D1499436

CONTEMPORARY MATHEMATICS FOR ELECTRONICS

CONTEMPORARY MATHEMATICS FOR ELECTRONICS

WILLIAM H. GOTHMANN

Senior Engineer
ISC Systems Corporation
Spokane, WA 99206

PRENTICE-HALL, INC., *Englewood Cliffs, New Jersey 07632*

Library of Congress Cataloging in Publication Data

Gothmann, William H., 1937–
    Mathematics for electronics.

    Includes index.
    1. Electronics—Mathematics.   I.   Title.
TK7835.G596   621.381'01'51   81–8497
ISBN 0-13-170274-2   AACR2

Editorial production supervision
and interior design by   James M. Chege

Cover design by: Tony Ferrara Studio

Manufacturing buyer: Gordon Osbourne

Printed in the United States of America

10   9   8   7   6   5   4   3   2   1

PRENTICE-HALL INTERNATIONAL, INC., *London*
PRENTICE-HALL OF AUSTRALIA PTY. LIMITED, *Sydney*
PRENTICE-HALL OF CANADA, LTD., *Toronto*
PRENTICE-HALL OF INDIA PRIVATE LIMITED, *New Delhi*
PRENTICE-HALL OF JAPAN, INC., *Tokyo*
PRENTICE-HALL OF SOUTHEAST ASIA PTE. LTD., *Singapore*
WHITEHALL BOOKS LIMITED, *Wellington, New Zealand*

To
June *and* **Leroy**

Whoever loves discipline loves knowledge
—Job

CONTENTS

Preface   xx

To the Reader   xxii

PART I   DIRECT-CURRENT MATHEMATICS   1

1 The Electronic Calculator   2

    1-1 Type of Arithmetic   *2*
        Algebraic notation   2
        RPN notation   3
        Which is better?   3
    1-2 Functions Available   *4*
        Data memory   4
        Program memory   4
    1-3 The Calculator for You   *5*

2 Introduction to Algebra   6

    2-1 A Number by Any Other Name   *6*
        Literals   7
    2-2 Algebraic Operations   *7*
        Exponentiation   7
        Order of operations   8
        Problems   10
    2-3 Terms Used in Algebra   *11*
        Expression of quantities   11
        Results of arithmetic operations   11
        Algebraic expressions   12
        Subscripts and superscripts   12
    2-4 Evaluating Algebraic Expressions   *13*
        Exponentiation   14
        Multiplication and division   14
        Addition and subtraction   15
        Problems   15

2-5   Signs of Grouping   *16*
    Exponents   17
    Problems   17

3   *Algebraic Addition and Subtraction*   19

3-1   Negative Numbers   *19*
    The number line   19
    Addition   20
    Subtraction   22
    Problems   24
3-2   Adding Algebraic Expressions   *24*
    Binomials, trinomials, and polynomials   25
    Problems   26
3-3   Subtracting Algebraic Expressions   *26*
    Monomials   27
    Binomials, trinomials, and polynomials   27
    Problems   28
3-4   Signs of Grouping   *28*
    Multiple groupings   29
    Problems   30

4   *Algebraic Multiplication and Division*   31

4-1   Signs in Multiplication   *31*
    Multiplying several numbers   33
    Problems   33
4-2   Terms with Exponents   *33*
    Problems   34
4-3   Multiplying Algebraic Expressions   *34*
    Monomials by monomials   35
    Polynomials by monomials   35
    Binomials by binomials   36
    Polynomials by binomials   37
    Polynomials by polynomials   37
    Multiple signs of groupings   38
    Problems   38
4-4   Arithmetic Fractions   *39*
    Fraction rules   39
    Reduction   40
    Multiplying two fractions   42
    Dividing two fractions   43
    Adding two fractions   43
    Subtracting two fractions   46
    Proper and improper fractions   47
    Problems   50
4-5   Signs in Division   *51*
    Problems   52

4-6   Terms With Exponents   *52*
    Problems   53
4-7   Dividing Algebraic Expressions   *53*
    Monomial division   54
    Polynomials divided by a monomial   54
    Polynomials divided by polynomials   55
    Problems   56

5 Equations   57

5-1   Fundamentals of Equations   *57*
    Operations used on equations   58
5-2   Solving Equations   *59*
    General procedures   59
    Fractional equations   62
    Problems   65
5-3   Forming Equations   *65*
    Problems   67
5-4   Literal Equations   *67*
    Problems   69
5-5   Proportionality   *70*
    Application to electronics   72
    Ratios   73
    Problems   74

6 Calculating Decimal Numbers Effectively   76

6-1   Resistor Values   *76*
    Values under 10Ω   77
    Problems   78
6-2   Expressing numbers   *78*
    Fixed notation   79
    Scientific notation   79
    Engineering notation   80
    Problems   81
6-3   Manipulating Decimal Numbers   *81*
    Addition   81
    Subtraction   82
    Multiplication   82
    Division   82
    Exponentiation   83
    Using the calculator   83
    Problems   83
6-4   Expressing Accuracy   *84*
    Significant digits   84
    Rounding off   85
    Problems   85
6-5   Calculator Errors   *86*
    Expressing errors   86

        Errors in manipulation   87
        Problems   89
6-6   Propagation of Errors   *90*
        Addition   90
        Subtraction   91
        Multiplication   92
        Division   93
        Summary   94
        Problems   94
6-7   Electronic Units of Measure   *95*
        Metric prefixes   95
        Conversion of units   96
        Dimensional analysis   98
        Problems   99

7 *Ohm's Law in Series Circuits*   101

7-1   Solving Problems Using Ohm's Law   *101*
        Problem analysis   102
        Problems   103
7-2   Power and Energy   103
        Power   104
        Energy   105
        Problems   106
7-3   Resistors in the Series Circuit   *107*
        Current   107
        Resistance   107
        Voltage   108
        Power   110
        Problems   111
7-4   The Loop Equation   *112*
        Problems   113
7-5   The Unloaded Voltage Divider   *114*
        Problems   115

8 *Factoring*   116

8-1   Powers of Monomials   *116*
        Problems   117
8-2   Roots of Monomials   *118*
        Problems   119
8-3   Prime Numbers   *119*
        Algebraic expressions   120
        Problems   120
8-4   The Common Factor   *120*
        Problems   122
8-5   The Perfect Square   *122*
        Problems   124

The square of differences   124
Problems   126

8-6   The Difference of Squares   *126*
Problems   127

8-7   General Trinomial Factoring   *127*
Trinomials with two positive coefficients   128
Problems   129
The negative middle term   129
Problems   130
The negative third term   130
Problems   132
Summary of signs and operations   132
Trinomials with three coefficients   133
Problems   135

8-8   Summary   *135*
Problems   136

9 **Algebraic Fractions**   137

9-1   Review of Fractions   *137*
9-2   Reducing Fractions   *137*
Problems   139
9-3   Multiplication and Division of Fractions   *140*
Division   141
Problems   141
9-4   Lowest Common Multiple   *142*
Problems   144
9-5   Addition and Subtraction   *144*
Problems   146
9-6   Complex Fractions   *147*
Problems   149
9-7   Fractional Equations   *149*
Conditional equations   150
Literal equations   152
Problems   153

10 **Parallel and Series–Parallel Circuits**   154

10-1   The Parallel Circuit   *154*
10-2   Voltage and Current of the Parallel Circuit   *154*
Problems   155
10-3   Resistance of a Parallel Circuit   *156*
Two parallel resistors   157
Identical parallel resistors   158
Problems   158
10-4   Power in a Parallel Circuit   *159*
Problems   159
10-5   The Balanced Bridge   *160*
Fault Location   161
Problems   163

10-6   The Loaded Voltage Divider   *164*
    Changing the reference point   166
    Problems   168

11   *Batteries and Conductors*   169

11-1   Battery Internal Resistance   *169*
    Problems   171
11-2   Multiple Batteries   *171*
    Batteries in parallel   172
    Problems   174
11-3   Conductors   *174*
    Wire gauge   175
    Diameter   176
    Area   176
    Resistance   177
    Ampacity   177
    Voltage drop   178
    Resistivity   178
    Effect of temperature   180
    Problems   181

12   *Graphs and Equations*   182

12-1   Introduction to Graphs   182
    Locating points on a graph   182
    Plotting experimental data   184
    Nonlinear axes   188
    Problems   189
12-2   Graphing Equations   *190*
    Independent and dependent variables   190
    Plotting equations   191
    Higher-order equations   193
    Solving equations   194
    Problems   196
12-3   Developing Equations from Graphs   *198*
    Problems   201

13   *Simultaneous Equations*   203

13-1   Introduction   *203*
13-2   Solution by Graph   *204*
    Problems   205
13-3   Solution by Addition   *207*
    Problems   209
13-4   Solution by Substitution   *209*
    Problems   210
13-5   Solution by Equality   *211*
    Problems   212

13-6   Three Variables   *212*
       Problems   215
13-7   Literal Equations   *215*
       Problems   216

14   *Determinants*   217

   14-1   The Determinant Matrix   *217*
       Two variables   217
       Three variables   218
       More than three variables   219
       Problems   219
   14-2   Cramer's Rule for Equations   *219*
       Two variables   219
       Three variables   221
       Minors   223
       Evaluating higher-order determinants   224
       Higher-order equations   225
       Problems   227
   14-3   Determinant Manipulations   *227*
       Determinants with zero value   228
       Interchanging rows or columns   228
       Factoring   229
       The zero quadrant   229
       Reduction theorem   230
       Problems   231
   14-4   Pivotal Reduction Method   *232*
       Matrix format   232
       Problems   236

15   *Network Analysis*   237

   15-1   The Loop Method   *237*
       The distribution problem   241
       Problems   243
   15-2   The Nodal Method   *244*
       The unbalanced bridge   247
       Problems   249
   15-3   Thévenin's Theorem   *251*
       Batteries in parallel   254
       Problems   254
   15-4   Norton's Theorem   *255*
       Finding the Norton equivalent   255
       Problems   257
   15-5   The Superposition Method   *257*
       Problems   261
   15-6   Wye–Delta Transformations   *261*
       Problems   264

16 Direct-Current Meters 265

16-1 The Voltmeter *265*
  Ohms/volt rating 266
  Multirange voltmeters 267
  Problems 270
16-2 The Ammeter *270*
  Multirange Ammeters 271
  Problems 275
16-3 The Ohmmeter *275*
16-4 The Volt-Ohm-Milliammeter *276*
16-5 Effects of Meters Upon a Circuit *276*
  Voltmeter loading 277
  Electronic voltmeters 280
  Ammeter insertion loss 280
  Problems 281

PART II ALTERNATING-CURRENT MATHEMATICS **283**

17 Exponents and Radicals 284

17-1 Laws of Exponents *284*
  Addition and subtraction 284
  Multiplication 285
  Division 285
  Exponentiation 286
  Problems 287
17-2 Fractional Exponents *288*
  Problems 289
17-3 The Radical Sign *289*
  Rational and irrational numbers 290
  Literals 290
  Numerical constants 291
  Complete monomials 292
  Problems 293
17-4 Addition and Subtraction *293*
  Problems 295
17-5 Multiplication *296*
  Problems 297
17-6 Division *297*
  Problems 298

18 Quadratic Equations 299

18-1 Introduction *299*
18-2 Graphing *299*
  The circle 300
  The ellipse 300
  The hyperbola 301

The parabola   301
Solving equations   303
Problems   304

18-3   Solution by Factoring   *304*
    Problems   306

18-4   Solution by Quadratic Equation   *306*
    Imaginary roots   308
    Problems   310

19   Triangles   311

19-1   Angles   *311*
    Types of angles   311
    Plotting angles   312

19-2   Units of Angular Measurement   *313*
    Sexagesimal system   313
    Radian system   316
    Grad system   317
    Calculator conversion   318
    Problems   318

19-3   Triangles   *319*
    Similar triangles   319
    The right triangle   320
    Problems   322

20   Trigonometry   325

20-1   Trigonometric Functions   *325*
    Sine, cosine, and tangent   327
    Problems   328

20-2   Finding Unknown Sides   *329*
    Problems   332

20-3   Finding Unknown Angles   *333*
    Problems   335

20-4   Functions of Large Angles   *336*
    Ranges of the functions   340
    Problems   341

20-5   Advanced Concepts   *341*
    Law of sines   342
    Problems   343
    Law of cosines   343
    Problems   344
    Secant, cosecant, and cotangent   344
    Problems   345
    Trigonometric identities   345
    Problems   346

21   Introduction to Alternating Current   347

21-1   Generating Alternating Current   *347*
21–2   The Sine Wave   *348*

Frequency   350
Period   351
Phase angle   351
Angular velocity   352
The sine-wave equation   352
Harmonics   355
Wavelength   356
Problems   357

21-3   Expression of AC Values   *358*
Peak measurements   358
Average amplitude   358
Root mean square   361
Problems   365

22   **Phasors**   367

22-1   Phasors and Sine Waves   *367*
RMS values   368
22-2   The Complex Field   368
Problems   372
22-3   Rectangular Coordinates   *372*
Adding phasors   373
Subtracting phasors   374
Multiplying phasors   375
Dividing phasors   378
Problems   379
22-4   Polar Coordinates   *380*
Polar to rectangular conversion   380
Rectangular to polar conversion   381
Addition and subtraction   382
Multiplication   383
Division   384
Exponentiation   384
Problems   384

23   **Logarithms**   386

23-1   Some Definitions   *386*
Calculating the source   387
Logarithms to base 10   388
Logarithms to base ϵ   389
Problems   389
23–2   Properties of Logarithms   *390*
Log of product   390
Log of quotient   392
Log of exponential   394

Changing bases   395
Problems   396
23-3   The Decibel   *397*
Problems   398

24   *Inductance in Series* RL *Circuits*   399

24-1   Inductors in Series and in Parallel   *399*
Problems   401
24-2   Mutual Inductance   *402*
Problems   403
24-3   Transformers   *403*
Problems   405
24-4   *RL* Discharge Curve   *406*
Problems   408
24-5   Inductive Reactance   *408*
Reactances in series   409
Reactances in parallel   410
Problems   412
24-6   Series *RL* Circuits   *413*
Problems   415
24-7   Ohm's Law in Series Reactive Circuits   *415*
Problems   416

25   *Capacitance in Series* RC *Circuits*   418

25-1   Capacitors in Series and in Parallel   *418*
Capacitors in parallel   419
Capacitors in series   419
Problems   420
25-2   *RC* Time Constant   *421*
Discharge   423
Problems   424
25-3   Capacitive reactances in Series and in Parallel   *424*
Capacitive reactances in Series   426
Capacitive reactances in parallel   426
Problems   427
25-4   Series *RC* Circuits   *427*
Problems   428
25-5   Ohm's Law in Series Capacitive Reactive Circuits   *429*
Problems   430

26   RLC *Circuits*   431

26-1   Series Circuits   *431*
Problems   434
26-2   Parallel Circuits   *434*
Problems   437
26-3   Series-Parallel Circuits   *437*
Problems   440

26-4 Series Resonance   *441*
      The Q of a circuit   442
      Problems   444
26-5 Parallel Resonance   *444*
      The Q of a parallel circuit   446
      Problems   446

PART III   ACTIVE-DEVICE MATHEMATICS   **447**

27   Computer Arithmetic   448

27-1 The Decimal Number System   *448*
27-2 The Binary Number System   *449*
      Converting binary to decimal   449
      Converting decimal to binary   451
      Problems   452
27-3 The Octal Number System   *452*
      Binary–octal conversions   452
      Converting octal to decimal   453
      Converting decimal to octal   454
      Problems   454
27-4 The Hexadecimal Number System   *455*
      Converting hexadecimal to decimal   455
      Converting decimal to hexadecimal   456
      Problems   456
27-5 Introduction to Boolean Algebra   *457*
27-6 The AND Operator   *457*
      AND gates   457
      AND laws   458
27-7 The OR Operator   *459*
      OR gates   459
      OR laws   460
27-8 The NOT Operator   *460*
      Inverters   460
      NOT laws   461
27-9 Laws of Boolean Algebra   *462*
      Laws of complementation   462
      AND laws   462
      OR laws   462
      Commutive laws   463
      Associative laws   463
      Distributive laws   464
      De Morgan's theorem   465
      Problems   467
27-10 Reducing Boolean Expressions   *467*
      Problems   468

Appendices

I   Table of Symbols   *469*
II Greek Alphabet   *471*
III International System of Units (SI) Conversion Factors   *473*

Answers to Odd-Numbered Problems   475

Index   495

PREFACE

The Model T Ford was a splendid machine. It represented that spark of ingenuity that has made the United States admired by the world. Like the Model T, electronics has undergone many changes since 1906 when Lee DeForest ushered us into this exciting age of electronics through his invention of the vacuum tube. Many of these changes have occurred during the last ten years. Remarkable advances in microprocessors, laser communications, and medical electronics have added new dimensions to traditional analog communications subjects. These changes have necessitated changes within the classroom, which, in turn, must be reflected in textbooks used within the classroom. The influence of contemporary electronics is twofold:

(a) It must be reflected in the manner in which we teach these subjects.
(b) It must be reflected in the subjects we teach.

It is my earnest expectation that this book answers these two challenges by providing a modern text suitable to lower-division electronics courses.

In attempting to meet these two challenges, the book has been divided into three parts, corresponding to those used within the classroom:

(a) Direct-Current Mathematics
(b) Alternating-Current Mathematics
(c) Active-Device Mathematics

This allows both the instructor and the student to see both the forest and the trees.

Part I, Direct-Current Mathematics, opens the scene by introducing the student to the calculator and the types and features available. This is an encounter of the first kind, for the calculator and its manipulation are discussed throughout the book as applications arise. The student is further schooled in the calculator in a rather unique chapter that discusses notation, accuracy, error analysis, and error propagation (Chapter 6). Other features of Part I include:

(a) Manipulation of grouping designators is discussed in detail in both Chapters 2 and 3, easing the student over this sometimes difficult hurdle.
(b) Fractions, another potential problem area, are introduced in Chapter 4 and reviewed and explained further in Chapter 9.
(c) Dimensional analysis is used both as a conversion tool and as a tool for evaluating the correctness of an equation (Chapter 6).

(d)   Kirchhoff's loops are introduced very early when single-variable problems are discussed (Chapter 7). This eases their use in simultaneous equations.

(e)   Traditional methods for solving simultaneous equations are discussed in Chapter 12. Chapter 13 then discusses determinants in detail, including Cramer's rule, matrix manipulation, and pivotal reduction, allowing students to use these most effective techniques on their calculators.

Part II, Alternating-Current Mathematics, introduces phasors very early (Chapter 22) so they can be used in solving ac problems. Other features include:

(a)   Quadratics are related to conics (Chapter 18).

(b)   Calculator solution methods are emphasized in trigonometry (Chapters 19 and 20).

(c)   Rms is applied to both sine and digital waveforms (Chapter 21).

(d)   Calculator solutions are emphasized in time-constant problems (Chapters 24 and 25).

Part III, Active-Device Mathematics, introduces both number systems (base 2, 8, 10, and 16) and boolean algebra to the students, permitting them to enter the digital world with their eyes wide open.

The publication of this book realizes a ten-year dream of mine: a series of basic electronic texts having the same order of subjects and approach covering the following:

(a)   Electronic theory, represented by *Electronics: A Contemporary Approach,* Prentice-Hall, 1980

(b)   Laboratory projects, included in the above text

(c)   Mathematics for electronics, represented by *Contemporary Mathematics for Electronics*

No longer will students and instructors have to juggle, skip, and interpret content of subjects, for these two books have the same chapter and section titles any time commonality exists. Further, the order of subjects is identical, although their content differ, for one explains the theory of how electronic devices work and the other the mathematical skills necessary to understand the subject. The math book, for example, covers factoring and algebraic fractions that are not covered by the theory book. The theory book covers dielectrics, which are not covered by the math book. Both books are designed to be used independently or together to suit the program of the school. It is my hope that the publication of these two texts will be an invaluable aid in understanding electronics, both to the student and to the instructor.

Finally, I wish to thank the many people who contributed to the publication of this work, especially Barbara Castle for the magnificent and monumental task of typing and proofreading the manuscript, and my wife, Myrna, for her active encouragement and many helpful suggestions for its improvement.

Spokane, Washington   BILL GOTHMANN

TO THE READER

Mathematics has been called the queen of science, and with good reason. It enables the student not only to know "how," but "how much." Suppose that you had just withdrawn some shekels from your vast underground money vault and purchased a brand-spanking-new airplane. A friend of yours then inquires: "What is this new airplane's cruising speed?" "Fast," you reply. But what does "fast" mean? To a car, 60 miles per hour (mph) is fast, but to an orbiting space satellite, 25,000 mph is fast. So what does your friend know about your airplane's cruising speed? Very little. He only knows that it is faster than a speeding snail and slower than a cow that jumps over the moon. However, by applying a number to the speed, 160 mph for example, your friend gains an additional level of understanding of the speed of your airplane. Thus, mathematics permits a more accurate description of any phenomenon.

This is true in electronics, also. Not only can we describe the fact that a stereo set is blasting away from the neighbor's house, but we can describe how much it is blasting in terms of decibels or watts. Not only can we describe how wiring a capacitor and an inductor in parallel permits selection of only one radio station from the radio band, we can calculate the frequency of that station by merely knowing the value of the capacitor in a unit called a farad and knowing the value of the inductor in a unit called a henry. Again, mathematics provides a deeper understanding of electronics.

It is obvious from the fact that you are reading this that you want to have this deeper level of understanding. But how can you attain it? After all, everyone says that mathematics is hard! Tell you what I'm going to do. I'm going to let you in on the best kept secret since the apple fell on Newton's head. Math isn't that hard—in fact, it can be fun; it can be challenging. It is like a whodunnit mystery and you are the detective. However, it is only easy if a few simple rules are followed and, because of my overly generous nature, I'm going to let you in on these rules:

1. Mathematics is a science in which the next step builds on the present step. If the present step is foggy, the next step will be pea soup.
   Never, never assume you will "catch on" in the next step when you don't understand this step. Learn this one first.
2. Mathematics is a science having many logical rules. However, there is only one way these rules can be learned—practice, practice, and more practice.

Never, never fail to do even one homework assignment. Math can only be learned by practice.

3. You are going to get stumped during the course of this study—we all do. What then?

   When you do not understand:
   (a) *Read the text carefully, then read it again.*
   (b) *Study the examples carefully.*
   (c) *Seek out other textbooks and see what they have to say.*
   (d) *Ask your instructor or fellow students to explain it.*
   (e) *Above all, don't go on until you solve the problem.*

By following these relatively simple steps, you, too, can unlock the mysteries and enjoy the challenge of solving electronic problems.

PART I

DIRECT-CURRENT MATHEMATICS

In this part of the book, we study the mathematics required to solve direct current (dc) circuit problems. The dc circuit is one in which electronic charges consistently flow in one particular direction, and is used extensively in such things as cars, television sets, and computers. Algebra will be the primary tool used for mathematical analysis and will form the foundation upon which Part II, Alternating-Current Mathematics, will build.

1

THE ELECTRONIC
CALCULATOR

The electronic calculator has emerged as the most useful tool in electronic mathematics. Because of its importance and because the user is faced with such a variety of them on the market, let us take a few minutes to discuss the features of these marvelous, miniature, mathematical machines.

1-1 TYPE OF ARITHMETIC

Hand calculators use one of two arithmetic systems: algebraic or reverse Polish notation (RPN). Each has its advantages and disadvantages and each its enthusiastic supporters.

Algebraic Notation

The algebraic machines compute a problem exactly as it is written on paper. For example, suppose that I work for 3 days and wish to calculate my wages. My rate of pay is $9.50 per hour and my work schedule requires me to work 7 hours

Key	Display	
(		
7	7	
+	7	
8	8	
+	8	
6	6	
)	6	
×	6	
9	9	
.	9.0	
5	9.5	
=	199.5	Final answer

FIGURE 1-1 Algebraic notation in a calculator.

2

on Monday, 8 hours on Tuesday, and 6 hours on Wednesday. I can calculate my wages as follows:

$$\text{wages} = (7 + 8 + 6) \times 9.50$$

On this type of calculator this problem would be computed by pressing buttons in the sequence shown in Fig. 1-1. The answer, $199.50, would then appear as 199.5 on the display. Thus, in algebraic notation, the problem is entered into the calculator precisely as it would be written on paper, from left to right. This has the advantage of simplicity but the disadvantage that no intermediate answers are displayed. For example, if we also wanted to know the total number of hours worked, we would have to do the problem in two steps: first calculate total hours, then multiply by wages. Further, the algebraic notation has the disadvantage of requiring keystrokes to enter the parentheses, whereas the RPN method eliminates these actions.

RPN Notation

In the RPN method, the problem is calculated not as it is written on paper, but as it would be calculated if the problem were done manually. In the foregoing problem, for example, we would normally add the hours to obtain total hours, then multiply this by 9.5 to obtain the final answer. Thus, on an RPN calculator we would depress the keys in the order shown in Fig. 1-2. Note that the function to be performed (+) is entered after the two numbers are entered that must be added (7 and 8). Note further that two fewer keystrokes are required and that each intermediate answer is displayed.

Which Is Better?

Either calculator, arithmetic or RPN, can be used effectively in electronics. The arithmetic type has the advantage of permitting data to be entered as they are written

Key	Display	
7	7	
ENTER	7	This key terminates entry of this number and allows entry of another
8	8	
+	15	Displays the sum of 7 and 8
6	6	
+	21	Displays the sum of 15 and 6
9	9	
.	9.	
5	9.5	
×	199.5	Final answer

FIGURE 1-2 Reverse Polish notation in a calculator.

on the page. The RPN system requires slightly fewer keystrokes and displays intermediate results.

1-2 FUNCTIONS AVAILABLE

There is a dazzling array of functions available on today's calculators. No calculator has all of them—indeed, this itself would be a disadvantage. However, let us select those that are most useful in electronics. Table 1-1 lists those that are required, highly desirable, and desirable.

TABLE 1-1 Calculator functions used in electronics.

Required functions:
$+$, $-$, $\times$, $\div$, sin, $\sin^{-1}$, cos, $\cos^{-1}$, tan, $\tan^{-1}$, log, $1/X$, Y^X, $\sqrt{X}$, π, exponential notation

Highly desirable functions:
10^X, engineering notation, radian–degree conversion, polar–rectangular conversion, data memory

Desirable features:
X^2, e^X, %, % change, Ln, program memory

Data Memory

Many calculators have a data memory into which any number may be placed for recall at a later time. Some have only one such memory location and thus can accommodate only one number at a time, whereas other calculators may have 25 or more. This allows intermediate calculations to be remembered without having to write them down. Having at least one data memory location is a highly desirable feature; having more than one is a desirable feature.

Further, some calculators permit numbers to be added to, subtracted from, multiplied by, or divided into these memory locations, with the result automatically placed back into the memory (wiping out its previous number). This memory arithmetic feature is another desirable option.

Program Memory

A second type of memory available on some machines is a program memory. This is a multilocation memory allowing the calculator to record every time a key is pushed. This record is known as a program. Let us assume that we put the calculator into the "program" mode and press the keys $\boxed{+}$, $\boxed{5}$, $\boxed{=}$, $\boxed{\text{STOP}}$ on an algebraic notation calculator. These four keystrokes will be remembered in the same order as they were entered. We shall now switch from "program" to "run" and enter the number 31.36. When we now press the key $\boxed{\text{GO}}$, it will cause the calculator to execute the sequence of keystrokes recorded in the program until it obtains the $\boxed{\text{STOP}}$, in which case it will display the latest result. By depressing $\boxed{\text{GO}}$, we have done the same thing as if we had pressed $\boxed{+}$, $\boxed{5}$, and $\boxed{=}$. Thus, the calculator will display 31.36 + 5 or 36.36. By entering different numbers and pressing $\boxed{\text{GO}}$, the calculator will add 5 to each. Therefore, a program memory records and stores a set of keystroke operations that can be used over and over again. Without the program memory, these keystrokes would all have to be entered each time a new number is to be treated.

Some calculators retain the information in their data and program memories when the calculator is turned off. In others, the information is lost. In some calculators, the program and/or data memories can be recorded on small magnetic cards that can be read at a later time.

1-3 THE CALCULATOR FOR YOU

Since each person is different, each may want a different calculator. If you are a "math nut," only the finest will satisfy you. However, if you find math uninteresting, you should probably buy the calculator that has the basic required features.

One final word. Keep in mind that if the unit should malfunction, the manufacturer probably has the only facility capable of repairing it. Therefore, choose a brand whose manufacturer will back its operation.

2

INTRODUCTION TO ALGEBRA

In this chapter we learn the basics of the branch of math called algebra. We first examine the symbols we use daily to express numbers. We then examine the operations of algebra only to discover that they include the same rules that we have known and loved since the second grade. We also study the vocabulary used in algebra. Finally, we evaluate some algebraic expressions by finding the exact number that they represent.

2-1 A NUMBER BY ANY OTHER NAME

In arithmetic, we express numbers as Arabic symbols. The number 9, for example, could represent 9 cents, 9 fleas on a camel's back, or 9 radio sets, for this symbol always represents 9 items. We are not going to change any of this in algebra; the number 9 still represents 9 items.

We are also accustomed to expressing unknown quantities in our language. "How many miles per gallon did you get from your 1936 Osenfeffer on your trip to Transylvania?" Mergatroid asks Fauntleroy. In this case Mergatroid is asking for a number. But does she know the number? Of course not, or she would not need to ask. However, the answer is a specific number even though she does not know what the exact number is.

What symbol did Mergatroid use to express this unknown number? "Miles per gallon." She could have asked, "What mileage did you get?" in which case she would be using the symbol "mileage" to express this unknown quantity. Thus, she selected an appropriate symbol to represent this unknown quantity: one that is understood by all participating in the conversation.

What Mergatroid has just done was to use algebra, a system in which symbols are used to represent unknown quantities. However, in algebra, we try to keep the length of the symbol fairly short; MPG or just M could easily have done the job. We all are told what the symbol is and what it represents before we start the game.

EXAMPLE 2-1 Select appropriate symbols for representing the ages of Sue, Bud, and Mark, respectively.

SOLUTION We could represent Sue's age by the symbol S, Bud's with a B, and Mark's with an M. Again, each symbol represents a specific number. If Sue is 20, the value of S is 20; if Bud is 19, the value of B is 19; and if Mark is 3 years old, M is 3.

6

Literals

These letters we have been using to represent unknown quantities are called literals, after a Latin word meaning "letter." Thus, in Example 2-1, *B*, *S*, and *M* are all literals. We therefore have two classes of symbols to represent numbers: literals, which represent unknown numbers, and Arabic numerals, which represent known quantities.

EXAMPLE 2-2 Select appropriate literals for representing miles driven, miles per gallon, and gallons of gasoline used.

SOLUTION We cannot use an *M* to represent both miles and miles per gallon, just as in English the term "miles" cannot represent both quantities. We could, however, use another letter, *d,* to represent distance (miles), use *M* for miles per gallon, and *G* for gallons.

2-2 ALGEBRAIC OPERATIONS

The commonly used designations for addition (+), subtraction (−), multiplication (×), and division (÷) can also be used in algebra. However, there are two additional symbols that are frequently used for multiplication:

1. *Multiplication can be shown by a dot* (·). *Thus,* 3 × 4 *can be shown as* 3 × 4 *or* 3 · 4.
2. *In multiplying a literal by a number, or a literal by another literal, the multiplication sign is usually omitted.*

For example, 3*B* represents 3 × *B*. If *B* is 23, then 3*B* would be 3 × 23 or 69. In addition, *BB* represents *B* × *B* or, if *B* is 23, 23 × 23 or 529.

EXAMPLE 2-3 The literal *R* represents electrical resistance measured in a quantity called an ohm. State three ways of expressing 40 multiplied by the *R* of a certain wire.

Solution 40*R*, 40 × *R*, or 40 · *R*. In algebra, the accepted practice is to place the number before any literals. Thus, *R*40 would be a valid number but would not be in the customary form.

EXAMPLE 2-4 State six ways of expressing miles (represented by *d*) multiplied by speed (represented by *s*).

SOLUTION The following forms are all correct: *sd*, *ds*, *s* × *d*, *d* × *s*, *s* · *d*, and *d* · *s*.

Exponentiation

The operation of exponentiation is also performed in arithmetic, but not as often as addition, subtraction, multiplication, or division. Some examples are 3^2 and 5^4. In each case the base (the 3 and 5) is raised to a power (the 2 and 4). This is easily performed by remembering that the base is put down the number of times specified by the power (or exponent), and the resulting numbers are multiplied, giving the answer.

EXAMPLE 2-5 Compute 4^6.

SOLUTION The 4 must be put down 6 times and these multiplied:

$$4 \times 4 \times 4 \times 4 \times 4 \times 4 = 4096$$

Thus, 4^6 is 4096.

EXAMPLE 2-6 Compute 5^{10}.

SOLUTION $5^{10} = 5 \times 5 \times 5 \times 5 \times 5 \times 5 \times 5 \times 5 \times 5 \times 5 = 9,765,625$.

Order of Operations

Not only must close attention be paid to the symbols used for the operations, but the order in which they are used must be observed. What does $3 + 4 \times 5$ mean? Does it mean we are to add 3 and 4, obtaining 7, then multiply this by 5, obtaining 35? Or does it mean we are to multiply 4 by 5, obtaining 20, then add this to 3, making 23? The two solutions are obtained because in the first case we did addition first and then multiplied, whereas in the second case we multiplied first, then added. In algebra and in arithmetic the latter case is the correct one and 23 the correct answer, for multiplication takes precedence over addition. Table 2-1 formally states the order or hierarchy in which operations are to be performed.

TABLE 2-1 Hierarchy of algebraic operations.

Highest	1. Exponentiation
	2. Unary operations
	3. Multiplication and division
Lowest	4. Addition and subtraction

The highest order within the hierarchy is an exponent. For example, the 5 in 2^5 is an exponent. To illustrate the exponential, let us consider the number 4^{-5}. The -5 would be considered as a single number and the 4 would be raised to the -5 power. In a statement such as $3 + 4^2$, the 4 is first squared (raised to the second power), then this result is added to the 3, obtaining 19.

EXAMPLE 2-7 Compare the results obtained in the following two expressions:

(a) $5 + 6^3$ (b) $6^3 + 5$

SOLUTION Because of the hierarchy of algebra, the exponentiation must be done before the addition. Therefore, in both cases, the 6 must be raised to the third power, then the 5 added. The answers will be the same:

(a) $5 + 6^3 = 5 + 216 = 221$
(b) $6^3 + 5 = 216 + 5 = 221$

The next highest priority is the unary operation. This involves only one number, such as the "+" in +3 and the "−" in −6.

The third priority in the hierarchy is a class of two operations: multiplication and division. Both have equal priority but are lower than the unary or exponentiation.

This can lead to ambiguities such as in the problem $12 \div 4 \times 3$, where, if we divide first and then multiply, we obtain

$$12 \div 4 = 3$$
$$3 \times 3 = 9$$

However, if we multiply first and then divide, we obtain

$$4 \times 3 = 12$$
$$12 \div 12 = 1$$

Which is the correct answer, 9 or 1? In terms of pure algebra, either is. However, with the advent of the computer, designers of computer languages had to define this type of expression more exactly, so another rule was added that stated:

When equal priorities in the algebraic hierarchy are encountered, the computer will process the expression from left to right.

In the example above, therefore, the 12 would first be divided by the 4, then this result multiplied by 3. Thus, in the computer age, 9 is the correct result.

EXAMPLE 2-8 Evaluate the expression $10 \div 5 \times 3$.

SOLUTION Since division and multiplication have the same priority, the expression should be processed from left to right.

$$10 \div 5 = 2 \qquad \text{Evalute the leftmost operation.}$$
$$2 \times 3 = 6 \qquad \text{Multiply this result by the 3.}$$

Thus, the answer is 6.

The fourth priority in the hierarchy is another class of two operations: addition and subtraction. Both have equal priority but are lower than multiplication and division. Since it makes no difference whether we add, then subtract or subtract, then add (both result in the same answer), no ambiguities occur.

EXAMPLE 2-9 Compute the expression $5 + 6 - 2$.

SOLUTION We can add the 5 and 6, obtaining 11, then subtract 2, giving 9. However, we can also subtract 2 from 6, obtaining 4, then add 5, and still get 9.

EXAMPLE 2-10 Compute the expression $3 + 4 \times 5^2 - 6 \div 3$.

SOLUTION The highest priority of this problem is exponentiation, so this must be done first, obtaining

$$3 + 4 \times 25 - 6 \div 3$$

Next, we must perform the multiplication and division:

$$3 + 100 - 6 \div 3$$
$$3 + 100 - 2$$

We can now compute the final answer:

$$3 + 100 - 2 = 101$$

EXAMPLE 2-11 Compute the expression $12 \div 6 + 5 \times 4 \times 3^2 + 2$.

SOLUTION

$12 \div 6 + 5 \times 4 \times 3^2 + 2$	Original problem
$= 12 \div 6 + 5 \times 4 \times 9 + 2$	Exponentiation complete
$= 2 + 5 \times 4 \times 9 + 2$	Division complete
$= 2 + 20 \times 9 + 2$	5×4 computed
$= 2 + 180 + 2$	20×9 computed
$= 182 + 2$	$2 + 180$ computed
$= 184$	Final answer

EXAMPLE 2-12 Compute the expression $4 \div 2 \times 9 + 6 \times 5^2 \div 3$.

SOLUTION

$4 \div 2 \times 9 + 6 \times 5^2 \div 3$	Original problem
$= 4 \div 2 \times 9 + 6 \times 25 \div 3$	Exponential complete
$= 2 \times 9 + 6 \times 25 \div 3$	$4 \div 2$ computed
$= 18 + 6 \times 25 \div 3$	2×9 computed
$= 18 + 150 \div 3$	6×25 computed
$= 18 + 50$	$150 \div 3$ computed
$= 68$	Final answer

Note that where a question of identical priorities occurred, the problem was worked from left to right.

PROBLEMS

State three ways to express each of the following:

2-1. The literal V multiplied by 26
2-2. The literal Q multiplied by 7
2-3. The literal C multiplied by 100
2-4. The literal G multiplied by 75

State six ways to express the multiplication of the following literals:

2-5. f and L
2-6. C and X
2-7. L and C
2-8. V and I

Evaluate the following arithmetic expressions:

2-9.	4^3	**2-17.**	$2 \times 3 + 4$	**2-25.**	$14 + 7^2$
2-10.	5^2	**2-18.**	$3 \times 5 + 6$	**2-26.**	$7 + 14^2$
2-11.	6^3	**2-19.**	$6 + 7 \times 2$	**2-27.**	$20 - 2^2$
2-12.	7^3	**2-20.**	$7 - 2 \times 3$	**2-28.**	$17 - 3^2$
2-13.	3^5	**2-21.**	$14 + 21 \div 7$	**2-29.**	3×4^5
2-14.	10^7	**2-22.**	$30 + 10 \div 2$	**2-30.**	5×6^3
2-15.	6^6	**2-23.**	$8 \div 2 + 2$	**2-31.**	$6^2 \times 7^3$
2-16.	2^{10}	**2-24.**	$20 \div 5 - 3$	**2-32.**	$4^4 \div 2^2$

2-33.	$32 \div 4 \times 2$		**2-35.**	$27 \div 3 \times 9 + 7$
2-34.	$55 \div 5 \times 6$		**2-36.**	$75 \div 5 \times 5 + 3$

2-37. $3 + 4 \div 2^2 + 6$

2-38. $5 - 2^2 + 6 \div 3 \times 2$

2-39. $27 \div 3 \times 3^2 + 6$

2-40. $84 \div 12 \times 4^2 - 4 \times 3^2$

2-41. $3 + 120 \times 5^2 \div 10 - 7 \times 2^2$

2-42. $4^2 \div 2^2 + 6^3 \times 3^2 - 2^2 \times 5$

2-43. $12^4 \div 2^2 \div 3^2 \times 6^3 - 2^{16}$

2-44. $1.63^2 \div 2^2 \times 4^2 - 3.12$

2-3 TERMS USED IN ALGEBRA

Before we travel any further down this road, let us examine several definitions of terms commonly used in algebra.

Expression of Quantities

The entire field of numbers can be divided into two parts: constants and variables.

A constant is any fixed, unchangeable number.

Some examples are 2, 3, 3.14159265, ¾, and $\sqrt{35}$. In each case, the number will never change otherwise it will no longer be the same number. This contrasts with the variable.

The variable is a quantity that may change as conditions change.

Miles per hour is such an example, for it will change depending upon whether the car is a Rolls Royce or a Honda, and whether the driving is uphill or downhill. Age is a second example, for my age this year will differ from my age last year, and my age next year. Other examples include voltage, resistance, current, quality factor, distance, and sum.

Variables or constants may be expressed in many ways: for example, spoken word, printed word, and letter.

When a variable or a constant is expressed as a letter, it is called a literal.

Examples of constant literals include π (the Greek letter pi, which represents the number 3.14159265) and ϵ (the Greek letter epsilon, which represents the number 2.718281828). Examples of variable literals include V (voltage), I (electrical current intensity), R (electrical resistance), and C (electrical capacitance).

Results of Arithmetic Operations

The result of an addition process is called a sum.

When we add $3 + 4 + 7$, the answer, 14, is called the sum. Similarly, the sum of A and B is $A + B$.

The result of a subtraction process is called a difference.

When we subtract 3 from 12, the difference is 9. Similarly, the difference between A and B is $A - B$.

The result of a multiplication process is called a product.

When we multiply 5 by 6, the product is 30. Similarly, the product of A and B is AB.

The result of a division process is called the quotient.

When we divide 36 by 12, the quotient is 3. Similarly, the quotient of A divided by B is A/B.

Algebraic Expressions

An algebraic expression is a representation of a number using constants and/or variables in any sequence of operation.

Examples include A^2, $2A + 3$, $R^2 + 3(R + 6R) - 27$, and A/B. The expression is usually composed of portions called terms, which are added or subtracted.

A term is a single quantity, power, quotient, constant, or a product of these that is bounded by plus sign or minus sign, or that occurs at the beginning or end of the expression.

In the algebraic expression $V^4 + 3V^3I + 4V^2I^2 - 6VI^3 + 7I^4$, there are five terms: V^4, $3V^3I$, $4V^2I^2$, $6VI^3$, and $7I^4$.

The numerical constant part of the term is called a coefficient.

If there is no numerical constant, the coefficient is assumed to be a 1. Thus, in the expression above, the coefficients are 1, 3, 4, 6, and 7.

Algebraic expressions may have any number of terms.

An algebraic expression with only one term is called a monomial.

Note that the "mono" part of the word means 1. The following are examples: $3R^2G$, Q, $45A^2B^2DE$, $56P^2W^2Q$, and $3.15I^2V^2$.

An algebraic expression with two terms is called a binomial.

Note that the "bi" part of the word represents the number 2, as in such words as bicycle, biplane, and bifurcated. Some examples are $R_1 + R_2$, $3V^2 + 6E^2$, and $R^2S + 3$.

An algebraic expression with three terms is called a trinomial.

Note that "tri" represents the number 3, as in such words as triangle and tricycle. Some examples are $X^2 + 2X + 1$, $3V_1^2 + 2V_1V_2 + 6V_2^2$, and $A + B + 3$.

An algebraic expression with more than three terms is called a polynomial.

Note that "poly" means "many," as in such words as polygon, polygamy, and polyunsaturated. Some examples are $X^3 + 4X^2 + 3X + 5$ and $3R^5 - 6R^4S + 7R^3S^2 + 2R^2S^3 - 65RS^4 + 33S^5$.

Subscripts and Superscripts

A subscript is a number or letter printed lower than the main character and is used to differentiate between values of similar objects.

The 5 in R_5, the d in C_d, and the 7 in L_7 are examples. If there were 20 students in a class, we could designate the social security number of John as S_1, that of Mary as S_2, and that of Bill as S_3, and so forth until we reach S_{20}. Note that each is a student, but each has a different social security number. In a similar manner, a schematic diagram for an electronic computer may have 78 different resistors, each with a different value. They would be designated as R_1 through R_{78}. R_1 may be 1000 ohms, whereas R_2 may be 200 ohms and R_3, 65,000 ohms. Note that there is no mathematical relationship between the subscript itself and the value of the component.

A superscript is a symbol printed above the main character.

Thus, the 2 in R^2, the 4 in 6^4, and the a in Z^a are examples.

Numerical or literal superscripts are used to represent exponents.

The three examples above, therefore, have the exponents 2, 4, and a.

There is, however, one other type of superscript, the single-quote mark, called a "prime." This is used in a manner identical to that used for the subscript, to differentiate between similar objects having different values. For example, R' (pronounced R prime) might represent the value of the first resistor, R'' (pronounced R double prime) the value of the second resistor, and R''' (pronounced R triple prime) the value of the third resistor. This system is not used very frequently in electronics, however.

Figure 2-1 summarizes the parts of an expression.

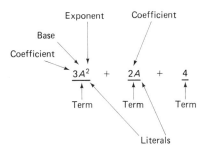

FIGURE 2-1 Parts of an algebraic expression.

2-4 EVALUATING ALGEBRAIC EXPRESSIONS

If the values of each of the literals of an expression is known, the exact value of the entire expression can be computed. This process is called evaluation and follows the priorities of operations listed in Section 2-2.

A word of caution is advisable here. The student should get into the habit of following a pattern in evaluation:

1. Write down the exact original expression.
2. Write the expression down a second time, substituting constants for each of the literals.
3. Perform the arithmetic, being careful (a) to document each step, and (b) not to do too much in one step.

Consider what happens when this procedure is not followed and a mistake occurs (and it will). The only choice the student has is to do the entire problem again. If the problem takes 4 hours to compute, another 4 hours must be expended, with the possibility of making a new mistake. However, if this procedure is followed, each step can be checked for errors, the exact error found, and the problem corrected. Many problems in electronics are quite lengthy and, by taking just a little more time to be orderly, much more time can be saved.

Exponentiation

Exponentiation has the highest priority (next to the unary) on the algebraic hierarchy. In evaluating such a term, if the base or the exponent is a literal, these values are substituted first. Then the base is written down the number of times specified by the exponent and these numbers multiplied.

EXAMPLE 2-13 Evaluate the term R^4 if R is 56.

SOLUTION Substituting yields

$$R^4 = 56^4 = 56 \times 56 \times 56 \times 56 = 9,834,496$$

EXAMPLE 2-14 Evaluate the expression C^D if C is 2.43 and D is 6.

SOLUTION Substituting for C and D, we obtain

$$C^D = 2.43^6 = 2.43 \times 2.43 \times 2.43 \times 2.43 \times 2.43 \times 2.43 = 205.89$$

Multiplication and Division

The next lower rung on the priority ladder is multiplication and division. Where there might be ambiguity, the operations are performed from left to right.

EXAMPLE 2-15 Evaluate RST if R is 16, S is 26.3, and T is 3.15.

SOLUTION Substituting yields

$$RST = 16 \times 26.3 \times 3.15 = 1325.52$$

EXAMPLE 2-16 Evaluate $R_1 \div R_2 \times R_3$ where R_1 is 30, R_2 is 20, and R_3 is 16.8.

SOLUTION Since there are two operations of equal priority, the problem must be worked left to right.

$$R_1 \div R_2 \times R_3 = 30 \div 20 \times 16.8 = 1.5 \times 16.8 = 25.2$$

Where exponentiation occurs, this must be done first.

EXAMPLE 2-17 Evaluate $C_1^2 \div C_2^3 \times C_3$, where C_1 is 30, C_2 is 2.59, and C_3 is 4.65.

SOLUTION Substituting, we obtain

$$\begin{aligned} C_1^2 \div C_2^3 \times C_3 &= 30^2 \div 2.59^3 \times 4.65 \\ &= 900 \div 17.37 \times 4.65 \\ &= 51.80 \times 4.65 \\ &= 240.88 \end{aligned}$$

Note that exponentiation was done before either the division or the multiplication.

Addition and Subtraction

The lowest priority in algebraic operations is addition and subtraction.

EXAMPLE 2-18   Evaluate $R_1 + R_2 - R_3$ if R_1 is 20, R_2 is 40, and R_3 is 16.

SOLUTION   Substituting the known values yields

$$R_1 + R_2 - R_3 = 20 + 40 - 16 = 44$$

EXAMPLE 2-19   Evaluate $3R^2 + 2R + 40$ if R is 33.2.

SOLUTION   Substituting for R, we obtain

$$
\begin{aligned}
3R^2 + 2R + 40 &= 3 \times 33.2^2 + 2 \times 33.2 + 40 \\
&= 3 \times 1102.24 + 2 \times 33.2 + 40 \\
&= 3306.72 + 66.4 + 40 \\
&= 3413.12
\end{aligned}
$$

EXAMPLE 2-20   Evaluate $3I_1^2 R_1 + 2I_2^2 R_2 + VI_3$ if R_1 is 20, R_2 is 30, I_1 is 5, I_2 is 3.9, I_3 is 2.6, and V is 4.1.

SOLUTION

$$
\begin{aligned}
&3I_1^2 R_1 + 2I_2^2 R_2 + VI_3 \\
&= 3 \times 5^2 \times 20 + 2 \times 3.9^2 \times 30 + 4.1 \times 2.6 \\
&= 3 \times 25 \times 20 + 2 \times 15.21 \times 30 + 4.1 \times 2.6 \\
&= 1500 + 912.6 + 10.66 \\
&= 2423.26
\end{aligned}
$$

Note that exponentiation was completed first, then multiplication, then addition.

PROBLEMS

Evaluate the following expressions:

2-45. R^3 if R is 3
2-46. R^s if R is 4, s is 9
2-47. C_1^D if C_1 is 3.16, D is 4
2-48. L^M if L is 5.96, M is 5
2-49. $3R_1 R_2 R_3$ if R_1 is 5, R_2 is 6.3, R_3 is 9.15
2-50. $3.6C_1 C_2 C_3$ if C_1 is 1.59, C_2 is 3.64, C_3 is 4.92
2-51. $R_1 \div R_2 \times R_3$ if R_1 is 6.21, R_2 is 2.15, R_3 is 7.8
2-52. $6V_1 \div V_s \times V_r \div V_t$ if V_1 is 13, V_s is 6.5, V_r is 7.2, V_t is 8.6
2-53. $3.2R_1^2 R_2$ if R_1 is 30, R_2 is 7
2-54. $4.6I_1^2 RF^3$ if I_1 is 9.6, R is 20, F is 2.31
2-55. $7.63W_1^2 W_3^3 W_4$ if W_1 is 5.3, W_3 is 4.1, W_4 is 7.3
2-56. $8.69P_1^2 P_2^4 P_3^9$ if P_1 is 2.13, P_2 is 4.96, P_3 is 1.615
2-57. $3R_1 + R_2 + R_3$ if R_1 is 5, R_2 is 6, R_3 is 9
2-58. $4R_1 + 2R_2^2 - 6$ if R_1 is 4, R_2 is 7
2-59. $3.16A_1 A_2 B + 6.95A_1 A_4 C$ if A_1 is 2, A_2 is 4, A_4 is 6, B is 2.13, C is 4.19
2-60. $3VI + I^2 R + V^2 \div R$ if V is 2.13, I is 4.9, R is 3.61
2-61. $3A^3 B^2 DL - 45ABD^2 + 63.19A^4 B^3 D^2 M$ if A is 3, B is 2.15, D is 2.6, L is 45, M is 2.97

2-62. $1.639ARB^2 + 2.75A^2RB - 7.32ARN^3$ if A is 1, R is 2.19, B is 7.32, N is 1.63

2-63. $3W_1^2W_2^3 + 2W_1W_2^4 + 4.9W_1^3W_2^3$ if W_1 is 2.6, W_2 is 1.32

2-64. $3R_1R_2^2 + 4R_1^2R_2 + 6R_2^3$ if R_1 is 3, R_2 is 4.3

2-5 SIGNS OF GROUPING

Several symbols are used to group terms together: parentheses, ( ); brackets, [ ]; braces, { }; and the vinculum or overscore, $\overline{}$. When these occur, the rules we have just applied to expressions should be applied to everything within the grouping, then the rules applied to the overall expression. For example, consider the expression $a^2 + b(c + d)$. The *(c + d)* should be treated as an expression and evaluated, then this result treated as a variable in the overall expression.

EXAMPLE 2-21   Evaluate the expression $3 + 20(6 + 9)$.

SOLUTION   The $(6 + 9)$ must be evaluated first:

$$
\begin{aligned}
3 + 20(6 + 9) &= 3 + 20 \times 15 && \text{(6 + 9) evaluated} \\
&= 3 + 300 && \text{Multiplication complete} \\
&= 303 && \text{Final answer}
\end{aligned}
$$

EXAMPLE 2-22   Evaluate the expression $45 + 8(9 + 6) + 7 \cdot \overline{6 + 5 \cdot 3}$.

SOLUTION   Again, the groupings should be treated as individual expressions and be evaluated first.

$$
\begin{aligned}
&45 + 8(9 + 6) + 7 \cdot \overline{6 + 5 \cdot 3} \\
&= 45 + 8 \times 15 + 7 \cdot \overline{6 + 15} && \text{(9 + 6) and } 5 \cdot 3 \text{ computed} \\
&= 45 + 120 + 7 \times 21 && 8 \times 15 \text{ and } \overline{6 + 15} \text{ computed} \\
&= 45 + 120 + 147 && 7 \times 21 \text{ computed} \\
&= 312 && \text{Final answer}
\end{aligned}
$$

There are many cases where a grouping occurs within another grouping. The inner grouping should be evaluated first by applying the rules of expressions. Then the outer grouping can be evaluated and, finally, the entire expression.

EXAMPLE 2-23   Evaluate the expression $5 + 3\,[6 \cdot 4 + 7(3 \cdot 4 + 9)]$.

SOLUTION   Note that the $(3 \cdot 4 + 9)$ grouping falls within the grouping bounded by brackets. This inner grouping must be evaluated first.

$$
\begin{aligned}
&5 + 3[6 \cdot 4 + 7(3 \cdot 4 + 9)] \\
&= 5 + 3[6 \cdot 4 + 7(12 + 9)] && 3 \cdot 4 \text{ computed} \\
&= 5 + 3[6 \cdot 4 + 7 \times 21] && (12 + 9) \text{ computed} \\
&= 5 + 3[24 + 147] && 6 \cdot 4 \text{ and } 7 \times 21 \text{ computed} \\
& && \text{because multiplication has} \\
& && \text{a higher priority than} \\
& && \text{addition} \\
&= 5 + 3 \times 171 && 24 + 147 \text{ computed} \\
&= 5 + 513 && 3 \times 171 \text{ computed} \\
&= 518 && \text{Final answer}
\end{aligned}
$$

Note that each grouping was computed to be a single number before combining with terms outside the grouping.

EXAMPLE 2-24   Evaluate $396 - [85 + 6 \cdot 2 - 4(3 \cdot 5 + 6)]$.

SOLUTION

$$396 - [85 + 6 \cdot 2 - 4(3 \cdot 5 + 6)]$$
$$= 396 - [85 + 6 \cdot 2 - 4(15 + 6)]$$
$$= 396 - [85 + 6 \cdot 2 - 4 \cdot 21]$$
$$= 396 - [85 + 12 - 84]$$
$$= 396 - 13$$
$$= 383$$

Note that the bracketed grouping was subtracted, whereas the $(3 \cdot 5 + 6)$ was multiplied by the 4.

Exponents

Exponents are, by their very nature, a grouping of themselves. There is, for example, a difference between 3^{5+2} and $3^5 + 2$. In the first case, the 2 is part of the superscript and, therefore, is to be added to the 5 to obtain the complete exponent. Thus, 3^{5+2} equals 3^7 or 2187. In the second case, the 2 is not part of the exponent, but should be added to 3^5. Thus, $3^5 + 2$ is equal to $243 + 2$ or 245. Again, note that whatever is superscripted is part of the exponent.

EXAMPLE 2-25   Evaluate $A^{A+B} + B$ if A is 3 and B is 2.

SOLUTION $A^{A+B} + B = 3^{3+2} + 2 = 3^5 + 2 = 245$.

EXAMPLE 2-26   Evaluate $R^{3T-S^2} + T^2$ if R is 2, S is 3, and T is 4.

SOLUTION   The entire exponent is a grouping and should be treated as an individual expression:

$R^{3T-S^2} + T^2$	
$= 2^{3 \cdot 4 - 3^2} + 4^2$	Substituted known values
$= 2^{12-9} + 4^2$	Computed $3 \cdot 4$ and 3^2
$= 2^3 + 4^2$	Computed $12 - 9$
$= 8 + 16$	Computed 2^3 and 4^2
$= 24$	Final answer

PROBLEMS

Evaluate the following expressions:

2-65. $7 + 6(3 + 2)$

2-66. $3.95 + 2.6(4.3 - 2.6)$

2-67. $14.95(2.1 + 3) - 6.2(4.9 + 7)$

2-68. $36.2(4.9 - 3.1) + 6(2.3 + 3.2) - 1.39(7.7 + 6.3 - 10.9)$

2-69. $2.19[3 - 2 + 6.3] - (4.9 + 7) + \overline{3.3 \cdot 7.2}$

2-70. $7.6 \cdot \overline{2.19 + 6.42} + 7.36 + \overline{3.25 + 9} - [2.19 - 1.36 + 7]$

2-71. $3 - (3 - \overline{3 - 2})$

2-72. $45 - 6[3 + \overline{9 - 7}]$

2-73. $\overline{2.6 - 3(4 + 2)} + 6(2 + 1)$

2-74. $[(43.2 - 7) + 6(2.19 + 7)]$

2-75. $[21.6(1.39 + 2 \cdot 3.65) + 7] \cdot [2.61(3.92 + 3)]$

2-76. $\{3.95 + 7.3(2.169 + 7.2 \cdot \overline{3.15 + 6.3})\}$

2-77. $2.19^{3+2} + 3.6$

2-78. $3.16^{7-2^2} - 20$

Assuming that A is 1, B is 2, C is 3, D is 4, E is 5, evaluate the following expressions:

2-79. $A + B(C + D)$

2-80. $AB + C(A + B)$

2-81. $3A(B + C) + 2(A + B)$

2-82. $3B(A + C) + 4(C + D + E)$

2-83. $10A\overline{B + C} - 2C(2E - 3) + 4[A + 3B]$

2-84. $12(E + 3) - \overline{6 + E + B} + 4[A + B + E]$

2-85. $A[B + C(A + B)]$

2-86. $3A[2B + 3C + (A + E)]$

2-87. $E - [C - (A - \overline{B - 1})]$

2-88. $\overline{3E - 4}[A + C(D - 2A)]$

2-89. $2B^2 - A(4A^2 - 3)$

2-90. $3A(C^3 + B)$

2-91. $4B^2(3C^3 - 2C + 4)$

2-92. $5C^3(4A - 2 + 6AB^2C) - 2A^2B^2$

2-93. $3E^2 + 4A(3B^2 - \overline{A^2 + 4})(A + 3)$

2-94. $\{4E^2 - 2[3E + 6A^2(E^2 - 3B^3)]\}(3E - 9)$

2-95. $3D^2 + 6\{4E^2 - 3[3C - (8A - \overline{4C - 3B})]\}$

2-96. $4E^3 + 3\{(6A + 3B)[3A^2 + \overline{B^2 - 2A} + D^2(3C - 2A)] + 4C^2\}$

2-97. $B^{C-A} + D$

2-98. $4C^{3A+B} - E$

2-99. B^{B^B}

2-100. $B^{2+C^2} - 2C^{A+2}$

2-101. $B^{B+2}(2C^{A+B} + 3C^2)$

2-102. $E^{B+1} - (2B^{B+1} + 3CD)$

3

ALGEBRAIC ADDITION
AND SUBTRACTION

A pessimist once declared, "There's nothing new under the sun." This could very well be the theme for this chapter, for it is merely an application of those principles the reader already knows about addition and subtraction. First, we examine the negative number system more closely, then apply this system to the addition and subtraction of algebraic expressions. Finally, we consider the effect of multiple signs of grouping on these processes.

3-1 NEGATIVE NUMBERS

It seems that every time our friendly mathematician runs into a brick wall, he invents a new system of mathematics to overcome it. In our early history, we counted in discrete units such as 3 saber-toothed tigers, 4 mammoths, or 6 dinosaurs. Later, to account for the parts of a mammoth or dinosaur he acquired through clubbing competitions, he invented the fractional system. Still later, Mr. Fahrenheit discovered that there were places in the world that fell below zero on his temperature scale. Further, an enterprising John Dillinger withdrew $20 from a bank where he did not have an account, resulting in a negative balance. For these and other reasons, our erudite mathematicians invented the negative number system.

This number system is used extensively in electronics: the grid of a vacuum tube must have a negative voltage on it for the tube to amplify properly. The output of an inverting operational amplifier is the opposite polarity from its input. That is, if we input $+5$ volts (V), it will output -5 V; if we input -2 V, it will output $+2$ V. Thus, it is essential that those in electronics become familiar with the negative number system.

The Number Line

The most straightforward illustration of our number system is the number line (Fig. 3-1). It looks very much like a horizontal thermometer scale. Positive numbers are to the right of the zero and negative numbers to the left.

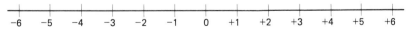

FIGURE 3-1   The number line.

19

Addition

When we add positive numbers on the number scale, we move to the right. Figure 3-2 illustrates the addition of +2 and +3. We start at +2 and move three places to the right, obtaining +5. This is shown in algebraic notation as

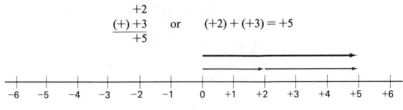

$$\begin{array}{r} +2 \\ (+)\,+3 \\ \hline +5 \end{array} \qquad \text{or} \qquad (+2)+(+3)=+5$$

FIGURE 3-2   Adding +3 to +2.

Note that there are three unary signs: the + before each of the three numbers; and one operation sign shown as a (+) on the left and an unparenthesized + on the right. The unary indicates the sign (positive or negative) of the individual number and the operation sign indicates that the process is addition, rather than subtraction, multiplication, or division.

In adding a negative number to any other number, we move to the left on the scale (Fig. 3-3). Here, a −3 is being added to a −2. To do this, we start at the −2 mark and move three places to the left, arriving at our destination, −5. Note that we moved to the left, for we were adding a negative number. In algebra, this addition is shown:

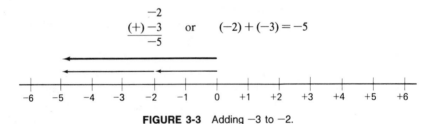

$$\begin{array}{r} -2 \\ (+)\,-3 \\ \hline -5 \end{array} \qquad \text{or} \qquad (-2)+(-3)=-5$$

FIGURE 3-3   Adding −3 to −2.

Again, each number has a unary sign (−) and there is one operation sign, (+), indicating addition.

We can also add −3 to a +2 (Fig. 3-4). We start at +2, go three jumps in the negative direction, and arrive at −1. Shown algebraically:

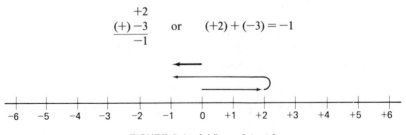

$$\begin{array}{r} +2 \\ (+)\,-3 \\ \hline -1 \end{array} \qquad \text{or} \qquad (+2)+(-3)=-1$$

FIGURE 3-4   Adding −3 to +2.

To complete this fearsome foursome we can add $+3$ to -2 (Fig. 3-5). Shown algebraically:

$$
\begin{array}{r}
-2 \\
(+)\ +3 \\
\hline
+1
\end{array}
\quad \text{or} \quad (-2) + (+3) = +1
$$

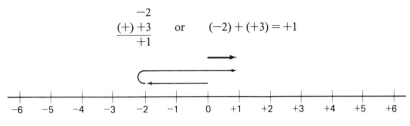

FIGURE 3-5 Adding $+3$ to -2.

What have we learned in our four illustrations?

$$(+2) + (+3) = +5$$
$$(-2) + (-3) = -5$$
$$(+2) + (-3) = -1$$
$$(-2) + (+3) = +1$$

We can summarize our findings as follows:

1. *To add numbers of like signs, add the absolute values and prefix the answer with the common sign.*

2. *To add numbers of unlike sign, find the difference of their absolute values and give the answer the sign of the larger.*

The absolute value of a number is that number prefixed with a positive sign. For example, the absolute value of both $+5$ and -5 is $+5$. Thus, statement 1 above indicates that when the two numbers have like signs (both are positive or both are negative), simply add the two values as if they were both positive numbers and give the answer the same sign as the original numbers. If, however, the two original numbers have differing signs (one positive and the other negative), statement 2 indicates that we must subtract the two as if they were both positive and give the answer the same sign as the larger of the two originally had.

EXAMPLE 3-1 Perform the following additions:

$$
\begin{array}{llllllll}
\text{(a)} & \begin{array}{r} +3 \\ (+)\ +9 \end{array} &
\text{(b)} & \begin{array}{r} -57 \\ (+)\ -23 \end{array} &
\text{(c)} & \begin{array}{r} +13 \\ (+)\ -12 \end{array} &
\text{(d)} & \begin{array}{r} -165 \\ (+)\ +126 \end{array}
\end{array}
$$

SOLUTION Problems (a) and (b) have like signs, so we have merely to add them and give the sign of the numbers to the answer:

$$
\begin{array}{llll}
\text{(a)} & \begin{array}{r} +3 \\ (+)\ \ +9 \\ \hline +12 \end{array} &
\text{(b)} & \begin{array}{r} -57 \\ (+)\ -23 \\ \hline -80 \end{array}
\end{array}
$$

Problems (c) and (d) have differing signs, and we must apply rule 2:

$$
\begin{array}{llll}
\text{(c)} & \begin{array}{r} +13 \\ (+)\ -12 \\ \hline +1 \end{array} &
\text{(d)} & \begin{array}{r} -165 \\ (+)\ +126 \\ \hline -39 \end{array}
\end{array}
$$

Note that in (c) the 13 is larger than the 12, so we gave the answer the same sign as the 13 originally had. In (d) the 165 is the larger, so we gave the answer its sign.

It should be noted that all calculators obey these rules. The only difficulty is how to enter a negative number. On most units, merely enter the 12, then press a $\boxed{\text{CHANGE SIGN}}$ key; this changes the $+12$ to a -12. The result can be manipulated with a variety of operations.

Many times, it is necessary to add a long list of positive and negative numbers. To do this manually, add all the positive numbers together, then add all the negative numbers together. Finally, subtract the negative numbers from the positive numbers. However, to add such a list on a calculator, it is not necessary first to segregate the positives and the negatives.

EXAMPLE 3-2   Add the following list of numbers: 26, -33, 10, 16, -27, 5, -49.

SOLUTION   To add them manually, segregate:

Positives	Negatives
26	-33
10	-27
16	-49
5	-109
57	

The final sum is: $(+57) + (-109) = -52$.

If the addition were done on a calculator, we would add 26 to -33, obtaining -7, this to 10, giving 3, and so forth.

Subtraction

Let us consider the subtraction problem $(+5) - (+3)$. In this example we are subtracting $(-)$ one positive number $(+3)$ from another $(+5)$. Another way of looking at this is: If I start at $+3$, how far do I have to go, and in what direction, to reach $+5$? If I move in the positive direction, the number is positive; if in the negative direction, the number is negative. Figure 3-6 illustrates the process; the answer is $+2$.

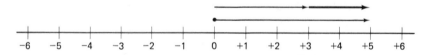

FIGURE 3-6   Subtracting $+3$ from $+5$.

Next, let us apply this analysis to the problem $(-5) - (-3)$. Remember, the question is: How far is it from the second number (-3) to the first number (-5)? We must move two units in the negative direction; therefore, the answer is -2 (**Fig. 3-7**).

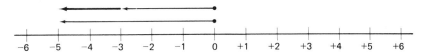

FIGURE 3-7 Subtracting −3 from −5.

We shall now try $(+5) - (-3)$ (Fig. 3-8). The answer is, of course, $(+8)$. Finally, we shall examine $(-5) - (+3)$ (Fig. 3-9), resulting in −8. Note that in each of these cases, we were trying to find the difference (distance) from the second arrowhead to the first.

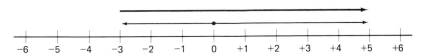

FIGURE 3-8 Subtracting −3 from +5.

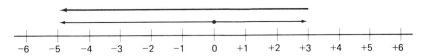

FIGURE 3-9 Subtracting +3 from −5.

What have we come up with?

(a) $(+5) - (+3) = +2$
(b) $(-5) - (-3) = -2$
(c) $(+5) - (-3) = +8$
(d) $(-5) - (+3) = -8$

Examine these results closely. Notice that we can obtain the identical answers by changing both the operation and sign of the subtrahend (the second number) as follows:

(a) $(+5) + (-3) = +2$
(b) $(-5) + (+3) = -2$
(c) $(+5) + (+3) = +8$
(d) $(-5) + (-3) = -8$

From this, we can devise the rule:

To algebraically subtract, change the sign of the subtrahend and add.

EXAMPLE 3-3 Subtract the second number from the first:

(a) 3, −10 (b) 7, +20 (c) −35, +6 (d) −10, −16

SOLUTION In each case, we must change the sign of the subtrahend (the second number) and add.

(a)
$$
\begin{array}{r}
+3 \\
(-)\,-10
\end{array}
\quad \text{change to} \quad
\begin{array}{r}
+3 \\
\underline{(+)\,+10} \\
+13 \quad \textit{Answer}
\end{array}
$$

(b)
$$
\begin{array}{r}
+7 \\
(-)\,+20 \\
\end{array}
$$
change to
$$
\begin{array}{r}
+7 \\
(+)\,-20 \\
\hline
-13 \quad \textit{Answer} \\
\end{array}
$$

(c)
$$
\begin{array}{r}
-35 \\
(-)\ +6 \\
\end{array}
$$
change to
$$
\begin{array}{r}
-35 \\
(+)\ -6 \\
\hline
-41 \quad \textit{Answer} \\
\end{array}
$$

(d)
$$
\begin{array}{r}
-10 \\
(-)\,-16 \\
\end{array}
$$
change to
$$
\begin{array}{r}
-10 \\
(+)\,+16 \\
\hline
+6 \\
\end{array}
$$

PROBLEMS

Add the following:

3-1. (a) 3, +5 (b) −2, +6 (c) −40, +35 (d) −86, −37
3-2. (a) 45, −36 (b) −72, −67 (c) −46, +92 (d) +37, +39
3-3. (a) +1.39, −7.62 (b) −8.76, +6.32 (c) −3.46, −10.92 (d) +8.62, +7.39
3-4. (a) +0.93, −7.215 (b) 24.62, +37.91 (c) −27.25, −2.36 (d) −3.917, +8.213
3-5. 35, −92, +6, +10, +13, −46
3-6. −72, −9, +6, −33, +10, +16
3-7. +17.32, −16.19, −4.27, −3.62, +4.78, +7.62
3-8. −1.315, +6.345, +4.468, −2.223, −7.621, −4.819

Subtract the second number from the first:

3-9. (a) +6, +9 (b) −7, +3 (c) −26, +62 (d) −73, −129
3-10. (a) −20, −16 (b) −37, −76 (c) −64, +29 (d) +739, −926
3-11. (a) +7.31, −3.68 (b) −2.87, +3.73 (c) −2.77, −7.63 (d) +9.31, +10.10
3-12. (a) +0.816, −6.315 (b) 17.36, +73.91 (c) −25.22, −36.17 (d) −3.215, +7.625

3-2 ADDING ALGEBRAIC EXPRESSIONS

In the real world, we never arithmetically add unlike quantities: miles to miles per gallon, sweet potatoes to monkeys, Japanese beetles to Chinese pheasants, but rather we first assure that the items are alike. This is also true in algebra, for we first assure the literals are identical and have identical powers, then we can add. This addition is performed by adding the coefficients.

To add like terms, add the coefficients.

EXAMPLE 3-4 Add the following:

(a) $6R^2LT$ and $-2R^2LT$ (b) RS and $7RS$ (c) $-75Z_2V$ and $-39Z_2V$

SOLUTION In each case the literal quantities are identical, so we can add the coefficients.

(a)
$$
\begin{array}{r}
6R^2LT \\
(+)\,-2R^2LT \\
\hline
4R^2LT \\
\end{array}
$$

(b)
$$
\begin{array}{r}
RS \\
(+)\,+7RS \\
\hline
8RS \\
\end{array}
$$

(c)
$$
\begin{array}{r}
-75Z_2V \\
(+)\ -39Z_2V \\
\hline
-114Z_2V \\
\end{array}
$$

Note in (b) that the coefficient is 1. Just for fun, let us assume that R is 10, L is 3, and T is 4 and examine the first addition problem.

$$6R^2LT = 6 \times 10^2 \times 3 \times 4 = 7200$$
$$-2R^2LT = -(2 \times 10^2 \times 3 \times 4) = -2400$$
$$4R^2LT = 4 \times 10^2 \times 3 \times 4 = 4800$$

But what can be done about the addition of unlike quantities? Assume that $2RL$ must be added to $2RC$. The result is $2RL + 2RC$, for the coefficients cannot be combined.

EXAMPLE 3-5   Add the following:

(a) $P, 2T$ (b) $4R, -7Z$ (c) $3V^2, 2V$

SOLUTION   In each case the literals or powers of the literals are not identical. The solutions are:

(a) $\begin{array}{r} PT \\ (+)\,2T \\ \hline PT + 2T \end{array}$ (b) $\begin{array}{r} 4R \\ (+)\,-7Z \\ \hline 4R - 7Z \end{array}$ (c) $\begin{array}{r} 3V^2 \\ (+)\,2V \\ \hline 3V^2 + 2V \end{array}$

Again, the only way the coefficients can be combined is if the literal part of the terms are identical in every manner.

Binomials, Trinomials, and Polynomials

The terms of any size expression can be combined if the literal portions are identical. Rearrange the like terms into columns and add as was done for monomials.

EXAMPLE 3-6   Add $4 + 3C$ to $7 + 9C$.

SOLUTION

$$\begin{array}{r} 4 + 3C \\ 7 + 9C \\ \hline 11 + 12C \end{array}$$

EXAMPLE 3-7   Add $5Z^2 - R^2$ to $-7Z^2 + R^2$.

SOLUTION

$$\begin{array}{r} 5Z^2 - R^2 \\ -7Z^2 + R^2 \\ \hline -2Z^2 + 0R^2 = -2Z^2 \end{array}$$

EXAMPLE 3-8   Add $4R^2S - 2T^2U + 3SU$ to $-5R + 6R^2S + 2SU + Q^2$.

SOLUTION   Rearrange to put like terms in column with like terms:

$$\begin{array}{l} 4R^2S - 2T^2U + 3SU \\ -5R + Q^2 +\ 6R^2S + 2SU \\ \hline -5R + Q^2 + 10R^2S - 2T^2U + 5SU \end{array}$$

Note that similar terms combined, but dissimilar terms are merely carried through to the final expression.

EXAMPLE 3-9 Add $1.3V_1V_2 - 6V_2^2 + V_1V_2V_3 - V_3$ to $4V_1V_3 - 7.6V_1V_2 + 6V_3 - 10V_1V_2$.

SOLUTION

$$
\begin{array}{l}
1.3V_1V_2 - 6V_2^2 + V_1V_2V_3 - \ \ V_3 \\
4V_1V_3 - 10\ \ V_1V_2 + 6V_3 \\
- 7.6V_1V_2 \\
\hline
4V_1V_3 - 16.3V_1V_2 - 6V_2^2 + V_1V_2V_3 + 5V_3
\end{array}
$$

Note that there were two V_1V_2 terms in the second expression.

PROBLEMS

Add the following:

3-13. $2R,\ 4R$

3-14. $4R^2,\ -3R^2$

3-15. $1.6QP,\ -3.7QP$

3-16. $-1.9LTP,\ -3.4LTP$

3-17. $2R,\ 3T$

3-18. $2RS,\ 4RST$

3-19. $-1.7LM,\ -3.9L^2M$

3-20. $1.935L^2T,\ 2.315L^2T$

3-21. $7Z + 4T,\ 8Z + 7T$

3-22. $4M_1M_2 + 7L_1L_2,\ -6M_1M_2 - 8L_1L_2$

3-23. $4.5Z_1Z_2 - 3.6Z_2Z_3,\ -2.1Z_1Z_3 - 7.6Z_2Z_3$

3-24. $39.62I_1L + 6I_2L,\ 25.2I_1T + 7.9I_2L$

3-25. $4A - 3B + C,\ 6A + 2B + 4AB + CD$

3-26. $-7.3Z_1^2Z_2 + 6.9Z_1Z_2^2 + 8.3Z_2^2,\ -2.3Z_1^2Z_2 + 2.3Z_1Z_2^2 + 7.9Z_2^2Z_2^2$

3-27. $-4L_1^2L_2^4 + 6L_1^4L_2^2L_3 + 7L_1L_2L_3^5,\ 6L_1^2L_2^2 + 7L_1^4L_2^2L_3$

3-28. $3.96f_1^2f_2^2 - 2.92f_1^2f_3 + 7.21f_1^2f_2 + 10.21f_1^2f_3^2,\ 4.56f_1^2f_2 - 2.15f_1^2f_3^2 + 6.72f_1^2\,f_3^2$

3-29.
$$
\begin{array}{l}
4I_1^2R_1 - 3I_2^2R_2 + \ \ 7I_1^2R_2 \\
7I_1^2R_1 + 4I_2^2R_2 - \ \ 6I_1^2R_2 \\
-3I_1^2R_1 + 7I_2^2R_2 - 13I_1^2R_2
\end{array}
$$

3-30.
$$
\begin{array}{l}
2VI_1 - 6VI_2 - 5VI_3 \\
-5VI_1 - 7VI_2 - 6VI_3 \\
6VI_1 - 6VI_2 - 9VI_3
\end{array}
$$

3-31.
$$
\begin{array}{l}
-2fL^2\ \ + 6f^2L + 3f^2L^2 \\
-5f^2L^2 + 7fL^2 - 2\,f^2L \\
3f^2L - 8fL^2 + 6\,f^2L^2
\end{array}
$$

3-32.
$$
\begin{array}{l}
3X_LX_C - 6X_LX_T + 7Z^2 \\
4Z^2 + 9X_LX_C - 7X_LX_T \\
-7X_LX_T - 3X_LX_C + 6Z^2
\end{array}
$$

3-3 SUBTRACTING ALGEBRAIC EXPRESSIONS

To subtract, change the sign of the subtrahend and add.

The subtrahend may be a monomial, binomial, trinomial, or polynomial. However, in each case its sign is changed. Furthermore, the sign of each term within the subtrahend expression is changed.

Monomials

When subtracting monomials, after the sign has been changed, the two may be combined only if their literal portions are identical, as in addition.

EXAMPLE 3-10   Subtract the following:

    (a) $13ab$ from $-4ab$ (b) $16a^2b$ from $12a^2b$ (c) R^2S^2 from $-2R^2S^2$

SOLUTION

(a)
$$\begin{array}{r} -4ab \\ (-)+13ab \\ \hline \end{array}$$
change to
$$\begin{array}{r} -4ab \\ (+)-13ab \\ \hline -17ab \end{array}$$

(b)
$$\begin{array}{r} +12a^2b \\ (-)+16a^2b \\ \hline \end{array}$$
change to
$$\begin{array}{r} +12a^2b \\ (+)-16a^2b \\ \hline -4a^2b \end{array}$$

(c)
$$\begin{array}{r} -2R^2S^2 \\ (-)+R^2S^2 \\ \hline \end{array}$$
change to
$$\begin{array}{r} -2R^2S^2 \\ (+)-R^2S^2 \\ \hline -3R^2S^2 \end{array}$$

Note that in each of the foregoing cases, the literal portions are identical.

EXAMPLE 3-11   Subtract the following:

    (a) $2ab^2$ from $4ab$ (b) $-4R_1R_2$ from $2R_1R_2^2$ (c) $7XZ$ from $-2XZM$

SOLUTION   In each of these cases the literal portions of the terms differ.

(a)
$$\begin{array}{r} +4ab \\ (-)+2ab^2 \\ \hline \end{array}$$
change to
$$\begin{array}{r} +4ab \\ (+)-2ab^2 \\ \hline 4ab-2ab^2 \end{array}$$

(b)
$$\begin{array}{r} +2R_1R_2^2 \\ (-)-4R_1R_2 \\ \hline \end{array}$$
change to
$$\begin{array}{r} +2R_1R_2^2 \\ (+)+4R_1R_2 \\ \hline 2R_1R_2^2+4R_1R_2 \end{array}$$

(c)
$$\begin{array}{r} -2XZM \\ (-)+7XZ \\ \hline \end{array}$$
change to
$$\begin{array}{r} -2XZM \\ (+)-7XZ \\ \hline -2XZM-7XZ \end{array}$$

Binomials, Trinomials, and Polynomials

For expressions with more than one term, the signs on all terms of the subtrahend must be changed, identical terms lined up in the same columns, and the terms added as in addition.

EXAMPLE 3-12   Subtract the following:

    (a) $3ab$ from $7a^2+2ab$

(b) $4L^2M - 2LM^2$ from $L^3 - 3L^2M + 7LM^2 - M^3$

(c) $-4X_1^3X_2 + 7X_1^2X_2^2 - 7X_1X_2^2 + 6X^4$ from $3X_2^4 + 7X_1^2X_2^2 + 2X^4$

SOLUTION

(a)

$7a^2 + 2ab$		$7a^2 + 2ab$
$(-) + 3ab$	change to	$(+) - 3ab$
		$7a^2 - ab$

(b)

$L^3 - 3L^2M + 7LM^2 - M^3$		$L^3 - 3L^2M + 7LM^2 - M^3$
$(-)\quad 4L^2M - 2LM^2$	change to	$(+) - 4L^2M + 2LM^2$
		$L^3 - 7L^2M + 9LM^2 - M^3$

Note that the signs of all subtrahend terms were changed.

(c)

$$\begin{array}{ll} & 3X_2^4 \qquad\qquad + 7X_1^2X_2^2 \qquad\qquad + 2X_2^4 \\ (-) & \quad -4X_1^3X_2 + 7X_1^2X_2^2 - 7X_1X_2^3 + 6X_2^4 \end{array} \quad \text{change to}$$

$$\begin{array}{ll} & 3X_2^4 \qquad\qquad + 7X_1^2X_2^2 \qquad\qquad + 2X_2^4 \\ (+) & \quad +4X_1^3X_2 - 7X_1^2X_2^2 + 7X_1X_2^3 - 6X_2^4 \\ \hline & 3X_2^4 + 4X_1^3X_2 + \qquad 0 + 7X_1X_2^3 - 4X_2^4 \\ = & 3X_2^4 + 4X_1^3X_2 + 7X_1X_2^3 - 4X_2^4 \end{array}$$

Note that the signs of all subtrahend terms were changed.

PROBLEMS

Subtract the following:

3-33. $3\,X^2$ from $7\,X^2$

3-34. $5Z_1^2$ from $-7Z_1^2$

3-35. $-A^2B$ from $3\,A^2B$

3-36. $-P^2WZ$ from $9\,P^2WZ$

3-37. $75\,AZ$ from $26\,A^2Z$

3-38. $-23\,X^2Y^2Z^4$ from $27\,X^3\,Y^2Z^4$

3-39. $7.313\,P^2Q^2RS$ from $-8.216\,P^2Q^2RS^2$

3-40. $-22.16\,A^2BR$ from $-16.23\,A^2B$

3-41. $35\,VI + 27\,P$ from $-26\,VI + 27\,P + 36I^2R$

3-42. $-26.31I_1R_1 + 32.19I_2R_2$ from $7.39I_1R_1 - 26.29I_2R_2 + 7.31I_1R_2 + 3.19I_2R_1$

3-43. $-2I^2Z + 7I^2R - 6I^2X_L$ from $35I^2X_C + 17I^2Z$

3-44. $A^2B + BC^2 - 35\,AD^3$ from $-2BC^2 + 6\,RS$

3-45. $3\,M^3N^2 - 2\,M^2N^3 + 7N^5$ from $5\,M^5 - 7\,M^4N^2 + 8\,M^3N^2 - 2N^5$

3-46. $7WX^2Z - 2WXZ^2 + WZ + X^2Z$ from $22\,WXZ^2 + W^2XZ + X^2Z$

3-47. $1.315f_1^2f_0 - 2.613f_1f_0^2$ from $3.153f_1^3 + 2.315f_1f_0^2 - f_0^3$

3-48. $28f_1X_C - 7.31f_2X_L + 7.98f_1Z$ from $f_1X_L + 2.15f_2X_L$

3-4 SIGNS OF GROUPING

When dealing with signs of grouping, the inner group is always calculated first. When adding, the signs of grouping can be removed and similar terms combined.

EXAMPLE 3-13 Perform the following additions:

(a) $(3A + C) + (A + 3C + D)$

(b) $(2W^2P^2 + 3WP^3 + 4P^4) + (-7W^4 - W^3P + 7WP^3 - 6P^4)$

SOLUTION Expressed as groups:

(a) $(3A + C) + (A + 3C + D)$

$\quad = 3A + C + A + 3C + D$

$\quad = 4A + 4C + D$

Note that when the parentheses are removed, the $+$ operation sign is also removed.

(b) $(2W^2P^2 + 3WP^3 + 4P^4) + (-7W^4 - W^3P + 7WP^3 - 6P^4)$

$\quad = 2W^2P^2 + 3WP^3 + 4P^4 - 7W^4 - W^3P + 7WP^3 - 6P^4$

$\quad = -7W^4 - W^3P + 2W^2P^2 + 10WP^3 - 2P^4$

Note that similar terms were combined.

When subtracting, the sign of each of each term of the subtrahend must first be changed, then the sign of grouping removed, and the expression treated as in addition. Consider the problem $4Z^3 - (4M^3 + Z^3)$. The minus sign is not a unary, but an indication of a subtraction operation. Because of this, the quantity $(4M^3 + Z^3)$ is a subtrahend, and all signs within it must be changed. Therefore,

$$4Z^3 - (4M^3 + Z^3) = 4Z^3 + (-4M^3 - Z^3)$$
$$= 4Z^3 - 4M^3 - Z^3$$
$$= 3Z^3 - 4M^3$$

EXAMPLE 3-14 Perform the following subtractions:

(a) $(3X^2 - X) - (2X^2 - 7X)$

(b) $(3W_1^2W_2 + 6W_1W_2^2 + W_2^3) - (7W_1^3 + 3W_1W_2^2 - 9W_2^3)$

SOLUTION

(a) $(3X^2 - X) - (2X^2 - 7X) = (3X^2 - X) + (-2X^2 + 7X)$

$\quad\quad\quad\quad\quad\quad\quad\quad\quad\quad = 3X^2 - X - 2X^2 + 7X$

$\quad\quad\quad\quad\quad\quad\quad\quad\quad\quad = X^2 + 6X$

Note that the signs of both subtrahend terms were changed.

(b) $(3W_1^2W_2 + 6W_1W_2^2 + W_2^3) - (7W_1^3 + 3W_1W_2^2 - 9W_2^3)$

$\quad = (3W_1^2W_2 + 6W_1W_2^2 + W_2^3) + (-7W_1^3 - 3W_1W_2^2 + 9W_2^3)$

$\quad = 3W_1^2W_2 + 6W_1W_2^2 + W_2^3 - 7W_1^3 - 3W_1W_2^2 + 9W_2^3$

$\quad = -7W_1^3 + 3W_1^2W_2 + 3W_1W_2^2 + 10W_2^3$

Multiple Groupings

When a grouping occurs within a grouping, the inner one must be treated first, being careful to change all the signs within a grouping preceded by a minus sign.

EXAMPLE 3-15   Perform the following operation: $3XY - [WX - (4XY + 9)]$

SOLUTION

$$
\begin{aligned}
&3XY - [WX - (4XY + 9)] &&\text{Original problem}\\
&= 3XY - [WX - 4XY - 9] &&\text{Inner grouping subtracted}\\
&= 3XY - WX + 4XY + 9 &&\text{Outer grouping subtracted}
\end{aligned}
$$

EXAMPLE 3-16   Perform the following operation:

$$21I^2R - \{36I^2Z + (25I^2R - 6) - [I^2R + 6I^2Z - (2I^2R + 7)]\} + 4I^2R$$

SOLUTION

$$
\begin{aligned}
&21I^2R - \{36I^2Z + (25I^2R - 6) - [I^2R + 6I^2Z - (2I^2R + 7)]\} + 4I^2R\\
&= 21I^2R - \{36I^2Z + 25I^2R - 6 - [I^2R + 6I^2Z - 2I^2R - 7]\} + 4I^2R\\
&= 21I^2R - \{36I^2Z + 25I^2R - 6 - I^2R - 6I^2Z + 2I^2R + 7\} + 4I^2R\\
&= 21I^2R - 36I^2Z - 25I^2R + 6 + I^2R + 6I^2Z - 2I^2R - 7 + 4I^2R\\
&= -I^2R - 30I^2Z - 1
\end{aligned}
$$

PROBLEMS

Perform the following operations:

3-49. $3 + (6 + 3)$

3-50. $17.3 + (16.9 + 12.1)$

3-51. $3AB + C + (2AB - 3C)$

3-52. $-6Y^2 + (3Y^2 + Y) + 6Y$

3-53. $(3VI - 6P + 4Q) + (2VI + 7P + 27Q)$

3-54. $(I_1^2 - 3I_2^2) + (-7I_2^2 + 6I_1^2)$

3-55. $25 - (7 + 3)$

3-56. $72 - (126 - 25)$

3-57. $17.63 - (21 - 2.75 - 3.12)$

3-58. $(123.9 - 28) - (117.3 - 26.1)$

3-59. $2X_L^2 - (7X_C^2 - Z^2)$

3-60. $3V_1 - (IR - 2V_1 + V_2)$

3-61. $7L_1 - (6M_1 - 3L_1 + 2L_2)$

3-62. $23M_1 - (6M_1 - 3M_1 + 3L_1 - 5L_2)$

3-63. $(W - X) - (W - Y) - (W - Z)$

3-64. $(C_1 - C_2 + C_3) - (2C_2 - 2C_3 + 2C_4) - (3C_3 - 3C_4 + 3C_5)$

3-65. $L_1 - [3L_2 - (13L_1 - 12L_2 - 7L_3)]$

3-66. $2R_1 - [3R_1 - (2R_2 - 3R_1)]$

3-67. $(8R_1^2 - 2R_2^2) - [(3R_1^2 - 4R_2^2) - R_2^2 - R_1^2]$

3-68. $[13Q_1 - (12Q_1 - 8Q_2 + Q_3) - 7] - 8Q_1 - (7Q_2 + 6Q_1)$

3-69. $(8C_1 + 7C_2 - 16C_3) - \{7C_1 - 7C_2 - [8C_1 - (6C_2 - C_3)]\}$

3-70. $(3V_1 - 2V_2) - (6V_1 + V_2) - [(3V_1 - 7V_2) + (6V_1 - 3V_2)]$

4

ALGEBRAIC MULTIPLICATION
AND DIVISION

In this chapter we examine the operations of multiplication and division. First, we look at numerical multiplication—that by which we have been calculating our periodic paychecks, parcel postage, and production profit. We then take up in order the three skills necessary for algebraic multiplication: the treatment of signs, the treatment of exponents, and the treatment of groupings. With these procedures in hand, we shall find that multiplication is as easy for us as for rabbits.

We then consider the operation of division, the most common of which is the arithmetic fraction. After marshaling our fractional skills, we duel with the signs of division, skirmish with exponents, and, finally conquer the division of algebraic expressions.

4-1 SIGNS IN MULTIPLICATION

Multiplication can best be pictured by remembering that it is successive addition. For example, 3×4 means that we are to add 4 three times, $4 + 4 + 4$. This can be represented on the number line as the addition of three line segments, traveling in the positive direction (Fig. 4-1). In a similar manner, $3 \times (-4)$ can be thought of as $(-4) + (-4) + (-4)$ and represented as the addition of three line segments in the negative direction (Fig. 4-2).

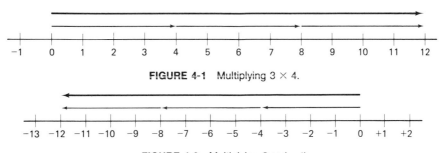

FIGURE 4-1 Multiplying 3×4.

FIGURE 4-2 Multiplying $3 \times (-4)$.

In multiplying $(-3) \times 4$, the result is seen to be the same as multiplying $4 \times (-3)$ or -12. However, when we represent $(-3) \times 4$ on the number line, we run into difficulties. We know that the result is -12. But how can we show this? One way is to recognize that $(-3) \times 4$ is the same as $-(3 \times 4)$. Therefore, we can find the result of 3 × 4 and merely change its sign. To do this on the number line, note that we must rotate the number 12 one-half turn to arrive at our answer, −12 (Fig. 4-3).

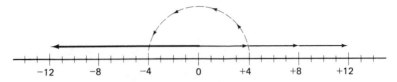

FIGURE 4-3 Multiplying (−3) × 4.

This operation can also be thought of as (−1)(3 × 4). Thus, we can multiply the 3 × 4 to obtain 12, then multiply this by −1, resulting in −12. Note that we have again rotated the result on the number line by one-half turn. This is a general principle we will use extensively in ac electronics:

To multiply any number by −1, rotate the number line by one-half turn.

We can use this principle in analyzing (−3) × (−4) (Fig. 4-4). The expression can be thought of as (−1)[(+3) × (−4)]. Thus, we first find (+3) × (−4), resulting in our favorite number, −12. We then multiply this result by −1, rotating the −12 by one-half turn to +12. Therefore, (−3) × (−4) = +12.

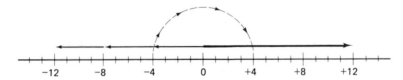

FIGURE 4-4 Multiplying (−3) × (− 4).

Let us now summarize our findings:

(a) (+3) × (+4) = +12
(b) (+3) × (−4) = −12
(c) (−3) × (+4) = −12
(d) (−3) × (−4) = +12

From this the rules of signs in multiplication follow:

1. *Multiplying numbers with like signs results in a positive product.*

2. *Multiplying numbers with unlike signs results in a negative product.*

EXAMPLE 4-1 Multiply:

(a) 13 × 16   (b) −12 × 64   (c) 72 × −35
(d) −26 × −33

SOLUTION Use rules (a) and (b) above:

(a) 13 × 16 = 208 (like signs)   (b) −12 × 64 = −768 (unlike signs)
(c) 72 × −35 = −2520 (unlike signs)   (d) −26 × −33 = 858 (like signs)

It should be noted that electronic calculators obey these rules for signs, making signed arithmetic much easier to do. Try it.

Multiplying Several Numbers

When several numbers are multiplied together, the sign can be determined according to the rule:

When multiplying several numbers, if the number of minus signs is even, the sign of the product is positive. If the number of minus signs is odd, the sign of the product is negative.

The rule makes sense if it is kept in mind that every pair of minus signs being multiplied together results in a plus sign. $(-1) \times (-1) = +1$, for example.

EXAMPLE 4-2 Multiply:

(a) $-6, -5, +3, +2, -5$
(b) $+7, -6, +5, -4, +3, -2, +1$

SOLUTION

(a) The number of minus signs is 3, an odd number. Therefore, the final product is negative:

$$(-6) \times (-5) \times (+3) \times (+2) \times (-5) = -900$$

(b) The number of minus signs is even, so the result is positive:

$$(+7) \times (-6) \times (+5) \times (-4) \times (+3) \times (-2) \times (+1) = +5040$$

PROBLEMS

Multiply:

4-1. (a) 12×36 (b) $(-5) \times (+63)$ (c) $35 \times (-27)$ (d) $(-73) \times (-32)$
4-2. (a) 21×22 (b) $(-39) \times (45)$ (c) $67 \times (-89)$ (d) $(-84) \times (-61)$
4-3. (a) 44×27 (b) $28 \times (-7)$ (c) $(-17) \times (-19)$ (d) $(-45) \times (72)$
4-4. (a) $(-5) \times (62)$ (b) $(-12) \times (-32)$ (c) $(+19) \times (-47)$ (d) 77×66
4-5. (a) 1.73×2.69 (b) $6.32 \times (-7.29)$ (c) $(-23.1) \times 17.6$
     (d) $(-3.16) \times 2.95$
4-6. (a) $176.2 \times (-23.14)$ (b) $(-73.16) \times 1.6$ (c) $(-5.36) \times 57.2$ (d) 27.19×1.6
4-7. $(-3) \times (6) \times (-5) \times (-7)$
4-8. $(36) \times (-4) \times (23) \times (76) \times (-2)$
4-9. $(-12) \times (-19) \times (-16) \times (-13)$
4-10. $(-6) \times (-12) \times (35) \times (-7)$
4-11. $(17.23) \times (-6.19) \times (-2.69)$
4-12. $(-2.16) \times (-7) \times (7.39) \times (-15.61)$

4-2 TERMS WITH EXPONENTS

Many times, terms that must be multiplied contain exponents. Identical literal terms can be combined by multiplication according to the following rules:

If two terms have identical bases, they can be multiplied by adding exponents.

This can be illustrated by using numerals. Let us multiply 3^2 by 3^4 using this rule:

$$3^2 \times 3^4 = 3^{2+4} = 3^6$$
$$9 \times 81 = 729$$

It should be emphasized that the exponents can only be combined if the bases are identical.

EXAMPLE 4-3 Multiply:

(a) a^2 by a^3 (b) R_1 by R_1^2 (c) L_1^3 by L_1^{-1} (d) V^a by V^b

SOLUTION In each case the exponents may be added, for in each problem the bases are identical.

(a) $a^2 \times a^3 = a^{2+3} = a^5$
(b) $R_1 \times R_1^2 = R_1^{1+2} = R_1^3$

Note that where no exponent is indicated, it is assumed to be 1.

(c) $L_1^3 \times L_1^{-1} = L_1^{3+(-1)} = L_1^2$
(d) $V^a \times V^b = V^{a+b}$

Where the bases are not identical, the terms must be left as a product.

EXAMPLE 4-4 Multiply:

(a) $b^2 \times b_1^2$ (b) $X_L \times X_C$ (c) $R_1 \times R_2$

SOLUTION In each case the bases are not identical. Therefore:

(a) $b^2 \times b_1^2 = b^2 b_1^2$ (b) $X_L \times X_C = X_L X_C$ (c) $R_1 \times R_2 = R_1 R_2$

PROBLEMS

Multiply, leaving the result in exponential format:

4-13.   (a) $3^2 \times 3^5$ (b) $4^3 \times 4^9$ (c) $5^7 \times 5^6$ (d) $7^7 \times 7^7$
4-14.   (a) $17^{12} \times 17^2$ (b) $12^2 \times 12^4$ (c) $11^7 \times 11^{12}$ (d) $25^2 \times 25^3$
4-15.   (a) $4^4 \times 4^{-2}$ (b) $7^7 \times 7^{-6}$ (c) $9^{-12} \times 9^{-10}$ (d) $12^{-12} \times 12^7$
4-16.   (a) $17.6^3 \times 17.6^{-4}$ (b) $12.1^{12} \times 12.1^{-17}$ (c) $14.6^9 \times 14.6^{-7}$ (d) $0.139^7 \times 0.139^{-12}$

Multiply:

4-17.   (a) $A^2 \times A^3$ (b) $X_1^2 \times X_1^8$ (c) $P^4 \times P^7$ (d) $V^4 \times V^7$
4-18.   (a) $R_1^2 \times R_1^4$ (b) $C_1^7 \times C_1^9$ (c) $C_7^7 \times C_7^6$ (d) $W^3 \times W^2$
4-19.   (a) $L_1^{-3} \times L_1^5$ (b) $R_1^7 \times R_1^{-10}$ (c) $M_3^{-4} \times M_3^{-7}$
4-20.   (a) $T_1^4 \times T_1^{-7}$ (b) $B_L^{-3} \times B_L^{-4}$ (c) $S^{-R} \times S^{-T}$

4-3 MULTIPLYING ALGEBRAIC EXPRESSIONS

In this section we discuss multiplying algebraic expressions: monomials by monomials, polynomials by monomials, binomials by binomials, polynomials by binomials, and polynomials by polynomials. These operations require that three procedures be used:

1. The sign of the product must be determined as described in Section 4-1.
2. Those literals with identical bases may be multiplied as described in Section 4-2.
3. The coefficients must be multiplied as described next.

Monomials by Monomials

Monomials may be multiplied according to the following rule:

When multiplying monomials, (1) multiply the coefficients using signed arithmetic, and (2) combine literal terms having identical bases by adding the exponents.

Again, those terms having differing bases are shown as products in the final expression.

EXAMPLE 4-5   Multiply $-3AB^2C$ by $9A^2CD$.

SOLUTION   The coefficients must first be multiplied, yielding $(-3) \times 9$ or -27. Next, examine the A literal. Both expressions have one, so they can be combined:

$$A \times A^2 = A^{1+2} = A^3$$

Next, examine the B literal. Since the second expression does not contain a B literal, the B will be carried through to the final expression. Now examine the C literal:

$$C \times C = C^{1+1} = C^2$$

Finally, the D literal will be carried through to the final product. In summary:

$$(-3AB^2C)(9A^2CD) = (-3)(+9)(A^{1+2}B^2C^{1+1}D) = -27A^3B^2C^2D$$

EXAMPLE 4-6   Multiply

    (a) $13X_LX_C^2$ by $2X_LX_C$ (b) $12AB^2C^{-2}$ by AB^2C^3
    (c) $16V_1^2V_2^3V_3^2$ by $-7V_1V_2^2V_3^{-4}$

SOLUTION

    (a) $(13X_LX_C^2)(2X_LX_C) = (2 \times 13)X_L^{1+1}X_C^{2+1} = 26X_L^2X_C^3$
    (b) $(12AB^2C^{-2})(AB^2C^3) = (12 \times 1)(A^{1+1}B^{2+2}C^{-2+3}) = 12A^2B^4C$
    (c) $(16V_1^2V_2^3V_3^2)(-7V_1V_2^2V_3^{-4}) = [16 \times (-7)](V_1^{2+1}V_2^{3+2}V_3^{2+(-4)}) = -112V_1^3V_2^5V_3^{-2}$

Polynomials by Monomials

The procedure for multiplying any expression by a monomial is an extension of the foregoing process:

To multiply any expression by a monomial, multiply each of the terms within the expression by the monomial and retain the operation signs of the expression.

Let us first apply this procedure to a purely numerical expression: $6(3 + 4 - 5)$. We must multiply each term within the expression by the monomial (the 6), retaining the operation signs ($+$ and $-$) within the parentheses:

$$6(3 + 4 - 5) = 6 \cdot 3 + 6 \cdot 4 - 6 \cdot 5)$$

This executes the foregoing algebraic rule. However, since the problem is numerical, we can compare the left side with the right side of the equals sign:

$$6(2) = 18 + 24 - 30$$
$$12 = 12$$

It works! Next, let us apply the procedure to some algebraic expressions. Note that each multiplication within the problem becomes a monomial multiplied by a monomial, retaining operation signs.

EXAMPLE 4-7 Multiply:

(a) $A(B + C)$
(b) $V_1 V_2 (XY - XZ - YZ)$
(c) $2P(P + Q)$
(d) $3P^2 Q^2 (PQ - P^2 Q + 3P^2 Q^2 + 4P^4)$

SOLUTION

(a) $A(B + C) = AB + AC$
(b) $V_1 V_2 (XY - XZ - YZ) = V_1 V_2 XY - V_1 V_2 XZ - V_1 V_2 YZ$
     Each term within the parentheses was multiplied by the $V_1 V_2$ term and operation signs ($-$ and $-$) were retained.
(c) $2P(P + Q) = 2P^2 + 2PQ$
     Exponents were treated as in monomial multiplication: Where identical bases occur, the exponents add.
(d) $3P^2 Q^2 (PQ - P^2 Q - 3P^2 Q^2 + 4P^4) = 3P^3 Q^3 - 3P^4 Q^3 - 9P^4 Q^4 + 12P^6 Q^2$
     Note that, since there were four terms between the parentheses, there were four terms in the final result.

Binomials by Binomials

To multiply a binomial by a binomial, each term of one expression must be multiplied by each term of the second expression. Some have referred to this as the FOIL system: First, Outer, Inner, Last. In the expression

$$Y = (A + B)(C + D)$$

the first terms, A and C, would be multiplied, then the outer terms, A and D, then the inner terms, B and C, then the last terms, B and D. The signs given each of the products would be those that considered the signs of each of the operands being multiplied.

EXAMPLE 4-8 Multiply $(A - B)(A + C)$.

SOLUTION Multiply the first: $A \cdot A = A^2$

The outer: $A \cdot C = AC$
The inner: $(-B)(A) = -AB$
The last: $(-B)(C) = -BC$

Thus, $(A - B)(A + C) = A^2 + AC - AB - BC$.

EXAMPLE 4-9 Multiply $(R_1^2 R_2 - R_2^2)(R_1 R_2^2 - R_1^2)$.

SOLUTION Multiply the first: $(R_1^2 R_2)(R_1 R_2^2) = R_1^3 R_2^3$

$$\text{Outer:} \quad (R_1^2 R_2)(-R_1^2) = -R_1^4 R_2$$
$$\text{Inner:} \quad (-R_2^2)(R_1 R_2^2) = -R_1 R_2^4$$
$$\text{Last:} \quad (-R_2^2)(-R_1^2) = R_1^2 R_2^2$$

Thus, $(R_1^2 R_2 - R_2^2)(R_1 R_2^2 - R_1^2) = R_1^3 R_2^3 - R_1^4 R_2 - R_1 R_2^4 + R_1^2 R_2^2.$

EXAMPLE 4-10 Multiply $(3PQ^2 - 4PR)(4PQ^2 + 5PR)$.

SOLUTION

$$(3PQ^2 - 4PR)(4PQ^2 + 5PR) = 12P^2Q^4 + 15P^2Q^2R - 16P^2Q^2R - 20P^2R^2$$
$$= 12P^2Q^4 - P^2Q^2R - 20P^2R^2$$

Note that similar terms were combined.

Polynomials by Binomials

This procedure also requires that each term of one expression be multiplied by each term of the second expression. The bookkeeping can become fairly extensive, so a system should be followed whereby no terms "fall through the crack" and are omitted. One such system is to work left to right on the left expression and left to right on the right expression. Assume that

$$Y = (A + B)(C + D + E)$$

Multiply as follows: AC, AD, AE, BC, BD, BE. Thus, $Y = AC + AD + AE + BC + BD + BE$.

EXAMPLE 4-11 Multiply $(V_1 + 2V_2)(2V_1 + V_2 - V_3)$.

SOLUTION

$$(V_1 + 2V_2)(2V_1 + V_2 - V_3)$$
$$= (V_1)(2V_1) + (V_1)(V_2) + (V_1)(-V_3) + (2V_2)(2V_1) + (2V_2)(V_2) + (2V_2)(-V_3)$$
$$= 2V_1^2 + V_1 V_2 - V_1 V_3 + 4V_1 V_2 + 2V_2^2 - 2V_2 V_3$$
$$= 2V_1^2 + 5V_1 V_2 - V_1 V_3 - 2V_2 V_3 + 2V_2^2$$

Polynomials by Polynomials

A similar procedure can be followed in multiplying a polynomial by another polynomial: work left to right in each expression. For example,

$$Z = (A + B + C)(D + E + F + G)$$

Multiply the terms in the following order:

$$Z = AD + AE + AF + AG + BD + BE + BF + BG + CD + CE + CF + CG$$

EXAMPLE 4-12 Multiply $Y = (4R_1^2 - R_2^2 + 5R_3)(6R_1^2 + 7R_2^2 - 4R_3 + R_4 + 5R_5)$.

SOLUTION

$$Y = (4R_1^2)(6R_1^2) + (4R_1^2)(7R_2^2) + (4R_1^2)(-4R_3) + (4R_1^2)(R_4)$$
$$+ (4R_1^2)(5R_5) + (-R_2^2)(6R_1^2) + (-R_2^2)(7R_2^2) + (-R_2^2)(-4R_3)$$
$$+ (-R_2^2)(R_4) + (-R_2^2)(5R_5) + (5R_3)(6R_1^2) + (5R_3)(7R_2^2)$$
$$+ (5R_3)(-4R_3) + (5R_3)(R_4) + (5R_3)(5R_5)$$

$$= 24R_1^4 + 28R_1^2R_2^2 - 16R_1^2R_3 + 4R_1^2R_4 + 20R_1^2R_5 - 6R_1^2R_2^2$$
$$- 7R_2^4 + 4R_2^2R_3 - R_2^2R_4 - 5R_2^2R_5 + 30R_1^2R_3 + 35R_2^2R_3$$
$$- 20R_3^2 + 5R_3R_4 + 25R_3R_5$$
$$= 24R_1^4 + 22R_1^2R_2^2 + 14R_1^2R_3 + 4R_1^2R_4 + 20R_1^2R_5 - 7R_2^4$$
$$+ 39R_2^2R_3 - R_2^2R_4 - 20R_3^2 + 5R_3R_4 + 25R_3R_5$$

Multiple Signs of Grouping

When signs of grouping appear, the inner grouping must be treated first.

EXAMPLE 4-13 Multiply $(A + B)[3X + 4A(A^2 - B^2)]$.

SOLUTION The inner grouping must be treated first as a monomial multiplied by a binomial.

$$(A + B)[3X + 4A(A^2 - B^2)] = (A + B)[3X + 4A^3 - 4AB^2] = 3AX + 4A^4$$
$$- 4A^2B^2 + 3BX + 4A^3B - 4AB^3$$

EXAMPLE 4-14 Multiply $[A^2 + 2B(A + B)][3A^2 - 4B(A - B)]$.

SOLUTION The inner groupings must be treated first, then the outer. Note that the inner groupings consist of a monomial multiplied by a binomial, and the resultant outer groupings a trinomial multiplied by a trinomial.

$$[A^2 + 2B(A + B)][3A^2 - 4B(A - B)]$$
$$= [A^2 + 2AB + 2B^2][3A^2 - 4AB + 4B^2]$$
$$= 3A^4 - 4A^3B + 4A^2B^2 + 6A^3B - 8A^2B^2 + 8AB^3 + 6A^2B^2 - 8AB^3 + 8B^4$$
$$= 3A^4 + 2A^3B - 6A^2B^2 + 8B^4$$

PROBLEMS

Multiply

4-21. (a) $7A^2 \times 8AB^3$ (b) $2L_1^2L_2 \times 4L_2^7$ (c) $45N_p^2N_s \times 17N_p^4N_s^2$

4-22. (a) $3fC^2 \times 4f^{-4}C^3P^3$ (b) $77f_1f_2^2 \times 35f_2^2f_3f_4^2$ (c) $45L_1^{-2}L_2^3L_3L_4^4 \times 56L_1^3L_2L_3^{-3}L_5^5$

4-23. (a) $17.3a^2b \times 1.82a^2C$ (b) $0.831L_1^3L_2 \times 1.316L_1^{-4}L_2^2L_3$ (c) $7.91Q_1^2Q_2 \times 8.61Q_1^{-4}Q_2$

4-24. (a) $2.31R^2S^{-2}T \times 4.63R^4S^{-2}T^{-4}U$ (b) $3.46A^5B \times 7.92A^3B^{-2}$
(c) $84.2M^6N \times 75.9M^{-3}N^{-4}$

4-25. (a) $7R_1^2R_2(3R_1^3 + 2R_2^3)$ (b) $17P_1^2(P_2^2 + P_3^3)$ (c) $105A^2R(B^2 + R^2)$

4-26. (a) $4AB(A + B)$ (b) $9RS(R^2 + 3T)$ (c) $15M_1^2M_2^4(3M_1 - 2M_2)$

4-27. (a) $2.1X^2Y^2A(WX^2Y - 3X^2Z)$ (b) $4.5L_1^2L_2^2(0.3L_1^2 - 4L_2^2)$

4-28. (a) $0.73A^2B^2(4.6A^2C^2 - 6.9A^2B^2)$ (b) $4.5T_1^2T_2^3(3RST_1^2 - 2.9S^2T_2^2U^2)$

4-29. (a) $(A + B)(R + S)$ (b) $(A^2 - B^2)(R^2 + B^2)$ (c) $(3A^3 + 6B^2)(4A^3 - 7B)$

4-30. (a) $(3R_1 + R_2)(4R_1 - R_1R_2)$ (b) $(7L + 3M)(5L - 6M)$ (c) $(2X_L + X_C)(X_1 - 2X_C)$

4-31. (a) $(7.3QP - 2.6BP^2)(4.9Q^2P - 7.3BP^2)$ (b) $(8.61X_LX_C - 2.96X_L^2X_C)(X_L - 3.6X_C)$
(c) $(2.39V_1^2 - 6.7V_2^2)(3.1V_1 + 2.7V_2)$

4.32. (a) $(A + B)^2$ (b) $(3A + C)^2$ (c) $(2A - 3B)^2$

4-33. $(A + B)(A + B + C)$

4-34. $(A - B)(2A - 3B + 6C)$

4-35. $(R - S)(R^2 + 2S - T)$

4-36. $(6M_1 - M_2)(M_1^2 - 2M_1M_2 + M_2^2)$

4-37. $(3X_L - X_C)(4X_L^2 - 2X_LX_C + 6X_C^2)$

4-38. $(2.6V_1 + V_2)(3.7V_1^3 - 7.9V_2^2 + 7.3V_3)$

4-39. $(0.319I_L - 3.16I_C)(I_R^2 - 4.6I_LI_C + I_C^2)$

4-40. $(A + B)(A + C)(2B - C)$

4-41. $(X_1 + X_2)^2(X_1 - X_2)$

4-42. $(R + S + T)(R + S + T + U)$

4-43. $(7 + 3R + 6T)(3 + 7R + 6T + 7RT)$

4-44. $(X_L - 3X_C + X_T)(X_L - 4X_C + X_T)$

4-45. $(V_1 + 7V_2 - 7V_3)(3V_1 - 6V_2 + 7V_3)$

4-46. $(17X_L^2 - X_LX_C + X_C^2)(49X_C^3 - 7X_L^2 + 5)$

4-47. $(3C_T - C_L + C_R)(C_TC_LC_R - 7C_M + C_L^4)$

4-48. $(1.63L_1 - 3.9L_2 + 7.8L_3)(2.95L_1 - 7.31L_2 - 3.15L_3)$

4-49. $(29.5W + 6Y + 7.39Z)(3.32W^2 + 7.31WY + 3.21YZ)$

4-50. $(A + B)[2A + B(2 + C)]$

4-51. $(3R - 2S)[4R - 3S(4R + 2S)]$

4-52. $(V_1V_2 - V_3)[2V_1(V_2 - V_3) + 3V_2(V_1 + V_2)]$

4-53. $4(X_1 + X_2) - 2(X_1^2 - X_2^2) + 6X(X_1 - X_2)$

4-54. $3V_1(V_1 + V_2)[V_1^2 - V_2^2(3 - V_1)]$

4-55. $5(WX - Y) - 6(WX - WZ) + 4W(WX - Y)$

4-56. $P_1(P_1 - P_2) - (3P_1 - P_2^2)(P_1 + P_2 + P_3)$

4-4 ARITHMETIC FRACTIONS

We are about to study the division of one algebraic expression by another. However, this division follows the same rules as arithmetic fractions. Therefore, a review of fractions is in essence an introduction to algebraic division.

Just what is a fraction? It can be thought of in two ways:

1. A fraction is a method of expressing parts of a whole whereby the numerator (the number above the vinculum) expresses the number of parts taken and the denominator (the number below the vinculum) expresses the number of parts necessary to form a whole.

2. A fraction is a division process whereby the numerator is divided by the denominator.

Consider the fraction 3/4. According to the first definition, we are taking three parts and it takes four parts to form a whole. According to the second definition, we are dividing 3 by 4, resulting in 0.75. Both points of view are correct.

Fraction Rules

Since a fraction is fundamentally a division process, it follows that multiplication and division of the parts of the fraction are easier to do than addition and subtraction. Of the five operations of addition, subtraction, multiplication, division, and exponentiation, only two can be done to both denominator and numerator, leaving the value of the fraction unchanged.

The value of a fraction remains unchanged when the numerator and denominator are both multiplied by a constant or both divided by a constant.

Consider the fraction 12/16. We can multiply both the numerator and denominator by any constant and its value remains unchanged (Table 4-1). We can also divide both numerator and denominator by a constant and its value remains unchanged

TABLE 4-1 Multiplying the fraction $^{12}\!/_{16}$ by a constant.

Constant	Operation	Fraction Becomes	Decimal Value
1	$\dfrac{12 \times 1}{16 \times 1}$	$\dfrac{12}{16}$	0.75
2	$\dfrac{12 \times 2}{16 \times 2}$	$\dfrac{24}{32}$	0.75
3	$\dfrac{12 \times 3}{16 \times 3}$	$\dfrac{36}{48}$	0.75
4	$\dfrac{12 \times 4}{16 \times 4}$	$\dfrac{48}{64}$	0.75
5	$\dfrac{12 \times 5}{16 \times 5}$	$\dfrac{60}{80}$	0.75

(Table 4-2). Note, however, that the remaining operations cannot be performed on the numerator and denominator (Table 4-3). In each case, the value of the fraction is changed.

In summary, because a fraction is a division process, only multiplication and division by a constant can be performed on the numerator and denominator if the value of the fraction is to remain unchanged.

TABLE 4-2 Dividing the fraction $^{12}\!/_{16}$ by a constant.

Constant	Operation	Fraction Becomes	Decimal Value
1	$\dfrac{12 \div 1}{16 \div 1}$	$\dfrac{12}{16}$	0.75
2	$\dfrac{12 \div 2}{16 \div 2}$	$\dfrac{6}{8}$	0.75
3	$\dfrac{12 \div 3}{16 \div 3}$	$\dfrac{4}{5.333}$	0.75
4	$\dfrac{12 \div 4}{16 \div 4}$	$\dfrac{3}{4}$	0.75
5	$\dfrac{12 \div 5}{16 \div 5}$	$\dfrac{2.40}{3.20}$	0.75

Reduction

A fraction is said to be reduced when its numerator and denominator can only be evenly divided by the number 1 and itself.

Since the numerator and denominator can be changed only by multiplication or division, reduction of a fraction must use these two operations.

Consider the fraction 9/12. It can be reduced by dividing both numerator and denominator by 3:

$$\frac{9}{12} = \frac{9/3}{12/3} = \frac{3}{4}$$

The 3 and 4 can only be divided evenly (that is, there is no remainder) by the

TABLE 4-3 Improper fractional operations.

Constant	Operation	Fraction Becomes	Decimal Value
Addition			
0	$\dfrac{12+0}{16+0}$	$\dfrac{12}{16}$	0.75
1	$\dfrac{12+1}{16+1}$	$\dfrac{13}{17}$	0.7647
2	$\dfrac{12+2}{16+2}$	$\dfrac{14}{18}$	0.7778
Subtraction			
0	$\dfrac{12-0}{16-0}$	$\dfrac{12}{16}$	0.75
1	$\dfrac{12-1}{16-1}$	$\dfrac{11}{15}$	0.7333
2	$\dfrac{12-2}{16-2}$	$\dfrac{10}{14}$	0.7143
Exponentiation			
1	$\dfrac{12^1}{16^1}$	$\dfrac{12}{16}$	0.75
2	$\dfrac{12^2}{16^2}$	$\dfrac{144}{256}$	0.5625
3	$\dfrac{12^3}{16^3}$	$\dfrac{1728}{4096}$	0.4219

number 1. Therefore, the fraction 3/4 is its reduced form. However, many times it is not apparent what number evenly divides into the numerator and denominator. When this occurs, we can factor both parts of the fraction. To factor is to find the series of prime numbers that, when multiplied together, result in the original number. A prime number is one that can be divided evenly only by itself and 1. Several rules may help:

1. If a number is even, it can be divided by 2.
2. If the sum of the digits of a number is evenly divisible by 3, the original number is divisible by 3. For example, the number 1728 is divisible by 3 because $1 + 7 + 2 + 8$ or 18 is divisible by 3.
3. Any number ending in 5 or 0 can be evenly divided by 5.

EXAMPLE 4-15 Factor:

(a) 36   (b) 270   (c) 174

SOLUTION

(a) 36 can be divided by both 2 and 3, yielding

$$36 = 2 \cdot 3 \cdot 6$$

But 6 can be divided by 2 and 3.

$$36 = 2 \cdot 3 \cdot 2 \cdot 3$$
$$36 = 2 \cdot 2 \cdot 3 \cdot 3$$

(b) 270 is divisible by 2 and 5:

$$270 = 2 \cdot 5 \cdot 27$$

27 is divisible by 3:

$$270 = 2 \cdot 5 \cdot 3 \cdot 9$$
$$270 = 2 \cdot 5 \cdot 3 \cdot 3 \cdot 3$$

(c) $174 = 3 \cdot 58$
$174 = 3 \cdot 2 \cdot 29$

We are now in a position to reduce any fraction. Find the prime numbers, then divide both numerator and denominator by common factors.

EXAMPLE 4-16 Reduce:

(a) $\dfrac{54}{135}$ (b) $\dfrac{18}{162}$ (c) $\dfrac{84}{140}$

SOLUTION

(a) $\dfrac{54}{135} = \dfrac{2 \cdot 3 \cdot 3 \cdot 3}{3 \cdot 3 \cdot 3 \cdot 5}$

Divide both numerator and denominator by $3 \cdot 3 \cdot 3$:

$$\dfrac{54}{135} = \dfrac{2}{5}$$

(b) $\dfrac{18}{162} = \dfrac{2 \cdot 3 \cdot 3}{2 \cdot 3 \cdot 3 \cdot 3 \cdot 3}$

Divide both numerator and denominator by $2 \cdot 3 \cdot 3$:

$$\dfrac{18}{162} = \dfrac{1}{3 \cdot 3} = \dfrac{1}{9}$$

(c) $\dfrac{84}{140} = \dfrac{2 \cdot 2 \cdot 3 \cdot 7}{2 \cdot 2 \cdot 5 \cdot 7} = \dfrac{3}{5}$

Multiplying Two Fractions

Multiplication is performed using the following rule:

To multiply two fractions, multiply their numerators to obtain the numerator of the result and multiply their denominators to obtain the denominator of the result.

EXAMPLE 4-17 Multiply:

(a) $\dfrac{3}{4} \times \dfrac{7}{8}$ (b) $\dfrac{4}{7} \times \dfrac{5}{9}$

SOLUTION

(a) $\dfrac{3}{4} \times \dfrac{7}{8} = \dfrac{3 \times 7}{4 \times 8} = \dfrac{21}{32}$

Note that the numerators, 3 and 7, were multiplied together to obtain the numerator of the result, 21, and the denominators, 4 and 8, to obtain the denominator, 32.

(b) $\dfrac{4}{7} \times \dfrac{5}{9} = \dfrac{4 \times 5}{7 \times 9} = \dfrac{20}{63}$

Dividing Two Fractions

The parts of a division problem are:

$$\dfrac{\text{dividend}}{\text{divisor}} = \text{quotient}$$

or dividend $\div$ divisor = quotient
or dividend divided by divisor = quotient.

To divide one fraction by another, invert the divisor and multiply.

EXAMPLE 4-18   Divide:

(a) $\dfrac{3}{4}$ by $\dfrac{7}{8}$

(b) $\dfrac{5}{32}$ by $\dfrac{21}{49}$

SOLUTION

(a) $\dfrac{3}{7} \div \dfrac{7}{8}$

$= \dfrac{3}{4} \times \dfrac{8}{7} = \dfrac{24}{28} = \dfrac{6}{7}$

Note that the 7/8 was inverted to 8/7, and the problem was treated as in multiplication.

(b) $\dfrac{5}{32} \div \dfrac{21}{49}$

$= \dfrac{5}{32} \times \dfrac{49}{21} = \dfrac{245}{672} = \dfrac{35}{96}$

Adding Two Fractions

In real life, we never add unlike items together: dollars and miles, for example. Similarly, in algebra we never add coefficients unless the literals are identical. Fractions also follow this rule:

To add fractions with identical denominators, add the numerators and give the result the common denominator.

Thus, if the denominators are identical, the numerators can be added.

EXAMPLE 4-19   Add:

(a) $\dfrac{1}{4} + \dfrac{1}{4}$ (b) $\dfrac{3}{8} + \dfrac{1}{8}$ (c) $\dfrac{9}{32} + \dfrac{5}{32}$

SOLUTION

(a) $\dfrac{1}{4}+\dfrac{1}{4}=\dfrac{1+1}{4}=\dfrac{2}{4}=\dfrac{1}{2}$

Note that the numerators were added. The result could then be reduced.

(b) $\dfrac{3}{8}+\dfrac{1}{8}=\dfrac{3+1}{8}=\dfrac{4}{8}=\dfrac{1}{2}$

(c) $\dfrac{9}{32}+\dfrac{5}{32}=\dfrac{9+5}{32}=\dfrac{14}{32}=\dfrac{7}{16}$

If the denominators are not identical, a new denominator must be assigned, one that is the lowest possible number. This number is called the lowest common denominator (LCD).

The lowest common denominator is the smallest number into which both denominators can be evenly divided.

The easiest way to find this LCD is to factor the two denominators and build an LCD from these factors by including all the factors of each denominator in the result. For example, let us assume that the two denominators are 54 and 60. Factoring, we obtain

$$54 = 2 \cdot 3 \cdot 3 \cdot 3 \qquad 60 = 2 \cdot 2 \cdot 3 \cdot 5$$

The LCD must contain the entire first denominator: $2 \cdot 3 \cdot 3 \cdot 3$.
Then analyze the second denominator. It contains two 2's, so

$$\text{LCD must contain } 2 \cdot 2 \cdot 3 \cdot 3 \cdot 3$$

It also contains one 3, so the LCD must contain one 3; in fact, it already does. Finally, the second denominator contains one 5, so

$$\text{LCD must contain } 2 \cdot 2 \cdot 3 \cdot 3 \cdot 3 \cdot 5$$

This, then, is the LCD and each denominator can be divided evenly into this number.

$$\text{LCD} = 2 \cdot 2 \cdot 3 \cdot 3 \cdot 3 \cdot 5 = 540$$

EXAMPLE 4-20 Find the LCDs:

(a) 20, 36   (b) 30, 48   (c) 10, 16, 24

SOLUTION

(a) $20 = 2 \cdot 2 \cdot 5$
$36 = 2 \cdot 2 \cdot 3 \cdot 3$

 1. Write down the first number:

$$\text{LCD must contain } 2 \cdot 2 \cdot 5$$

 2. Analyzing the 36, the LCD must contain $2 \cdot 2$—it does.
 3. The LCD must contain $3 \cdot 3$—it does not. Therefore,

$$\text{LCD} = 2 \cdot 2 \cdot 5 \cdot 3 \cdot 3 = 180$$

(b) $30 = 2 \cdot 3 \cdot 5$
$48 = 2 \cdot 2 \cdot 2 \cdot 2 \cdot 3$
1. LCD must contain $2 \cdot 3 \cdot 5$.
2. LCD must contain four 2's.

LCD must contain $2 \cdot 2 \cdot 2 \cdot 2 \cdot 3 \cdot 5$

3. LCD must contain one 3—it does.
4. LCD $= 2 \cdot 2 \cdot 2 \cdot 2 \cdot 3 \cdot 5 = 240$
(c) $10 = 2 \cdot 5$
$16 = 2 \cdot 2 \cdot 2 \cdot 2$
$24 = 2 \cdot 2 \cdot 2 \cdot 3$
1. LCD must contain $2 \cdot 5$.
2. Analyzing 16, it must contain $2 \cdot 2 \cdot 2 \cdot 2$; therefore,

LCD must contain $2 \cdot 2 \cdot 2 \cdot 2 \cdot 5$

3. Analyzing 24, it must contain three 2's—it does. It must contain one 3. Therefore,

LCD $= 2 \cdot 2 \cdot 2 \cdot 2 \cdot 3 \cdot 5 = 240$

Having the lowest common denominators, let us now redirect our attention to the problem of addition.

To add two fractions:

(a) *Find their LCD.*
(b) *Multiply the first fraction by the number 1, chosen so that the resulting denominator is the LCD.*
(c) *Multiply the second fraction by the number 1, chosen so that its denominator is the LCD.*
(d) *Add the resulting numerators.*
(e) *Give the result the LCD.*

EXAMPLE 4-21 Add $1/12 + 1/15$.

SOLUTION The LCD is $2 \cdot 2 \cdot 3 \cdot 5 = 60$.

$$\frac{1}{12} + \frac{1}{15}$$
$$= \left(\frac{1}{12} \times \frac{5}{5}\right) + \left(\frac{1}{15} \times \frac{4}{4}\right)$$
$$= \frac{5}{60} + \frac{4}{60} = \frac{9}{60} = \frac{3}{20}$$

Note that the 12 must be multiplied by 5 to match the LCD. However, if we multiply the denominator by 5, we must multiply the numerator by 5. Thus, we have effectively multiplied the fraction by 5/5 or 1. Similarly, 15 must be multiplied by 4 to reach the LCD and, since the denominator is multiplied by 4, the numerator must be so multiplied. Again, we have multiplied the fraction by 1 (the 4/4), leaving it unchanged. We can now add the numerators because the denominators are identical. The result happens to be reducible.

Subtracting Two Fractions

Algebraic subtraction follows a rule very similar to that of addition:

To subtract algebraic fractions with identical denominators, subtract the numerators and give the result the common denominator.

EXAMPLE 4-22 Subtract:

(a) $\dfrac{1}{4}$ from $\dfrac{3}{4}$ (b) $\dfrac{2}{7}$ from $\dfrac{5}{7}$ (c) $\dfrac{6}{17}$ from $\dfrac{16}{17}$

SOLUTION

(a) $\dfrac{3}{4} - \dfrac{1}{4} = \dfrac{3-1}{4} = \dfrac{2}{4} = \dfrac{1}{2}$

(b) $\dfrac{5}{7} - \dfrac{2}{7} = \dfrac{5-2}{7} = \dfrac{3}{7}$

(c) $\dfrac{16}{17} - \dfrac{6}{17} = \dfrac{16-6}{17} = \dfrac{10}{17}$

When the denominators are not identical, we must proceed as follows:

To subtract two fractions:

(a) *Find their LCD.*
(b) *Multiply the first fraction by the number 1, chosen so that the resulting denominator is the LCD.*
(c) *Multiply the second fraction by the number 1, chosen so that its denominator is the LCD.*
(d) *Subtract the resulting numerators.*
(e) *Give the result the LCD.*

EXAMPLE 4-23 Subtract 7/15 from 11/12.

SOLUTION

The LCD is $2 \cdot 2 \cdot 3 \cdot 5 = 60$.

$$\frac{11}{12} - \frac{7}{15}$$
$$= \left(\frac{11}{12} \times \frac{5}{5}\right) - \left(\frac{7}{15} \times \frac{4}{4}\right)$$
$$= \frac{55}{60} - \frac{28}{60}$$
$$= \frac{55-28}{60} = \frac{27}{60} = \frac{9}{20}$$

EXAMPLE 4-24 Subtract 17/28 from 43/48.

SOLUTION The LCD is $7 \cdot 2 \cdot 2 \cdot 3 \cdot 2 \cdot 2 = 336$.

$$\frac{43}{48} - \frac{17}{28}$$

$$= \left(\frac{43}{48} \times \frac{7}{7}\right) - \left(\frac{17}{28} \times \frac{12}{12}\right)$$

$$= \frac{301}{336} - \frac{204}{336}$$

$$= \frac{301 - 204}{336} = \frac{97}{336}$$

Proper and Improper Fractions

Fractions with values greater than 1 can be expressed in either of two forms: proper or improper. A fraction is said to be proper if the fractional part of the number is less than 1. It is said to be improper if the fractional portion is greater than 1. For example, assume that our friend Mergatroid baked two pies and divided each into six pieces. Then, assume that our friend Fauntleroy, returning from a hard day at the race track, spied the pies and ate one piece. We could express the number of pies left to serve for dinner as:

$$\frac{11}{6} \quad \text{(improper)} \quad \text{or} \quad 1\frac{5}{6} \quad \text{(proper)}$$

In the preceding paragraphs our discussion of fractions presumes that they are all in improper form. If we encounter a proper fraction, it can be put in improper form by multiplying the whole number by the denominator, then adding this result to the numerator for a final numerator.

EXAMPLE 4-25 Convert to improper fractions:

(a) $1\frac{7}{8}$ (b) $6\frac{11}{16}$ (c) $4\frac{27}{28}$

SOLUTION

(a) $1\frac{7}{8} = \frac{(1 \times 8) + 7}{8} = \frac{15}{8}$

(b) $6\frac{11}{16} = \frac{(6 \times 16) + 11}{16} = \frac{107}{16}$

(c) $4\frac{27}{28} = \frac{(4 \times 28) + 27}{28} = \frac{139}{28}$

An improper fraction can be converted to a proper fraction by dividing the numerator by the denominator. The quotient becomes the whole number and the remainder the new numerator.

This operation can also be done on a calculator by dividing the numerator by the denominator. That part of the result to the left of the decimal point becomes the whole number. If the fractional part of the result is then multiplied by the denominator, the new numerator is obtained.

EXAMPLE 4-26 Convert 365/27 to a proper fraction.

SOLUTION We do this in two ways:

(a) Performing it manually yields

$$365 \div 27 = 13 \qquad \text{with a remainder of } 14$$

Therefore,

$$\frac{365}{27} = 13\frac{14}{27}$$

(b) Performing the operation on a calculator yields

$$365 \div 27 = 13.51851852$$

Next, subtract out the 13 (the whole number):

$$13.51851852 - 13 = 0.51851852$$

Multiply this result by the denominator (27):

$$0.51851852 \times 27 = 14.0000000$$

The remainder is, therefore, 14. The final result is

$$\frac{365}{27} = 13\frac{14}{27}$$

EXAMPLE 4-27 Convert 1473/53 to a proper fraction:

SOLUTION We obtain the result in two ways:

(a) Perform it manually:

$$1473 \div 53 = 27 \qquad \text{with a remainder of } 42$$

Therefore,

$$\frac{1473}{53} = 27\frac{42}{53}$$

(b) Perform it on a calculator:

$$\frac{1473}{53} = 27.79245283$$

The whole number is 27.

$$27.79245283 - 27 = 0.79245283$$
$$0.79245283 \times 53 = 42.0000000$$

The numerator is 42 and the final result is

$$\frac{1473}{53} = 27\frac{42}{53}$$

EXAMPLE 4-28 Perform the indicated operations:

(a) $1\frac{7}{8} + 3\frac{3}{4}$ (b) $6\frac{7}{9} - 3\frac{1}{5}$ (c) $5\frac{11}{12} \times 2\frac{7}{30}$ (d) $6\frac{3}{4} \div 7\frac{3}{7}$

SOLUTION

(a) There are two ways to add proper fractions: (1) add the whole numbers and then add the fractions, or (2) convert both to improper fractions, add, then convert back to a proper fraction.

1.
$$1\frac{7}{8} + 3\frac{3}{4} = (1+3) + \left(\frac{7}{8} + \frac{3}{4}\right)$$
$$= 4 + \left(\frac{7}{8} + \frac{6}{8}\right)$$
$$= 4 + \frac{13}{8}$$
$$= 4 + 1\frac{5}{8}$$
$$= 5\frac{5}{8}$$

2.
$$1\frac{7}{8} + 3\frac{3}{4}$$
$$= \frac{15}{8} + \frac{15}{4}$$
$$= \frac{15}{8} + \frac{30}{8} = \frac{45}{8} = 5\frac{5}{8}$$

(b) Subtraction can also be done these two ways:

1.
$$6\frac{7}{9} - 3\frac{1}{5} = (6-3) + \left(\frac{7}{9} - \frac{1}{5}\right)$$
$$= 3 + \left(\frac{35}{45} - \frac{9}{45}\right)$$
$$= 3 + \left(\frac{26}{45}\right)$$
$$= 3\frac{26}{45}$$

2.
$$6\frac{7}{9} - 3\frac{1}{5}$$
$$= \frac{61}{9} - \frac{16}{5} = \frac{305}{45} - \frac{144}{45}$$
$$= \frac{161}{45} = 3\frac{26}{45}$$

(c) Multiplication can only be done using improper fractions:

$$5\frac{11}{12} \times 2\frac{7}{30}$$
$$= \frac{71}{12} \times \frac{67}{30} = \frac{4757}{360} = 13\frac{77}{360}$$

(d) As in multiplication, division requires improper fractions:

$$6\frac{3}{4} \div 7\frac{3}{7}$$
$$= \frac{27}{4} \div \frac{52}{7}$$
$$= \frac{27}{4} \times \frac{7}{52} = \frac{189}{208}$$

PROBLEMS

Reduce:

4-57. (a) $\dfrac{4}{16}$ (b) $\dfrac{12}{36}$ (c) $\dfrac{14}{28}$

4-58. (a) $\dfrac{12}{52}$ (b) $\dfrac{48}{52}$ (c) $\dfrac{45}{75}$

4-59. (a) $\dfrac{96}{152}$ (b) $\dfrac{120}{3600}$ (c) $\dfrac{210}{11,760}$

4-60. (a) $\dfrac{150}{25,200}$ (b) $\dfrac{10,010}{1,541,540}$ (c) $\dfrac{280}{43,120}$

Multiply:

4-61. (a) $\dfrac{1}{2} \times \dfrac{7}{8}$ (b) $\dfrac{6}{7} \times \dfrac{1}{3}$ (c) $\dfrac{4}{7} \times \dfrac{1}{7}$

4-62. (a) $\dfrac{2}{3} \times \dfrac{3}{8}$ (b) $\dfrac{7}{16} \times \dfrac{14}{15}$ (c) $\dfrac{6}{7} \times \dfrac{13}{16}$

4-63. (a) $\dfrac{12}{17} \times \dfrac{4}{7}$ (b) $\dfrac{8}{9} \times \dfrac{16}{23}$ (c) $\dfrac{7}{17} \times \dfrac{7}{17}$

4-64. (a) $\dfrac{25}{31} \times \dfrac{17}{65}$ (b) $\dfrac{31}{32} \times \dfrac{12}{53}$ (c) $\dfrac{97}{100} \times \dfrac{84}{1321}$

4-65. (a) $\dfrac{27}{37} \times \dfrac{84}{121}$ (b) $\dfrac{64}{67} \times \dfrac{48}{51}$ (c) $\dfrac{121}{563} \times \dfrac{85}{91}$

4-66. (a) $\dfrac{57}{91} \times \dfrac{74}{75}$ (b) $\dfrac{121}{361} \times \dfrac{12}{13}$ (c) $\dfrac{141}{1371} \times \dfrac{84}{2671}$

Divide:

4-67. (a) $\dfrac{3}{4} \div \dfrac{4}{7}$ (b) $\dfrac{7}{8} \div \dfrac{3}{4}$ (c) $\dfrac{4}{7} \div \dfrac{1}{3}$

4-68. (a) $\dfrac{3}{7} \div \dfrac{6}{7}$ (b) $\dfrac{8}{9} \div \dfrac{5}{12}$ (c) $\dfrac{6}{13} \div \dfrac{11}{15}$

4-69. (a) $\dfrac{1}{5} \div \dfrac{3}{7}$ (b) $\dfrac{1}{5} \div \dfrac{8}{9}$ (c) $\dfrac{11}{12} \div \dfrac{1}{3}$

4-70. (a) $\dfrac{1}{7} \div \dfrac{1}{5}$ (b) $\dfrac{1}{12} \div \dfrac{4}{9}$ (c) $\dfrac{1}{16} \div \dfrac{1}{16}$

4-71. (a) $\dfrac{12}{13} \div \dfrac{12}{17}$ (b) $\dfrac{46}{51} \div \dfrac{12}{13}$ (c) $\dfrac{3}{26} \div \dfrac{5}{52}$

4-72. (a) $\dfrac{78}{83} \div \dfrac{7}{16}$ (b) $\dfrac{121}{136} \div \dfrac{6}{7}$ (c) $\dfrac{8}{19} \div \dfrac{9}{16}$

Add:

4-73. (a) $\dfrac{1}{3} + \dfrac{1}{2}$ (b) $\dfrac{3}{4} + \dfrac{1}{12}$ (c) $\dfrac{3}{5} + \dfrac{3}{10}$

4-74. (a) $\dfrac{3}{8} + \dfrac{3}{16}$ (b) $\dfrac{7}{8} + \dfrac{1}{64}$ (c) $\dfrac{4}{9} + \dfrac{3}{48}$

4-75. (a) $\dfrac{3}{10} + \dfrac{7}{15}$ (b) $\dfrac{7}{11} + \dfrac{17}{33}$ (c) $\dfrac{8}{11} + \dfrac{1}{16}$

4-76. (a) $\dfrac{7}{9} + \dfrac{13}{15}$ (b) $\dfrac{5}{12} + \dfrac{15}{27}$ (c) $\dfrac{9}{16} + \dfrac{7}{30}$

4-77. (a) $\dfrac{13}{20} + \dfrac{12}{25}$ (b) $\dfrac{19}{27} + \dfrac{7}{48}$ (c) $\dfrac{8}{9} + \dfrac{13}{52}$

4-78. (a) $\dfrac{59}{64} + \dfrac{11}{80}$ (b) $\dfrac{17}{45} + \dfrac{16}{27}$ (c) $\dfrac{19}{35} + \dfrac{13}{69}$

Subtract:

4-79. (a) $\dfrac{3}{16} - \dfrac{1}{8}$ (b) $\dfrac{5}{12} - \dfrac{1}{3}$ (c) $\dfrac{7}{8} - \dfrac{1}{2}$

4-80. (a) $\dfrac{13}{15} - \dfrac{7}{15}$ (b) $\dfrac{11}{18} - \dfrac{1}{6}$ (c) $\dfrac{13}{20} - \dfrac{3}{10}$

4-81. (a) $\dfrac{4}{9} - \dfrac{1}{15}$ (b) $\dfrac{11}{12} - \dfrac{1}{12}$ (c) $\dfrac{6}{7} - \dfrac{3}{28}$

4-82. (a) $\dfrac{17}{24} - \dfrac{6}{52}$ (b) $\dfrac{11}{48} - \dfrac{5}{30}$ (c) $\dfrac{17}{35} - \dfrac{8}{21}$

Perform the following:

4-83. (a) $3\dfrac{7}{8} + 2\dfrac{3}{4}$ (b) $12\dfrac{6}{7} + 4\dfrac{12}{21}$

4-84. (a) $5\dfrac{7}{12} + 3\dfrac{11}{18}$ (b) $14\dfrac{3}{8} + 3\dfrac{5}{12}$

4-85. (a) $6\dfrac{7}{8} - 2\dfrac{1}{3}$ (b) $5\dfrac{3}{4} - 2\dfrac{1}{15}$

4-86. (a) $4\dfrac{3}{4} - 2\dfrac{1}{12}$ (b) $6\dfrac{7}{15} - 3\dfrac{7}{30}$

4-87. (a) $3\dfrac{3}{4} \times 2\dfrac{7}{8}$ (b) $8\dfrac{2}{9} \times 1\dfrac{1}{3}$

4-88. (a) $2\dfrac{5}{7} \times 1\dfrac{3}{5}$ (b) $4\dfrac{7}{12} \times 2\dfrac{6}{7}$

4-89. (a) $3\dfrac{2}{3} \div 1\dfrac{3}{5}$ (b) $6\dfrac{7}{15} \div 4\dfrac{3}{4}$

4-90. (a) $6\dfrac{7}{8} \div 2\dfrac{1}{3}$ (b) $8\dfrac{2}{9} \div 2\dfrac{5}{7}$

4-5 SIGNS OF DIVISION

The signs of division follow the same rules as those of multiplication:

> *Dividing numbers of like sign results in a positive quotient. Dividing numbers of unlike sign results in a negative quotient.*

To visualize these rules, let us examine each case separately.

When a positive number, 20 for example, is divided by another positive number, 4 for example, we are asking, "How many 4's are there in 20?" The answer, of course, is 5—a positive 5. Similarly, when we divide a −20 by −4, we are asking, "How many −4's are there in −20?" For example, "How many people would I owe $4 each in order to owe a total of $20?" The answer is, of course, 5. To summarize, a positive number divided by a positive number yields a positive quotient; a negative number divided by a negative number yields a positive quotient.

Consider a temperature scale. Starting at 0°, how many times would the temperature have to rise by 4° in order to reach −20°? It is obvious that the temperature must fall, not rise, five times in order to reach −20°. Thus, the answer is −5, not +5, indicating that the direction the temperature must go is opposite to the rise of 4°. Thus, dividing −20 by +4 yields −5. In a similar manner, if the temperature starts at 0°, how many times must it fall by 4° to reach +20°? Again, the answer is that it must change five times, but in a direction opposite to falling. We indicate this direction by affixing the minus sign to the quotient. Therefore, 20 divided by −4 yields −5.

EXAMPLE 4-29   Divide:

(a) $5 \div -7$ (b) $-20 \div 6$ (c) $40 \div 11$ (d) $-75 \div -31$

SOLUTION

(a) $5 \div -7 = -0.7143$
(b) $-20 \div 6 = -3.333$
(c) $40 \div 11 = 3.636$
(d) $-75 \div -31 = 2.419$

PROBLEMS

Divide:

4-91. (a) $26 \div 42$ (b) $75 \div (-63)$ (c) $(-125) \div (-76)$
4-92. (a) $125 \div 63$ (b) $(-26) \div 3$ (c) $20 \div (-4)$
4-93. (a) $(-27) \div (-60)$ (b) $(-72) \div 27$ (c) $120 \div (-36)$
4-94. (a) $125 \div 36$ (b) $(-128) \div (-23)$ (c) $(-72) \div 23$

4-6 TERMS WITH EXPONENTS

Identical literals can be combined by division according to the following rule:

If two literals have identical bases, they can be divided by subtracting the exponent of the denominator from that of the numerator and retaining the literal.

This can be illustrated using numbers. Let us divide 3^5 by 3^2 using this rule:

$$\frac{3^5}{3^2} = 3^{5-2} = 3^3$$

$$\frac{243}{9} = 27$$

It should be emphasized that exponents can only be combined if the literals are identical.

EXAMPLE 4-30   Divide:

(a) $A^3 \div A^3$ (b) $R_1^2 \div R_1$ (c) $L_1^3 \div L_1^{-1}$ (d) $V^a \div V^b$
(e) $1 \div C_1^3$

SOLUTION

(a) $\dfrac{A^3}{A^3} = A^{3-3} = A^0 = 1$

Note that any number (except 0) divided by itself is 1. Therefore, any number (except 0) raised to the 0 power is 1.

(b) $\dfrac{R_1^2}{R_1} = R_1^{2-1} = R_1^1 = R_1$

Note that when the exponent is not explicitly stated, it is assumed to be 1.

(c) $\dfrac{L_1^3}{L_1^{-1}} = L^{3-(-1)} = L_1^4$

(d) $\dfrac{V^a}{V^b} = V^{a-b}$

(e) $\dfrac{1}{C_1^3} = \dfrac{C_1^0}{C_1^3} = C_1^{0-3} = C_1^{-3}$

Note that when a literal is moved from the denominator to the numerator, the sign of its exponent is changed.

Where the literals differ, the exponents cannot be combined.

EXAMPLE 4-31 Divide:

(a) $b^2 \div b_1^2$ (b) $X_L \div X_C$ (c) $R_1 \div R_2$

SOLUTION In each case, the literals are not the same and therefore cannot be combined. The answers are:

(a) $\dfrac{b^2}{b_1^2}$ (b) $\dfrac{X_L}{X_C}$ (c) $\dfrac{R_1}{R_2}$

PROBLEMS

Divide, leaving the result in exponential format:

4-95. (a) $3^5 \div 3^2$ (b) $4^9 \div 4^3$ (c) $5^7 \div 5^6$ (d) $7^7 \div 7^7$

4-96. (a) $17^{12} \div 17^2$ (b) $12^4 \div 12^2$ (c) $11^7 \div 11^{12}$ (d) $25^3 \div 25^2$

4-97. (a) $4^4 \div 4^{-2}$ (b) $7^7 \div 7^{-6}$ (c) $9^{-12} \div 9^{-11}$ (d) $12^{-12} \div 12^7$

4-98. (a) $17.6^3 \div 17.6^{-4}$ (b) $12.1^{12} \div 12.1^{-17}$ (c) $14.6^9 \div 14.6^{-7}$ (d) $0.139^7 \div 0.139^{-12}$

Divide:

4-99. (a) $A^3 \div A^2$ (b) $X_1^8 \div X_1^2$ (c) $P^7 \div P^4$ (d) $V^7 \div V^4$

4-100. (a) $R_1^4 \div R_1^2$ (b) $C_1^9 \div C_1^7$ (c) $C_7^7 \div C_7^6$ (d) $W^3 \div W^2$

4-101. (a) $L_1^{-3} \div L_1^5$ (b) $R_1^7 \div R_1^{-10}$ (c) $M_3^{-4} \div M_3^{-7}$ (d) $Q_1^{-3} \div Q_1^{-4}$

4-102. (a) $T_1^4 \div T_1^{-7}$ (b) $B_L^{-3} \times B_L^{-4}$ (c) $S^{-R} \times S^{-T}$ (d) $V^a \div V^c$

4-7 DIVIDING ALGEBRAIC EXPRESSIONS

In this section we discuss the techniques for dividing algebraic expressions: monomials by monomials, polynomials by monomials, and polynomials by polynomials. Later, in Chapter 8, we will add the skill of factoring to our division arsenal.

Monomial Division

In dividing a monomial by a monomial, we must consider it in two steps:

1. The coefficients are first divided as in ordinary division. The quotient becomes the coefficient for the answer.
2. Each literal is then divided by the corresponding literal within the divisor. Where a literal is missing, it can be considered as being raised to the zero power.

EXAMPLE 4-32   Divide $26X^3Y^{-2}Z$ by $12W^2X^2Y^{-6}Z$.

SOLUTION

$$\frac{26X^3Y^{-2}Z}{12W^2X^2Y^{-6}Z} = \left(\frac{26}{12}\right)\left(\frac{W^0}{W^2}\right)\left(\frac{X^3}{X^2}\right)\left(\frac{Y^{-2}}{Y^{-6}}\right)\left(\frac{Z^1}{Z^1}\right)$$
$$= 2.167\,W^{0-2}X^{3-2}Y^{-2-(-6)}Z^{1-1}$$
$$= 2.167\,W^{-2}X^1Y^4Z^0$$
$$= 2.167\,W^{-2}XY^4$$

Note that, since the numerator contained no W literal, W^0 was substituted. Further, in the answer, the "1" was dropped from X^1, a common procedure, and Z^0 was dropped because it is equal to 1.

EXAMPLE 4-33   Divide $4.71a^2b^3c^{-3}$ by $3.5ab^4d^2$.

SOLUTION

$$\frac{4.71a^2b^3c^{-3}}{3.5ab^4d^2} = \left(\frac{4.71}{3.5}\right)a^{2-1}b^{3-4}c^{-3}d^{0-2}$$
$$= 1.346ab^{-1}c^{-3}d^{-2}$$

Note that the c term can be looked at as $c^{-3} \div c^0$, or c^{-3-0}, or c^{-3}. Thus, it is carried into the answer unchanged.

Polynomials Divided by a Monomial

When binomials or polynomials are divided by a monomial, each term of the numerator must be divided by the denominator.

EXAMPLE 4-34   Divide $36A^2C + 4A^3C^2$ by $2AC$.

SOLUTION

$$\frac{36A^2C + 4A^3C^2}{2AC} = \frac{36A^2C}{2AC} + \frac{4A^3C^2}{2AC} = 18A + 2A^2C$$

EXAMPLE 4-35   Divide $35.9BC^2D - 7.6BD^2 + 6C^2D^4$ by $3.56BCD^2$.

SOLUTION

$$\frac{35.9BC^2D - 7.6BD^2 + 6C^2D^4}{3.56BCD^2}$$

$$= 10.08CD^{-1} - 2.135C^{-1} + 1.685B^{-1}CD^2$$

Note that in both Examples 4-34 and 4-35, each term of the polynomial was divided by the monomial.

Polynomials Divided by Polynomials

Polynomials can be divided by polynomials using a procedure similar to arithmetic long division. It first requires that both the divisor and the dividend terms be arranged according to the descending power of a particular literal. For example, if we arranged the terms of the following expression in descending order of X,

$$3X - 4X^4 + 6 - 5X^2 = -4X^4 - 5X^2 + 3X + 6$$

then, following this rearrangement, we must divide the first term of each divisor into the first term of each dividend.

EXAMPLE 4-36   Divide $22X + 13X^2 + 6 + 3X^3$ by $4X + 6 + X^2$.

SOLUTION   Rearrange and divide:

$$
\begin{array}{r}
3X + 1 \\
X^2 + 4X + 6 \overline{)3X^3 + 13X^2 + 22X + 6} \\
\underline{3X^3 + 12X^2 + 18X} \\
X^2 + 4X + 6 \\
\underline{X^2 + 4X + 6}
\end{array}
$$

We first divide X^2 into $3X^3$, yielding the quotient, $3X$. Multiplying the divisor by $3X$ results in the expression $3X^3 + 12X^2 + 18X$. Subtracting this expression from the dividend results in $X^2 + 4X + 6$.

Dividing the X^2 of the divisor into the X^2 of the remaining dividend results in a quotient of $+1$. Again, multiplying the divisor by this quotient and subtracting this from the remaining dividend results in a remainder of 0. Therefore, our quotient is $3X + 1$ and our remainder is 0.

EXAMPLE 4-37   Divide $7AB^3 - 6A^2B^2 + 9B^4 + 6A^4 + 11A^3B$ by $B + 2A$.

SOLUTION   Rearrange terms according to the A variable and divide:

$$
\begin{array}{r}
3A^3 + 4A^2B - 5AB^2 + 6B^3 \\
2A + B \overline{)6A^4 + 11A^3B - 6A^2B^2 + 7AB^3 + 9B^4} \\
\underline{6A^4 + 3A^3B} \\
8A^3B - 6A^2B^2 \\
\underline{8A^3B + 4A^2B^2} \\
-10A^2B^2 + 7AB^3 \\
\underline{-10A^2B^2 - 5AB^3} \\
12AB^3 + 9B^4 \\
\underline{12AB^3 + 6B^4} \\
3B^4
\end{array}
$$

Note that in each case the quotient was determined by dividing the first term of the divisor into the first term of the remaining dividend. Finally, $3B^4$ was not divided by $2A + B$, because it would have resulted in a negative power of A in the quotient. Thus, the remainder is $3B^4$.

PROBLEMS

Divide:

4-103. (a) $6A^2B^3C$ by $2ABC$ (b) $30A^4B^2C$ by $5AB^2C$

4-104. (a) $32A^2CD$ by $16A^2B^2C$ (b) $2Z_1^2Z_2^3Z_3$ by $3Z_1Z_2Z_3^2$

4-105. (a) $3.19R^2S^3TV$ by $2.16RSV^2$ (b) $7.69M^4N^2PQ^4$ by $3.69M^2N^2P^3Q$

4-106. (a) $12.32M^3NP^2$ by $3.69MN^2P$ (b) $31.69F^2G^3HJ^2$ by $2.16FGHJ$

4-107. (a) $(3^2 + 3^3)$ by 3 (b) $(2^6 + 2^5 - 2^3)$ by 2^2

4-108. (a) $(1.6^6 + 1.6^4)$ by 1.6^2 (b) $(2.39^3 - 2.39^2)$ by 2.39

4-109. (a) $3X^2 + 4X$ by X (b) $2Y^3 - 6Y^2 + 2Y$ by $2Y$

4-110. (a) $12P^2 - 6PQ$ by $3P$ (b) $7P^4Q^2 + 6P^2Q + PQ$ by $2PQ$

4-111. $3X^3Y^2 + 4X^2Y^3 + 2XY^4 + Y^5$ by $2X^2Y$

4-112. $48M^3 + 36M^2N - 32MN^2 + 12N^3$ by $4MN^2$

4-113. $3.69R^3 + 7.31R^2S - 3.69RS^2$ by $2.13R^2S$

4-114. $7.31W^5T^2V + 7.69WT^3V^2N - 3.15W^2TN^2$ by $2.16WT^2$

4-115. $3X^3 - 4X^2 + 10X - 3$ by $3X - 1$

4-116. $X^3 + 7X - 6$ by $X - 2$ (*Hint:* Assume that the dividend is $X^3 + 0X^2 + 7X - 6$.)

4-117. $4A^3B^2 + 6A^2B^3 + A^4B + 4AB^4$ by $A + 2B$

4-118. $3X^3Y^2Z + 3X^2Y^3Z + X^4YZ + 6XY^4Z$ by $X + 2Y$

4-119. $16X^3Y + 13X^2Y^2 - 6Y^4 + 6X^4 - 2XY^3$ by $2X^2 + 3Y^2 + 4XY$

4-120. $13A^2 + 18 - A^3 + 3A^4$ by $A^2 - 2A + 6$

5

EQUATIONS

In this chapter we consider the great and mighty equals sign (=) and how it can be used to solve algebraic problems. We first study equality—what it is—then consider how we can maintain this equality by playing with terms on either side of the equals sign. We then consider methods of solving equations and, most important, how to express a problem in the universal language of mathematics. Finally, we consider some very real situations requiring the manipulation of equations.

5-1 FUNDAMENTALS OF EQUATIONS

An equation can be defined as two algebraic expressions separated by an equals sign. For example:

$$3V - V + 6 = 15V - 5V - 10 \qquad (5\text{-}1)$$

In this statement, we are saying that the expression to the left of the equals sign has the same numerical value (equal to) the expression to the right of the equals sign. It happens that this statement is true for only one unique value of V: 2. Substituting 2 into the equation yields

$$3V - V + 6 = 15V - 5V - 10$$
$$3 \cdot 2 - 2 + 6 = 15 \cdot 2 - 5 \cdot 2 - 10$$
$$10 = 10$$

Therefore, for the value of 2, the equation is a true statement. However, for the value of 3,

$$3V - V + 6 = 15V - 5V - 10$$
$$3 \cdot 3 - 3 + 6 = 15 \cdot 3 - 5 \cdot 3 - 10$$
$$12 \neq 20$$

(The symbol $\neq$ means "is not equal to.") Considering all the possible numbers our mind could generate, only one of these numbers, 2, meets the condition of equality. For this reason, this type of equation is called a conditional equation. It is true only under a specified set of conditions, in this case when V assumes the value of 2.

The identity equation, however, is true for all conditions. For example:

$$3X + 5 = 3X + 5$$

For $X = 1$:

$$3X + 5 = 3X + 5$$
$$3 \cdot 1 + 5 = 3 \cdot 1 + 5$$
$$8 = 8$$

For $X = 6$:

$$3X + 5 = 3X + 5$$
$$3 \cdot 6 + 5 = 3 \cdot 6 + 5$$
$$23 = 23$$

In fact, this equation will be true for any value of X. Therefore, the left expression is the identical twin of the right expression. In electronics, we are usually trying to solve for one particular value of a variable. Therefore, we use the conditional equation much more frequently than the identity equation.

Operations Used on Equations

We have used five arithmetic operations: addition, subtraction, multiplication, division, and exponentiation. It turns out that all five can be used with equations, as long as:

Any operation performed on the left side of an equation must also be performed on the right side in order to maintain equality.

Thus, let us add $5V + 3$ to both sides of Eq. (5-1):

$$3V - V + 6 = 15V - 5V - 10$$
$$5V + 3 + 3V - V + 6 = 15V - 5V - 10 + 5V + 3$$

Substituting 2 for V yields

$$5 \cdot 2 + 3 + 3 \cdot 2 - 2 + 6 = 15 \cdot 2 - 5 \cdot 2 - 10 + 5 \cdot 2 + 3$$
$$23 = 23$$

The equality still holds.

Next, let us subtract $5V + 3$ from both sides:

$$3V - V + 6 - (5V + 3) = 15V - 5V - 10 - (5V + 3)$$
$$3V - V + 6 - 5V - 3 = 15V - 5V - 10 - 5V - 3$$

Substitute 2 for V:

$$3 \cdot 2 - 2 + 6 - 5 \cdot 2 - 3 = 15 \cdot 2 - 5 \cdot 2 - 10 - 5 \cdot 2 - 3$$
$$-3 = -3$$

The equality holds.

Next, we multiply both sides by $5V + 3$:

$$(3V - V + 6)(5V + 3) = (15V - 5V - 10)(5V + 3)$$

Substitute 2 for V:

$$(3 \cdot 2 - 2 + 6)(5 \cdot 2 + 3) = (15 \cdot 2 - 5 \cdot 2 - 10)(5 \cdot 2 + 3)$$
$$10 \times 13 = 10 \times 13$$
$$130 = 130$$

The equality again holds.

Next, we divide each side by $5V + 3$:

$$\frac{3V - V + 6}{5V + 3} = \frac{15V - 5V - 10}{5V + 3}$$

Substitute 2 for V:

$$\frac{3 \cdot 2 - 2 + 6}{5 \cdot 2 + 3} = \frac{15 \cdot 2 - 5 \cdot 2 - 10}{5 \cdot 2 + 3}$$

$$\frac{10}{13} = \frac{10}{13}$$

$$0.7692 = 0.7692$$

Again, the equality holds.

Each side can also be raised to a power and the equality remains unchanged. Let us square our favorite equation:

$$(5V - V + 6)^2 = (15V - 5V - 10)^2$$

Substitute 2 for V:

$$10^2 = 10^2$$

$$100 = 100$$

Thus, the equality holds.

Summarizing, as long as we operate the same on each side of the equation, the condition of the equality remains unchanged.

5-2 SOLVING EQUATIONS

In this section, we consider techniques for solving conditional equations. We limit our discussion to:

1. Those conditional equations that have a single variable.
2. That variable is raised to no power above 1.

Some examples of such equations include:

(a) $X + 3X = 2 - 4X$

(b) $\dfrac{X}{3} + 3X = \dfrac{16X + 3}{7}$

(c) $\dfrac{4 + X}{3X} = \dfrac{3 - X}{4X}$

General Procedures

We can state the following as a general procedure for solving all types of equations. It may not be the shortest procedure for a particular equation, but it will consistently find the solution:

(a) Multiply both sides by the denominators of fractions to remove those denominators.

(b) Remove all parentheses by adding, subtracting, or multiplying. Combine similar terms.

(c) Add or subtract to move all terms containing the variable to the left of the equation and those not containing the variable to the right of the equation.

(d) Combine terms by adding or subtracting.

(e) Divide both sides by the coefficient of the variable.

(f) Check the solution by substitution.

We use this procedure for solving all equations in this section and in the sections that follow.

EXAMPLE 5-1 Solve for V:

$$3V - 5V + 6 = 7V - 10$$

SOLUTION

Step (a) Unnecessary.

Step (b) Unnecessary.

Step (c) Subtract $7V$ from each side of the equation to get all terms containing V's on the left:

$$3V - 5V + 6 - 7V = \cancel{7V} - 10 - \cancel{7V}$$
$$3V - 5V + 6 - 7V = -10$$

We must next subtract 6 from each side to move the 6 to the right side:

$$3V - 5V + \cancel{6} - 7V - \cancel{6} = -10 - 6$$
$$3V - 5V - 7V = -10 - 6$$

We have now satisfied step (c).

Step (d) Combine terms:

$$-9V = -16$$

Step (e) Divide both sides by -9:

$$-9V = -16$$
$$\frac{-9V}{-9} = \frac{-16}{-9}$$
$$V = 1.778$$

Step (f) Check the solution by substitution into the original equation:

$$3V - 5V + 6 = 7V - 10$$
$$3(1.778) - 5(1.778) + 6 = 7(1.778) - 10$$
$$2.444 = 2.446$$

The difference is due to the rounding of the answer to four places. Whereas the answer actually was 1.777777778, we assumed that it was 1.778. If we had used all 10 places, the check becomes

$$3(1.777777778) - 5(1.777777778) + 6 = 7(1.777777778) - 10$$
$$2.444444444 = 2.444444450$$

Before moving on, let us reexamine step 3. Note that to move a $+7V$ to the other side of the equation, we must subtract $7V$. Conversely, to move a $-7V$ to the other side, we must add $7V$. Thus:

To move a positive term to the opposite side of the equation, subtract that term from both sides. To move a negative term to the other side of the equation, add that term to both sides.

EXAMPLE 5-2 Solve for P:

$$3P + 13P - 3 = 12P + 6 - 9P + 7$$

SOLUTION Before applying steps (a) through (e), we can save some trouble by combining similar terms:

$$3P + 13P - 3 = 12P + 6 - 9P + 7$$
$$16P - 3 = 3P + 13$$

Step (a) Unnecessary.
Step (b) Unnecessary.
Step (c) Move terms:

$$16P - 3 = 3P + 13$$

Subtract $3P$:

$$16P - 3 - 3P = 3P + 13 - 3P$$

Add 3:

$$16P - 3 - 3P + 3 = 3P + 13 - 3P + 3$$

Step (d) Combine similar terms:

$$13P = 16$$

Step (e) Divide by coefficient of variable:

$$\frac{13P}{13} = \frac{16}{13}$$
$$P = 1.231$$

Step (f) Substitution:

$$3P + 13P - 3 = 12P + 6 - 9P + 7$$
$$3(1.231) + 13(1.231) - 3 = 12(1.231) + 6 - 9(1.231) + 7$$
$$16.70 = 16.69$$

Again, since we expressed the answer to four places, the check is slightly off in the fourth place.

EXAMPLE 5-3 Solve for X_C:

$$3(X_C - 5) + 5X_C = 2(X_C - 3) - 6$$

SOLUTION

Step (a) Unnecessary.
Step (b) Remove parentheses and combine terms:

$$3(X_C - 5) + 5X_C = 2(X_C - 3) - 6$$
$$3X_C - 15 + 5X_C = 2X_C - 6 - 6$$
$$8X_C - 15 = 2X_C - 12$$

Step (c)   Move terms:
Subtract $2X_C$:

$$8X_C - 15 - 2X_C = 2X_C - 12 - 2X_C$$

Add 15:

$$8X_C - 15 - 2X_C + 15 = 2X_C - 12 - 2X_C + 15$$

Step (d)   Combine similar terms:

$$6X_C = 3$$

Step (e)   Divide by 6:

$$X_C = \frac{3}{6} = 0.5000$$

Step (f)   Check the result by substitution:

$$3(0.5 - 5) + 5 \times 0.5 = 2(0.5 - 3) - 6$$
$$-11.00 = -11.00$$

Note that the check is perfectly accurate because 0.5000 is the exact answer.

Fractional Equations

Equations stated in fractional form may also be solved using the foregoing procedure. However, we must first multiply each side of the equation by each denominator. If, after multiplying through by denominators and reducing, we still have a variable raised to a power greater than $1(X^2, A^3, B^2, \text{etc.})$, the procedures discussed in Chapter 18 must be used. If, however, the highest power of the variable is 1, the procedures in this chapter may be used.

EXAMPLE 5-4   Solve for R:

$$\frac{R - 2}{3} = \frac{R + 5}{6}$$

SOLUTION

Step (a)   Multiply through by denominators:

$$\frac{R - 2}{\cancel{3}}(\cancel{3})(6) = \frac{R + 5}{\cancel{6}}(3)(\cancel{6})$$

Step (b)   Remove parentheses:

$$6R - 12 = 3R + 15$$

Step (c)   Move terms:

$$6R - 12 + 12 - 3R = 3R + 15 + 12 - 3R$$
$$3R = 27$$

Step (d)   Combine terms (unnecessary).
Step (e)   Divide by the coefficient:

$$\frac{\cancel{3}R}{\cancel{3}} = \frac{27}{3}$$
$$R = 9$$

Step (f) Check the solution:

$$\frac{R-2}{3} = \frac{R+5}{6}$$
$$\frac{9-2}{3} = \frac{9+5}{6}$$
$$\frac{7}{3} = \frac{7}{3}$$

The solution is $R = 9$.

EXAMPLE 5-5 Solve for V:

$$\frac{V+5}{6} + 2V = \frac{V-2}{7} + 6V + 3$$

SOLUTION

Step (a) Multiply through by denominators. We shall do this in two steps: Multiply by 6:

$$\frac{V+5}{\cancel{6}}(\cancel{6}) + 2V(6) = \frac{V-2}{7}(6) + 6V(6) + 3(6)$$

Multiply by 7:

$$(V+5)(7) + 2V(6)(7) = \frac{V-2}{\cancel{7}}(6)(\cancel{7}) + 6V(6)(7) + 3(6)(7)$$

Step (b) Remove parentheses:

$$7V + 35 + 84V = 6V - 12 + 252V + 126$$

Combine terms:

$$91V + 35 = 258V + 114$$

Step (c) Move terms:

$$91V + 35 - 258V - 35 = 258V + 114 - 258V - 35$$

Step (d) Combine:

$$-167V = 79$$

Step (e) Divide by the coefficient:

$$\frac{-167V}{-167} = \frac{79}{-167}$$
$$V = -0.4731$$

Step (f) Check by substitution:

$$\frac{V+5}{6} + 2V = \frac{V-2}{7} + 6V + 3$$

$$\frac{-0.4731 + 5}{6} + 2(-0.4731) = \frac{-0.4731 - 2}{7} + 6(-0.4731) + 3$$

$$-0.1917 = -0.1919$$

The slight difference is due to the roundoff error of the calculator. The correct solution to four places is

$$V = -0.4731$$

EXAMPLE 5-6 Solve for I:

$$\frac{3}{I+4} = \frac{6}{4I-5}$$

SOLUTION

Step (a)   Multiply by denominators. We shall do it in two steps. First, multiply by $I + 4$:

$$\frac{3}{I+4}\,(I+4) = \frac{6}{4I-5}\,(I+4)$$

$$3 = \frac{6(I+4)}{4I-5}$$

Then, multiply by $4I - 5$:

$$3(4I-5) = \frac{6(I+4)}{4I-5}\,(4I-5)$$

Step (b)   Remove parentheses:

$$12I - 15 = 6I + 24$$

Step (c)   Move terms:

$$12I - 15 + 15 - 6I = 6I + 24 + 15 - 6I$$

Step (d)   Combine terms:

$$6I = 39$$

Step (e)   Divide by the coefficient:

$$\frac{6I}{6} = \frac{39}{6}$$

$$I = 6.500$$

Step (f)   Check:

$$\frac{3}{I+4} = \frac{6}{4I-5}$$

$$\frac{3}{6.5+4} = \frac{6}{4(6.5)-5}$$

$$0.2857 = 0.2857$$

The solution is $I = 6.500$.

PROBLEMS

Solve for the variable:

5-1. $V + 5 = 6$

5-2. $P + 7 = -3$

5-3. $3X - 4 = 7$

5-4. $5T - 2 = 73$

5-5. $3S + 7 = 7S - 2$

5-6. $4Q + 5Q - 6 = -2Q + 7Q + 7$

5-7. $-6.321A + 7.393 = 7.415A + 6.015$

5-8. $36.32 - 7.513I = 2.615I - 3.149$

5-9. $3.159I - 7.563 + 7.312I = 2.569I$

5-10. $3.64 - 7.2V + 7.319 = 0$

5-11. $3(V - 2) = 2(V + 6)$

5-12. $7(Q + 10) = -3(-Q - 20)$

5-13. $10(P + 6) - 2P = 4 + 6P + 5(P - 1)$

5-14. $2S - 3(S + 6) = 4S + 6$

5-15. $3R + 6(R - 2) = 7(R + 5) + 6(R + 2)$

5-16. $X_C - 2(X_C - 5) = 4X_C - 7(X_C + 5)$

5-17. $5R_1 - 7 + 3(R_1 - 6) - 7(R_1 + 3) = 0$

5-18. $7B_L - 2(6 + B_L - 7B_L) = 3(B_L + 6)$

5-19. $2.319(G - 6.215) + 7.316G = 12.71G$

5-20. $3.416(R - 7.213) = 2.169R + 7(2.592R)$

5-21. $\dfrac{R + 6}{3} = \dfrac{R - 7}{8}$

5-22. $\dfrac{Q + 3}{7} = \dfrac{Q - 10}{2}$

5-23. $\dfrac{I - 3}{7} - 3 = \dfrac{I + 6}{2}$

5-24. $\dfrac{V + 6}{2} - 4V = (V - 3)3$

5-25. $\dfrac{13X_L - 3}{7} - \dfrac{X_L + 3}{2} = 5$

5-26. $\dfrac{7R + 6}{10} - 2 = \dfrac{R + 7}{3} + 7R$

5-27. $\dfrac{17L - 6}{2} + 6L = 2L + 3$

5-28. $\dfrac{2R - 6}{3} = \dfrac{12R - 5}{7} + 6R$

5-29. $\dfrac{2.951R + 7.314}{2.615} = \dfrac{7.215R - 6.142}{2.015} - 3.149$

5-30. $\dfrac{-3.156P + 3.149}{12.63} = \dfrac{7.461P}{2.159} + 2.169$

5-31. $\dfrac{17}{P + 3} = \dfrac{2}{6 + P}$

5-32. $\dfrac{3}{V + 2} = \dfrac{2}{V + 7}$

5-33. $\dfrac{7}{I + 6} = \dfrac{2}{4L + 9}$

5-34. $\dfrac{6}{3R - 7} = \dfrac{17}{2R + 6}$

5-35. $\dfrac{1.715}{6.412R - 7.315} = \dfrac{4.316}{2.817R + 2.615}$

5-36. $\dfrac{7.315}{2.816R + 7.196} = \dfrac{14.31}{26.32R - 27.19}$

5-3 FORMING EQUATIONS

There are many occasions when an equation is stated in words and must be translated into algebra. It should be emphasized that algebra is a universal language—it is not dependent on English, French, or Russian. Thus, our task is that of a translator, changing an English statement to a universally understood language, algebra. In so doing, we, like the translator, look for words in one language that can express those of the second. Table 5-1 is an English–algebraic dictionary of these key words.

The skill of forming equations is a very valuable one for the technician. Any problem must be first translated into algebra before it can be solved. Thus, the forming process is essential to reaching a solution. This process consists of five steps:

(a) Identify the variables and assign them a letter name.
(b) Identify the values given.
(c) Analyze carefully, looking for the key words.
(d) Write the equation.
(e) Solve for the unknown, if required.

TABLE 5-1 Translating English to algebra.

English	Algebra
Sum	$+$
Added	$+$
Increase	$+$
Plus	$+$
Less	$-$
Difference	$-$
Remainder	$-$
Subtract	$-$
Decrease	$-$
Minus	$-$
Product	$\times$
Multiply	$\times$
Times	$\times$
Double	$2\times$
Triple	$3\times$
Quotient	$\div$
Divide	$\div$
Square	n^2
Cube	n^3
Square root	$\sqrt{\ }$
Cube root	$\sqrt[3]{\ }$
Inverse	$1/n$
Equal	$=$
Is	$=$

EXAMPLE 5-7 In a transistor circuit, the voltage across the load resistor is the product of the collector current and the load resistance. If the current is 0.005 A and the resistance is 1000 Ω, what is the voltage?

SOLUTION

Step (a)   Identify the variables.

Let us go through the problem in detail. The first phrase, "in a transistor circuit," tells us nothing about values or numbers, so we ignore it. The phrase "the voltage across the load resistor" contains seven words; however, only one can have a number attached to it—the voltage. We call this voltage V. Next, the key word "is" appears. Here is our equals sign. We now have "$V =$." But what does V equal? The "product," another key word, tells us that it is a multiplying process of two or more numbers. We now have to look for these numbers. The only nouns that can be assigned numbers in the remainder of the sentence are "current" and resistance." Thus, voltage (V) equals current (I) times resistance (R):

$$V = IR$$

We now have our equation. The next sentence gives us some numerical values to be used in the equation: current (I) is 0.005 A; resistance (R) is 1000 Ω. Thus, substituting our equation we obtain

$$
\begin{aligned}
V &= IR \\
&= (0.005)(1000) \\
&= 5.000
\end{aligned}
$$

EXAMPLE 5-8 In John Metermotor's model IRA voltmeter, the scale voltage is equal to the sum of the movement resistance and series resistance divided by the full-scale sensitivity. What is the equation for scale voltage?

SOLUTION The variables and their key words are: voltage, equal, sum, movement resistance, series resistance, divided by, sensitivity. Arranged in equation form:

$$V = \frac{R_{\text{MOV}} + R_{\text{SER}}}{\text{sens.}}$$

PROBLEMS

5-37. The power dissipated by a type 2N2222 transistor is the product of its voltage and current. If the voltage is 5.69 V and the current is 0.013 A, what is the power?

5-38. The total inductance of a series circuit is the sum of the series inductances. If inductors of 3 H, 0.50 H, and 2 H are connected in series, what is the total inductance?

5-39. Cliff Trueheart's heartbeat increases by 20 beats/min to 95 beats when his true love approaches. What is his heartbeat when she is absent from his presence?

5-40. Two resistors connected in parallel have an equivalent resistance that is their product divided by their sum. If resistances of 1000 Ω and 2500 Ω are connected in parallel, what is the equivalent resistance?

5-41. Henry Worthmuch purchased two diodes, one costing twice that of the other. If Henry paid a total of $0.75, what is the cost of each diode?

5-42. The total time it takes to form one frame on a display terminal is equal to the product of the time it takes to scan one line and the number of lines per frame. If the time it takes to scan one line is 63 μs and the total number of lines per frame is 262, how long does it take to form one frame?

5-43. The collector current of Klugeco's Hifail transistor is equal to the sum of the leakage current and the product of gain and base current. If the leakage current is 1 μA, the gain is 200, and the base current 20 μA, what is the collector current?

5-44. Skills Sharpshooter gets paid an amount per radio he repairs plus an hourly rate. For any overtime hours, his hourly rate doubles but his piece rate remains fixed. Develop an equation expressing Skill's pay based upon the number of straight-time hours, number of overtime hours, number of radios repaired, hourly (straight-time) rate, and piece rate.

5-45. The inverse of total capacitance of capacitors connected in parallel is equal to the sum of the inverses of each capacitor. Develop a formula for the inverse of total capacitance.

5-46. The formula for finding the area of Granny Gordon's garden is the product of half the width and the sum of the two parallel sides. If her garden is 20 m wide and the parallel sides are 15 m and 25 m, what is the total area?

5-4 LITERAL EQUATIONS

Literal equations are those containing several letters (unknowns). The following are examples:

$$A = \pi r^2 \qquad Q = \frac{f_0}{f_{\text{BW}}} \qquad Z = \sqrt{X^2 + R^2}$$

In many cases a literal equation can be found from a reference book. However, the unknown we have to find may not be to the left of the equal sign. We may know that $Q = f_0/f_{BW}$—but we know Q and f_0 and must solve for f_{BW}. Thus, we must transpose the equations; that is, we must solve for a different variable. The procedure for doing this is the same as that presented in Section 5-2. There are cases when this procedure requires more lengthy calculations than necessary. However, you will eventually arrive at the answer.

EXAMPLE 5-9   Solve for b_1:

$$A = \frac{1}{2}(b_1 + b_2)h$$

SOLUTION

Step (a)   Multiply by denominators:

$$(2)A = \frac{1}{2}(b_1 + B_2)h(2)$$
$$2A = (b_1 + b_2)h$$

Step (b)   Remove parentheses:

$$2A = b_1h + b_2h$$

Step (c)   Move terms:

$$2A - b_1h - 2A = b_1h + b_2h - b_1h - 2A$$
$$-b_1h = b_2h - 2A$$

Step (d)   Combine terms (unnecessary).

Step (e)   Divide by the coefficient:

$$\frac{-b_1h}{-b_1} = \frac{b_2h - 2A}{-b_1h}$$
$$h = \frac{b_2h - 2A}{-b_1h}$$

The minus sign in the denominator can be removed by multiplying both numerator and denominator by (-1):

$$\frac{(b_2h - 2A)(-1)}{(-b_1h)(-1)}$$
$$= \frac{2A - b_2h}{b_1h}$$

In some cases it may be necessary to factor, or un-multiply, in step (e) to get the variable by itself. The following example illustrates the difficulty.

EXAMPLE 5-10   Solve for R_1:

$$V_1 = I_1R_1 + V_2 + I_2R_1$$

SOLUTION

Step (a)   Unnecessary.
Step (b)   Unnecessary.
Step (c)   Move terms:

$$V_1 - I_1R_1 - I_2R_1 - \cancel{V_1} = \cancel{I_1R_1} + V_2 + \cancel{I_2R_1} - \cancel{I_1R_1} - \cancel{I_2R_1} - V_1$$
$$-I_1R_1 - I_2R_1 = V_2 - V_1$$

Step (d) Unnecessary.

Step (e) Divide by the coefficient:

This illustrates the difficulty. There is no single coefficient of the variable R_1. To make a single coefficient, we must un-multiply (factor) such that

$$-I_1R_1 - I_2R_1 = V_2 - V_1$$
$$R_1(-I_1 - I_2) = V_2 - V$$
$$\frac{R_1(\cancel{-I_1 - I_2})}{(\cancel{-I_1 - I_2})} = \frac{V_2 - V_1}{-I_1 - I_2}$$
$$R_1 = \frac{V_2 - V_1}{-I_1 - I_2}$$

Although this is correct, we can get it in better form by multiplying both numerator and denominator by (-1), yielding

$$R_1 = \frac{V_1 - V_2}{I_1 + I_2}$$

PROBLEMS

Solve for the variables indicated on the right:

5-47. $V = IR$ (I)

5-48. $L_1 = mL_k$ (m)

5-49. $V = \dfrac{X^2}{2C_m}$ (C_m)

5-50. $P = I^2R$ (R)

5-51. $h_{fe} = -\dfrac{1}{1 + h_{21}}$ (h_{21})

5-52. $T_A = \dfrac{T_M}{1 - E}$ (E)

5-53. $D = \dfrac{E_d L_l}{2 E_a A}$ (E_a)

5-54. $r_i = r_e + r_b(1 - a)$ (a)

5-55. $\dfrac{f_2}{f_1} = \dfrac{1 - m}{n + 1}$ (n)

5-56. $V = \dfrac{SaRT}{0.049}$ (T)

5-57. $Y = SX + b$ (S)

5-58. $Y_{12} = R_e + j\omega C$ (C)

5-59. $G = \dfrac{K}{\omega(1 + \omega^2 T^2)}$ (T^2)

5-60. $E_{cc} = E_c - I_c R_c$ (I_c)

5-61. $L_1 = \dfrac{m}{1 - m^2} L_K$ (L_K)

5-62. $E_0 = \dfrac{R_{in} + R_e}{R_{in}} E_{in}$ (R_{in})

5-63. $A_i = \dfrac{a}{1 - a}$ (a)

5-64. $A_i = \dfrac{-(r_m + r_b)}{r_b + r_c + r_l}$ (r_b)

5-65. $A_i = \dfrac{h_{21} y_l}{h_{22} + y_l}$ (y_l)

5-66. $A = \dfrac{\mu Z_2}{r_p + Z_2}$ (Z_2)

5-67. $A_v = \dfrac{-ar_l}{r_e + r_b(1 - a)}$ (r_b)

5-68. $C_1 = \dfrac{1 - m^2}{m_1} C_2$ (C_2)

5-69. $L = \dfrac{L_s}{1 + \omega^2 L_s C_d}$ (L_s)

5-70. $E_Q = 0.41(E_B - E_Y) + 0.48(E_R - E_Y)$ (E_Y)

5-71. $r_0 = r_c + r_e - r_m + \dfrac{r_e(r_m - r_e)}{r_g + r_b + r_e}$ (r_m)

5-72. $A_i = \dfrac{-a}{a + r_e/r_c}$ (r_e)

5-73. $a = \dfrac{0.049 V}{A_1 + A_2 + A_3}\left(\dfrac{1}{T_M} - \dfrac{1}{T_E}\right)$ (A_1)

5-74. Prob. 5-73 (T_M)

5-5 PROPORTIONALITY

Let us assume you had a job that paid by the hour, $100 per hour. If you worked 1 h, you would expect to receive $100; if you worked 5 h, you would expect to receive $500. We could, therefore, say that your amount of pay is directly proportional to the hours worked. We express this algebraically as

$$P = kh$$

where P is total amount of pay, h the number of hours worked, and k is called the constant of proportionality. In this case, k is the rate of pay per hour. Another way of expressing this is

$$P \propto h$$

where the symbol $\propto$ represents "is directly proportional to." By stating that $P \propto h$, we mean that doubling h doubles P; tripling h triples P; halving h halves P.

There are many cases when proportionality occurs. Assuming a constant speed, the time it takes to travel from point A to point B is directly proportional to distance. Thus, doubling the distance doubles the transit time; halving the distance halves the travel time. Express this as

$$t = ks \quad \text{or} \quad t \propto s$$

where t represents the time, s the distance, and k is the constant of proportionality. In this case k is the speed.

EXAMPLE 5-11 Assuming a constant rate of speed, what is the effect upon time of decreasing the distance to $\frac{1}{3}$ that of the original distance?

SOLUTION Assume that t_0 is the initial time and t_1 is the new rate:

$$t_0 \propto S$$
$$t_1 \propto \frac{1}{3}S$$
$$\propto \frac{1}{3}t_0$$

Thus, the time is $\frac{1}{3}$ of the original time.

By providing a unique set of values, the constant of proportionality can be computed. If, for example, we know that the amount of pay is directly proportional to hours worked and that 4 h of work yielded $2000, we can compute k:

$$P = kh$$
$$2000 = k \times 4$$
$$k = \$500 \text{ per hour}$$

EXAMPLE 5-12 A crazy collie christened Calamity wags his tail at a rate that is directly proportional to how fast you pat his head. Determine the constant of proportionality if he wags his tail 80 times/min while his head is patted 25 times/min.

SOLUTION If W represents wags and p represents pats:

$$W = kp$$
$$80 = k \times 25$$
$$k = 3.20$$

Not only can elements be directly related to each other, but other relationships can exist. The area of a circle is directly proportional to the square of the radius:

$$A = kr^2 \quad \text{or} \quad A \propto r^2$$

Thus, doubling the radius causes the area to increase by a factor of 4:

$$A_D = k(2r)^2$$
$$= 4kr^2$$

EXAMPLE 5-13 What is the effect upon the area of a circle if the radius is increased five times?

SOLUTION

$$A_0 \propto r^2$$
$$A_1 \propto (5r)^2$$
$$\propto 25r^2$$

Thus, the area is increased 25 times.

Proportionalities can also be inverse. For example, the time it takes to travel from Seattle to Juneau is inversely proportional to speed. That is, the greater the speed, the less the time. Express this as

$$t = k\frac{1}{s} \quad \text{or} \quad t \propto \frac{1}{s}$$

Thus, doubling the speed cuts the time in half; halving the speed doubles the travel time.

EXAMPLE 5-14 What effect will traveling five times the speed have upon travel time?

SOLUTION

$$t_0 \propto \frac{1}{s}$$
$$t_1 \propto \frac{1}{5s}$$
$$\propto \frac{1}{5}\left(\frac{1}{s}\right)$$
$$\propto \frac{1}{5}t_0$$

Thus, the travel time is $\frac{1}{5}$ of what it was originally.

EXAMPLE 5-15 The amount of light illuminating a surface is inversely proportional to the square of the distance from the light source. If the distance is increased by 50%, what is the effect upon the surface light?

SOLUTION Let L be light and s be distance.

$$L_0 \propto \frac{1}{s^2}$$

$$L_1 \propto \frac{1}{(1.5s)^2}$$

$$\propto \frac{1}{2.25s^2}$$

$$\propto 0.4444L_0$$

Thus, light at the increased distance is 0.4444 times that of the initial condition.

Application to Electronics

Proportionalities exist everywhere in electronics. Some of these are:

1. Current (I) is directly proportional to voltage (V) if resistance is held constant.
2. If voltage is held constant, power (P) is directly proportional to current.
3. If voltage is held constant, current is inversely proportional to resistance (R).
4. If resistance is held constant, power is proportional to the square of the current.
5. Impedance (Z) of a transformer secondary is inversely proportional to the square of the turns ratio (n).

EXAMPLE 5-16 What is the effect upon current if voltage is increased by 5%, assuming constant resistance?

SOLUTION Using relationship 1:

$$I_0 \propto V$$
$$I_1 \propto 1.05V$$
$$\propto 1.05I_0$$

Thus, current increases by 5%.

EXAMPLE 5-17 When power is 30 W and current is 7.39 A, what is the constant of proportionality?

SOLUTION Using relationship 2:

$$P = kI$$
$$30 = k \times 7.39$$
$$k = 4.060$$

EXAMPLE 5-18 If resistance is decreased to 75% of its original value, what is the effect upon current?

SOLUTION Using relationship 3:

$$I_0 \propto \frac{1}{R}$$

$$I_1 \propto \frac{1}{0.75R}$$
$$\propto \frac{1}{0.75} \frac{1}{R}$$
$$\propto 1.333 I_0$$

Thus, current increases by 33.33%.

EXAMPLE 5-19 If the turns ratio of a transformer triples, what is the effect upon secondary impedance?

SOLUTION Using relationship 5:

$$Z_0 \propto \frac{1}{n^2}$$
$$Z_1 \propto \frac{1}{(3n)^2}$$
$$\propto \frac{1}{9} \frac{1}{n^2}$$
$$\propto 0.1111 Z_0$$

Thus, the impedance is 0.1111 of its original value.

Ratios

If two sets of proportions with the same k are divided by each other, the constant of proportionality cancels out, leaving a *ratio*. For example, knowing that on this particular job total pay is proportional to hours worked, we obtain

$$P_1 = kh_1$$
$$P_2 = kh_2$$
$$\frac{P_1}{P_2} = \frac{h_1}{h_2}$$

Note that we are assuming the same constant of proportionality; thus, we are assuming that the hourly rate of pay remains unchanged. This ratio is a very useful relationship.

EXAMPLE 5-20 If I worked 20 h and received $5280, how much would I earn in 76 h?

SOLUTION

$$\frac{P_1}{P_2} = \frac{h_1}{h_2}$$

Substituting the first set of conditions (5280 and 20) for P_1 and h_1 and the second set of conditions (P_2, which I do not know, and 76) for P_2 and h_2:

$$\frac{5280}{P_2} = \frac{20}{76}$$
$$P_2 = \$20{,}064$$

Thus, for 76 h at the same rate of pay, I would receive $20,064.

EXAMPLE 5-21 Within a circuit having a current of 5 A, the voltage if 76 V. If voltage were increased to 89 V, how much is current if resistance is constant?

SOLUTION For the first set of conditions:

$$I_1 = kV_1$$

For the second set:

$$I_2 = kV_2$$

Divide:

$$\frac{I_1}{I_2} = \frac{V_1}{V_2}$$

Substitute values:

$$\frac{5}{I_2} = \frac{76}{89}$$
$$I_2 = 5.855 \text{ A}$$

EXAMPLE 5-22 For a particular circuit with constant resistance, the power is 7.693 W and the current is 27.29 A. If current were decreased to 21.31 A, what would the power be?

SOLUTION

$$P_1 = kI_1^2$$
$$P_2 = kI_2^2$$
$$\frac{P_1}{P_2} = \frac{I_1^2}{I_2^2}$$
$$\frac{7.693}{P_2} = \frac{27.29^2}{21.31^2}$$
$$P_2 = 4.691 \text{ W}$$

PROBLEMS

5-75. Assuming a constant rate of speed, what is the effect upon time if distance is increased by 20%?

5-76. Assuming a constant rate of speed, what is the effect upon time if distance is decreased to 27% of its initial value?

5-77. Ringo the rattler rattles his rattles at a rate that is inversely proportional to the distance between him and his opponent. If his opponent cuts his distance in half, what happens to Ringo's rattle rate?

5-78. Wiley Willard wolf woos Winnie wolf by wailing woefully at a rate that is proportional to the distance between him and Winnie. If Winnie wends her way 20 times farther away from Willard, what happens to Willard's wailing rate?

5-79. Moving a light source from 20 m away from a surface to 16 m away has what effect upon the surface light?

5-80. If voltage increases 20%, what is effect upon current if resistance is held constant?

5-81. If current decreases by 27%, what is effect upon power if voltage is held constant?

5-82. If resistance increases by 23%, what is effect upon current if voltage is constant?

5-83. If current is doubled, what happens to power if resistance is constant?

5-84. If the turns ratio decreases to 62% of its original value, what happens to the impedance of the transformer secondary?

5-85. If current is 20 A and voltage 60 V, what is the constant of proportionality when resistance is constant?

5-86. If resistance is 76,000 Ω when current is 0.031 A, what is the constant of proportionality if voltage is constant?

5-87. If Jimmy worked 7 hours for 63 peanuts, how many peanuts would he earn in 23 hours?

5-88. On her birthday, King Katua of Krakatoa presents Queen Quastach with gold which has a weight that is inversely proportional to her age. If she received 500 kg of gold when she was 38, how much will she receive when she reaches 56?

5-89. The thermal resistance of a transistor heat sink is inversely proportional to its area. If 1 in.² has a thermal resistance of 125°C/W, how large a heat sink will be necessary for a thermal resistance of 10°C/W?

5-90. Frequency is inversely related to wavelength. If the frequency is 2000 Hz at a wavelength of 150,000 m, what is the wavelength of a frequency of 15,273 Hz?

5-91. If the power is 27 W at 2.39 A, what will the power be at a current of 5.32 A? The resistance is constant.

5-92. If the current is 20 A at a resistance of 27,000 Ω, what is the current at a resistance of 1000 Ω? The voltage is constant.

6

CALCULATING DECIMAL NUMBERS
EFFECTIVELY

In this chapter we consider numbers. What are the different ways a number can be expressed? How do we express accuracy in numbers? What does the electronic calculator do to our accuracy, and how many figures must we read to express the number accurately? We also consider a method to convert from any system of units to any other system and we apply this method to the analysis of algebraic formulas.

6-1 RESISTOR VALUES

Since resistors come in all values from 0.1 to 27,000,000 Ω (electrical sizes), they provide an ideal vehicle for analyzing numbers. Figure 6-1 illustrates the markings found on the carbon composition resistor—the most commonly used in electronic systems. These markings consist of five bands. The two left bands yield two digits

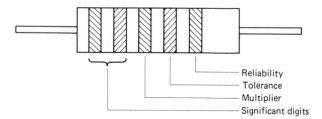

FIGURE 6-1 Resistor color bands.

according to Table 6-1; the third band, a multiplier; the fourth band, a tolerance; and the fifth band, a reliability code. Assume that a resistor has the following color bands: red, violet, yellow, gold, and orange. From Table 6-1, the red and violet bands yield 27. The yellow band tells us to add four zeros, making 270,000. Thus, the resistor is 270,000 Ω. The gold band tells us that the tolerance of this resistor is ±5% and that its actual value lies between 270,000 minus 5% of 270,000, and 270,000 plus 5% of 270,000. Therefore, the actual resistor value is between 256,500 and 283,500. The fifth band tells us that 0.01% of these units failed after 100 h of operation.

> **EXAMPLE 6-1** Determine the value of a resistor with green, blue, red, silver, and red bands.

SOLUTION According to Table 6-1, green and blue yield 56. The red band adds two zeros, making 5600, and the tolerance is 10%. The actual value is 5600 ± 10% of 5600 or 5040 to 6160 Ω. The 100-h reliability test resulted in 0.1% of the units failing.

TABLE 6-1 Resistor color code.

Color	Significant Figure	Multiplier	Tolerance (%)	100-h Reliability (%)
Black	0	1		
Brown	1	10		1.0
Red	2	10^2		0.1
Orange	3	10^3		0.01
Yellow	4	10^4		0.001
Green	5	10^5		
Blue	6	10^6		
Violet	7	10^7		
Gray	8	10^8		
White	9	10^9		
Gold		0.1	5	
Silver		0.01	10	
No color			20	No test made

EXAMPLE 6-2 Determine the value of a resistor with brown, black, and black bands.

SOLUTION Brown and black yields 10. The third band, black, tells us to add zero zeros, still giving us 10. The missing fourth band tells us this is a 20% resistor and that its value lies between 8.0 and 12.0 Ω. The missing fifth band indicates that no reliability test was made.

EXAMPLE 6-3 What color bands would a resistor of 1200 Ω, 5%, have?

SOLUTION From Table 6-1, 12 gives brown and red, two zeros another red, and 5% a gold band. Thus, the colors are brown, red, red, and gold.

Values Under 10 Ω

Values less than 10 Ω use a gold or silver multiplier (third) band. A gold band indicates that the first two digits are to be multiplied by 0.1; a silver band indicates multiplication by 0.01.

EXAMPLE 6-4 What is the value of a resistor having orange, orange, silver, and gold color bands?

SOLUTION The first two bands produce 33. The third band indicates that we are to multiply this by 0.01, giving 0.33 Ω for the nominal value. The tolerance is 5%. Therefore, its actual value lies between 0.3135 and 0.3465 Ω. No reliability test was made.

EXAMPLE 6-5 What resistor values do bands of brown, green, gold, and gold represent?

SOLUTION The first two bands represent 15. The third band represents a multiplier of 0.1, giving 1.5 Ω, and the fourth band a 5% tolerance. Therefore, the value lies between 1.425 and 1.575 Ω.

EXAMPLE 6-6   What bands does a 4.7-Ω 5% resistor have?

SOLUTION   The 47 gives yellow and violet. We must multiply this by 0.1 to obtain the 4.7, so the next band is gold. The tolerance band is also gold. The resistor has the bands yellow, violet, gold, and gold.

PROBLEMS

Find the minimum, maximum, and nominal values and, where applicable, the results of the reliability test for the following resistors:

6-1.   Brown, orange, red, gold, brown
6-2.   Green, brown, orange, silver, red
6-3.   Violet, green, brown, gold
6-4.   Violet, orange, yellow
6-5.   Red, red, red
6-6.   Brown, black, brown, silver
6-7.   Blue, red, orange, gold
6-8.   Red, yellow, orange, silver
6-9.   Orange, orange, gold, gold
6-10.   Red, black, silver, gold
6-11.   Orange, white, silver
6-12.   Gray, red, gold

What are the color bands of the following resistors?

6-13.   130 Ω, 5% tolerance, no test
6-14.   2400 Ω, 10% tolerance, no test
6-15.   43,000 Ω, 5% tolerance, 0.1% reliability
6-16.   910,000 Ω, 10% tolerance, 0.1% reliability
6-17.   3.6 Ω, 5% tolerance, no test
6-18.   0.18 Ω, 10% tolerance, no test
6-19.   5.6 Ω, 5% tolerance, 0.001% reliability
6-20.   0.75 Ω, 10% tolerance, 1% reliability

6-2   EXPRESSING NUMBERS

As can be seen of resistors, electronics uses quantities requiring a large string of zeros. Very small currents also require multiple zeros. However, the powers of 10 can be expressed in two ways, as shown in Table 6-2: decimal format and exponential format. In this way, very small and very large numbers can be conveniently represented. There are three systems in common use for expressing these numbers: fixed, scientific, and engineering. Most calculators use two or three of these systems of notation, so we should be familiar with them.

TABLE 6-2   Exponential format.

Decimal Format	Exponential Format
0.000,001	10^{-6}
0.000,01	10^{-5}
0.0001	10^{-4}
0.001	10^{-3}
0.01	10^{-2}
0.1	10^{-1}
1	10^{0}
10	10^{1}
100	10^{2}
1000	10^{3}
10,000	10^{4}
100,000	10^{5}
1,000,000	10^{6}

Fixed Notation

In fixed notation the number is expressed just as it was in Section 6-1, without a power of 10. The following are examples:

$$1.39 \qquad 0.0001763$$
$$0.13642 \qquad 1,700,000,000.0$$
$$0.000,000,000,000,139$$

As can be seen, this works quite well as long as the numbers are not too small or too large. When this occurs, scientific or engineering notation is used.

Scientific Notation

Scientific notation expresses a number as a single digit (1 through 9) to the left of the decimal point, the point, then the rest of the significant digits, multiplied by 10 raised to a power. The following is an example:

$$3.146 \times 10^{12}$$

This means 3.146 multiplied by 10^{12}, that is, 3.146 with the decimal point moved 12 places to the right. In fixed notation:

$$3.146 \times 10^{12} = 3,146,000,000,000$$

Note that the number 12, the exponent, represents the number of places the decimal point must be moved to convert back to fixed notation. If this exponent is positive, the point is moved to the right; if negative, the point is moved to the left.

EXAMPLE 6-7 Express the following numbers in fixed-point notation:

(a) 4.59×10^6 (b) 2.13×10^{-7}

SOLUTION

(a) The point must be moved six places to the right.

$$4.59 \times 10^6 = 4,590,000$$

(b) The point must be moved seven places to the left.

$$2.13 = 0.000,000,213$$

To convert a fixed-point number to scientific notation, the process is reversed. Move the point to the place where a single digit 1 through 9 appears to the left of the point. If the point was moved left, use a positive exponent, its value indicating the number of places the point was moved. If the point was moved to the right, use a negative exponent, its value representing the number of places moved.

EXAMPLE 6-8 Express the following in scientific notation:

(a) 2,135.6   (b) −7.1563   (c) 0.01395   (d) −63,490

SOLUTION

(a) $2{,}135.6 = 2.1356 \times 10^3$
(b) $-7.1563 = -7.1563 \times 10^0$
(c) $0.01395 = 1.395 \times 10^{-2}$
(d) $-63{,}490 = -6.3490 \times 10^4$

Note that the sign of the value is carried through in scientific notation.

Engineering Notation

Engineering notation is by far the most common system used in electronics. It is similar to scientific notation except that:

(a) The exponent must be a multiple of 3.
(b) The numbers 1 through 999 may therefore appear to the left of the point.

Some examples are:

$$3.169 \times 10^3 \qquad 71.56 \times 10^6$$
$$23.165 \times 10^{-6} \qquad 831.6 \times 10^{-9}$$

To convert a number from engineering notation to fixed notation, move the point to the right for a positive exponent or the left for a negative exponent.

EXAMPLE 6-9 Express in fixed notation:

(a) 31.65×10^3 (b) -673.1×10^{-3} (c) -79.6×10^6

SOLUTION

(a) $31.65 \times 10^3 = 31{,}650$
(b) $-673.1 \times 10^{-3} = -0.6731$
(c) $-79.6 \times 10^6 = -79{,}600{,}000$

To convert a number from fixed into engineering notation, move the point such that between one and three digits appear to the left of the point and the exponent is evenly divisible by 3. Zero is also a proper exponent.

EXAMPLE 6-10 Convert to engineering notation:

(a) 3150   (b) $-21{,}619$ (c) 0.001396   (d) -0.0000003776

SOLUTION In each case, move the point in blocks of three digits until between one and three digits appear to the left of the point, the most left being nonzero.

(a) $3150 = 3.150 \times 10^3$
(b) $-21{,}619 = -21.619 \times 10^3$
(c) $0.001396 = 1.396 \times 10^{-3}$
(d) $-0.0000003776 = -377.6 \times 10^{-9}$

Note that we had to move three blocks of three digits to the right.

PROBLEMS

Express in scientific and engineering notation:

6-21. (a) 31.69 (b) 21,673

6-22. (a) 743.9 (b) 67,615,721

6-23. (a) 0.0139 (b) −0.00001364

6-24. (a) −0.0000159 (b) 0.007823

6-25. (a) −765,317 (b) 8,472,000

6-26. (a) 976,000,000 (b) −84,300,000,000

6-27. (a) 53,670,000 (b) 34,672

6-28. (a) 3.156 (b) 0.157

Express in fixed notation:

6-29. (a) 37.6×10^3 (b) 734.1×10^0

6-30. (a) 765.2×10^{-9} (b) -3.19×10^{-2}

6-31. (a) -7.316×10^6 (b) 3.87×10^4

6-32. (a) 5.15×10^4 (b) -6.32×10^6

6-33. (a) -7.96×10^1 (b) 8.76×10^{-2}

6-34. (a) 9.28×10^7 (b) -3.98×10^{-7}

6-3 MANIPULATING DECIMAL NUMBERS

Numbers in any of the three notations can be added, subtracted, multiplied, and divided. Usually, it is necessary only to enter the numbers and the operation into the calculator and ask it to manipulate the decimal point. However, a basic understanding of these manipulations aids in fully using the power of the calculator.

Addition

Numbers can be added by hand only if their exponents of 10 are identical. If they are not, make them identical by moving the point. Finally, add the two numbers and affix the common exponent.

EXAMPLE 6-11 Add 6.319×10^2 to 1316.7.

SOLUTION Since the first number has a 2 for its 10^2 term, we shall make the second number have a 2 also:

$$1316.7 = 13.167 \times 10^2$$

We can now add:

$$
\begin{array}{r}
13.167 \times 10^2 \\
+ \quad 6.319 \times 10^2 \\
\hline
19.486 \times 10^2
\end{array}
$$

This is the same as saying that "13.167 apples plus 6.319 apples equals 19.486 apples." Note that numbers can be added only if their exponents of 10 are identical; apples can only be added to apples. Convert to engineering notation:

$$19.486 \times 10^2 = 1.9486 \times 10^3$$

EXAMPLE 6-12 Add $3.195 \times 10^5 + 26.16 \times 10^6$.

SOLUTION Modify the first operand to make the exponent 6:

$$3.195 \times 10^5 = 0.3195 \times 10^6$$

I had trouble with this type of manipulation until I realized that if we multiply the 10^5 by 10^1, making 10^6, we must divide 3.195 by the same 10 in order to maintain the same value. So line up the decimal points and add:

$$\begin{array}{r} 0.3195 \times 10^6 \\ + \underline{26.16 \quad \times 10^6} \\ 26.4795 \times 10^6 \end{array}$$

Subtraction

Subtracting is similar to adding except, once we have converted to apples, we subtract instead of add.

EXAMPLE 6-13 Subtract 0.3195 from 76.21×10^{-1}.

SOLUTION Convert 0.3195:

$$0.3195 = 3.195 \times 10^{-1}$$

Subtract:

$$\begin{array}{r} 76.21 \quad \times 10^{-1} \\ - \underline{3.195 \times 10^{-1}} \\ 73.015 \times 10^{-1} \end{array}$$

Convert the answer to engineering notation:

$$73.015 \times 10^{-1} = 7.3015 \times 10^0$$

Note that subtraction cannot be performed by hand unless the exponents are identical. The calculator does this all for us.

Multiplication

Multiplication of decimal numbers follows the rules for exponents: to multiply numbers with the same base, add exponents. In this case, to multiply powers of 10, add the exponents.

EXAMPLE 6-14 Multiply:

(a) 2.16×10^3 by 7.39×10^{-6}
(b) 76.15 by 3.15×10^4

SOLUTION

(a) $\quad (2.16 \times 10^3) \times (7.39 \times 10^{-6})$
$\quad = (2.16 \times 7.39) \times (10^3 \times 10^{-6})$
$\quad = 15.9624 \times 10^{3-6}$
$\quad = 15.9624 \times 10^{-3}$

(b) $\quad (76.15) \times (3.15 \times 10^4)$
$\quad = (76.15 \times 10^0) \times (3.15 \times 10^4)$
$\quad = (76.15 \times 3.15) \times (10^0 \times 10^4)$
$\quad = 239.8725 \times 10^{0+4}$
$\quad = 239.8725 \times 10^4$
$\quad = 2.398725 \times 10^6$

Division

Division also follows the rules of exponents for those powers of 10: they are subtracted.

EXAMPLE 6-15 Divide:

(a) 36.15×10^6 by 7.39×10^3

(b) -273.7×10^{-3} by 7.315×10^{-6}

SOLUTION

(a) $$\frac{36.15 \times 10^6}{7.39 \times 10^3}$$

$$= \frac{36.15}{7.39} \times \frac{10^6}{10^3}$$

$$= 4.892 \times 10^{6-3}$$

$$= 4.892 \times 10^3$$

(b) $$\frac{-273.7 \times 10^{-3}}{7.315 \times 10^{-6}}$$

$$= \frac{-273.7}{7.315} \times \frac{10^{-3}}{10^{-6}}$$

$$= -37.42 \times 10^{-3-(-6)}$$

$$= -37.42 \times 10^3$$

Exponentiation

Exponentiation follows the "power to a power" rule of exponents:

$$(a^m)^n = a^{m \times n}$$

EXAMPLE 6-16 Compute $(3.76 \times 10^7)^3$.

SOLUTION

$$(3.76 \times 10^7)^3 = (3.76)^3 \times (10^7)^3 = 53.16 \times 10^{21}$$

Note that each part of the number is raised to the power.

Using the Calculator

Most calculators have an exponent key, labeled $\boxed{\text{EEX}}$ or $\boxed{\text{EXP}}$. Using it greatly simplifies data entry of the very large and very small numbers encountered in electronics. To enter 13,000,000,000, for example, enter $\boxed{1}$, $\boxed{3}$, $\boxed{\text{EEX}}$ $\boxed{9}$. The $\boxed{\text{EEX}}$ $\boxed{9}$ means "$\times 10^9$" and saves entering the long string of zeros. The key can also be used for entering negative exponents by entering $\boxed{\text{CHS}}$ or $\boxed{+/-}$ following the entry of the exponent. Thus, 73.6×10^{-9} would be entered as $\boxed{7}$ $\boxed{3}$ $\boxed{.}$ $\boxed{6}$ $\boxed{\text{EEX}}$ $\boxed{9}$ $\boxed{\text{CHS}}$.

PROBLEMS

Compute using the calculator's exponent key and express in engineering notation:

6-35. (a) $(3.15 \times 10^5) + (7.39 \times 10^6)$

(b) $(73.16 \times 10^6) + (317.7 \times 10^3)$

(c) $(0.0317) + (3.76 \times 10^{-2})$

(d) $(87.61 \times 10^3) + (7.869 \times 10^4)$

6-36. (a) $(0.315 \times 10^{-3}) + (76.17 \times 10^{-4})$

(b) $(735.7 \times 10^9) + (118.215 \times 10^7)$

(c) $(83.16 \times 10^{-3}) + (7.315 \times 10^{-2})$

(d) $(0.013 \times 10^{-6}) + (3.169 \times 10^{-8})$

6-37. (a) $(73.15 \times 10^3) - (8.21 \times 10^2)$

(b) $(84.16 \times 10^4) - (2159 \times 10^2)$

(c) $(7777 \times 10^{-3}) - (0.01316 \times 10^0)$

(d) $(849.6 \times 10^6) - (7.21 \times 10^8)$

6-38. **(a)** $(1.315 \times 10^{-4}) - (66.315 \times 10^{-2})$
(b) $(72.73 \times 10^{-7}) - (0.813 \times 10^{-5})$
(c) $(437.6 \times 10^{-2}) - (1.316 \times 10^{1})$
(d) $(789.0 \times 10^{3}) - (1.319 \times 10^{5})$

6-39. **(a)** $(778.3 \times 10^{-4}) \times (998.8 \times 10^{-4})$
(b) $(802.2 \times 10^{-3}) \times (7.976 \times 10^{2})$
(c) $(-194.6 \times 10^{-6}) \times (53.89 \times 10^{-3})$
(d) $(7.249 \times 10^{3}) \times (0.765 \times 10^{-4})$

6-40. **(a)** $(8.326 \times 10^{9}) \times (-12.01 \times 10^{-7})$
(b) $(0.9785 \times 10^{-3}) \times (620.5 \times 10^{4})$
(c) $(66.98 \times 10^{-3}) \times (79.64 \times 10^{-5})$
(d) $(0.9935 \times 10^{-9}) \times (4.718 \times 10^{+7})$

6-41. **(a)** $(427.8 \times 10^{-7}) \div (5.862 \times 10^{3})$
(b) $(0.432 \times 10^{-2}) \div (7.321 \times 10^{4})$

(c) $(91.91 \times 10^{4}) \div (-2.966 \times 10^{5})$
(d) $(-7486 \times 10^{3}) \div (438.2 \times 10^{-4})$

6-42. **(a)** $(260.0 \times 10^{-6}) \div (-205.5 \times 10^{4})$
(b) $(926.9 \times 10^{9}) \div (624.8 \times 10^{0})$
(c) $(-1.700 \times 10^{12}) \div (0.09182 \times 10^{-4})$
(d) $(-4.569 \times 10^{-6}) \div (49.49 \times 10^{8})$

6-43. **(a)** $(533.1 \times 10^{4})^{5}$
(b) $(-5.092 \times 10^{6})^{2}$
(c) $(65.59 \times 10^{-3})^{-3}$
(d) $(9.817 \times 10^{-4})^{6}$

6-44. **(a)** $(0.06075 \times 10^{-3})^{4}$
(b) $(-42.59 \times 10^{-6})^{3}$
(c) $(0.05859 \times 10^{4})^{-3}$
(d) $(8.151 \times 10^{-6})^{3.2}$

6-4 EXPRESSING ACCURACY

Most of the numerical values we use in electronics are obtained from some measuring
instrument: a voltmeter, ammeter, or ohmmeter, for example. However, each of these
instruments has only a certain number of digits that can be used to express the
number.

Significant Digits

Assume that I just read the voltage from the power outlet just to the right of
my desk. "The voltage is 120 volts," I declare confidently. But examine this statement.
Just what are the outer limits this voltage could be? Every meter has an error. If
my meter had an error of ±10 V, the actual reading lies between 110 and 130 V: if
my meter had an error of ±0.01 V, the actual reading lies between 119.99 and 120.01
V. Just how can I communicate this to you in the number 120? The answer lies in
the number of digits I give you. If I express the value as 120.0, the scientist should
assume that the value is accurate to four significant digits; that is, all four digits
are accurate and the fifth digit is not. Thus, the actual value is 120.0 ± 0.05 V. If
I express the value as 120.000, the number is accurate to six significant digits and
the seventh is not accurate. Thus, the actual value is 120.000 ± 0.0005. Note that
the term "significant digits" indicates how many digits are accurate and that the
next digit to the right is ±5.

In very small numbers, the leading zeros are not considered significant digits.
The number 0.000,015,2 is considered significant to three places, for the zeros in
front of the 152 merely tell where the point is being placed, not the accuracy of
the reading. Thus, the number represents 0.000,015,2 ± 0.000,000,05. On the other
hand, the number 0.000,015,200 is considered accurate to five places. The numbers
to the left of 152 are not considered significant, but those to the right are considered
significant, for they contribute to the expression of accuracy. In this case, the actual
value is 0.000,015,200 ± 0.000,000,000,5 and the value lies between 0.000,015,199,5
and 0.000,015,200,5.

EXAMPLE 6-17 How many significant digits do each of the following numbers contain?

(a) 87.95   (b) 6.800   (c) 799.338   (d) 0.0094
(e) 0.0009464   (f) 0.005196000

SOLUTION In each case count the number of digits from right to left starting with the leftmost nonzero digit.

(a) 4   (b) 4   (c) 6   (d) 2   (e) 4   (f) 7

EXAMPLE 6-18 What are the highest and lowest possible values that each of the following numbers express?

(a) 300.9   (b) 0.382   (c) 161,869.9   (d) 0.0038420   (e) −0.0137000

SOLUTION In each case the highest and lowest values are ±5 in the digit to the right of the least significant digit.

(a) The lowest value is 300.9 − 0.05 and the highest, 300.9 + 0.05. Therefore, the answer is 300.85 to 300.95.
(b) 0.382 ± 0.0005 or 0.3815 to 0.3825
(c) 161,869.9 ± 0.05 or 161,869.85 to 161,869.95
(d) 0.003,842,0 ± 0.000,000,05 or 0.003,841,95 to 0.003,842,05
(e) −0.013,700,0 ± 0.000,000,05 or −0.013,700,05 to −0.013,699,95

Rounding Off

Sometimes a calculator will give many digits of supposedly significant data, but we wish to express the accuracy only to a certain number of digits. Let us assume that we calculated a voltage of 3.16961523 and wish to express this number to three significant digits. Rather than calling it 3.16, we should examine the next digit, the 9, as to whether it is 5 or more, and if it is, increase our third digit by 1, making the result 3.17.

EXAMPLE 6-19 Round off the following numbers to four places:

(a) 30.5612   (b) 0.00716543   (c) −840.9631   (d) 0.0008137614

SOLUTION

(a) We want to keep four places, so examining the fifth place, the 1, we see that it is less than five. Therefore, the correct answer is 30.56.
(b) 0.007165
(c) In this case the fifth place is 6. Therefore, we must add one to the fourth place, making −841.0.
(d) 0.0008138
    Calculators automatically do this rounding off for us.

PROBLEMS

How many significant digits do each of the following numbers contain?

6-45. (a) 7.315 (b) −70.0 (c) 85.931
6-46. (a) −8.41 (b) 13 (c) 84.3164
6-47. (a) 0.02950 (b) −0.00243 (c) 4.22410

6-48. (a) 0.01436 (b) −0.00263000 (c) −0.058902
6-49. (a) 768 (b) 0.345 (c) 3.100
6-50. (a) 131.5 (b) 1.203 (c) 994

What are the highest and lowest numbers that each of the following expresses?

6-51. (a) 1.3 (b) 1.30 (c) 1.300
6-52. (a) 1.3000 (b) 1.30000 (c) 1.300000
6-53. (a) 438.08 (b) 438.083 (c) 438.0836
6-54. (a) −3.16 (b) −3.160 (c) −3.1600
6-55. (a) 251.3×10^4 (b) 6.435×10^{-5} (c) 5.763×10^2
6-56. (a) 650.0×10^3 (b) 2.11×10^{-6} (c) 9.787×10^{-2}

Round off as indicated:

6-57. Two places: (a) 3.156 (b) 4.7923 (c) 7.176
6-58. Four places: (a) −7.1693 (b) 8.42178 (c) −0.0135000
6-59. Three places: (a) 3.1451 (b) 0.74349 (c) 7.76543
6-60. Two places: (a) 9.22 (b) 0.55633 (c) 0.0615143
6-61. Three places: (a) −1.2637 (b) 9.917 (c) 0.81193
6-62. Three places: (a) 0.502286 (b) −7.78843 (c) 506.371

6-5 CALCULATOR ERRORS

The hand-held calculator has eased much of the burden of tedious hand calculations. However, with this easing has come an increasing dependency upon it. In this section we find that (a) we can introduce errors into our calculations by the way we operate the device, and (b) the limitations of the unit itself introduce errors into our calculations.

Expressing Errors

With the advent of the hand-held calculator, it has become more important to understand the problem of errors. There are two methods of expressing errors: absolute and relative. Assume a voltage reading of 0.53 V and that this is accurate to two significant places. The absolute error represents the error voltage, in this case 0.005 V. Similarly, if a meter read 25 V ± 10 V, the absolute error is 10 V.

The problem of absolute errors is that it does not express how good (or bad) a reading is. An absolute voltage error of 5 V on a 15-V reading is pretty inaccurate, for the actual voltage lies between 5 and 15 V. On the other hand, an absolute error of 5 V on a 1,000,000-V reading is very accurate, for the actual value lies between 999,995 and 1,000,005 V.

Therefore, a second way of expressing error is in terms of relative error. The relative error is the percent of the reading that the absolute error represents. It can be computed as

$$E_{REL} = \frac{E_{ABS}}{\text{reading}} \times 100$$

where E_{REL} is the relative error and E_{ABS} the absolute error. Using our previous example, a reading of 15 V with a possible error of 5 V represents a relative error of

$$E_{REL} = \frac{5}{15} \times 100 = 33.33\%$$

An absolute error of 5 V on 1,000,000 V represents

$$E_{REL} = \frac{5}{1,000,000} \times 100 = 0.0005\%$$

EXAMPLE 6-20 Compute the absolute and relative errors:

(a) 600 ± 20 V   (b) 27 ± 3 V   (c) 831.6

SOLUTION

(a) $E_{ABS} = 20$

$$E_{REL} = \frac{E_{ABS}}{reading} \times 100$$

$$= \frac{20}{600} \times 100 = 3.333\%$$

(b) $E_{ABS} = 3$ V

$$E_{REL} = \frac{E_{ABS}}{reading} \times 100$$

$$= \frac{3}{27} \times 100 = 11.11\%$$

(c) The absolute error is ± 0.05, since the number has four significant places.

$$E_{REL} = \frac{E_{ABS}}{reading} \times 100$$

$$= \frac{0.05}{831.6} \times 100 = 0.006013\%$$

EXAMPLE 6-21 Jason Joscowitz had a 12-in. ruler graduated in ⅛-in. increments and used it to measure the upper left cuspid of his pet gorilla, Epios. If Jason was very careful, measuring to one division and Epios's tooth measured ½ in., what are the absolute and relative errors?

SOLUTION The absolute error is plus or minus one division on the ruler, or $\pm \frac{1}{8}$ in. The relative error is

$$E_{REL} = \frac{E_{ABS}}{reading} \times 100$$

$$= \frac{\frac{1}{8}}{\frac{1}{2}} \times 100 = 25\%$$

Errors in Manipulation

Try the following on your calculator: add 3.15×10^{-10} to 1.00. What is your result? When I tried it, I obtained 1.000000000. But what happened to my 3.15×10^{-10}—it disappeared? This is only one example of how errors can be introduced into calculations. In the next few paragraphs, we examine several ways of introducing errors into our solutions. The 10-digit calculator will be used as an example. However, the same types of errors are present regardless of the number of places displayed.

1. *Adding or subtracting a very large number and a very small number.* We have seen that adding a very large number, 1.000, to a very small number, 3.15 $\times$ 10^{-10}, results in an error. The correct solution is

$$1.000 + 3.15 \times 10^{-10} = 1.000{,}000{,}000{,}315$$

However, since the calculator can only display 10 digits, everything to the right of the tenth digit is lost. This type of error is called roundoff error, for the calculator rounds off to 10 places.

This error is also present in subtraction. Subtracting 3.15 $\times$ 10^{-10} from 1.00 results in

$$1.00 - 3.15 \times 10^{-10} = 0.999{,}999{,}999{,}685$$

However, since the calculator can only express 10 places, the displayed result is $999.9999997 \times 10^{-3}$ in engineering notation or 1.000000000 in fixed notation. Thus, accuracy is again lost to roundoff error.

This type of error is always present. However, it can be minimized by first adding or subtracting all the small numbers, then proceeding to the larger numbers.

EXAMPLE 6-22 Add the following numbers: 4.93 $\times$ 10^{-10}, 4.46 $\times$ 10^{-10}, 4.75 $\times$ 10^{-10}, 3.95 $\times$ 10^{-10} and 1.000.

SOLUTION We first put the numbers in ascending order, then compare the result of adding the numbers in ascending order with that obtained by adding in descending order. The ordered list becomes

$$3.95 \times 10^{-10}$$
$$4.46 \times 10^{-10}$$
$$4.75 \times 10^{-10}$$
$$4.93 \times 10^{-10}$$
$$1.000$$

Adding the list top to bottom results in 1.000,000,002. Adding the list from the bottom to the top results in 1.000,000,000, for each time a small number is added to a large number, roundoff error occurs. The actual answer is 1.000,000,001,809 and, as can be seen, the first answer correctly expresses this number to the tenth place.

2. *Subtracting two almost equal numbers.* Let us assume that a measurement yields 1.02 and is accurate to ± 0.01. This is a relative accuracy of $\pm 1\%$. Next, let us assume that a second measurement results in 1.01 and is also accurate to ± 0.01. Subtracting these two numbers results in a difference of 0.01 ($1.02 - 1.01 = 0.01$). However, because of the tolerance problem, the first number could be as high as 1.03 and the second number as low as 1.00. Subtracting these two numbers results in 0.03. Again, the first number could be as low as 1.01 and the second as high as 1.02. Subtracting these yields -0.01. Thus, our final answer is 0.01 ± 0.02, a relative error of 200%. Whereas we started with numbers accurate to $\pm 1\%$, we ended up with a result accurate to $\pm 200\%$.

But how can this problem be avoided? First, wherever possible, avoid subtracting almost-equal numbers. Second, express the two numbers as accurately as possible before the subtraction, recognizing that a great deal of accuracy will be lost as a result of the process. Third, where accuracy must be known, calculate the resulting absolute error by adding the absolute errors of the two operands. In the example above, adding the absolute error of the first to that of the second results in

$$0.01 + 0.01 = 0.02$$

Therefore, the result will have an absolute error of 0.02. This can then be converted to a relative error.

EXAMPLE 6-23   Compute the error in subtracting $5.133 \pm 1\%$ from $5.246 \pm 1\%$.

SOLUTION   The nominal answer is

$$5.246 - 5.133 = 0.113$$

The absolute error of the operands are

$$5.246 \times 0.01 = 0.05246$$
$$5.133 \times 0.01 = \underline{0.05133}$$
$$\text{solution} = 0.10379$$

This represents a relative error in the solution of

$$E_{\text{REL}} = \frac{E_{\text{ABS}}}{\text{reading}} \times 100$$
$$= \frac{0.10379}{0.113} \times 100 = 91.85\%$$

The relative error is, therefore, $\pm 91.85\%$.

EXAMPLE 6-24   Compute the error in subtracting $0.1730 \pm 4\%$ from $0.9138 \pm 2\%$.

SOLUTION   The nominal answer is

$$0.9138 - 0.1730 = 0.7408$$

The absolute errors are

$$0.1730 \times 0.04 = 0.00692$$
$$0.9138 \times 0.02 = \underline{0.018276}$$
$$\text{solution} = 0.025196$$

This represents a relative error of

$$E_{\text{REL}} = \frac{E_{\text{ABS}}}{\text{reading}} \times 100$$
$$= \frac{0.025196}{0.7408} \times 100 = 3.401\%$$

PROBLEMS

Compute the absolute and relative errors:

6-63.   (a) $852.1 \pm 8\%$ (b) $6.29 \pm 5\%$

6-64.   (a) $0.6530 \pm 7\%$ (b) $1.15 \pm 6\%$

6-65.   (a) 3.28 ± 0.27 (b) 0.289 ± 0.003

6-66.   (a) 0.409 ± 0.08 (b) 637 ± 29

6-67.   (a) -3.87 ± 4 (b) $-3.87 \pm 4\%$

6-68.   (a) 2.398 ± 9 (b) $2.398 \pm 9\%$

Add the following numbers in ascending and descending order. Which is more accurate?

6-69. 4; 4; 4; 4; 4; 4; 4; 1.00×10^{10}

6-70. 0.44, 0.45, 0.46, 0.47, 0.48, 0.49, 1.00×10^9

6-71. The following is the expression for the sine of 30° accurate to 13 places. The perfectly accurate answer is 0.500. Perform the sum left to right and right to left and compare answers. (*Note:* 5! means 5 factorial. This indicates that you are to multiply by every integer between 5 and 1 inclusive—5 · 4 · 3 · 2 · 1.)

$$\sin 30° = \frac{\pi}{6} - \frac{(\pi/6)^3}{3!} + \frac{(\pi/6)^5}{5!} - \frac{(\pi/6)^7}{7!} + \frac{(\pi/6)^9}{9!} - \frac{(\pi/6)^{11}}{11!} + \frac{(\pi/6)^{13}}{13!} + \cdots$$

6-72. The expression for ϵ (a constant having a value of 2.718,281,828,459) is

$$\epsilon = 1 + \frac{1}{1} + \frac{1}{2!} + \frac{1}{3!} + \frac{1}{4!} + \frac{1}{5!} + \cdots$$

Sum all the terms first left to right, then right to left up to and including the fifteenth term and compare the results.

Compute the difference, absolute error, and relative error:

6-73. $(4.13 \pm 0.01) - (4.44 \pm 0.01)$

6-74. $(6.8015 \pm 0.0012) - (6.8056 \pm 0.0042)$

6-75. $(0.109 \pm 0.0065) - (0.120 \pm 0.0078)$

6-76. $(3207.5 \pm 3\%) - (3225 \pm 5\%)$

6-77. $(7.886 \pm 2\%) - (7.800 \pm 3\%)$

6-78. $(28.136 \pm 0.3\%) - (28.100 \pm 0.5\%)$

6-6 PROPAGATION OF ERRORS

Each time we add, subtract, multiply, or divide, we carry along with us an error. The question is whether this error is significant or not, and we cannot know this unless we get a feeling for how much error is being propagated.

Addition

The absolute error of an addition process is the sum of the absolute errors of the operands. Assuming that E_A and E_B are errors of A and B, respectively, we have

$$(A + E_A) + (B + E_B) = (A + B) + (E_A + E_B)$$

EXAMPLE 6-25 Find the sum and the maximum error when adding the following: 5.14 ± 0.13 and 7.32 ± 0.22.

SOLUTION The nominal sum is

$$5.14 + 7.32 = 12.46$$

The absolute error of the result is the sum of the individual absolute errors:

$$E_{SUM} = E_{N1} + E_{N2}$$
$$= 0.13 + 0.22 = 0.35$$

We can convert this to a relative error:

$$E_{REL} = \frac{E_{ABS}}{\text{measured}} \times 100$$

$$= \frac{0.35}{12.46} \times 100 = 2.81\%$$

EXAMPLE 6-26   Find the sum and the maximum error:

$$(4.83 \pm 6\%) + (172.03 \pm 0.001\%) + (515.3 \pm 0.1) + (660.4 \pm 0.5)$$

SOLUTION

The sum is

$$4.83 + 172.03 + 515.3 + 660.4 = 1352.56$$

The absolute errors are

$$
\begin{array}{lr}
4.83 \times 6\% & = 0.2898 \\
172.03 \times 0.001\% & = 0.0017203 \\
& 0.1 \\
& \underline{0.5} \\
\text{Total} & 0.8915203
\end{array}
$$

Thus, the answer is 1352.56 ± 0.892 or a relative error of

$$E_{REL} = \frac{E_{ABS}}{\text{measured}} \times 100$$

$$= \frac{0.892}{1352.56} \times 100 = 0.066\%$$

Note that, even though one operand had a relative error of 6%, it had little effect upon the error of the sum.

Subtraction

The absolute error of a subtraction process is the sum of the absolute errors of the operands:

$$(A \pm E_A) - (B \pm E_B) = (A - B) \pm (E_A + E_B)$$

EXAMPLE 6-27   Find the absolute and relative errors of $(3.769 \pm 0.07) - (3.142 \pm 0.04)$.

SOLUTION   The difference is

$$3.769 - 3.142 = 0.627$$

The resultant error is

$$E_{ABS} = 0.07 + 0.04 = 0.11$$

Express in relative terms:

$$E_{REL} = \frac{E_{ABS}}{\text{measured}} \times 100$$

$$= \frac{0.11}{0.627} \times 100 = 17.54\%$$

EXAMPLE 6-28 Find the absolute and relative errors:

$$(510 \pm 2.5\%) - (400 \pm 4\%)$$

SOLUTION The nominal difference is

$$510 - 400 = 110$$

The absolute errors are

$$
\begin{array}{ll}
510 \times 2.5\% = 12.75 \\
400 \times 0.04 = \underline{16.0} \\
\text{Result} \qquad \quad 28.75
\end{array}
$$

Thus, the answer is 110 ± 28.75 or $110 \pm 26.14\%$.

Multiplication

In multiplication, the relative error of the product is the sum of the relative errors of the operands. This can be shown as follows:

$$(A + E_A)(B + E_B) = AB + AE_B + BE_A + E_A E_B$$

We can assume that $E_A E_B$ is very small. Dropping this and subtracting out *AB*, the resulting error is

$$E_{\text{RESULT}} = AE_B + BE_A$$

We can now compute the relative error by dividing by the result:

$$E_{\text{REL}} = \frac{AE_B + BE_A}{AB} = \frac{E_B}{B} + \frac{E_A}{A}$$

This can be recognized as the sum of the relative errors of the operands.

EXAMPLE 6-29 Compute the relative and absolute errors:

$$(55.4 \pm 2\%)(872.9 \pm 3\%)$$

SOLUTION The nominal product is

$$55.4 \times 872.9 = 48,358.66$$

The relative errors are

$$
\begin{array}{l}
E_{\text{PRODUCT}} = E_1 + E_2 \\
\qquad \quad = 2 + 3 \\
\qquad \quad = 5\%
\end{array}
$$

Thus, the solution is $48,359 \pm 5\%$ or $48,359 \pm 2418$.

EXAMPLE 6-30 Compute the relative and absolute errors: $(275.4 \pm 2.9)(0.5914 \pm 0.0009)$.

SOLUTION The nominal solution is

$$275.4 \times 0.5914 = 162.87156$$

The relative errors are

$$E_{\text{REL1}} = \frac{E_{\text{ABS}}}{\text{measured}} \times 100$$

$$= \frac{2.9}{275.4} \times 100$$

$$= 1.053\%$$

$$E_{\text{REL2}} = \frac{E_{\text{ABS}}}{\text{measured}} \times 100$$

$$= \frac{0.0009}{0.5914} \times 100$$

$$= 0.152\%$$

The total error is

$$E_{\text{T}} = E_1 + E_2$$

$$= 1.053 + 0.152$$

$$= 1.21\%$$

Thus, the solution is $162.9 \pm 1.2\%$ or 162.9 ± 2.0.

Division

The relative error of a division process is the sum of the relative errors of the dividend and the divisor. This can be shown as a division process:

$$\frac{A + E_{AA}}{B + E_{AB}} = \frac{A}{B} + E_{AR}$$

where E_{AA}, E_{AB}, and E_{AR} are the absolute errors of A, B, and the result, respectively. Solve for E_{AR}:

$$E_{AR} = \frac{A + E_{AA}}{B + E_{AB}} - \frac{A}{B}$$

$$= \frac{AB + BE_{AA} - AB - AE_{AB}}{B(B + E_{AB})}$$

The relative error of the result can be found by dividing by the result, A/B:

$$E_{RR} = \frac{BE_{AA} - AE_{AB}}{(A/B)(B + E_{AB})B}$$

$$= \frac{BE_{AA} - AE_{AB}}{A(B + E_{AB})}$$

$$= \frac{BE_{AA}}{A(B + E_{AB})} - \frac{AE_{AB}}{A(B + E_{AB})}$$

However,

$$\frac{B}{B + E_{AB}} \approx 1 \quad \text{and} \quad \frac{A}{A} = 1 \quad \text{and} \quad B + E_{AB} \approx B$$

Therefore,

$$E_{RR} = \frac{E_{AA}}{A} - \frac{E_{AB}}{B}$$

This equation states that the relative error of a division result is the difference between the relative errors of the operands. Note that if both E_{AA} and E_{AB} are positive, the statement is true. However, if E_{AA} is positive and E_{AB} negative, the equation says that the relative error of a division process is the sum of the errors of the two operands. Unless we know for certain the polarities of the relative errors, we have to assume that they add.

EXAMPLE 6-31 What are the absolute and relative errors of dividing $(4.67 \pm 2\%)$ by $(5.41 \pm 8\%)$?

SOLUTION The nominal result is

$$\frac{4.67}{5.41} = 0.8632$$

The relative errors must be summed:

$$E_{RR} = E_{RA} + E_{RB}$$
$$= 2 + 8 = 10\%$$

Therefore, the solution is $0.8632 \pm 10\%$ or 0.8632 ± 0.0863.

EXAMPLE 6-32 Analyze the quotient of $(997.2 \pm 0.3) \div (241.9 \pm 1.6)$.

SOLUTION

$$\frac{997.2}{241.9} = 4.122$$
$$E_{RR} = E_{RA} + E_{RB}$$
$$= \left(\frac{E_{AA}}{A} + \frac{E_{AB}}{B} \right) \times 100$$
$$= \left(\frac{0.3}{997.2} + \frac{1.6}{241.9} \right) \times 100$$
$$= 0.692\%$$

The solution is $4.122 \pm 0.692\%$ or 4.122 ± 0.029.

Summary

The following summarizes the propagation of errors:

Addition and subtraction: The absolute error of the result is the sum of the absolute errors of the operands. Multiplication and division: The relative error of the result is the sum of the relative errors of the operands.

PROBLEMS

Find the nominal answer, the maximum absolute error, and maximum relative error:

6-79. $(7.80 \pm 0.3) + (6.06 \pm 0.6)$ **6-84.** $(70.2 \pm 5\%) + (9.19 \pm 2\%)$

6-80. $(8.315 \pm 0.005) + (50.86 + 0.06)$ **6-85.** $(6.652 \pm 0.3\%) + (2.338 \pm 0.1\%)$

6-81. $(0.00710 \pm 0.00005) + (0.00776 \pm 0.00009)$ **6-86.** $(0.1100 \pm 0.2\%) + (0.7353 \pm 0.6\%)$

6-82. $(0.44 \pm 0.02) + (0.86 \pm 0.09)$ **6-87.** $(97.59 \pm 0.01) - (29.67 \pm 0.07)$

6-83. $(3.57 \pm 2\%) + (5.96 \pm 3\%)$ **6-88.** $(841.3 \pm 0.7) - (781.6 \pm 1.5)$

6-89.	$(2387 \pm 6\%) - (3325 \pm 12\%)$	**6-96.**	$(0.7183 \pm 0.01\%)\,(1.107 \pm 0.17\%)$
6-90.	$(0.151 \pm 2\%) - (0.290 \pm 7\%)$	**6-97.**	$(3713 \pm 5\%)\,(2.75 \pm 1\%)$
6-91.	$(392 \pm 7)\,(531 \pm 12)$	**6-98.**	$(0.5055 \pm 0.06\%)\,(988.98 \pm 0.001\%)$
6-92.	$(7.77 \pm 0.07)\,(53.34 \pm 0.65)$	**6-99.**	$(16.55 \pm 0.03) \div (499 \pm 3)$
6-93.	$(1.834 \pm 0.012)\,(0.9190 \pm 0.0001)$	**6-100.**	$(3.519 \pm 0.016) \div (0.0889 \pm 0.0006)$
6-94.	$(0.0340 \pm 0.002)\,(0.0228 \pm 0.080)$	**6-101.**	$(9.94 \pm 1\%) \div (0.05250 \pm 0.06\%)$
6-95.	$(1.55 \pm 1\%)\,(56.4 \pm 5\%)$	**6-102.**	$(393.60 \pm 0.01\%) \div (404 \pm 0.7\%)$

6-7 ELECTRONIC UNITS OF MEASURE

In electronics we use very, very small units, 10^{-12} F, for example, and very, very large units, 10^{12} Ω, for example. Because of this, the metric system of prefixes is especially valuable to us. In this section we examine this system and how to convert from one system of units to another. Finally, we examine a system for conversion that not only can be done blindly, but is very useful in analyzing the validity of a formula.

Metric Prefixes

The system of metric prefixes is built around engineering notation. Table 6-3 lists the prefixes used and their meanings. These prefixes would be attached to electronic units to express very large or very small units. Thus, 1000 V is 1 kilovolt (kV); 1,000,000 V is 1 megavolt (MV); 0.001 V is 1 millivolt (mV).

To change a unit into proper metric engineering notation, move the decimal point to the left or the right in groups of three until the desired notation is obtained. If the point is moved to the right, move upward on Table 6-3. If the point is moved to the left, move downward on Table 6-3.

EXAMPLE 6-33 Convert the following electrical quantities to proper metric notation:

(a) 6,520,000 Ω
(b) 0.00153 s
(c) 0.000,000,045 A
(d) 4561 V

TABLE 6-3 Metric prefixes.

Decimal	Exponential	Prefix	Symbol
0.000,000,000,000,000,001	10^{-18}	atto	a
0.000,000,000,000,001	10^{-15}	femto	f
0.000,000,000,001	10^{-12}	pico	p
0.000,000,001	10^{-9}	nano	n
0.000,001	10^{-6}	micro	μ
0.001	10^{-3}	milli	m
1	10^{0}	(units)	
1000	10^{3}	kilo	k
1,000,000	10^{6}	mega	M
1,000,000,000	10^{9}	giga	G
1,000,000,000,000	10^{12}	tera	T
1,000,000,000,000,000	10^{15}	peta	P
1,000,000,000,000,000,000	10^{18}	exa	E

SOLUTION

(a) The point must be moved to the left two groups of three, obtaining 6.52. Thus, starting at the units position on the table, move down two places, arriving at "mega." The correct answer is, therefore, 6.52 megohms (MΩ).

(b) The point must be moved to the right one group of three digits, obtaining 1.53. This means moving up from the units place on the table by one line. The answer is 1.53 milliseconds (ms).

(c) 45 nanoamps (nA)

(d) 4.561 kilovolts (kV)

Most electronic formulas use unit dimensions and it is therefore necessary to convert from metric notation to units. This is easily done by using the exponential format from the table. Thus, 1 pA is 10^{-12} A; 4.59 MHz is 4.59×10^6 Hz.

EXAMPLE 6-34 Convert the following to units:

(a) 5.63 kΩ   (b) 36.9 μA   (c) 415.6 mV

SOLUTION In each case, multiply the unit by the exponential shown in Table 6-3.

(a) 5.63×10^3 Ω

(b) 36.9×10^{-6} A

(c) 415.6×10^{-3} V

Conversion of Units

I firmly believe that the more tools a technician has in his repertoire, the better off he is. However, sometimes a tool is so superior as to make the use of any other tool extremely awkward. The calculator is such a tool when compared to the slide rule. The method of conversion discussed in this section is also such a tool. Study it carefully and you will find that you, too, will discard other conversion methods in favor of this one. It allows you to convert blindly—there is no need to figure out whether to multiply or divide by a conversion factor.

Let us assume that you have 4.5 in. and must convert this to centimeters (cm). Assume you have memorized that there are 2.54 cm in 1 in. Start by stating that "X cm =" on the left of an equation. Then put down the given units on the right, "4.5 in.". So far we have

$$X \text{ cm} = 4.5 \text{ in.}$$

Note that the units (cm and in.) are placed beside the values (X and 4.5). We must now "cancel" the "in." But to do this we must have "in." in a denominator. Therefore, construct a fraction out of the statement "2.54 cm = 1 in.," putting the "1 in." in the denominator and the "2.54 cm" in the numerator. This gives us

$$X \text{ cm} = \frac{4.5 \text{ in.}}{1} \times \frac{2.54 \text{ cm}}{1 \text{ in.}}$$

We can now cancel the units, leaving just cm (the desired final unit) in the numerator:

$$X \text{ cm} = \frac{4.5 \;\cancel{\text{in.}}}{1} \times \frac{2.54 \text{ cm}}{1 \;\cancel{\text{in.}}}$$

Now that we have the right unit in the numerator, merely do the arithmetic: multiply 4.5 by 2.54. Thus,

$$X \, cm = \frac{4.5 \, in.}{1} \times \frac{2.54 \, cm}{1 \, in.}$$
$$= (4.5 \times 2.54) \, cm$$
$$= 11.43 \, cm$$

EXAMPLE 6-35 Convert 3.54 m³ to gallons knowing only the following:

(a) 1 British dry butt (Bdb) = 126 gal
(b) 1 Bdb = 0.57281 m³

SOLUTION

$$X \, gal = \frac{3.54 \, m^3}{1} \times \frac{1 \, Bdb}{0.57281 \, m^3} \times \frac{126 \, gal}{1 \, Bdb}$$

Step 1: Write down "X gal =":
Step 2: Write down the number we are converting from (3.54 m³).
Step 3: Pick out a conversion factor that has m³ in its denominator so that we can cancel it out (1 Bdb = 0.57281 m³).
Step 4: Pick out another factor that will cancel out British dry butts (1 Bdb = 126 gal).
Step 5: Having converted to the right units, do the arithmetic.

$$X \, gal = \frac{3.54 \times 126}{0.57281} \, gal = 778.7 \, gal$$

Note that the necessity for canceling units is what determines whether the numerical conversion factor goes in the numerator or the denominator.

EXAMPLE 6-36 On the planet K-9, the following units are used for measurements of sound intensity:

1 arf = 3.69 woofs
1 arf = 7.63 ruffs
1 ruff = 4.39 yelps

How many yelps are there in 4.63 woofs?

SOLUTION

$$X \, yelps = \frac{4.63 \, woofs}{1} \times \frac{1 \, arfs}{3.69 \, woofs} \times \frac{7.63 \, ruff}{1 \, arf} \times \frac{4.39 \, yelps}{1 \, ruff}$$
$$= 42.03 \, yelps$$

As can be seen, the system can be used for converting any unit into another unit.

EXAMPLE 6-37 How many microvolts are there in 2.798 kV?

SOLUTION The conversion factors can be obtained from Table 6-3.

$$X\mu V = \frac{2.798 \text{ kV}}{1} \times \frac{1000 \text{ V}}{1 \text{ kV}} \times \frac{1\mu V}{10^{-6} \text{ V}}$$
$$= 2.798 \times 10^9 \mu V$$

Dimensional Analysis

The foregoing method of conversion of units is quite valuable in analyzing the validity of formulas. For example, let us examine the formula for finding the volume of a sphere:

$$V = \tfrac{4}{3}\, \pi r^3$$

where V is the volume and r is the radius of the sphere. If we substitute the dimensions into the right side and discard any constants (the 4, 3, and π), we can find whether the formula is correct in its dimensions. This substitution can be in actual units (meters, feet, centimeters) or in just the categories of units (length, mass, time). Substitute actual units:

$$V = (m)^3 = m^3$$

We are saying that volume can be measured in cubic meters. Thus, the dimensions of the formula are correct.

Whenever a formula is presented, get in the practice of analyzing the units. In this way, any literals that have been accidentally misplaced will be detected.

EXAMPLE 6-38 The following loop equation was written for a certain electrical circuit. Knowing that $V = IR$ (volts equals amperes times ohms), analyze the equation for validity.

$$V = 3.6IR + 2.3V + 7.8I^2R$$

SOLUTION The $3.6IR$ is fine, for we have volts on the left side of the equation and we know that IR is the same as volts. The next term, $2.3V$, is also correct, for we are still adding volts to volts. The last term, however, is not correct, for volts is IR and we have I^2R; thus, we actually have ampere-volts instead of volts. Since we can only add volts to volts, this last term is wrong.

EXAMPLE 6-39 Dame Desidra Drusula Dempsy's demesne's deed declares its dimensions as

$$A = \frac{1}{2}(d_1 + d_2)d_3 + \frac{\pi d_3^2}{4} + \frac{1}{2}d_4d_5 + \frac{1}{8}d_6$$

where A is the area of the estate and the d's represent linear measurements. In court, Dame Dempsy's dashing lawyer, Jonathon Jurist, disproved the document using dimensional analysis. How did he do it?

SOLUTION He analyzed each of the terms by substituting m (meters) for each of the d's and omitting the constants:

$$A = (m + m)(m) + m^2 + m \cdot m + m$$
$$= m^2 + m^2 + m^2 + m$$

Note that adding meters to meters results in meters in the first term. Note further that the last term, meters, cannot be added to the rest of the terms, which are in meters² (area). Therefore, Jonathon Jurist was able to prove to the court that the formula was incorrect.

PROBLEMS

Convert to proper metric notation:

6-103. (a) 136,590,000 Ω (b) 17,570 V
6-104. (a) 0.0216 s (b) 6206 A
6-105. (a) 0.00829 A (b) 0.000,093,7 A
6-106. (a) 411,000,000 Ω (b) 59,400 Ω
6-107. (a) 22,900,000,000 Hz (b) 0.000,000,000,071 F
6-108. (a) 0.07 s (b) 0.00872 A

Convert to unit measure:

6-109. (a) 4.96 MΩ (b) 5.42 kΩ
6-110. (a) 31.2 ms (b) 7.67 ms
6-111. (a) 81.35 ns (b) 105 pF
6-112. (a) 41.3 μs (b) 4.39 GΩ

Note: Problems 6-113 through 6-118 involve real units.

6-113. How many square feet are there in 1 rod² if (a) 1 acre = 160 rods², (b) 1 acre = 4860 yd², (c) 1 yd² = 9 ft².

6-114. How many cubic centimeters are there in 1 gal if (a) 1 ft³ = 7 gal, (b) 1 in. = 2.54 cm, (c) 1 ft = 12 in.

6-115. How many grains are there in 1 short ton if (a) 1 gram = 0.0352739 ounce (avoir.), (b) 1 gram = 15.4324 grains, (c) 1 short ton = 2000 lb (avoir.), (d) 1 lb = 16 ounce (avoir.).

6-116. How many pennyweights are there in 1 pound (avoir.) if (a) 1 scruple = 0.833333 pennyweight, (b) 1 scruple = 20 grains, (c) 1 ounce (avoir.) = 437.5 grains, (d) 1 ounce (avoir.) = 28.349527 grams, (e) 1 gram = 0.00220462 lb (avoir.).

6-117. How many British nails in 1 mile if (a) 1 British nail = 5.715 cm, (b) 1 m = 100 cm, (c) 1 m = 1.093614 yd, (d) 1 link = 0.22 yd, (e) 1 link = 0.01 chain, (f) 1 mile = 80 chains.

6-118. How many British noggins in 1 cubic centimeter (cc) if (a) 1 noggin = ⅓₂ gal, (b) 1 dram (fluid) = 0.0078125 pint, (c) 8 pints = 1 gal, (d) 1 dram = 60 minims, (e) 1 minim = 0.0020833 oz (fluid), (f) 1 oz (fluid) = 29.5737 cc.

6-119. How many milliseconds in 30 seconds?
6-120. How many milohms in 20 GΩ?
6-121. How many nanoamperes in 3.5 kA?
6-122. How many microvolts in 3.79 MV?
6-123. Knowing that $P = I^2R$, $P = VI$, $P = V^2/R$, which of the following are invalid formulas?

(a) $P_T = P_1 + P_2 + I^2(R_1 + R_2)$

(b) $P_T = (I_1 + I_2)^2(R_1 + R_2 + R_3) + \dfrac{V^2}{R_1 + R_2}$

(c) $P_T = \dfrac{(V_1 + V_2)^2}{R_1 + R_2} + I_1(I_2 + I_3)^2R$

(d) $P_T = P_1 + P_2^2$

(e) $P_T = P_1 + (V_1 + V_2)I + (V_3 + V_4)I$

(f) $P_T = (I_1 + I_2)(I_3 + I_4)R + \dfrac{(V_1 + V_2)^2 + P}{R}$

(g) $P_T = \dfrac{V_1^2 + V_2^2}{R} + I_1^2 + I_2^2 R$

6-124. Knowing that $V = IR$, $V = P/I$, and $V = \sqrt{PR}$, which of the following are invalid formulas?

(a) $V_T = I_1 (R_1 + R_2) + V_1 + I_2 (R_2 + R_3)$

(b) $V_T = I_1 R_1 + \dfrac{P_1 + P_2}{I_2} + \sqrt{(P_3 + P_4)(R_5 + R_6)}$

(c) $V_T = P_1 R_1 + I_2 R_2 + (I_3 + I_4)R_4$

(d) $V_T = \dfrac{P_1 - P_2}{I} + \dfrac{I_2 R_2}{4}$

(e) $V_T = \dfrac{P_1 + P_2}{I_1 - I_2} + \dfrac{1}{2} V_2 - \sqrt{P_4 (R_4 + R_5)}$

(f) $V_T = V_2 - V_3 + I^2 (R_1 + R_2) + \dfrac{P_1 + P_2}{I}$

(g) $V_T = \dfrac{V_2 - V_3}{2} + \dfrac{1}{2} \left(\sqrt{P_2 R_2} + \dfrac{P_3}{I_2} \right)$

7

OHM'S LAW IN SERIES CIRCUITS

In this chapter we consider a brief formula relating voltage, current, and resistance: Ohm's law. This law is the most important formula that exists in the mathematical analysis of electricity and yet, its author, Georg Simon Ohm, devoted his entire life to its development.

7-1. SOLVING PROBLEMS USING OHM'S LAW

Ohm's law states that voltage *(V)* is equal to current *(I)* multiplied by resistance *(R):*

$$V = IR$$

The formula has three variables and, knowing any two, we are able to solve for the third. Thus, it has three forms:

$$V = IR \qquad I = \frac{V}{R} \qquad R = \frac{V}{I}$$

It is important to realize that the variables of Ohm's law must all be applied to the same portion of the circuit. That is, the voltage across a particular circuit or portion of a circuit is equal to the current through that circuit or portion multiplied by the resistance of that circuit or portion.

EXAMPLE 7-1 If current is 3 A and voltage 7 V, what is resistance?

SOLUTION We know both current and voltage and, examining the foregoing three equations, we see that the third formula is the one having both current and voltage to the right of the equals sign. Selecting this formula yields

$$R = \frac{V}{I} = \frac{7}{3} = 2.333 \ \Omega$$

EXAMPLE 7-2 The current through a certain 5-kΩ resistor is 6 mA. What is the voltage across the resistor?

SOLUTION The unknown value is voltage and the known values are current and resistance. Examining the three Ohm's law formulas, we see that the first one has the unknown on the left and known values on the right. Therefore,

$$V = IR = 6 \times 10^{-3} \times 5 \times 10^3 = 30 \ \text{V}$$

Problem Analysis

Mathematics is an orderly subject and those who use it must adopt this order to gain its full benefits. For this reason, the reader should practice the following procedure in solving circuit problems:

(a) Draw a circuit diagram. This is an invaluable aid in visualizing the invisible electronic forces within a circuit.
(b) Put the circuit values on the diagram, assigning the unknowns as literals.
(c) State the formula to be used.
(d) Substitute the known values into the formula.
(e) Solve for the unknown values.
(f) Above all, practice neatness. Mathematics is a language of order.

EXAMPLE 7-3 A certain soldering iron draws 435 mA from its 117-V source. What is the resistance of the iron?

SOLUTION The schematic is shown in Figure 7-1. Note that both known and unknown values are shown. Next, state the formula used and the values substituted into the formula:

$$R = \frac{V}{I} = \frac{117}{0.435} = 269.0 \ \Omega$$

The problem of units in Ohm's law can be greatly simplified by recognizing that:

$$\text{volts} = \text{mA} \times \text{k}\Omega \quad \text{and} \quad \text{volts} = \mu\text{A} \times \text{M}\Omega$$

This can be proven by recognizing that mA ($I \times 10^{-3}$) multiplied by kΩ ($R \times 10^3$) leaves IR. Similarly, μA ($I \times 10^{-6}$) multiplied by MΩ ($R \times 10^6$) is also IR. Thus, we can substitute these units directly into the formulas.

EXAMPLE 7-4 The input resistance to a certain transistor is 15 MΩ. How much current does the transistor input draw if 7.6 V is applied?

SOLUTION Figure 7-2 shows the schematic diagram. We can compute the results in μA, for MΩ is used:

$$I = \frac{V}{R} = \frac{7.6}{15} = 0.5067 \ \mu\text{A}$$
$$= 506.7 \ \text{nA}$$

EXAMPLE 7-5 A feedback resistor in an amplifier has a resistance of 20 kΩ and draws 1.315 mA. What is its voltage drop?

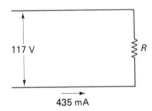

117 V   R

435 mA

FIGURE 7-1 Example 7-3.

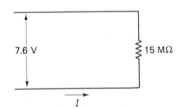

7.6 V   15 MΩ

I

FIGURE 7-2 Example 7-4.

SOLUTION Use units of kΩ, mA, and volts:

$$V = IR = 1.315 \times 20 = 26.30 \text{ V}$$

PROBLEMS

Solve for the unknown values:

	V	I	R
7-1.	2.6 V	20 A	
7-2.	3.9 V	70 A	
7-3.	23.1 V	3.15 A	
7-4.	4.31 V	0.39 A	
7-5.	76.1 V		4.31 Ω
7-6.	444 V		73.6 Ω
7-7.	7.21 V		8.23 Ω
7-8.	175 kV		590.6 Ω
7-9.		880 A	403.8 Ω
7-10.		6.56 A	42.68 Ω
7-11.		53.1 A	9.693 Ω
7-12.		400.8 A	0.8757 Ω
7-13.	144.7 V	2.499 mA	
7-14.	19.82 V	5.615 mA	
7-15.	8.47 V	8.617 μA	
7-16.	2.026 V	2.809 μA	
7-17.	7.721 V		7.003 kΩ
7-18.	72.55 V		3.39 kΩ
7-19.	4.33 V		343 kΩ
7-20.	38.80 V		6.796 MΩ
7-21.		57.3 mA	44.45 kΩ
7-22.		1.744 μA	9.526 MΩ
7-23.		71.4 mA	14.12 kΩ
7-24.		724 μA	3.82 MΩ

7-25. A bias resistor of 1 MΩ has 5 V across it. What is its current?

7-26. A ballast resistor draws 5 A and has 4 V across it. What is its resistance?

7-27. A meter shunt resistor has 50 mV across it and draws 15 A. What is its resistance?

7-28. An integrated circuit draws 100 mA and has a resistance of 50 Ω. What is its voltage drop?

7-29. A 5-kΩ collector resistor has 5.93 V across it. What is its current?

7-30. Jack Ford's 1946 flivver draws 400 A from its 12-V battery when Jack stomps on the starter. What is the resistance of the starter?

7-31. Dumb Duncan stuck his tongue into a 117-V light socket. If the resistance of his tongue was 2 kΩ, how much current did his jolting tongue draw?

7-32. Witch Haggard Hermoine's 1000-year-old mansion, Dreadful Dacha, has a resistance of 5 MΩ from its highest gable to the ground. How much current will the ghoulish house draw when struck by 100 MV of lightning?

7-2. POWER AND ENERGY

We use the terms "power" and "energy" daily in our conversations. But just what do each mean? They are both precise scientific terms that we, as electronics experts, must learn and use.

Power

Power is best described by realizing that 1 horsepower equals 550 ft-lb/s. That is, if strong man Gibraltar Gossamer were to lift 550 lb 1 ft every second, he would be exerting 1 horsepower (hp). In electronics, power is measured in watts (W), there being 746 W in 1 hp. We can calculate power by knowing that

$$P = VI$$

That is, power is voltage multiplied by current. Combining this equation with Ohm's law results in the various equation forms shown in Table 7-1. Although the table looks formidable, I recommend that you memorize only those shown in boxes. The other forms are derived from these primary formulas. Note that there are a total of four variables: P, V, I, and R.

TABLE 7-1 Forms of Ohm's law and the power law.

$\boxed{P = VI}$	$\boxed{P = I^2 R}$
$V = \dfrac{P}{I}$	$R = \dfrac{P}{I^2}$
$I = \dfrac{P}{V}$	$I = \sqrt{\dfrac{P}{R}}$
$\boxed{P = \dfrac{V^2}{R}}$	$\boxed{V = IR}$
$R = \dfrac{V^2}{P}$	$I = \dfrac{V}{R}$
$V = \sqrt{PR}$	$R = \dfrac{V}{I}$

EXAMPLE 7-6 A 50-W light bulb is operated from a 117-V source. What is its current and resistance while it is on (Fig. 7-3)?

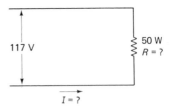

117 V

50 W
R = ?

I = ?

FIGURE 7-3 Example 7-6.

SOLUTION We must first identify the variables we know: power and voltage. Knowing this, we can solve for the other two quantities: current and resistance. Let us first solve for current by selecting that equation that relates power, voltage, and current.

$$P = VI$$
$$50 = 117 \times I$$
$$I = 427.4 \text{ mA}$$

We can next solve for resistance. Rather than depend upon a computed result (current), it is better practice to use those quantities that were given (power and voltage).

$$P = \frac{V^2}{R}$$
$$50 = \frac{117^2}{R}$$
$$R = 273.8 \ \Omega$$

We can check the two results by relating them to a given value:

$$V = IR$$
$$= 0.4274 \times 273.8$$
$$= 117.0 \ V$$

It checks.

EXAMPLE 7-7   A transistor draws 20 mA and has 10 V across it. What is its resistance and what power must it dissipate (Fig. 7-4)?

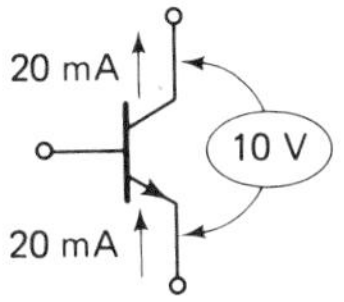

FIGURE 7-4   Example 7-7.

SOLUTION   The known variables are voltage and current; the unknown variables are resistance and power. Relate *V*, *I*, and *R*:

$$V = IR$$
$$10 = 20 \times R$$
$$R = 0.5 \ k\Omega = 500 \ \Omega$$

Relate *V*, *I*, and *P*:

$$P = VI$$
$$= 10 \times 20 = 200 \ mW$$

Energy

Energy can be described very easily in electrical terms. If we allow a 100-W light bulb to burn for 3 days, we should expect to pay the power company more than if we allow it to burn for only 3 h. This is because we pay for energy consumed, not power. Energy *(W)* is, therefore, power *(P)* multiplied by time *(t)*:

$$W = Pt$$

We pay the power company for the energy we have consumed in kilowatt-hours: power (kilowatts) multiplied by the number of hours these kilowatts are expended.

EXAMPLE 7-8   An electric dryer is rated at 12.5 kW. If it takes ½ h to dry a load of clothes, and electricity costs 5 cents/kWh, what is the cost of drying one load?

SOLUTION   We must first compute the energy expended. We can then multiply this by the cost per kWh.

$$W = Pt$$
$$= 12{,}500 \times 0.5$$
$$= 6250 \text{ Wh}$$
$$= 6.25 \text{ kWh}$$
$$\text{cost} = \text{kWh} \times \frac{\text{cost}}{\text{kWh}}$$
$$= 6.25 \times \frac{0.05}{1}$$
$$= \$0.3125 \text{ per load}$$

EXAMPLE 7-9   A clock consumes 2 W of 117 V AC electrical energy. What is the cost of operating the clock for 1 year if electrical energy costs 3 cents/kWh?

SOLUTION

$$W = Pt$$
$$= 2\,\cancel{W} \times \frac{24 \text{ h}}{1 \cancel{\text{day}}} \times \frac{365 \cancel{\text{days}}}{1 \text{ yr}} \times \frac{\text{kW}}{1000 \cancel{W}}$$
$$= 17.5 \text{ kWh/yr}$$
$$\frac{\text{cost}}{\text{yr}} = \frac{\text{kWh}}{\text{yr}} \times \frac{\text{cost}}{\text{kWh}}$$
$$= 17.5 \times 0.03 = \$0.525 \text{ per year}$$

PROBLEMS

7-33. How much current do each of the following 117-V lamps draw? What is the resistance of each?   **(a)** 50 W   **(b)** 60 W   **(c)** 100 W   **(d)** 500 W

7-34. What is the power consumed and resistance of each of the following 28-V lamps? **(a)** type 307, 670 mA   **(b)** type 313, 170 mA   **(c)** type 1251, 230 mA   **(d)** type 1819, 40 mA

7-35. A 7408 integrated circuit draws 16 mA from its 5-V source. What is its power and resistance?

7-36. A 4193 integrated circuit draws 1.6 mA from its 10-V source. What is its power and resistance?

7-37. What is the maximum current a 1-W 1-kΩ resistor can draw? What is its maximum voltage?

7-38. What is the maximum current a 50-W 10-Ω resistor can draw? What is its maximum voltage?

7-39. What is the power consumed by and voltage across a 1-mA meter having a resistance of 150 Ω?

7-40. How much power is consumed by a 40-MΩ insulator across 2000 V? What is the current drawn?

7-41. A transistor must not consume more than 500 mW. What is the maximum current it may draw at 5 V? at 10 V? at 15 V?

7-42. A 12-V 55-Ah battery draws 55 A. What power is it producing across what resistance?

Compute the missing electrical quantities:

	V	I	R	P
7-43.	93 V	811 mA		
7-44.	5.5 V	9.76 μA		
7-45.	36.8 V		4.02 kΩ	
7-46.	69.7 V		560 mΩ	

7-47.	206 mV		111 W
7-48.	26.9 V		9.76 W
7-49.		910 mA	143 kΩ
7-50.		7.43 μA	8.52 kΩ
7-51.		3.94 A	3.56 W
7-52.		112 mA	722 mW
7-53.		8.59 kΩ	745 mW
7-54.		591 kΩ	340 μW

7-55. What is the cost of operating a 100-W 117-V light bulb for its rated 750 h if electrical energy costs 4 cents/kWh?

7-56. What is the revenue received by a power company per year from its 1-MW power plant if it receives an average of 1 cent/kWh?

7-57. What is the yearly cost of leaving on outside lamps totaling 720 W for an average of 8 h per night? Electricity costs 3 cents/kWh.

7-58. What is the yearly cost of a 25-W refrigerator bulb if it is on for 5 min per day? Electricity costs 3 cents/kWh.

7-3 RESISTORS IN THE SERIES CIRCUIT

In the series circuit, resistances are connected head to tail (Fig. 7-5), whereas in the parallel circuit, all the heads are connected together and all the tails are connected together (Fig. 7-6). In this section we investigate the series circuit—currents, resistances, voltages, and power—using Ohm's law.

FIGURE 7-5 The series circuit.   **FIGURE 7-6** The parallel circuit.

Current

Current in a series circuit is the same throughout its length (Fig. 7-7). This is comparable to a water pipe: the water flowing out of the pipe is equal to the water flowing into the pipe, regardless of its length.

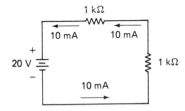

FIGURE 7-7 Current in a series circuit.

Resistance

The total resistance of a series circuit is equal to the sum of the individual resistors:

$$R_t = R_1 + R_2 + R_3 + \ldots$$

In Fig. 7-7, for example, the total resistance as seen by the 20-V supply is

$$R_t = R_1 + R_2 = 1 + 1 = 2 \text{ k}\Omega$$

EXAMPLE 7-10 Resistances of 1 kΩ, 2 kΩ, 5 kΩ, and 10 kΩ are connected in series. What is the total resistance of the circuit?

SOLUTION The total resistance is the sum of the individual resistances:

$$R_t = R_1 + R_2 + R_3 + R_4$$
$$= 1 + 2 + 5 + 10 = 18 \text{ k}\Omega$$

Note that as long as the units are the same, they can be added.

EXAMPLE 7-11 Resistances of 1 kΩ, 340 Ω, and 50 kΩ are connected in series. What is the total resistance?

SOLUTION We shall do the problem in kΩ:

$$R_t = R_1 + R_2 + R_3$$
$$= 1 + 0.34 + 50 = 51.34 \text{ k}\Omega$$

Voltage

Kirchhoff's voltage law states:

The sum of all the voltages around a complete circuit is equal to zero volts.

Let us analyze Fig. 7-8 in detail. The total resistance the supply must drive is

$$R_t = R_1 + R_2 + R_3$$
$$= 10 + 4 + 6 = 20 \text{ k}\Omega$$

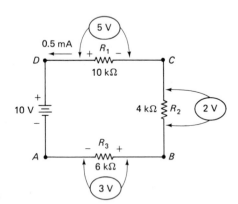

FIGURE 7-8 Series voltages.

Knowing total resistance, we can calculate total current:

$$I = \frac{V_s}{R_t} = \frac{10}{20} = 0.5 \text{ mA}$$

We can now calculate each of the voltages a voltmeter would see across each of the resistors. One of the very useful characteristics of Ohm's law is that it applies to any circuit or portion of a circuit. Let us apply it to R_1. We know the resistance, 10 kΩ, and the current, 0.5 mA. Let us calculate the voltage:

$$V = IR = 0.5 \times 10 = 5 \text{ V}$$

Thus, a voltmeter would read 5 V when placed across R_1, in the polarity shown.

Let us next calculate the voltages across R_2 and R_3:

$$V_2 = I_2 R_2 = 0.5 \times 4 = 2 \text{ V}$$
$$V_3 = I_3 R_3 = 0.5 \times 6 = 3 \text{ V}$$

Finally, we shall apply Kirchhoff's voltage law. Pretend you are a voltmeter with one lead on point A. We shall move the second lead. Starting at point A, moving counterclockwise (ccw) through R_3, we will see a voltage increase, from 0 V to +6 V. Continuing through R_2, our voltmeter will increase from 6 V to 8 V. Thus, with one lead on A and the other on C, we read 8 V. Continuing through R_1, we shall read 10 V as we move our lead from C to D. We are now effectively across the supply. As we next move our lead from D to A, we shall see a decrease in voltage of 10 V. Thus, assigning voltage increases as positive and voltage decreases as negative, we have

$$\begin{aligned} 0 &= V_s + V_1 + V_2 + V_3 \\ &= -10 + 5 + 2 + 3 \\ &= 0 \end{aligned}$$

Note that the law applies if we assume that (a) voltage increases are positive and decreases are negative, or (b) voltage increases are negative and decreases are positive. In the foregoing illustration, we traveled ccw using assumption (a). If we had traveled clockwise (cw), we could have used assumption (b).

EXAMPLE 7-12 Compute the voltage drops, total current, and total resistance of Fig. 7-9.

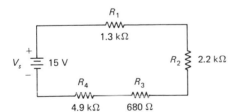

FIGURE 7-9 Example 7-12.

SOLUTION Computing total resistance and current:

$$\begin{aligned} R_t &= R_1 + R_2 + R_3 + R_4 \\ &= 1.3 + 2.2 + 0.68 + 4.9 \\ &= 9.08 \text{ k}\Omega \end{aligned}$$

$$I_t = \frac{V_s}{R_t} = \frac{15}{9.08} = 1.652 \text{ mA}$$

We can calculate each of the voltage drops by Ohm's law: V_1 (voltage across R_1), V_2 (voltage across R_2), V_3 (voltage across R_3), and V_4 (voltage across R_4).

$$V_1 = I_t R_1 = 1.652 \times 1.3 = 2.148 \text{ V}$$
$$V_2 = I_t R_2 = 1.652 \times 2.2 = 3.634 \text{ V}$$
$$V_3 = I_t R_2 = 1.652 \times 0.68 = 1.123 \text{ V}$$
$$V_4 = I_t R_4 = 1.652 \times 4.9 = 8.095 \text{ V}$$

We can use Kirchhoff's voltage law to check our answers:

$$0 = V_s + V_1 + V_2 + V_3 + V_4$$
$$= 15 - 2.148 - 3.634 - 1.123 - 8.095$$
$$= 0$$

Power

Power in a series circuit is additive. That is, the total power is equal to sum of the power dissipated in the individual resistors.

$$P_t = P_1 + P_2 + P_3 + \ldots$$

EXAMPLE 7-13 Compute the power dissipated in the circuit of Fig. 7-9.

SOLUTION We shall obtain the answer in two ways:

(a) Total power equals total voltage multiplied by total current.

$$P_t = V_s I_t$$
$$= 15 \times 1.652$$
$$= 24.78 \text{ mW}$$

(b) Total power is the sum of the power dissipated by each resistor.

$$P_1 = V_1 I_t$$
$$= 2.148 \times 1.652 = 3.548 \text{ mW}$$
$$P_2 = V_2 I_t$$
$$= 3.634 \times 1.652 = 6.003 \text{ mW}$$
$$P_3 = V_3 I_t$$
$$= 1.123 \times 1.652 = 1.855 \text{ mW}$$
$$P_4 = V_4 I_t$$
$$= 8.095 \times 1.652 = 13.373 \text{ mW}$$
$$P_t = P_1 + P_2 + P_3 + P_4$$
$$= 3.548 + 6.003 + 1.855 + 13.373$$
$$= 24.78 \text{ mW}$$

Note that the two methods yield identical results.

EXAMPLE 7-14 Three resistors are in series with a 28-V supply: 300 Ω, 500 Ω, and 600 Ω. Compute the power dissipated by each resistor and total power.

SOLUTION Compute the total resistance:

$$R_t = R_1 + R_2 + R_3$$
$$= 300 + 500 + 600 = 1400 \text{ Ω}$$

Compute the current:

$$I = \frac{V}{R} = \frac{28}{1.4} = 20 \text{ mA}$$

Compute the individual power dissipations:

$$P_{300} = I^2 R$$
$$= (0.020)^2 (300) = 120 \text{ mW}$$
$$P_{500} = I^2 R$$
$$= (0.020)^2 (500) = 200 \text{ mW}$$
$$P_{600} = (0.020)^2 (600) = 240 \text{ mW}$$

The total power is the sum:

$$P_t = P_{300} + P_{500} + P_{600}$$
$$= 120 + 200 + 240 = 560 \text{ mW}$$

PROBLEMS

Compute the total resistance of the following series-connected resistors:

	R_1	R_2	R_3	R_4
7-59.	92.63 Ω	359 Ω	11.37 Ω	697.8 Ω
7-60.	7.07 kΩ	4.74 kΩ	7.86 Ω	13.88 kΩ
7-61.	8.39 Ω	838 Ω	4.12 kΩ	97.6 Ω
7-62.	43.1 kΩ	62.1 kΩ	38.5 kΩ	9.724 kΩ
7-63.	9.49 mΩ	4.29 Ω	8.81 kΩ	8.48 MΩ
7-64.	9.75 mΩ	5.10 Ω	1.56 kΩ	4.62 MΩ

7-65. The divider string for a meter has 1.8-kΩ, 3.5-kΩ, 9-kΩ, and 35-kΩ resistors in series. What is the total resistance?

7-66. The 0.2-Ω internal resistance of a battery is in series with two wires, each having 150-mΩ resistance, and two lamps each having 5-Ω resistance. What is the total resistance of the circuit?

Problems 7-67 through 7-72 refer to Fig. 7-10. Compute the voltage drop across each of the resistors.

	R_1	R_2	R_3	V_s
7-67.	178 Ω	653 Ω	376 Ω	16 V
7-68.	7.40 kΩ	7.99 kΩ	4.40 kΩ	45 V
7-69.	39.24 kΩ	23.47 kΩ	96.03 kΩ	42 V
7-70.	9.00 MΩ	463 kΩ	13.08 kΩ	20 V
7-71.	8.12 MΩ	639 kΩ	75.3 kΩ	23 V
7-72.	180 mΩ	4.05 Ω	8.41 Ω	6 V

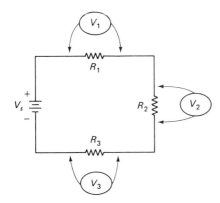

FIGURE 7-10 Problems 7-67 through 7-72.

7-73. Two resistors, 5 kΩ and 7.5 kΩ, are in series with a 12-V supply. Compute the voltage drops across the resistors.

7-74. A 12-V battery feeds a 25-Ω load over two wires each having 180-mΩ resistance. What is the voltage drop across each of the wires and the 25-Ω load?

Compute the power dissipated by each of the resistors and the total power.

7-75. The circuit of Prob. 7-67.
7-76. The circuit of Prob. 7-68.
7-77. The circuit of Prob. 7-69.
7-78. The circuit of Prob. 7-70.

7-4 THE LOOP EQUATION

Not only can resistors be connected in series, but voltage sources can also be so connected. When the two are connected such that the currents generated by each supply aid each other, they are said to be connected series-aiding (Fig. 7-11). When the currents oppose each other, they are said to be connected series-opposing (Fig. 7-12). In both, Kirchhoff's voltage equation can be used to analyze the current by constructing an equation. This equation, called a loop equation, can be derived by traveling around a complete circuit. If a voltage supply lead is positive as we enter the supply, call it positive. If negative, call it negative. Call all resistor drops negative *IR* drops. Current direction is in the same direction as we are traveling if the answer comes out positive. It is opposite to the direction we are traveling if the answer is negative.

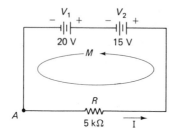

FIGURE 7-11   Series aiding supplies.   FIGURE 7-12   Series opposing supplies.

EXAMPLE 7-15   Compute the current in Fig. 7-11.

SOLUTION   The loop equation is also called a mesh equation. Starting from point *A* in the direction indicated by the arrow:

$$-5I + 15 + 20 = 0$$
$$I = \frac{35}{5} = 7 \text{ mA}$$

The current flows in the same direction as the mesh arrow, for our answer was positive.

EXAMPLE 7-16   Compute the current in Fig. 7-12.

SOLUTION Starting at point A, travel in the direction shown:

$$-20 + 15 - 5I = 0$$
$$-5I = 5$$
$$I = -1 \text{ mA}$$

Actual current flow is 1 mA in a direction opposite to that shown by the arrow, for our answer is negative.

EXAMPLE 7-17 Compute the current and voltage drops in Fig. 7-13.

SOLUTION Starting at point A, travel in the direction of M:

$$-20 - 5I + 15 - 2I - 1.5I - 17 - 3I = 0$$
$$-11.5I = 22$$
$$I = -1.913 \text{ mA}$$

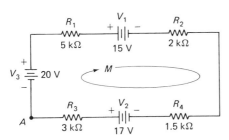

FIGURE 7-13 Example 7-17.

PROBLEMS

Problems 7-79 through 7-84 refer to Fig. 7-14. Compute total current for the series circuit shown. A negative value indicates a polarity opposite to that shown in the figure.

	R_1	R_2	R_3	V_1	V_2	V_3
7-79.	15 kΩ	72 kΩ	56 kΩ	24 V	16 V	68 V
7-80.	7.1 kΩ	6.2 kΩ	4.5 kΩ	50 V	13 V	70 V
7-81.	2.15 kΩ	256 Ω	817 Ω	−8.8 V	−2.3 V	7.7 V
7-82.	4.2 MΩ	560 kΩ	515 kΩ	−7.6 V	−7.7 V	5.6 V
7-83.	5.1 Ω	1.67 Ω	501 mΩ	1.86 V	−2.63 V	3.56 V
7-84.	5.1 kΩ	3.0 kΩ	44.3 kΩ	−8.65 V	918 mΩ	41 V

Problems 7-85 through 7-88 refer to Fig. 7-15. Compute the current, voltage drops, and total power dissipated.

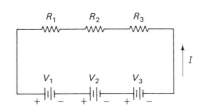

FIGURE 7-14 Problems 7-79 through 7-84.

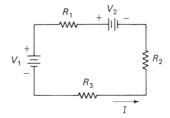

FIGURE 7-15 Problems 7-85 through 7-88.

	V_1	V_2	R_1	R_2	R_3
7-85.	37 V	85 V	66 kΩ	87 kΩ	43 kΩ
7-86.	9.1 V	−7.3 V	103 MΩ	691 MΩ	225 MΩ
7-87.	−5.27 V	9.24 V	14.9 kΩ	3.69 kΩ	1.36 kΩ
7-88.	−7.6 V	5.1 V	167 Ω	53 Ω	72 Ω

7-5 THE UNLOADED VOLTAGE DIVIDER

A voltage divider receives a supply voltage and distributes it as various voltages to a load (Fig. 7-16). Most electronic circuits contain one or more dividers for providing needed supply current or changing the voltage level of a signal. Thus, it is a very important circuit.

If less than 10% of the resistor string current flows through the load, the divider is said to be unloaded; if greater than 10%, it is said to be loaded. The current flowing through the string resistors is called bleeder current and will usually be specified within the problem.

By changing the reference point, both negative and positive voltages may be fed to the load (Fig. 7-17). Note that the circuits of Figs. 7-16 and 7-17 are identical. Only that point defined as 0 V to the load has changed.

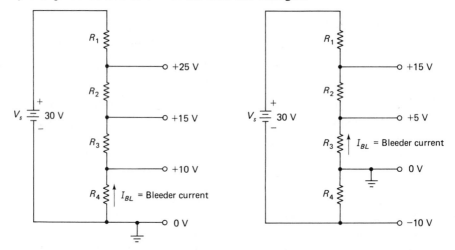

FIGURE 7-16 The unloaded voltage divider.   **FIGURE 7-17** Changing the reference.

EXAMPLE 7-18 Compute the string resistors and their individual power dissipations in Fig. 7-16 for a bleeder current of 6 mA.

SOLUTION There is 10-V difference between the 0-V and 10-V line and the current is given as 6 mA. Therefore,

$$R_4 = \frac{V}{I} = \frac{10}{6} = 1.667 \text{ k}\Omega$$
$$P_4 = VI = 10 \times 6 = 60 \text{ mW}$$

Compute R_2 and R_3:

$$R_3 = \frac{V}{I} = \frac{15 - 10}{6} = 833 \ \Omega$$
$$P_3 = VI = (15 - 10)(6) = 30 \ \text{mW}$$
$$R_2 = \frac{V}{I} = \frac{25 - 15}{6} = 1.667 \ \text{k}\Omega$$
$$P_2 = VI = (25 - 15)(6) = 60 \ \text{mW}$$

The voltage across R_1 is the difference between the supply voltage, 30 V, and the voltage between the 0-V and 25-V terminals, 25 V. Thus,

$$R_1 = \frac{V}{I} = \frac{30 - 25}{6} = 833 \ \Omega$$
$$P_1 = VI = (30 - 25)(6) = 30 \ \text{mW}$$

EXAMPLE 7-19 Compute the string resistors of Fig. 7-17 if the bleeder current is 50 mA.

SOLUTION

$$R_4 = \frac{V}{I} = \frac{10}{50} = 200 \ \Omega$$
$$P_4 = VI = 10 \times 50 = 500 \ \text{mW}$$
$$R_3 = \frac{V}{I} = \frac{5 - 0}{50} = 100 \ \Omega$$
$$P_3 = VI = 5 \times 50 = 250 \ \text{mW}$$
$$R_2 = \frac{V}{I} = \frac{15 - 5}{50} = 200 \ \Omega$$

The voltage across R_1 is the difference between the 30-V supply and the voltage between the -10-V output and the $+15$-V output.

$$R_1 = \frac{V}{I} = \frac{30 - [15 - (-10)]}{50} = 100 \ \Omega$$
$$P = VI = 5 \times 50 = 250 \ \text{mW}$$

PROBLEMS

Compute the resistance of and power dissipated by the string resistors of the dividers of Problems 7-89 through 7-94.

7-89. $V_s = 20$ V; $I_{BL} = 16$ mA; outputs: 6 V, 12 V, and 15 V
7-90. $V_s = 60$ V; $I_{BL} = 50$ mA; outputs: 5 V, 20 V, and 40 V
7-91. $V_s = 75$ V; $I_{BL} = 20$ mA; outputs: 10 V, 16 V, and 48 V
7-92. $V_s = 45$ V; $I_{BL} = 15$ mA; outputs: -30 V, -20 V, and $+5$ V
7-93. $V_s = 25$ V; $I_{BL} = 100$ mA; outputs: -15 V, -5 V, and $+5$ V
7-94. $V_s = 55$ V; $I_{BL} = 125$ mA; outputs: -20 V, -15 V, -8 V, $+5$ V, $+8$ V, $+15$ V

8

FACTORING

Factors and factoring are studied for two prime purposes. First, factoring is essential in reducing algebraic expressions. Second, factoring is one of the easiest methods of solving algebraic equations.

In preceding chapters we have found that several numbers can be multiplied together to obtain a solution. These numbers are called factors and in this chapter we study factors in detail. Quite a bit of attention is devoted to a process called factoring, where the answer we referred to is broken into its component factors. To make the objective clearer, perhaps we should call this process un-multiplying.

8-1 POWERS OF MONOMIALS

We start the chapter by discussing the powers and roots of monomials, for this skill is required in some of the factoring processes we shall use.

When monomials are raised to a power, each part of the monomial is raised to that power.

$$(abc)^n = a^n b^n c^n$$

EXAMPLE 8-1 Compute:

(a) $(4xy^2)^2$ (b) $(5x^4y^3z)^2$

SOLUTION

(a) $(4xy^2)^2 = (4)^2(x)^2(y^2)^2 = 16x^2y^4$
(b) $(5x^4y^3z)^2 = (5)^2(x^4)^2(y^3)^2(z)^2$
$$= 25x^8y^6z^2$$

Note that a power is raised to a power by multiplying the exponents as in (b) above, where $(x^4)^2$ is x^8.

We should state at this point that the square of a negative number (that is, raising it to a power of 2) results in a positive number. Examine the following:

$$(-2)^2 = (-2)(-2) = +4$$
$$(-3)^2 = (-3)(-3) = +9$$

In fact, raising a negative number to any even power results in a positive number:

$$(-2)^4 = (-2)(-2)(-2)(-2)$$
$$= [(-2)(-2)][(-2)(-2)]$$
$$= [+4][+4] = +16$$
$$(-3)^6 = [(-3)(-3)][(-3)(-3)][(-3)(-3)]$$
$$= [+9][+9][+9]$$
$$= +729$$

Whereas raising a negative number to an even power results in a positive result, raising a negative number to an odd power results in a negative result. Examine the following:

$$(-2)^3 = (-2)(-2)(-2) = [(-2)(-2)](-2)$$
$$= [+4](-2) = -8$$
$$(-3)^7 = [(-3)(-3)][(-3)(-3)][(-3)(-3)](-3)$$
$$= [+9][+9][+9](-3)$$
$$= -2187$$

EXAMPLE 8-2 Compute:

(a) $(-5A^3B^4C)^3$ (b) $(-6.3M^3N^4P^2Z^4)^3$

SOLUTION

(a) $(-5A^3B^4C)^3 = (-5)^3(A^3)^3(B^4)^3(C)^3$
$$= -125A^9B^{12}C^3$$

(b) $(-6.3M^3N^4P^2Z^4)^3 = (-6.3)^3(M^3)^3(N^4)^3(P^2)^3(Z^4)^3$
$$= -250.05M^9N^{12}P^6Z^{12}$$

Again, note that each part (or factor) is raised to the power. Note further that raising a negative number to an odd power results in a negative answer, whereas raising a negative number to an even power results in a positive number. The calculator automatically computes the proper sign. This process is true for any power.

EXAMPLE 8-3 Compute $(-4.4A^2B^2C)^5$.

SOLUTION

$$(-4.4A^2B^2C)^5 = -1649.16A^{10}B^{10}C^5$$

PROBLEMS

Compute the following:

8-1. (a) $(3AB)^2$ (b) $(4CD^2)^2$

8-2. (a) $(7xy)^2$ (b) $(9x^2y^2)^2$

8-3. (a) $(3AC)^3$ (b) $(4A^2B^3)^2$

8-4. (a) $(2x^2yz^3)^2$ (b) $(5M^3N^4P)^2$

8-5. (a) $(-5R^2S^3)^2$ (b) $(-7R_4R_2^2)^3$

8-6. (a) $(-7P^3R^2)^3$ (b) $(-2S_1^2S_2^3)^3$

8-7. (a) $(4.19x^2y^3)^4$ (b) $(-5.62W^4P^2)^5$

8-8. (a) $(-2.64R_1^2S_1^2)^4$ (b) $(-3.16P_1^2Q^2)^3$

8-9. (a) $(-1.315V_1V_2^2)^7$ (b) $(-1.0936X_1^2X_2^3)^9$

8-10. (a) $(-2.153C_1^2C_2^3)^5$ (b) $(-4.193L_1^4L_2^2L_3)^7$

The square root of a number, *A*, is that number which when multiplied by itself results in *A*. For example, the square root of 225 is 15, for when 15 is multiplied by itself, the result is 225.

$$15 \times 15 = 225$$

There are two symbols in mathematics used to represent the square root, the radical sign and the fractional power. The square root of 225 can be represented as

$$\sqrt{225} \quad \text{or} \quad \sqrt[2]{225} \quad \text{or} \quad (225)^{1/2}$$

In each case, the 2 represents how many numbers must be multiplied to obtain the original number. In the example above, the number 15 must be operated on twice to obtain 225.

When the radical sign is used without a number ($\sqrt{}$), the square root is meant ($\sqrt[2]{}$).

We should at this point get a little more precise. The square root of a positive number results in two answers, one positive and one negative. For example, the square root of 225 is

$$\sqrt{225} = +15 \quad \text{and} \quad \sqrt{225} = -15$$

Proof: $(+15)(+15) = 225$
$(-15)(-15) = 225$

This is expressed as

$$\sqrt{225} = \pm 15$$

indicating that the square root of 225 is plus or minus 15.

The square root of a monomial is found by finding the square root of each element within the monomial.

EXAMPLE 8-4 Find $\sqrt{4X^2 Y^4 Z^6}$.

SOLUTION

$$\begin{aligned}
\sqrt{4X^2 Y^4 Z^6} &= (\sqrt{4})(\sqrt{X^2})(\sqrt{Y^4})(\sqrt{Z^6}) \\
&= (\pm 2)(X^2)^{1/2}(Y^4)^{1/2}(Z^6)^{1/2} \\
&= (\pm 2)(X)(Y^2)(Z^3) \\
&= \pm 2XY^2 Z^3
\end{aligned}$$

Note that the easiest way to find the square root of a literal raised to a power is by expressing the square root in its fractional form. This allows the rules of fractions to be used in computing the final exponent. The exponent of *Z* was computed as

$$(Z^6)^{1/2} = Z^{(6 \times 1/2)} = Z^3$$

EXAMPLE 8-5 Find $\sqrt{6X^4 Y^2 Z^4}$.

SOLUTION

$$\begin{aligned}
\sqrt{6X^4 Y^2 Z^4} &= (\sqrt{6})(X^4)^{1/2}(Y^2)^{1/2}(Z^4)^{1/2} \\
&= \pm 2.449 X^2 Y Z^2
\end{aligned}$$

PROBLEMS

Find the square root:

8-11. (a) $4A^2B^2$ (b) $49C^4D^2$

8-12. (a) $25A^6B^2$ (b) $100X_1^2X_2^2$

8-13. (a) $16C_1^{10}C_2^4$ (b) $36D_1^8D_2^6D_3^4$

8-14. (a) $121E_1^2E_2^6E_3^2$ (b) $144V_1^4V_2^8V_3^{16}$

8-15. (a) $3A^2B^4$ (b) $7A^2B^4C^2D^6$

8-16. (a) $17R_1^4R_2^6R_3^2$ (b) $20X_1^2X_2^4X_3^6$

8-3 PRIME NUMBERS

In this chapter we study methods of breaking an algebraic expression into its factors, each of which is a prime number. A prime number is one that is evenly divisible only by 1 and the number itself.

EXAMPLE 8-6 Find all the prime numbers between 1 and 10 inclusive.

SOLUTION 1, 2, 3, 5, and 7 are prime numbers, for they can only be divided evenly (without a remainder) by 1 and the number itself. However, 4 is not prime, for it can be divided by 2; 6 is not, for it can be divided by 2 and 3; 8 is not, for it can be divided by 2; 10 is not, for it can be divided by 2 and 5.

Any number can be "unmultiplied" to obtain a product of prime numbers. This process is called factoring and each of the prime numbers is called a factor. A numerical integer constant can be broken into its factors by a trial-and-error process. There are several tricks that can be used to tell if it contains certain numbers:

1. All even numbers can be evenly divided by 2.
2. If the sum of the digits with a number is evenly divisible by 3, the number itself is evenly divisible by 3.
3. A number ending in 5 can be evenly divided by 5.

EXAMPLE 8-7 Factor 2520.

SOLUTION Since it is even, it can be divided by 2:

$$2520 = 2 \cdot 1260$$

Divide again by 2:

$$2520 = 2 \cdot 2 \cdot 630$$

Divide again by 2:

$$2520 = 2 \cdot 2 \cdot 2 \cdot 315$$

The 315 ends in 5, so it can be divided by 5:

$$2520 = 2 \cdot 2 \cdot 2 \cdot 5 \cdot 63$$

The sum of 6 and 3 in 63 is 9, which is evenly divisible by 3. Therefore, 63 is evenly divisible by 3:

$$2520 = 2 \cdot 2 \cdot 2 \cdot 5 \cdot 3 \cdot 21$$

The number 21 is also divisible by 3:

$$2520 = 2 \cdot 2 \cdot 2 \cdot 5 \cdot 3 \cdot 3 \cdot 7$$

Since 7 is prime, we have the prime factors of 2520. Rearranging in ascending order, we obtain

$$2520 = 2 \cdot 2 \cdot 2 \cdot 3 \cdot 3 \cdot 5 \cdot 7$$

EXAMPLE 8-8   Factor 30,030.

SOLUTION

$$30,030 = 2 \cdot 15,015$$
$$30,030 = 2 \cdot 5 \cdot 3003$$
$$30,030 = 2 \cdot 5 \cdot 3 \cdot 1001$$

The 1001 cannot be divided by 2, 3, or 5. Thus, we shall try higher prime numbers in ascending order: 7, 11, 13, 17, 19, 23. . . .

$$30,030 = 2 \cdot 5 \cdot 3 \cdot 7 \cdot 143$$
$$30,030 = 2 \cdot 5 \cdot 3 \cdot 7 \cdot 11 \cdot 13$$

These are all prime numbers. Rearranging yields

$$30,030 = 2 \cdot 3 \cdot 5 \cdot 7 \cdot 11 \cdot 13$$

Algebraic Expressions

We can also find the prime factors in an algebraic expressions. Although these prime factors can be expressed as products of literals, they are usually "unmultiplied" as far as necessary to reduce the exponent to its lowest value. For example, $20X^2Y^4$ would be left as it is instead of

$$20X^2Y^4 = 2 \cdot 2 \cdot 5 \cdot X \cdot X \cdot Y \cdot Y \cdot Y \cdot Y$$

In this form, $20X^2Y^4$, it is considered prime.

The next sections discuss four different models for factoring:

1. A common factor
2. The perfect square
3. Difference of squares
4. A trial-and-error method for trinomials

Using just these four methods, almost all factoring situations can be easily treated.

PROBLEMS

Reduce to prime factors:

8-17. (a) 140 (b) 60 (c) 240
8-18. (a) 180 (b) 100 (c) 75
8-19. (a) 264 (b) 936 (c) 204
8-20. (a) 374 (b) 450 (c) 840

8-21. (a) 3648 (b) 1377 (c) 5148
8-22. (a) 3780 (b) 8280 (c) 9280
8-23. (a) 238,425 (b) 110,352 (c) 193,545
8-24. (a) 8,473,905 (b) 3,504,000 (c) 8,469,846

8-4   THE COMMON FACTOR

This common factor technique should always be the first one attempted when considering binomials and larger expressions. In it, each term is examined to see if there is a literal or a constant that is common to all terms. The model expression is

$$aX + aY + aZ = a(X + Y + Z)$$

Note that the a is common to all three terms of the expression. This a is then placed on the outside of an expression, each term of which has been divided by the a.

Example 8-9 Factor:

(a) $3X + 3Y$
(b) $4X^2 + 8Y^2$
(c) $7X^2 + 49X^2Y$
(d) $3X^2Y^2 + 3X^2Z + 6X^2Y^3Z^2$

SOLUTION

(a) $3X + 3Y$

The two terms have a common "3." Place this on the outside of the parentheses:

$$3X + 3Y = 3(X + Y)$$

We can prove the correctness of this solution by multiplying the right expression, obtaining the original expression:

$$3(X + Y) = 3X + 3Y$$

Again, it should be emphasized that factoring is unmultiplying. Thus, the proof is remultiplication.

(b) $4X^2 + 8Y^2$

We can completely factor the expression:

$$4X^2 + 8Y^2 = (2 \cdot 2 \cdot X \cdot X) + (2 \cdot 2 \cdot 2 \cdot Y \cdot Y)$$

Note that there are common terms, 2 and 2. Place these on the outside of the parentheses, leaving what is left inside:

$$4X^2 + 8Y^2 = (2 \cdot 2)[(X \cdot X) + (2 \cdot Y \cdot Y)]$$
$$= 4(X^2 + 2Y^2)$$

Proof:

$$4(X^2 + 2Y^2) = 4X^2 + 8Y^2$$

(c) $7X^2 + 49X^2Y$

Factor completely:

$$7X^2 + 49X^2Y = 7 \cdot X \cdot X + 7 \cdot 7 \cdot X \cdot X \cdot Y$$

Factor out the common terms:

$$(7 \cdot X \cdot X)[(1) + (7 \cdot Y)] = 7X^2(1 + 7Y)$$

Note that factoring a $7X^2$ out of a $7X^2$ yields 1. This can be shown by assuming that the original $7 \cdot X \cdot X$ is $7 \cdot X \cdot X \cdot 1$:

$$7X^2 + 49X^2Y = 7 \cdot X \cdot X \cdot 1 + 7 \cdot 7 \cdot X \cdot X \cdot Y$$
$$= (7 \cdot X \cdot X)[(1) + (7 \cdot Y)]$$
$$= 7X^2(1 + 7Y)$$

Proof:

$$7X^2(1 + 7Y) = 7X^2 + 49X^2Y$$

(d) $3X^2Y^2 + 3X^2Z + 6X^2Y^3Z^2$

Factor completely:

$$3X^2Y^2 + 3X^2Z + 6X^2Y^3Z^2 = (3 \cdot X \cdot X \cdot Y \cdot Y) + (3 \cdot X \cdot X \cdot Z)$$
$$+ (2 \cdot 3 \cdot X \cdot X \cdot Y \cdot Y \cdot Y \cdot Z \cdot Z)$$

Select the common factors:

$$(3 \cdot X \cdot X)[(Y \cdot Y) + (Z) + (2 \cdot Y \cdot Y \cdot Y \cdot Z \cdot Z)]$$
$$= 3X^2(Y^2 + Z + 2Y^3Z^2)$$

Proof:

$$3X^2(Y^2 + Z + 2Y^3Z^2) = 3X^2Y^2 + 3X^2Z + 6X^2Y^3Z^2$$

Again, it should be emphasized that, when factoring, all common factors should be treated before any other method is used.

PROBLEMS

Factor:

8-25. $2A + 2B$

8-26. $5X + 5Y$

8-27. $4R_1 + 8R_2$

8-28. $6V_1 + 12V_2$

8-29. $25S_1 + 165S_1S_2$

8-30. $8A_1^2B_1^3 + 12A_1B_1^4$

8-31. $14X^5Y^4 + 7X^2Y^2$

8-32. $16M^2N^4P + 8MN^4$

8-33. $9P_L^2 + 18P_L$

8-34. $3P_1P_2^2P_3^3 + P_2^4P_3$

8-35. $9V_1 + 3V_1V_2$

8-36. $12X^2Y + 6XY$

8-37. $18R_1^2R_2 + 12R_1R_2^2$

8-38. $20V_1V_2^2V_3^3 + 50V_1^3V_2^2V_3$

8-39. $3A + 3B + 3C$

8-40. $4V_1 + 8V_2 + 6V_3$

8-41. $16V_1V_2^2 + 18V_1^2V_2 + 12V_1V_2$

8-42. $26P_1^2P_2^3P_3 + 2P_1^2P_2^6 + 6P_1^3P_3^2P_4$

8-43. $48E_1^2E_2^3V_1 + 20E_2^3V_1^3V_2^2 + 40E_2^2V_2^2$

8-44. $44S_1^2S_2^2S_3^2 + 33S_1S_2 + 22S_1^4S_3S_2^2$

8-45. $20A^2B^3C + 4AB^3C^2D + 6A^3B^2E + B^2C^2$

8-46. $75X_1^2X_2^3X_3 + 55X_2^4X_3^5 + 70X_1^4X_3^5$

8-5   THE PERFECT SQUARE

It should first be emphasized that the perfect-square method is a trial-and-error process. Upon trying it, if it works, the factors are found. However, if it does not work, another method must be tried.

The model for this method is

$$a^2 + 2ab + b^2 = (a + b)^2$$

This model can be proved by

$$(a + b)^2 = (a + b)(a + b) = a^2 + 2ab + b^2$$

Although the model can easily be verified, using the model is more involved.

It can be summarized as follows:

(a) Arrange the terms according to the model, with the higher-order terms (those with the highest exponents) on the outside.
(b) Take the square root of the first term.
(c) Take the square root of the last term.
(d) Multiply these two roots together, then multiply by 2.
(e) If the product matches the middle term of the original expression, that expression is a perfect square according to the model. If it does not match, the expression cannot be factored as a perfect square.
(f) If it is a perfect square, the square root found in step (b) is the a of the model and that of step (c) the b of the model.

EXAMPLE 8-10   Factor $X^2 + 4X + 4$.

SOLUTION   There is no common factor, so we shall see if it is a perfect square.

Step (a)   The expression is already correctly arranged.
Step (b) $\sqrt{X^2} = X$
Step (c) $\sqrt{4} = 2$
Step (d)   Two times the product $= 2(2X) = 4X$.
Step (e)   This product does match the $4X$ of the original expression. Therefore, it is a perfect square.
Step (f)   The X from step (b) becomes the a and the 2 from step (c) becomes the b in the model. Therefore,

$$X^2 + 4X + 4 = (X + 2)^2$$

Proof:

$$(X + 2)^2 = (X + 2)(X + 2) = X^2 + 4X + 4$$

EXAMPLE 8-11   Factor $16X^2 + 25Y^2 + 40XY$.

SOLUTION   There is no common factor.

Step (a) $16X^2 + 40XY + 25Y^2$
Step (b) $4X$
Step (c) $5Y$
Step (d) $40XY$
Step (e)   It is a perfect square.
Step (f) $16X^2 + 40XY + 25Y^2 = (4X + 5Y)^2$
           Proof: Multiplying out the right side results in the left side of the equation.

EXAMPLE 8-12   Factor $49M^4N^2 + 9X^4Y^2 + 42X^2YM^2N$.

SOLUTION   There is no common factor.

Step (a) $9X^4Y^2 + 42X^2YM^2N + 49M^4N^2$
Step (b) $3X^2Y$
Step (c) $7M^2N$
Step (d) $42X^2YM^2N$
Step (e)   It is a perfect square.
Step (f) $9X^4Y^2 + 42X^2YM^2N + 49M^4N^2 = (3X^2Y + 7M^2N)^2$
           Proof: Multiplying out the right side results in the left side of the equation.

EXAMPLE 8-13   Factor $4X_1^2 + 6X_1X_2 + 9X_2^2$.

SOLUTION   There is no common factor.

Step (a) $4X_1^2 + 6X_1X_2 + 9X_2^2$
Step (b) $2X_1$
Step (c) $3X_2$
Step (d) $12X_1X_2$
Step (e)   The expression is not a perfect square. Therefore, if it can be factored, some other method must be used.

EXAMPLE 8-14   Factor $25X^4Y^2 + 50X^3YZ + 25X^2Z^2$.

SOLUTION   There is a common factor:

$$25X^4Y^2 + 50X^3YZ + 25X^2Z^2 = 25X^2(X^2Y^2 + 2XYZ + Z^2)$$

We can now try the perfect-square method on the expression within the parentheses:

Step (a) $X^2Y^2 + 2XYZ + Z^2$
Step (b) XY
Step (c) Z
Step (d) $2XYZ$
Step (e)   It is a perfect square.
Step (f) $X^2Y^2 + 2XYZ + Z^2 = (XY + Z)^2$
  The final solution is

$$25X^4Y^2 + 50X^3YZ + 25X^2Z^2 = 25X^2(XY + Z)^2$$

PROBLEMS

Factor:

8-47. $A^2 + 10AB + 25B^2$

8-48. $4R^2 + 4RS + S^2$

8-49. $25B^2 + 20BC + 4C^2$

8-50. $9P_1^2 + 12P_1P_2 + 4P_2^2$

8-51. $25E_1^2 + 30E_1E_2 + 9E_2^2$

8-52. $49C_1^2 + 56C_1C_2 + 16C_2^2$

8-53. $16A_1^2 + 24A_1A_2 + 9A_2^2$

8-54. $49X^2Y^2 + 84XYZ + 36Z^2$

8-55. $25A^2B^4 + 30AB^2CD + 9C^2D^2$

8-56. $81M^4N^6 + 90M^2N^3P^2Q + 25P^4Q^2$

8-57. $9R_1^2R_2^2 + 16R_3^2 + 24R_1R_2R_3$

8-58. $40MPQ + 16P^2Q^2 + 25M^2$

8-59. $AX^2 + 2AXY + AY^2$

8-60. $2BC^2 + 4BCD + 2BD^2$

8-61. $18X^2Y + 12XY^2 + 2XY^3$

8-62. $27P_1^3 + 36P_1^2P_2 + 12P_1P_2^2$

8-63. $28M^3N + 28M^2N^2 + 7MN^3$

8-64. $200R_1^2R_2R_3 + 80R_1R_2^2R_3^2 + 125R_1^3$

8-65. $180A^3B^4C^2D + 50ABC^4D^2 + 162A^5B^7$

8-66. $36C_1^3C_2^2C_3 + 64C_1C_3^3 + 96C_1^2C_2C_3^2$

The Square of Differences

In the preceding model we have used for the perfect square, we have assumed that these two square roots were added, then multiplied:

$$(a + b)^2 = (a + b)(a + b) = a^2 + 2ab + b^2$$

If we now substitute $-b$ for the b on the left side of the equation:

$$[a + (-b)]^2 = (a - b)^2 = (a - b)(a - b)$$
$$= a^2 - 2ab + b^2$$

Note that the only effect of changing the sign of one of the square roots is to change the sign of the middle term. The first and last terms will always be positive because, whenever either a negative or a positive number is squared, the result is positive.

$$(+a)^2 = +a^2 \qquad (+b)^2 = +b^2$$
$$(-a)^2 = +a^2 \qquad (-b)^2 = +b^2$$

Squared terms will always be positive.

We can use this information in our factoring. Express the foregoing equations in reverse:

$$a^2 - 2ab + b^2 = (a - b)^2$$

The method for using this model is almost identical to that stated in the preceding subsection:

(a) Arrange the terms according to the model with the higher-order terms (those with the highest exponents) on the outside.
(b) Take the square root of the first term.
(c) Take the square root of the last term.
(d) Multiply these two roots together, then multiply by 2.
(e) If the product matches the middle term of the original expression (omit the sign), that expression is a perfect square. If it does not match, it is not a perfect square.
(f) If it is a perfect square, the square root found in step (b) is the a of the model and that of step (c) is the b of the model.
(g) If the middle term of the original expression is positive, the roots are added according to the model: $a^2 + 2ab + b^2 = (a + b)^2$. If the middle term of the original expression is negative, the roots are subtracted according to the model: $a^2 - 2ab + b^2 = (a - b)^2$.

EXAMPLE 8-15 Factor $4W^2 - 12WX + 9X^2$.

SOLUTION There is no common factor.

Step (a) $4W^2 - 12WX + 9X^2$
Step (b) $2W$
Step (c) $3X$
Step (d) $12WX$
Step (e) It is a perfect square.
Step (f) Since the middle term, $-12WX$, is negative, the roots are subtracted: $4W^2 - 12WX + 9X^2 = (2W - 3X)^2$.

EXAMPLE 8-16 Factor $25R_1^2R_2^4 - 40R_1R_2^2R_3R_4 + 16R_3^2R_4^2$.

SOLUTION There is no common factor.

Step (a) $25R_1^2R_2^4 - 40R_1R_2^2R_3R_4 + 16R_3^2R_4^2$
Step (b) $5R_1R_2^2$
Step (c) $4R_3R_4$
Step (d) $40R_1R_2^2R_3R_4$
Step (e) It is a perfect square.
Step (f) $25R_1^2R_2^4 - 40R_1R_2^2R_3R_4 + 16R_3^2R_4^2 = (5R_1R_2^2 - 4R_3R_4)^2$

EXAMPLE 8-17 Factor $36V_1^2V_2^4 + 4V_1^4 - 24V_1^3V_2^2$.

SOLUTION There is a common factor:

$$36V_1^2V_2^4 + 4V_1^4 - 24V_1^3V_2^2 = 4V_1^2(9V_2^4 + V_1^2 - 6V_1V_2^2)$$

Applying the foregoing procedure to the expression within the parentheses:

Step (a) $V_1^2 - 6V_1V_2^2 + 9V_2^4$
Step (b) V_1
Step (c) $3V_2^2$
Step (d) $6V_1V_2^2$
Step (e) It is a perfect square.
Step (f) $(V_1 - 3V_2^2)^2$

The entire solution is

$$36V_1^2V_2^4 + 4V_1^4 - 24V_1^3V_2^2 = 4V_1^2(V_1 - 3V_2^2)^2$$

PROBLEMS

Factor:

8-67. $P^2 - 10PQ + 25Q^2$

8-68. $16R_1^2 - 24R_1R_2 + 9R_2^2$

8-69. $25Z_1^2 - 20Z_1Z_2 + 4Z_2^2$

8-70. $49T_1^2T_2^4 - 28T_1T_2^2T_3 + 4T_3^2$

8-71. $4X^4 - 8X^3 + 4X^2$

8-72. $18X^3y - 4X^2y^2 + 8Xy^3$

8-73. $64L_1^4L_2^4L_3^2 - 96L_1^3L_2^2L_3^3 + 36L_1^2L_3^4$

8-74. $625I_1^6I_2^4 - 1500I_1^5I_2^5 + 900I_1^4I_2^6$

8-6 THE DIFFERENCE OF SQUARES

One of the most interesting models for factoring is that of the difference of squares. Note what happens to the middle terms in the following multiplication:

$$\begin{array}{r} a + b \\ a - b \\ \hline -ab - b^2 \\ a^2 + ab \\ \hline a^2 \quad\quad - b^2 \end{array}$$

Note that the *ab* terms drop out, leaving just the difference of squares. We can use this property to find factors of the difference between squares according to the model

$$a^2 - b^2 = (a + b)(a - b)$$

The procedure can be stated as follows:

(a) If the signs of the two terms differ, the difference of squares method can be tried.
(b) Take the square root of the first term.
(c) Take the square root of the last term.
(d) Form a product of the difference and the sum of the roots.

EXAMPLE 8-18 Factor $4A^2 - C^2$.

SOLUTION

Step (a) It is a difference of squares.

Step (b) $2A$
Step (c) C
Step (d) $(2A - C)(2A + C)$

EXAMPLE 8-19   Factor $81R_1^2R_2^2 - 64R_3^4$.

SOLUTION

Step (a)   It is a difference of squares.
Step (b) $9R_1R_2$
Step (c) $8R_3^2$
Step (d) $(9R_1R_2 - 8R_3^2)(9R_1R_2 + 8R_3^2)$

Do not forget to check for monomial factors before using this method.

EXAMPLE 8-20   Factor $12L_1^4L_2 - 27L_1^2L_2^3$.

SOLUTION   At first glance, the problem does not appear to be either a perfect square or a difference of squares. However, after factoring the common terms, we obtain

$$12L_1^4L_2 - 27L_1^2L_2^3 = 3L_1^2L_2(4L_1^2 - 9L_2^2)$$

Apply the rules for difference of squares to the expression within the parentheses:

Step (a)   The expression is a difference of squares.
Step (b) $2L_1$
Step (c) $3L_2$
Step (d) $(2L_1 - 3L_2)(2L_1 + 3L_2)$

The final solution is

$$12L_1^4L_2 - 27L_1^2L_2^3 = 3L_1^2L_2(2L_1 - 3L_2)(2L_1 + 3L_2)$$

PROBLEMS

Factor:

8-75. $B^2 - N^2$

8-76. $16X_1^2 - 25X_2^2$

8-77. $4L_1^2 - 9L_2^2$

8-78. $64V_1^2 - 81V_2^2$

8-79. $E_1^2 - E_2^2$

8-80. $225R_1^2 - 36R_2^2$

8-81. $100Z_1^2Z^2 - 81Z_3^4$

8-82. $64I_1^4I_2^2 - 49I_3^6$

8-83. $36P_1^2P_2^{10} - 121P^{26}$

8-84. $196A^{10}B^4 - 225C^6D^3$

8-85. $320X_1^4 - 125X_1^2X_2^2$

8-86. $200Q_1^5 - 288Q_1^3Q_2^4$

8-87. $343L_1^4L_2^5L_3^6 - 448L_1^2L_2L_3^8$

8-88. $2352V_1^4V_2^3V_3^2 - 2028V_1^2V_2^7$

8-89. $a^2 + 2ab + b^2 - c^2 - 2cd - d^2$

8-90. $\dfrac{1}{4a^2} - C^2$

8-91. $\dfrac{1}{(a+b)^2} - \dfrac{1}{(c+d)^2}$

8-92. $\dfrac{4X_1X_2^4}{X_1^2 + 2X_1X_2 + X_2^2} - \dfrac{4X_1X_2^2}{4X_1^2 - 12X_1X_2 + 9X_2^2}$

8-7   GENERAL TRINOMIAL FACTORING

The foregoing methods of factoring trinomials are special cases of the method illustrated in this section. These special cases provide solution to certain trinomials with first and last terms that are perfect squares. But what if these terms are not perfect squares? Does this mean that the trinomial cannot be factored? On the contrary, it could very well be possible to find prime factors for the expression.

Trinomials with Two Positive Coefficients

To use the general trinomial method, we should first examine the process of multiplication of binomials. In the following illustration, both numerical and literal coefficients are used:

$$
\begin{array}{ll}
2X + Y & aX + Y \\
3X + Y & bX + Y \\
\hline
+2XY + Y^2 & aXY + Y^2 \\
(2 \cdot 3)X^2 + \quad 3XY & abX^2 + \quad bXY \\
\hline
2 \cdot 3X^2 + (3 + 2)XY + Y^2 & abX^2 + (a + b)XY + Y^2
\end{array}
$$

Let us examine what happened to the coefficients, a and b. Note that, in the solution, the coefficient of the X^2 term is the product of the coefficients a and b, whereas in the middle term it is the sum of the same two coefficients, a and b. Consider the expression

$$12M^2 + 7MN + N^2 = (aM + N)(bM + N)$$

If it can be factored according to the foregoing pattern, the 12 must be the product of two numbers, a and b, and the 7 must be the sum of those same two numbers, a and b. By a trial-and-error process, we can find these two numbers. We know that 12 can be found by multiplying the following a's and b's:

$$
\begin{array}{ll}
12 = a \cdot b & 12 = 4 \times 3 \\
12 = 1 \times 12 & 12 = 6 \times 2 \\
12 = 2 \times 6 & 12 = 12 \times 1 \\
12 = 3 \times 4 &
\end{array}
$$

Since the middle term is 7, what products of 12 yield a sum of 7?

$$
\begin{array}{ll}
1 + 12 & \text{No} \\
2 + 6 & \text{No} \\
3 + 4 & \text{Yes, we have a solution.}
\end{array}
$$

The numbers 3 and 4 satisfy the two conditions:

1. Their product is 12.
2. Their sum is 7.

The solution to the problem is

$$12M^2 + 7MN + N^2 = (3M + N)(4M + N)$$

Proof:

$$(3M + N)(4M + N) = 12M^2 + 3MN + 4MN + N^2 = 12M^2 + 7MN + N^2$$

Thus, for expressions of the form

$$gX^2 + hXY + Y^2$$

the following method can be followed:

(a) Arrange the terms in descending powers.
(b) Select the coefficient of the first term.
(c) Find all its possible factors.
(d) Select those factors whose sum is the coefficient of the middle term.
(e) Construct the product using these factors as coefficients of the X term.

EXAMPLE 8-21   Factor $10V_1^2 + 7V_1V_2 + V_2^2$.

SOLUTION

Step (a) $10V_1^2 + 7V_1V_2 + V_2^2$
Step (b)   10
Step (c)   10, 1; 2, 5
Step (d) $10 + 1 = 11$
            $2 + 5 = 7$ This is the correct sum.
Step (e) $10V_1^2 + 7V_1V_2 + V_2^2 = (2V_1 + V_2)(5V_2 + V_2)$

EXAMPLE 8-22   Factor $16P_1P_2 + P_2^2 + 48P_1^2$.

SOLUTION

Step (a) $48P_1^2 + 16P_1P_2 + P_2^2$
Step (b)   48
Step (c)   1, 48; 2, 24; 3, 16; 4, 12; 6, 8
Step (d)   The only pair whose sum is 16 is 4, 12.
Step (e) $16P_1P_2 + P_2^2 + 48P_1^2 = (4P_1 + P_2)(12P_1 + P_2)$

PROBLEMS

Factor:

8-93. $6V_1^2 + 5V_1V_2 + V_2^2$
8-94. $4V_1^2 + 5V_1V_2 + V_2^2$
8-95. $8I^2 + 6IJ + J^2$
8-96. $9Z_1^2 + 10Z_1Z_2 + Z_2^2$
8-97. $24P^2 + 14PQ + Q^2$
8-98. $25Y_1^2 + 10Y_1Y_2 + Y_2^2$

8-99. $30Q^2 + 13QR + R^2$
8-100. $27X_L^2 + 12X_LX_C + X_C^2$
8-101. $36V_s^2 + 15V_sV_p + V_p^2$
8-102. $40A^2 + 13AB + B^2$
8-103. $60D_1^2 + 17D_1D_2 + D_2^2$
8-104. $75R_1^2 + 20R_1R_2 + R_2^2$

The Negative Middle Term

When the middle term of a trinomial is negative and the last term positive, the factors will both contain minus signs:

$$4X^2 - 5XY + Y^2 = (4X - Y)(X - Y)$$

This occurs because the last term of the trinomial (Y^2) is formed by multiplying two negative numbers $(-Y$ and $-Y)$, whereas the middle term is formed by the sum of two negative terms: (a) $(4X)(-Y) = -4XY$, and (b) $(-Y)(X) = -XY$. As far as the coefficients are concerned, however, the middle term is still the sum of the factors of the first term:

$$5 = 4 + 1$$

Thus, we can use the same procedure on the following problems that we have been using.

EXAMPLE 8-23   Factor $6B^2 - 5B + 1$.

SOLUTION

Step (a) $6B^2 - 5B + 1$
Step (b)   6
Step (c)   6, 1; 2, 3
Step (d) $2 + 3 = 5$
Step (e) $6B^2 - 5B + 1 = (2B - 1)(3B - 1)$

Note that both factors were assigned minus signs, since the sign of the middle term of the trinomial is negative.

EXAMPLE 8-24   Factor $-6X_1 X_2 + 8X_1^2 + X_2^2$

SOLUTION

Step (a) $8X_1^2 - 6X_1 X_2 + X_2^2$
Step (b)   8
Step (c)   1, 8; 2, 4
Step (d) $2 + 4 = 6$
Step (e) $8X_1^2 - 6X_1 X_2 + X_2^2 = (2X_1 - X_2)(4X_1 - X_2)$

Again, the factors both have minus signs because the middle term's coefficient is negative, -6.

PROBLEMS

Factor:

8-105. $6X^2 - 5X + 1$
8-106. $10X^2 - 11X + 1$
8-107. $12L_1^2 - 7L_1 L_2 + L_2^2$
8-108. $20M_1^2 - 12M_1 M_2 + M_2^2$
8-109. $18P^2 - 19PQ + Q^2$

8-110. $25P_1^2 - 26P_1 P_2 + P_2^2$
8-111. $30Y_1^2 - 13Y_1 Y_2 + Y_2^2$
8-112. $48Q_1^2 + Q_2^2 - 19Q_1 Q_2$
8-113. $-15M_A M_B + M_B^2 + 50M_A^2$
8-114. $-16R_1 R_2 + R_2^2 + 28R_1^2$

The Negative Third Term

Let us now examine a trinomial such as $10X^2 + 3XY - Y^2$. Note that the last term is negative. Its factors are $5X - Y$ and $2X + Y$. Observe how the middle term and the last term are formed:

$$
\begin{array}{cc}
\begin{array}{r}
5X - Y \\
2X + Y \\
\hline
5XY - Y^2 \\
(5 \cdot 2)X^2 - 2XY \\
\hline
10X^2 + 3XY - Y^2
\end{array}
&
\begin{array}{r}
aX - Y \\
bX + Y \\
\hline
aXY - Y^2 \\
abX^2 - \\
\hline
abX^2 + (a - b)XY - Y^2
\end{array}
\end{array}
$$

Examine this illustration carefully to determine how each coefficient is formed. As in the preceding subsection, the coefficient of the first term is the product of the coefficients of the first terms of the factors. Note that the second term now contains the difference between the coefficients. Thus, when the third term of a trinomial is negative, the middle term contains the difference between the coefficients. Our procedure for factoring trinomials now becomes:

(a) Arrange the terms in descending powers.
(b) Select the coefficient of the first term.
(c) Find all its possible factors.
(d) If the third term of the trinomial is negative, select those factors whose difference is the middle term.
(e) If the third term of the trinomial is positive, select those factors whose sum is the coefficient of the middle term.
(f) Construct the product using these factors as coefficients of the first terms.

EXAMPLE 8-25   Factor $16A^2 + 6AB - B^2$.

SOLUTION

Step (a) $16A^2 + 6AB - B^2$
Step (b)   16
Step (c)   1, 16; 2, 8; 4, 4
Step (d) $8 - 2 = 6$
Step (e)   Not applicable.
Step (f) $16A^2 + 6AB - B^2 = (8A - B)(2A + B)$

We know that one of the factors is a sum and the other a difference, but which is which? Keep in mind how the middle term of the trinomial is formed. In this case, it is the difference of $(8A)(B)$ and $(2A)(B)$. For this difference to be positive, the $8AB$ must be positive and the $2AB$ negative. This can only happen if the operations are as shown.

EXAMPLE 8-26   Factor $-N_2^2 + 20N_1^2 + 8N_1N_2$.

SOLUTION

Step (a) $20N_1^2 + 8N_1N_2 - N_2^2$
Step (b)   20
Step (c)   1, 20; 10, 2; 4, 5
Step (d) $10 - 2 = 8$
Step (e)   Not applicable.
Step (f) $(10N_1 - N_2)(2N_1 + N_2)$

If the sign of both the second and the third terms of the trinomial is negative, the procedure is the same: One of the factors is a sum, the other a difference. The sum and difference signs are assigned to make the second term of the trinomial negative.

EXAMPLE 8-27   Factor $20R^2 - RS - S^2$.

SOLUTION

Step (a) $20R^2 - RS - S^2$
Step (b)   20
Step (c)   1, 20; 2, 10; 4, 5

Step (d) $5 - 4 = 1$
Step (e) Not applicable.
Step (f) $(4R - S)(5R + S)$

The minus sign was assigned to make the product $5RS$ negative and $4RS$ positive. That way, $-5RS + 4RS = -RS$ and the middle term is obtained.

EXAMPLE 8-28 Factor $48X^2 - Y^2 - 13XY$.

SOLUTION

Step (a) $48X^2 - 13XY - Y^2$
Step (b) 48
Step (c) 1, 48; 2, 24; 3, 16; 4, 12; 6, 8
Step (d) $16 - 3 = 13$
Step (e) Not applicable.
Step (f) $(16X + Y)(3X - Y)$

PROBLEMS

Factor:

8-115. $6W^2 + WX - X^2$

8-116. $8M^2 + 7MN - N^2$

8-117. $10R_1^2 + 9R_1R_2 - R_2^2$

8-118. $16Z_1^2 + 15Z_1Z_2 - Z_2^2$

8-119. $24L_1^2 + 10L_1L_2 - L_2^2$

8-120. $24Q_1^2 + 23Q_1Q_2 - Q_2^2$

8-121. $30L_1^2 + L_1L_2 - L_2^2$

8-122. $-L^2 + 36M^2 + 9LM$

8-123. $4Z_1^2 - 3Z_1Z_2 - Z_2^2$

8-124. $6L^2 - 5LM - M^2$

8-125. $-P^2 - PQ + 20Q^2$

8-126. $G^2 - 48F^2 + 13FG$

8-127. $16M^2N^2 + 6MNP - P^2$

8-128. $18V_1^2V_2^4 - V_3^4 - 7V_1V_2^2V_3^2$

8-129. $-10C_1^2C_2^3C_3 + 24C_1^4C_2^6 + C_3^2$

8-130. $12C_1^3 - 2C_1^2C_2 - 2C_1C_2^2$

8-131. $32R_1^3R_2^2 - 8R_1^2R_2^3 - 4R_1R_2^4$

8-132. $100N_1^4N_2 - 5N_1^3N_2^2 - 5N_1^2N_2^3$

Summary of Signs and Operations

Table 8-1 summarizes what we have studied concerning the signs of the factors and how the middle term is obtained for each different sign combination. If the first sign of the trinomial is not positive, a (-1) can be factored out of it and the foregoing methods applied. The table is evident if we multiply out each combination of factors:

1. $(aX + bY)(cX + dY) = (ac)X^2 + (bc + ad)XY + (bd)Y^2$
2. $(aX + bY)(cX - dY) = (ac)X^2 + (bc - ad)XY - (bd)Y^2$
3. $(aX - bY)(cX - dY) = (ac)X^2 - (bc + ad)XY + (bd)Y^2$
4. $(aX - bY)(cX + dY) = (ac)X^2 - (bc - ad)XY - (bd)Y^2$

TABLE 8-1 Factor operations.

1st	2nd	3rd	Signs of Factors	Middle Term Found by:
+	+	+	+, +	Addition
+	+	−	+, −	Subtraction
+	−	+	−, −	Addition
+	−	−	+, −	Subtraction

Above the "1st 2nd 3rd" header spans: TRINOMIAL TERMS

Trinomials with Three Coefficients

Consider the trinomial $6X^2 + 13XY + 6Y^2$. Its factors are $(3X + 2Y)$ and $(2X + 3Y)$. Let us multiply these factors and study closely how each of the coefficients is determined.

$$
\begin{array}{r}
3X + 2Y \\
2X + 3Y \\
\hline
9XY + 6Y^2 \\
6X^2 + 4XY \\
\hline
6X^2 + 13XY + 6Y^2
\end{array}
$$

Note that the coefficient of the X^2 term is found by multiplying the coefficients of the X terms. Similarly, the coefficient of the Y^2 term is found by multiplying the coefficients of the Y terms. Thus, our process of factoring this trinomial is as follows:

(a) Arrange the terms in descending powers.
(b) Select the coefficient of the first term.
(c) Find all its possible factors.
(d) Select the coefficient of the last term.
(e) Find all its possible factors.
(f) By trial and error, try all the combinations of (c) for the coefficients of X and those of (e) for the coefficients of Y until a combination is found that yields the middle coefficient of the trinomial.

EXAMPLE 8-29 Factor $6Y^2 + 23XY + 20Y^2$.

SOLUTION

Step (a) $6Y^2 + 23XY + 20Y^2$
Step (b) 6
Step (c) 1, 6; 2, 3
Step (d) 20
Step (e) 1, 20; 2, 10; 4, 5
Step (f) We shall try 1, 6 for the X coefficient, and all other combinations for the Y term:

1. $(1X + 1Y)(6X + 20Y)$ Middle term $26XY$
2. $(1X + 2Y)(6X + 10Y)$ Middle term $22XY$
3. $(1X + 4Y)(6X + 5Y)$ Middle term $29XY$
4. $(1X + 20Y)(6X + 1Y)$ Middle term $121XY$
5. $(1X + 10Y)(6X + 2Y)$ Middle term $62XY$
6. $(1X + 5Y)(6X + 4Y)$ Middle term $34XY$

Note that we had to try each combination in both positions. For example, we tried 1, 20 and 20, 1. Next, move to the 2, 3 combination of the X coefficient.

7. $(2X + 1Y)(3X + 20Y)$ Middle term $43XY$
8. $(2X + 2Y)(3X + 10Y)$ Middle term $26XY$
9. $(2X + 4Y)(3X + 5Y)$ Middle term $22XY$
10. $(2X + 20Y)(3X + 1Y)$ Middle term $62XY$
11. $(2X + 10Y)(3X + 2Y)$ Middle term $34XY$
12. $(2X + 5Y)(3X + 4Y)$ Middle term $23XY$

Finally, on the very last trial, we found the correct answer:

$$6X^2 + 23XY + 20Y^2 = (2X + 5Y)(3X + 4Y)$$

In Example 8-29 we could have skipped over trials 1, 2, 5, 6, 8, 9, 10, and 11 if we had been more observant. Note that the trinomial cannot be evenly divided by any prime number: 2, 3, 5, 7, and so forth. Therefore, each of the binomial factors must be incapable of being divided by any prime number. Thus, had we been more observant, we would only have had to try combinations 3, 4, 7, and 12. Furthermore, it seems that most problems work out that the factors are near to each other. Thus, it is a good strategy to try factors of 4, 5 before 2, 10 or 1, 20. Had we done this in the preceding answer, we would have found the answer after one or two trials.

EXAMPLE 8-30   Factor $14X^2 + 41XY + 15Y^2$.

SOLUTION

Step (a) $14X^2 + 41XY + 15Y^2$
Note that it cannot be reduced.
Step (b)   14
Step (c)   1, 14; 2, 7
Step (d)   15
Step (e)   1, 15; 3, 5
Step (f)   We shall try the closer factors first: 2, 7 and 3, 5.

$$(2X + 3Y)(7X + 5Y) \quad \text{Middle term } 31XY$$
$$(2X + 5Y)(7X + 3Y) \quad \text{Middle term } 41XY$$

We have the correct answer on the second trial. Thus,

$$14X^2 + 41XY + 15Y^2 = (2X + 5Y)(7X + 3Y)$$

Using the principles illustrated in Table 8-1, we can handle any combination of signs.

EXAMPLE 8-31   Factor $10A^2 - 3AB - 18B^2$.

SOLUTION

Step (a) $10A^2 - 3AB - 18B^2$
Step (b)   10
Step (c)   1, 10; 2, 5
Step (d)   18
Step (e)   1, 18; 2, 9; 6, 3
Step (f)   We shall try the 2, 5 and 6, 3 combinations first. Note that the middle term is obtained by subtraction.

$$(2A \quad 6B)(5A \quad 3B) \quad \text{No need to try. It is factorable.}$$
$$(2A \quad 3B)(5A \quad 6B) \quad \text{Middle term } 3AB$$

We know this is the right combination and that one has to be added and the other subtracted. Looking at the middle term of the trinomial, we see that it must be negative: $-3AB$. Therefore, we assign the signs to make the middle term negative. The solution is $10A^2 - 3AB - 18B^2 = (2A - 3B)(5A + 6B)$.

EXAMPLE 8-32   Factor $24V_2^2 + 11V_1^2 - 50V_1V_2$.

SOLUTION

Step (a) $11V_1^2 - 50V_1V_2 + 24V_2^2$
Step (b)   11
Step (c)   1, 11
Step (d)   24
Step (e)   1, 24; 2, 12; 3, 8; 4, 6
Step (f) $(V_1 - 4V_2)(11V_1 - 6V_2)$ Middle term $50V_1V_2$
           Correct on the first trial.

PROBLEMS

Factor:

8-133. $28N^2 + 29NP + 6P^2$

8-134. $12Z_1^2 + 32Z_1Z_2 + 21Z_2^2$

8-135. $54V_1^2 + 75V_1V_2 + 25V_2^2$

8-136. $12P^2 + 23PQ + 10Q^2$

8-137. $15W^2 + 43WN + 30N^2$

8-138. $28W^2 + 57WN + 14N^2$

8-139. $11V_1^2 + 23V_1V_2 + 12V_2^2$

8-140. $15I_1^2 + 92I_1I_2 + 12I_2^2$

8-141. $6Z^2 - 5ZR - 25R^2$

8-142. $12R_1^2 - 16R_1R_2 + 5R_2^2$

8-143. $12L^2 + 4LM - 21M^2$

8-144. $18R^2 - 37RS + 15S^2$

8-145. $63T_1 + 31T_1T_2 - 10T_2^2$

8-146. $12S_1^2 - 55S_1S_2 + 28S_2^2$

8-147. $13L^2 - 48LM - 16M^2$

8-148. $15f_0^2 + 71f_0f_s - 20f_s^2$

8-149. $27A_3^2 - 66A_3A_4 + 35A_4^2$

8-150. $30R_1^2 - 121R_1R_2 + 56R_2^2$

8-8   SUMMARY

Table 8-2 lists the general factoring models we have discussed in this chapter. Although four models are listed, the perfect square and difference of squares are special cases of the general trinomial methods. In any event, the common factor should always be tried first, then the perfect square and difference of squares, and finally, the general trinomial solution. Much mathematical manipulation can be avoided if this approach is used.

TABLE 8-2   Factoring models.

Section	Name	Models
8.4	Common factor	$aX + aY = a(X + Y)$
8.5	Perfect square	$X^2 + 2XY + Y^2 = (X + Y)^2$
		$X^2 - 2XY + Y^2 = (X - Y)^2$
8.6	Difference of squares	$X^2 - Y^2 = (X + Y)(X - Y)$
8.7	General trinomial	$abX^2 + (a + b)XY + Y^2 = (aX + Y)(bX + Y)$
		$abX^2 - (a + b)XY + Y^2 = (aX - Y)(bX - Y)$
		$abX^2 + (a - b)XY - Y^2 = (aX - Y)(bX + Y)$
		$acX^2 + (ad + bc)XY + bdY^2 = (aX + bY)(cX + dY)$

EXAMPLE 8-33   Factor $8AB^2 + 2A^3 - 8A^2B$.

SOLUTION   First try the common factor method:

$$8AB^2 + 2A^3 - 8A^2B = 2A(4B^2 + A^2 - 4AB)$$

The term within the parentheses can be rearranged and treated as a perfect square:

$$8AB^2 + 2A^3 - 8A^2B = 2A(A^2 - 4AB + 4B^2) = 2A(A - 2B)^2$$

EXAMPLE 8-34 Factor $2M^3 - 2MN^2 + M^2N - N^3$.

SOLUTION The problem may be unfactorable. First, however, let us factor $2M$ out of the first two terms and N out of the last two:

$$2M^3 - 2MN^2 + M^2N - N^3 = 2M(M^2 - N^2) + N(M^2 - N^2)$$

Next, factor $(M^2 - N^2)$ out of $2M(M^2 - N^2)$ and out of $N(M^2 - N^2)$:

$$= (M^2 - N^2)(2M + N)$$

But $M^2 - N^2$ is the difference of squares:

$$= (M + N)(M - N)(2M + N)$$

EXAMPLE 8-35 Factor $2A^3C + 4A^2BC - 2AC^3 + 2AB^2C$.

SOLUTION $2AC$ is common:

$$= 2AC(A^2 + 2AB - C^2 + B^2)$$

Rearrange:

$$= 2AC(A^2 + 2AB + B^2 - C^2)$$

Factor $A^2 + 2AB + B^2$:

$$= 2AC[(A + B)^2 - C^2]$$

Note that the expression within the brackets is a difference of squares:

$$= 2AC(A + B + C)(A + B - C)$$

PROBLEMS

Factor:

8-151. $2B^3C + 2B^2C^2$

8-152. $6R_1^2R_2^2 + 9R_1R_2^3 + 3R_1R_2^2R_3$

8-153. $8V_1^6V_2 + 4V_1^2V_2^2V_3^3$

8-154. $35ST^3U^2 + 35ST^2U^3 + 140STU^4$

8-155. $28A^3B + 28A^2B^2 + 7AB^3$

8-156. $24Z_1^2Z_2^3 + 4Z_1^3Z_2^2 + 36Z_1Z_2^4$

8-157. $3R_1^6R_2R_3 + 6R_1^4R_2^2R_3 + 3R_1^2R_2^3R_3$

8-158. $-84A_1^3A_2^2 + 63A_1A_2^3 + 28A_1^5A_2$

8-159. $36A_1^2A_2^4 + 25A_3^2 - 60A_1A_2^2A_3$

8-160. $3B^5 - 12A^4BC^2 - 12A^2B^3C$

8-161. $M^3 - MN^2$

8-162. $4N_1^3 - N_1N_2^2$

8-163. $FG_1^4 - 2F^3$

8-164. $12H^4J^5 - 3H^2JK^2$

8-165. $24B^3 + 4BC^2 + 20B^2C$

8-166. $7Q_1^4 - 35Q_1^3Q_2 + 42Q_1^2Q_2^2$

8-167. $12R_1^3R_2 + 10R_1^2R_2^2 + 2R_1R_2^3$

8-168. $6V_s^5V_0^2 + 9V_s^3V_0^3 + 3V_sV_0^4$

8-169. $12M^5T^6 + 9MT^2U^2 - 21M^3T^4U$

8-170. $300C_1^4C_2 + 10C_1^2C_2C_4^2 - 110C_1^3C_2C_4$

8-171. $12Z_1^3 - 22Z_1^2Z_2 + 6Z_1Z_2^2$

8-172. $70R_1^3R_2 + 21R_1^2R_2^2 - 28R_1R_2^3$

8-173. $-9R_1^3R_L^2 + 54R_1^4R_L - 45R_1^2R_L^3$

8-174. $54M_1^4M_2 + 9M_1^3M_2^2 - 30M_1^2M_2^3$

The following are for those who enjoy a challenge. Factor:

8-175. $378E_1^3E_2 - 63E_1^2E_2^2 - 588E_1E_2^3$

8-176. $128N_1^3N_2 + 592N_1^2N_2P_1 - 240N_1N_2P_1^2$

8-177. $L_s^3 - L_sL_T^2 - L_pL_s^2 + L_pL_T^2$

8-178. $A^4 + 2A^3B + 2A^2B^2 + 2AB^3 + B^4$

8-179. $A^3 + 2A^2B + 2AB^2 + B^3$

8-180. $3N^4 - 3N^3P - 3N^2P^2 + 3NP^3$

8-181. $4A^4C^5 - 16A^2B^2C$

8-182. $A^4 - 2A^2B^2 + B^4$

8-183. $2V_1^4 - 2V_1^3V_2 - 2V_1^2V_2^2 + 2V_1V_2^3$

8-184. $4R_P^3 + 2R_P^2R_Q - 7R_PR_Q^2 - 2R_Q^3$

9

ALGEBRAIC FRACTIONS

In Chapter 4 we studied the basic principles of fractions and in Chapter 8 we examined the process of factoring. In this chapter we apply this hitherto useless skill of factoring to algebraic fractions: addition, subtraction, multiplication, and division. By the time we finish this chapter, we will be able to solve equations expressed as fractions. In fact, if you peek into Chapter 10, you will see that these fractional equations are used to solve parallel resistor problems. Thus, although we have spent a long time circumnavigating the barn, we shall finally get to ride the horse.

9-1 REVIEW OF FRACTIONS

We learned in Chapter 4 the following definition of a fraction:

> *A fraction is a division process whereby the numerator (the number on top) is divided by the denominator (the number "down under").*

In this chapter we deal with algebraic fractions: fractions with literals in either the numerator, denominator, or both. All the principles and procedures that we discussed about numerical fractions apply to algebraic fractions. Thus, the following is fundamental to our dealing with any type of fraction:

> *The value of a fraction remains unchanged when the numerator and the denominator are both multiplied by or both divided by the same constant or variable.*

Notice that fractional rules allow only multiplication and division, not addition, subtraction, or exponentiation. It is this principle that gives students the most difficulty in studying fractions. However, remember that a fraction is fundamentally a division process and therefore can only be modified in form by multiplication or division.

9-2 REDUCING FRACTIONS

Let us review, for a minute, the reduction techniques used in arithmetic fractions. Consider the fraction

$$\frac{30}{36}$$

We can first factor its numerator, then its denominator, then divide out identical factors:

$$\frac{30}{36} = \frac{2 \cdot 3 \cdot 5}{2 \cdot 2 \cdot 3 \cdot 3} = \frac{\cancel{2} \cdot 3 \cdot 5}{\cancel{2} \cdot 2 \cdot 3 \cdot 3} = \frac{\cancel{2} \cdot \cancel{3} \cdot 5}{\cancel{2} \cdot 2 \cdot \cancel{3} \cdot 3}$$

In the first step, we crossed out the 2's, recognizing that

$$\frac{2 \cdot 3 \cdot 5}{2 \cdot 2 \cdot 3 \cdot 3} = \frac{2}{2} \cdot \frac{3 \cdot 5}{2 \cdot 3 \cdot 3} = 1 \cdot \frac{3 \cdot 5}{2 \cdot 3 \cdot 3}$$

Note that the process of crossing out the 2 is one of dividing 2 by 2, yielding 1. In a similar manner, we crossed out the 3's, recognizing that 3 divided by 3 is 1.

But what is reduction, then? It is a process of finding those factors that can be divided by themselves to obtain 1. This also applies in algebra. We can factor the numerator, then the denominator, then divide out identical factors, recognizing that a term divided by that same term is 1[1]. Consider the following fraction:

$$\frac{2A^2C}{AC}$$

We can use the same reduction technique for this algebraic fraction that we have used for arithmetic fractions. First, factor:

$$\frac{2 \cdot A \cdot A \cdot C}{A \cdot C}$$

Next, divide out identical terms:

$$\frac{2 \cdot \cancel{A} \cdot A \cdot \cancel{C}}{\cancel{A} \cdot \cancel{C}} = \frac{2 \cdot A}{1} = 2A$$

EXAMPLE 9-1 Reduce:

$$\frac{24R_1^2R_2^3R_3}{6R_1R_2^4R_3}$$

SOLUTION We can first factor, then divide out like terms:

$$\frac{\cancel{2} \cdot 2 \cdot 2 \cdot \cancel{3} \cdot \cancel{R_1} \cdot R_1 \cdot \cancel{R_2} \cdot \cancel{R_2} \cdot \cancel{R_2} \cdot \cancel{R_3}}{\cancel{2} \cdot \cancel{3} \cdot \cancel{R_1} \cdot \cancel{R_2} \cdot \cancel{R_2} \cdot \cancel{R_2} \cdot R_2 \cdot \cancel{R_3}} = \frac{4R_1}{R_2}$$

This can be recognized as a division process of a monomial by a monomial. However, by looking at it as a factoring problem, we can apply this technique to more complex problems.

EXAMPLE 9-2 Reduce:

$$\frac{4X^2 - 4Y^2}{6X^2 + 12XY + 6Y^2}$$

SOLUTION The problem cannot be reduced directly, but must be factored:

[1]   This statement is true for all values of the term except zero.

$$\frac{4X^2 - 4Y^2}{6X^2 + 12XY + 6Y^2} = \frac{2 \cdot 2(X+Y)(X-Y)}{2 \cdot 3(X+Y)(X+Y)} = \frac{2(X-Y)}{3(X+Y)}$$

Note that the terms 2 and $X + Y$ appeared in both the numerator and denominator and therefore can be divided out.

EXAMPLE 9-3 Reduce:

$$\frac{6M^2P^2Q - 3MNP^2Q - 3N^2P^2Q}{12M^2P^3Q + 18MNP^3Q + 6N^2P^3Q}$$

SOLUTION The term $3P^2Q$ can be factored out of the numerator and $6P^3Q$ can be factored out of the denominator.

$$\frac{3P^2Q(2M^2 - MN - N^2)}{6P^3Q(2M^2 + 3MN + N^2)} = \frac{3P^2Q(2M+N)(M-N)}{6P^3Q(2M+N)(M+N)} = \frac{M-N}{2P(M+N)}$$

EXAMPLE 9-4 Reduce:

$$\frac{16L_1^3 - 4L_1L_2^2}{4L_1^3L_2 + 6L_1^2L_2^2 + 2L_1L_2^3}$$

SOLUTION

$$= \frac{4L_1(4L_1^2 - L_2^2)}{2L_1L_2(2L_1^2 + 3L_1L_2 + L_2^2)}$$
$$= \frac{4L_1(2L_1 - L_2)(2L_1 + L_2)}{2L_1L_2(2L_1 + L_2)(L_1 + L_2)}$$
$$= \frac{2(2L_1 - L_2)}{L_2(L_1 + L_2)}$$

PROBLEMS

Reduce:

9-1. $\dfrac{3A^2}{6AB^2}$

9-2. $\dfrac{12Z_1^2Z_2^2}{4Z_1Z_2^3Z_3}$

9-3. $\dfrac{32N_1^2N_2^4}{26N_1N_2^2N_3}$

9-4. $\dfrac{132R_1R_2^2R_4}{24R_1R_3R_4^2}$

9-5. $\dfrac{3R_1^3 + 3R_1^2R_2}{R_1R_2}$

9-6. $\dfrac{3L_1^2L_2 + 3L_1L_2^2}{4L_1^2L_2 - 4L_1L_2^2}$

9-7. $\dfrac{12P^3Q^3 + 18P^2Q^4}{9P^4Q^2 - 6P^3Q^3}$

9-8. $\dfrac{12V_1^3V_2^3 - 12V_1^2V_2^2V_3}{4V_1^4V_2^3V_3 + 4V_1^3V_2^4V_3^3}$

9-9. $\dfrac{48I_1^5I_2^2 - 48I_1^4I_2^3}{108I_1^2I_2^4 + 36I_1I_2^5}$

9-10. $\dfrac{3P_1^3P_2^2P_4^2 - 4P_1P_2^3P_3P_4^2}{36P_1^2P_2^4P_4 - 12P_1P_3^3P_4^3}$

9-11. $\dfrac{A^2 - B^2}{A^2 + 2AB + B^2}$

9-12. $\dfrac{4P_1^2 - P_2^2}{4P_1^2 - 4P_1P_2 + P_2^2}$

9-13. $\dfrac{X^4 - X^2Y^2}{X^4 + 2X^3Y + X^2Y^2}$

9-14. $\dfrac{M^3N - MN^3}{M^4N^2 - M^2N^4}$

9-15. $\dfrac{3R_1^3 + 6R_1^2R_2 + 3R_1R_2^2}{2R_1^3R_2^3 + 4R_1^2R_2^4 + 2R_1R_2^5}$

9-16. $\dfrac{12M^3N - 10M^2N^2 - 12MN^3}{54M^4N + 72M^3N^2 + 24M^2N^3}$

In our study of arithmetic fractions in Chapter 4, we found the following:

To multiply fractions, multiply the numerators to obtain the numerator of the result and multiply the denominators to obtain the denominator of the result.

The same rule applies to algebraic fractions.

EXAMPLE 9-6 Multiply:

$$\frac{2A}{C} \cdot \frac{4AD}{E}$$

SOLUTION The numerator of the result is found by multiplying the numerators of the operands:

$$(2A)(4AD) = 8A^2D$$

The denominator of the result is found by multiplying the denominators together:

$$(C)(E) = CE$$

Therefore,

$$\frac{2A}{C} \cdot \frac{4AD}{E} = \frac{8A^2D}{CE}$$

EXAMPLE 9-6 Multiply:

$$\frac{R_1 + R_2}{S^2} \cdot \frac{R_1 - R_2}{R_1R_2}$$

SOLUTION Multiply the numerators:

$$(R_1 + R_2)(R_1 - R_2) = R_1^2 - R_2^2$$

Multiply the denominators:

$$(S^2)(R_1R_2) = S^2R_1R_2$$

Therefore,

$$\frac{R_1 + R_2}{S^2} \cdot \frac{R_1 - R_2}{R_1R_2} = \frac{R_1^2 - R_2^2}{R_1R_2S^2}$$

Multiplying can be greatly simplified if the numerator and denominator of each fraction are first factored. In this manner, dividing out can occur before the multiplication.

EXAMPLE 9-7 Multiply and reduce where possible:

$$\frac{9V_1^2 + 3V_1}{6V_1^2 + 5V_1 + 1} \cdot \frac{2V_1^2 + V_1}{V_1 + 1}$$

SOLUTION Each expression within the problem can first be factored:

$$= \frac{3V_1(3V_1+1)}{(3V_1+1)(2V_1+1)} \cdot \frac{V_1(2V_1+1)}{V_1+1}$$

$$= \frac{3V_1^2}{V_1+1}$$

EXAMPLE 9-8 Multiply and reduce where possible:

$$\frac{X^2-9}{3X^2+10X+3} \cdot \frac{X^2+6X+9}{3X^2+9X} \cdot \frac{4X^2+12X}{3X^2-8X-3}$$

SOLUTION Factoring yields

$$\frac{(X+3)(X-3)}{(3X+1)(X+3)} \cdot \frac{(X+3)(X+3)}{3X(X+3)} \cdot \frac{4X(X+3)}{(3X+1)(X-3)} = \frac{4(X+3)^2}{3(3X+1)^2}$$

Division

Division requires that the divisor (the fraction following the division sign) be inverted. The process then becomes one of multiplication.

EXAMPLE 9-9 Divide:

$$\frac{14B^2N}{6L} \div \frac{7QN}{3LB}$$

SOLUTION Invert the divisor and multiply:

$$\frac{14B^2N}{6L} \cdot \frac{3LB}{7QN} = \frac{B^3}{Q}$$

EXAMPLE 9-10 Divide:

$$\frac{A^2+2A+1}{3A^2+A} \div \frac{A^2-1}{3A^2-2A-1}$$

SOLUTION Invert, factor, and multiply:

$$= \frac{A^2+2A+1}{3A^2+A} \cdot \frac{3A^2-2A-1}{A^2-1}$$

$$= \frac{(A+1)(A+1)}{A(3A+1)} \cdot \frac{(3A+1)(A-1)}{(A+1)(A-1)}$$

$$= \frac{A+1}{A}$$

PROBLEMS

Multiply (do not attempt to reduce):

9-17. $\dfrac{3}{A} \cdot \dfrac{4}{B}$

9-18. $\dfrac{4A}{C} \cdot \dfrac{2A}{3BC}$

9-19. $\dfrac{R_1R_2}{R_3} \cdot \dfrac{R_1^2}{R_3}$

9-20. $\dfrac{2P_1P_2}{3P_3P_4} \cdot \dfrac{4P_1^2P_2}{5P_3^2P_4}$

9-21. $\dfrac{A+B}{2} \cdot \dfrac{3}{A-B}$

9-22. $\dfrac{4R_1+R_2}{5} \cdot \dfrac{R_1}{3R_2}$

9-23. $\dfrac{V_0 + V_s}{2V_0} \cdot \dfrac{V_0 - V_s}{3V_0}$

9-24. $\dfrac{T_L - 3}{4T_P} \cdot \dfrac{T_L + 3}{T_L T_P}$

9-25. $\dfrac{3I_1 - 2I_2}{2I_1} \cdot \dfrac{4I_1 + 3I_2}{I_1 + I_2}$

9-26. $\dfrac{4A^2 B + B^3}{2AB} \cdot \dfrac{3AB^2 +}{A + .}$

9-27. $\dfrac{L_1 + M}{L_1 - M} \cdot \dfrac{2L_1}{L_1 + M}$

9-28. $\dfrac{3P + Q}{2P + 3Q} \cdot \dfrac{P + Q}{P - Q}$

9-29. $\dfrac{4.3R_1 + 6.2R_2}{R_1} \cdot \dfrac{2.3R_1 + 6.1R_2}{R_1 + R_2}$

9-30. $\dfrac{A + C}{A + D} \cdot \dfrac{B + C}{B + D}$

9-31. $\dfrac{3A + B + C}{A + C} \cdot \dfrac{2A + 3B}{B}$

9-32. $\dfrac{4A + B}{C} \cdot \dfrac{2A + 3B + 4C}{A + B + 2C}$

Multiply and reduce where possible:

9-33. $\dfrac{3}{A} \cdot \dfrac{7}{6B}$

9-34. $\dfrac{4R_1 R_2}{6R_3} \cdot \dfrac{2R_1 R_3}{3R_2 R_3}$

9-35. $\dfrac{3L_1 L_2}{2L_1 L_3} \cdot \dfrac{6L_1 L_2 L_3}{4L_1^2 L_3}$

9-36. $\dfrac{V_1 + V_2}{3V_3} \cdot \dfrac{2V_1 V_3}{3V_1 + 3V_2}$

9-37. $\dfrac{V_1 - V_3}{V_1 - V_2} \cdot \dfrac{V_1^2 - V_2^2}{3V_1 + 3V_2}$

9-38. $\dfrac{3V_1 - 9}{V_1^2 - V_1 - 6} \cdot \dfrac{2V_1 + 2}{9}$

9-39. $\dfrac{4L + 4}{L - 6} \cdot \dfrac{3L^2 - 18L}{2L^2 + 5L + 3}$

9-40. $\dfrac{5L + 5M}{3L + M} \cdot \dfrac{12L^2 + 10LM + 2M^2}{10L + 5M}$

9-41. $\dfrac{R^2 - R - 6}{R^2 - 4} \cdot \dfrac{R^2 - 5R + 6}{3R - 9}$

9-42. $\dfrac{2P^2 - 5P + 2}{2P^2 - 3P - 2} \cdot \dfrac{4P^2 - 1}{2P^2 + P}$

9-43. $\dfrac{3I}{4R} \cdot \dfrac{R^2 + R}{R^2} \cdot \dfrac{3IR^2}{4I^2}$

9-44. $\dfrac{16V}{3I} \cdot \dfrac{12I}{3V} \cdot \dfrac{7I^2}{6V^2} \cdot \dfrac{6I}{16IV}$

9-45. $\dfrac{Q - 5}{Q^2 - 1} \cdot \dfrac{Q + 1}{Q^2 - 6Q + 5} \cdot \dfrac{Q^2 - 4Q - 5}{Q^2 - Q}$

9-46. $\dfrac{X_L^2 - X_c^2}{X_L^2 - 2X_L X_c + X_c^2} \cdot \dfrac{X_L^2 - X_L X_c}{X_c^2 - X_L X_c} \cdot \dfrac{X_L^2 X_c^2 + X_L X_c^2}{X_L + X_c}$

Perform the indicated operation and reduce, where possible:

9-47. $\dfrac{3}{4} \div \dfrac{7}{8}$

9-48. $\dfrac{25}{36} \div \dfrac{4}{9}$

9-49. $\dfrac{33}{48} \div \dfrac{11}{16}$

9-50. $\dfrac{7}{16} \div \dfrac{3}{8}$

9-51. $\dfrac{3AB}{6C} \div \dfrac{7B}{3C}$

9-52. $\dfrac{3R_1 R_2}{4V_3} \div \dfrac{2R_1 R_3}{10R_2 R_3}$

9-53. $\dfrac{3LM}{2} \div \dfrac{5L^2 M^2}{2T}$

9-54. $\dfrac{6MN^2}{3NP} \div \dfrac{4M^2}{3PQ}$

9-55. $\dfrac{L + M}{P} \div \dfrac{3L + 3M}{P^2}$

9-56. $\dfrac{R_1 + 2R_2}{R_3} \div \dfrac{R_1 R_3 + 2R_2 R_3}{R_3^2 + R_3}$

9-57. $\dfrac{L + M}{L^2 - M^2} \div \dfrac{L^2 + 2LM + M^2}{3L + 3M}$

9-58. $\dfrac{NA^2 - P}{NP} \div \dfrac{N^2 A^4 - P^2}{NA + P}$

9-59. $3\left[\dfrac{M_1 + M_2}{M_1^2 - M_2^2} \cdot \dfrac{M_1 - M_2}{3M_1 + 3M_2} \right] \div \dfrac{M_1 - M_2}{M_1 + M_2}$

9-60. $(A^2 - B^2) \div \left[\dfrac{A + B}{A^2 + 4AB + 4B^2} \cdot \dfrac{A^2 - 4B^2}{A + 2B} \right]$

9-4 LOWEST COMMON MULTIPLE

An understanding of the lowest common multiple is essential to algebraic addition of fractions. A multiple is any number containing a particular factor. For example,

12 is a multiple of 3; that is, 12 contains 3 as one of its factors. Similarly, 25, 30, and 60 are all multiples of 5. In a similar manner, $3ab$ is a multiple of a. $6A(B + C)$, $(B + C)^2$, and $(B + C)(B + D)$ are all multiples of $(B + C)$.

A common multiple is a number that contains two particular factors. For example, 30 is a common multiple of 5 and 3. That is, the number 30 contains both a 3 and a 5 as one of its factors. The numbers 45, 60, and 120 are also common multiples of 3 and 5. In a similar manner, $3A^2B^2$ is a common multiple of A and B. The expressions $4X^2Y(X + Y^2)^3(X - Y)$ and $3Y(X + Y^2)$ are both common multiples of Y and $X + Y^2$.

Although there are an infinite variety of multiples of a particular set of numbers, there is only one that is the lowest common multiple. For example, although 24, 48, and 2448 are multiples of 4 and 6, the number 12 has the distinction of being the very lowest in value that contains both 4 and 6 as one of its factors. It is, therefore, the lowest common multiple (LCM). Other examples are:

(a)   10 is the LCM of 2, 5.
(b)   12 is the LCM of 2, 3, 4.
(c)   60 is the LCM of 3, 4, 5.
(d)   8 is the LCM of 2, 4, 8.
(e) $3AB$ is the LCM of 3, A, AB.

The LCM can easily be found by factoring each of the set of values given and selecting the minimum number of factors necessary to contain the factors of each value.

EXAMPLE 9-11   Find the LCM of 25 and 35.

SOLUTION   First factor each value:

$$25 = 5 \times 5$$
$$35 = 7 \times 5$$

The solution must contain 5×5 in order to be evenly divisible by 25. So far we have

$$LCM_{25} = 5 \cdot 5$$

Furthermore, the LCM must contain 5×7 to be evenly divisible by 35. But we already have the required 5 in $5 \cdot 5$. Therefore, we merely have to multiply by a 7.

$$LCM_{25,35} = 5 \cdot 5 \cdot 7$$

Multiply this out:

$$LCM = 175$$

EXAMPLE 9-12   Find the LCM of 8, 20, and 30.

SOLUTION

$$8 = 2 \cdot 2 \cdot 2$$
$$20 = 2 \cdot 2 \cdot 5$$
$$30 = 2 \cdot 3 \cdot 5$$

$$LCM_8 = 2 \cdot 2 \cdot 2$$

For 20, we already have the 2 · 2, so we have only to multiply by 5.

$$\text{LCM}_{8,20} = 2 \cdot 2 \cdot 2 \cdot 5$$

For the 30, we already have the 2 and 5, so we must multiply by the 3.

$$\text{LCM}_{8,20,30} = 2 \cdot 2 \cdot 2 \cdot 3 \cdot 5$$

Final solution:

$$\text{LCM}_{8,20,30} = 120$$

EXAMPLE 9-13 Find the LCM of X, X^2, and Y^2.

SOLUTION

$$X = X$$
$$X^2 = X \cdot X$$
$$Y^2 = Y \cdot Y$$
$$\text{LCM of } X = X$$
$$\text{LCM of } X, X^2 = X \cdot X$$
$$\text{LCM of } X, X^2, Y^2 = X \cdot X \cdot Y \cdot Y$$
$$= X^2 Y^2$$

EXAMPLE 9-14 Find the LCM of $(a - b)(a + b)$, $(a + b)^2$, and $(a - b)(a + b)^2$.

SOLUTION

$$(a - b)(a + b) = (a - b)(a + b)$$
$$(a + b)^2 = (a + b)(a + b)$$
$$(a - b)^2(a + b) = (a - b)(a - b)(a + b)$$
$$\text{LCM of } (a - b)(a + b) = (a - b)(a + b)$$
$$\text{LCM of } (a - b)(a + b) \quad \text{and} \quad (a + b)^2 = (a - b)(a + b)(a + b)$$
$$\text{LCM of all three} = (a - b)(a - b)(a + b)(a + b)$$
$$= (a - b)^2(a + b)^2$$

PROBLEMS

Find the LCM:

9-61. (a) 3, 6 (b) 4, 6, 8 (c) 2, 4, 6
9-62. (a) 3, 7 (b) 3, 4, 9 (c) 6, 6, 2
9-63. (a) 7, 8 (b) 6, 8 (c) 4, 8
9-64. (a) 12, 15 (b) 8, 12 (c) 8, 15
9-65. (a) 6, 8, 12 (b) 7, 9, 12 (c) 11, 12, 15
9-66. (a) 15, 20 (b) 20, 24, 30 (c) 30, 33, 39
9-67. (a) 12, 15, 20, 21 (b) 20, 22, 30, 33
9-68. (a) 24, 36, 40, 44 (b) 26, 28, 30, 36
9-69. (a) 30, 60, 96, 121 (b) 48, 54, 56, 60

9-70. (a) 6, 24, 30, 45, 75 (b) 48, 52, 80, 90, 96
9-71. (a) X, X^2, X^3 (b) X, XY, Y^2
9-72. (a) WX^2, X^2y, Wy (b) PR, R^2S, R^4
9-73. $X + y$, $X - y$, $X^2(X + y)$
9-74. $(a + b)^2$, $a + c$, $(a + c)(a + b)$
9-75. WN, WAN^2, N^3
9-76. $R + S - t$, $R + S + t$, $R + S$
9-77. $(L + M)^3$, $L + M$, L
9-78. $3R^2$, $R + S$, $3R^2 + S$

9-5 ADDITION AND SUBTRACTION

Algebraic addition is very frequently required for combining resistor values in a parallel circuit. The mathematical technique is identical to that of arithmetic fractions:

(a) Find their lowest common denominator (LCD).

(b) Multiply the first fraction by the number 1, chosen so that the resulting denominator is the LCD.

(c) Multiply the second fraction by the number 1 chosen so that its denominator is the LCD.

(d) Add the resulting numerators.

(e) Give the result the LCD.

(f) Reduce, if possible.

EXAMPLE 9-15 Add:

$$\frac{A}{B} + \frac{2A}{BC}$$

SOLUTION The LCD is found by using the lowest-common-multiple techniques of Section 9-4. Thus, the LCD is BC. In order for the denominator of the first fraction to be BC, the entire fraction must be multiplied by C/C. The second fraction is then left alone.

$$\frac{A}{B} + \frac{2A}{BC} = \frac{A}{B} \cdot \frac{C}{C} + \frac{2A}{BC}$$
$$= \frac{AC}{BC} + \frac{2A}{BC}$$
$$= \frac{AC + 2A}{BC}$$

EXAMPLE 9-16 Add:

$$\frac{3PQ}{8} + \frac{7PQ}{2} - \frac{3PQ}{6}$$

SOLUTION The LCD must be formed from

$$8 = 2 \cdot 2 \cdot 2$$
$$2 = 2$$
$$6 = 2 \cdot 3$$
$$LCD = 2 \cdot 2 \cdot 2 \cdot 3 = 24$$

$$\frac{3PQ}{8} \cdot \frac{3}{3} + \frac{7PQ}{2} \cdot \frac{12}{12} - \frac{3PQ}{6} \cdot \frac{4}{4} = \frac{9PQ}{24} + \frac{84PQ}{24} - \frac{12PQ}{24}$$
$$= \frac{9PQ + 84PQ - 12PQ}{24}$$
$$= \frac{81PQ}{24} = \frac{27PQ}{8}$$

Subtraction is treated in the same manner as was done for polynomials.

EXAMPLE 9-17 Add

$$\frac{1}{2R+2} + \frac{6}{3R-3} - \frac{1}{1-R^2}$$

SOLUTION The denominators factored are:

$$2R + 2 = 2(R + 1)$$
$$3R - 3 = 3(R - 1)$$
$$1 - R^2 = -1(R + 1)(R - 1)$$

Thus, the LCD is

$$2 \cdot 3 \cdot (-1)(R + 1)(R - 1) = -6(R + 1)(R - 1)$$

Change to LCDs:

$$\frac{1}{2(R + 1)} \cdot \frac{-3(R - 1)}{-3(R - 1)} + \frac{6}{3(R - 1)} \cdot \frac{-2(R + 1)}{-2(R + 1)} - \frac{1}{(R + 1)(R - 1)(-1)} \cdot \frac{+6}{+6}$$

$$= \frac{-3(R - 1)}{-6(R + 1)(R - 1)} + \frac{-12(R + 1)}{-6(R + 1)(R - 1)} - \frac{6}{-6(R + 1)(R - 1)}$$

$$= \frac{-3R + 3 - 12R - 12 - 6}{-6(R + 1)(R - 1)}$$

$$= \frac{-15R - 15}{-6(R + 1)(R - 1)}$$

$$= \frac{-15(R + 1)}{-6(R + 1)(R - 1)}$$

$$= \frac{5}{2(R - 1)}$$

PROBLEMS

Add or subtract:

9-79. $\dfrac{R_1}{R_2} + \dfrac{2R_1}{3R_2}$

9-80. $\dfrac{3V}{2R} + \dfrac{7V}{6R}$

9-81. $\dfrac{V^2}{3R} + \dfrac{2V^2}{6R}$

9-82. $\dfrac{7P}{8Q} + \dfrac{6P}{5Q}$

9-83. $\dfrac{3M}{4N} + \dfrac{6N}{11N}$

9-84. $\dfrac{6S}{5T} - \dfrac{3S}{8T}$

9-85. $\dfrac{13R_1R_2}{24R_3} - \dfrac{7R_1R_2}{18R_3}$

9-86. $\dfrac{26L_1}{L_2} - \dfrac{5}{4L_2}$

9-87. $\dfrac{1}{a} + \dfrac{5}{b} - \dfrac{6}{c}$

9-88. $\dfrac{3}{R_1} - \dfrac{6}{R_2} - \dfrac{7}{R_2}$

9-89. $\dfrac{6}{L_1} - \dfrac{7}{L_2} - \dfrac{3}{L_1L_2}$

9-90. $\dfrac{10}{3M} - \dfrac{9}{4M} - \dfrac{8}{6M}$

9-91. $\dfrac{3V_1^2}{R_1} + \dfrac{6V_1^2}{R_2} - \dfrac{7V_1^2}{R_3}$

9-92. $\dfrac{8V_1V_2}{V_3} - \dfrac{6V_2}{V_1} - \dfrac{7V_2}{V_1V_3}$

9-93. $\dfrac{3R_1 - R_2}{6} + \dfrac{2R_1 + R_2}{5}$

9-94. $\dfrac{2V_1 + 3V_2}{6V_T} - \dfrac{3V_1 - 2V_2}{4V_T}$

9-95. $\dfrac{2P_N - 7P_R}{6P_N} - \dfrac{4P_N - 6P_R}{3P_R}$

9-96. $\dfrac{3L_1 - 6L_2}{5L_3} + \dfrac{4L_1 - 3L_2}{6L_3}$

9-97. $\dfrac{P_L}{R_1 + 3} + \dfrac{6P_L}{R_1^2 - 9}$

9-98. $\dfrac{L}{L - 2} - \dfrac{L}{L^2 - 3L + 2}$

9-99. $\dfrac{V_1}{V_1 - 3} + \dfrac{3V_1}{V_1^2 - 9}$

9-100. $\dfrac{3V + 2}{V + 2} + \dfrac{5V + 2}{V^2 + 2V}$

9-101. $\dfrac{I_1}{I_1 + 2} + \dfrac{4}{I_1} - \dfrac{I_1}{I_1 - 2}$

9-102. $\dfrac{A}{A + B} + \dfrac{3}{A} - \dfrac{3B}{A^2 + AB}$

9-103. $\dfrac{M}{M+3} + \dfrac{6M}{M^2-9} + \dfrac{2M}{M-3}$

9-105. $\dfrac{3}{T-3} + \dfrac{3}{T^2-7T+12} + \dfrac{4}{3T-12}$

9-104. $\dfrac{2S+1}{S+2} - \dfrac{S-1}{S-3} + \dfrac{9S+3}{S^2-S-6}$

9-106. $\dfrac{1}{3W+9} - \dfrac{2}{2W-8} + \dfrac{20}{6W^2-6W-72}$

9-6 COMPLEX FRACTIONS

A complex fraction is a fraction having a numerator or denominator that itself is a fraction. Usually, such an expression must be reduced to a simple fraction: one without a fraction in either the numerator or denominator. The procedure for this reduction consists of one primary rule:
Start with the simplest expression and work to the more complex.

EXAMPLE 9-18 Simplify:

$$\frac{\dfrac{a+b}{a-b} + a + b}{a+b-\dfrac{a+b}{a-b}}$$

SOLUTION We overhaul the numerator first:

$$\frac{a+b}{a-b} + a + b = \frac{a+b+a^2-b^2}{a-b}$$

Next, we manipulate the denominator:

$$a+b-\frac{a+b}{a-b} = \frac{a^2-b^2-a-b}{a-b}$$

We now combine the two results. Note that this is a division process:

$$\frac{\dfrac{a^2-b^2+a+b}{a-b}}{\dfrac{a^2-b^2-a-b}{a-b}} = \frac{a^2-b^2+a+b}{a-b} \div \frac{a^2-b^2-a-b}{a-b}$$

$$= \frac{a^2-b^2+a+b}{a-b} \cdot \frac{a-b}{a^2-b^2-a-b}$$

$$= \frac{a^2-b^2+a+b}{a^2-b^2-a-b}$$

EXAMPLE 9-19 Simplify:

$$\frac{\dfrac{R+S}{R} + \dfrac{R+S}{S}}{\dfrac{\frac{R}{S}+1}{R+S} + \dfrac{\frac{S}{R}+1}{R+S}}$$

SOLUTION We work over the numerator first. Note how important it is in this type of problem to keep very close track of where you are. Numerator:

$$\frac{R+S}{R}+\frac{R+S}{S}=\frac{S(R+S)+R(R+S)}{RS}$$

$$=\frac{S^2+RS+R^2+RS}{RS}$$

$$=\frac{R^2+2RS+S^2}{RS}$$

$$=\frac{(R+S)^2}{RS}$$

Next, we compute the fractions within the denominator:

$$\frac{R}{S}+1=\frac{R+S}{S}$$

$$\frac{S}{R}+1=\frac{R+S}{R}$$

Perform the left division within the denominator (note that division and multiplication must be done before addition and subtraction):

$$\frac{\dfrac{R}{S}+1}{R+S}=\frac{R+S}{S}\div R+S$$

$$=\frac{\cancel{R+S}}{S}\cdot\frac{1}{\cancel{R+S}}$$

$$=\frac{1}{S}$$

Next, perform the right division within the denominator:

$$\frac{\dfrac{S}{R}+1}{R+S}=\frac{R+S}{R}\div R+S$$

$$=\frac{\cancel{R+S}}{R}\cdot\frac{1}{\cancel{R+S}}$$

$$=\frac{1}{R}$$

Complete the denominator:

$$\frac{\dfrac{R}{S}+1}{R+S}+\frac{\dfrac{S}{R}+1}{R+S}$$

$$=\frac{1}{S}+\frac{1}{R}$$

$$=\frac{R+S}{RS}$$

Divide the numerator by the denominator:

$$\frac{\dfrac{R+S}{R}+\dfrac{R+S}{S}}{\dfrac{\dfrac{R}{S}+1}{R+S}+\dfrac{\dfrac{S}{R}+1}{R+S}}$$

$$=\frac{(R+S)^2}{RS}\div\frac{R+S}{RS}$$

$$=\frac{(R+S)^2}{\cancel{RS}}\cdot\frac{\cancel{RS}}{\cancel{R+S}}$$

$$=R+S$$

PROBLEMS

Simplify:

9-107. $\dfrac{\dfrac{3}{A}+\dfrac{4}{B}}{5}$

9-108. $\dfrac{\dfrac{R}{S}+\dfrac{S}{R}}{RS}$

9-109. $\dfrac{\dfrac{M}{L}+\dfrac{L}{M}}{M^2+L^2}$

9-110. $\dfrac{\dfrac{2}{2S}+\dfrac{3}{3R}}{\dfrac{R}{S}+1}$

9-111. $\dfrac{\dfrac{1}{R}+2}{\dfrac{\dfrac{1}{R}}{\dfrac{1}{R}+1}+\dfrac{1}{R}}$

9-112. $\dfrac{\dfrac{M+1}{M-1}+\dfrac{M-1}{M+1}}{\dfrac{M+1}{M-1}-\dfrac{M-1}{M+1}}$

9-113. $\dfrac{\dfrac{\dfrac{1}{C}+1}{C}+1}{\dfrac{\dfrac{1}{C}-1}{C}-1}$

9-114. $\dfrac{\dfrac{A+B}{A-B}-1-\dfrac{A-B}{A+B}}{\dfrac{A+B}{A-B}+1+\dfrac{A-B}{A+B}}$

9-115. $\dfrac{\dfrac{1}{B+3}+\dfrac{1}{B-2}+\dfrac{2-B}{B^2+B-6}}{\dfrac{15-5B}{B^2-9}+1}$

9-116. $\dfrac{\dfrac{3}{X-2}-\dfrac{5}{X+3}-\dfrac{15}{X^2+X-6}}{\dfrac{3}{X-2}-\dfrac{12}{X^2-4}}$

9-7 FRACTIONAL EQUATIONS

In this section we discuss two major types of fractional equations: (a) those conditional equations having a unique solution, and (b) literal equations. Both are used extensively in electronics.

Conditional Equations

Many equations we encounter have fractional numbers as a part of them. For example, consider the following two equations:

$$\frac{3}{4}A + \frac{6}{7}B = C$$

$$\frac{7}{9}X + \frac{3}{3}Y = 26$$

In each case, the fractions may be eliminated by multiplying the entire equation by each denominator.

EXAMPLE 9-20 Eliminate the fraction from the equation:

$$A = \frac{1}{2}(B_1 + B_2)H$$

SOLUTION Multiply the entire equation by the denominator (2).

$$2A = 2\left(\frac{1}{2}\right)(B_1 + B_2)(H)$$
$$= (B_1 + B_2)(H)$$

EXAMPLE 9-21 The formula for parallel inductors is as follows. Eliminate the fraction.

$$\frac{1}{L_T} = \frac{1}{L_1} + \frac{1}{L_2}$$

SOLUTION Multiply the equation by each of the denominators:

$$\frac{1}{L_T}(L_T)(L_1)(L_2) = \frac{1}{L_1}(L_T)(L_1)(L_2) + \frac{1}{L_2}(L_T)(L_1)(L_2)$$
$$L_1 L_2 = L_T L_2 + L_T L_1$$

This same procedure applies to all fractional equations: Eliminate the fractions, then solve the equation.

EXAMPLE 9-22 Solve for X in each equation:

(a) $\dfrac{1}{X} + \dfrac{1}{3} + 5 = 0$

(b) $\dfrac{1}{X+3} + \dfrac{1}{X-6} = 0$

(c) $\dfrac{1}{X} + \dfrac{1}{3X} + \dfrac{3}{5X} = 10$

(d) $\dfrac{X+3}{5} + X = 0$

SOLUTION

(a) $\dfrac{1}{X} + \dfrac{1}{3} + 5 = 0$

$$\frac{3}{3X} + \frac{X}{3X} + \frac{15X}{3X} = 0$$

$$3 + X + 15X = 0$$

$$3 + 16X = 0$$

$$X = -\frac{3}{16}$$

(b) $\dfrac{1}{X+3} + \dfrac{1}{X-6} = 0$

$$\frac{1}{X+3}(X-6)(X+3) + \frac{1}{X-6}(X+3)(X-6) = 0$$

$$X - 6 + X + 3 = 0$$

$$2X = 3$$

$$X = 1.5$$

(c) $\dfrac{1}{X} + \dfrac{1}{3X} + \dfrac{3}{5X} = 10$

$$\frac{1}{X}(3 \cdot 5 \cdot X) + \frac{1}{3X}(3 \cdot 5 \cdot X) + \frac{3}{5X}(3 \cdot 5 \cdot X) = 10\,(3 \cdot 5 \cdot X)$$

$$15 + 5 + 9 = 150X$$

$$X = \frac{29}{150}$$

(d) $\dfrac{X+3}{5} + X = 0$

$$\frac{X+3}{5}(5) + X(5) = 0$$

$$X + 3 + 5X = 0$$

$$X = -0.5$$

Note that in each case, the entire equation was multiplied through by each of the denominators.

EXAMPLE 9-23 The inverse of the total resistance of a parallel circuit is equal to the sum of the inverses of the individual resistors. If resistors of 47 kΩ, 75 kΩ, and 100 kΩ are in parallel, what is the total resistance?

SOLUTION We must first develop the equation. The inverse of a quantity is 1 over that quantity. Thus, the inverse of total resistance is $1/R_T$. The equation is then

$$\frac{1}{R_T} = \frac{1}{R_1} + \frac{1}{R_2} + \frac{1}{R_3}$$

Substitute the known values:

$$\frac{1}{R_T} = \frac{1}{47} + \frac{1}{75} + \frac{1}{100}$$

In this case, it is easier to add the inverses on the calculator:

$$\frac{1}{R_T} = 0.021277 + 0.013333 + 0.01000$$

$$= 0.044610$$

$$R_T = \frac{1}{0.044610} = 22.42 \text{ k}\Omega$$

Literal Equations

Literal equations are those having letters to represent values. They are also known as formulas. The following are examples:

$$A = \pi r^2 \qquad \frac{1}{R_T} = \frac{1}{R_1} + \frac{1}{R_2}$$

Many times, it is necessary to solve for a quantity that is buried deeply within a complex literal equation. It may look difficult, but solving for such a quantity is fairly simple if the procedure explained in Chapter 5 is used:

(a) Multiply both sides by the denominators of fractions to remove these denominators.
(b) Remove all parentheses by adding, subtracting, or multiplying. Combine similar terms.
(c) Add or subtract to move all terms containing the variable to the left of the equation and those not containing the variable to the right of the equation.
(d) Combine terms by adding or subtracting.
(e) Divide both sides of the coefficient of the variable.

We have studied the application of this procedure for all except complex fractions. To deal with these, we must spend considerable effort on step 1, converting it to a simple fraction.

EXAMPLE 9-24 Solve for A.

$$Y = \frac{\dfrac{1}{A} + \dfrac{1}{B} + \dfrac{1}{C}}{B + C + \dfrac{1}{AC}}$$

SOLUTION We must first convert to a simple fraction:

$$Y = \frac{\dfrac{BC + AC + AB}{ABC}}{\dfrac{ABC + AC^2 + 1}{AC}}$$

$$= \frac{BC + AC + AB}{AB\cancel{C}} \cdot \frac{A\cancel{C}}{ABC + AC^2 + 1}$$

$$= \frac{BC + AC + AB}{AB^2C + ABC^2 + B}$$

Now we can complete step (a):

$$Y(AB^2C + ABC^2 + B) = BC + AC + AB$$
$$AB^2CY + ABC^2Y + BY = BC + AC + AB$$

Complete the remaining steps:

$$AB^2CY + ABC^2Y - AC - AB = BC - BY$$
$$A(B^2CY + BC^2Y - C - B) = BC - BY$$
$$A = \frac{BC - BY}{B^2CY + BC^2Y - C - B}$$

Eliminate the fraction:

9-117. $V = \frac{4}{3}\pi R^3$

9-118. $V = \pi\frac{h}{3}(r_1^2 + r_1 r_2 + r_2^2)$

9-119. $V = \frac{\pi}{3}r^2 h$

9-120. $V = \frac{1}{6}H(S_0 + 4S_1 + S_2)$

9-121. $N = D\frac{PB}{48EI}(3L^2 - 4B^2) - \frac{M_1 L^2}{16EI}$

9-122. $F = \frac{3P}{4}\left(\frac{L}{L_1} - \frac{1}{3}\right)$

Solve for the unknown value:

9-123. $\frac{3}{X} + \frac{4}{X} = 5$

9-124. $\frac{2}{L_1} - \frac{3}{L_1} = 4$

9-125. $\frac{5}{L+1} + \frac{6}{L+7} = 0$

9-126. $\frac{6}{L-2} + \frac{7}{L+3} = 0$

9-127. $\frac{M+3}{6} + \frac{M+2}{7} = 3$

9-128. $\frac{R_1 + 6}{3} + \frac{R_1 - 3}{6} = 7$

9-129. $\frac{3k-2}{4} - \frac{5k-2}{6} = 3$

9-130. $\frac{3A-6}{5} - \frac{4A+3}{3} - 2A = 1$

9-131. $\frac{R_T - 2}{3} - \frac{R_T - 6}{4} - 7R_T = 4$

9-132. Quickie Dickey can assemble a computer in 120 h and Slick Sam can build the same computer in 150 h. How many hours will it take to build the computer if both work on it? (*Hint:* What part of the total job does each man complete in 1 h? What part of the job will they complete in 1 h if both work?)

9-133. The inverse of total capacitance for capacitors in series is equal to the sum of the inverses of the two capacitors. If the two capacitors are 2 F and 4 F, what is the total capacitance?

9-134. Solve for h and T.

$$\frac{\frac{1}{h} + \frac{3}{T}}{\frac{2}{h} - \frac{1}{T}} = 1$$

9-135. Solve for C_T and R_1.

$$\frac{C_T}{C_1 - C_2} = \frac{R_T}{\frac{1}{R_1} + \frac{1}{R_2}}$$

9-136. Solve for Z_1.

$$\frac{\frac{3}{Z_1} + \frac{1}{Z_2}}{\frac{R_1 R_T}{R_1 + R_T} + 1} = 1$$

9-137. Solve for C_1.

$$\frac{\frac{C_3}{C_2} + \frac{C_3}{C_1}}{\frac{C_3}{C_2} - \frac{C_3}{C_1} + 1} = C_3$$

10

PARALLEL AND SERIES–PARALLEL CIRCUITS

Parallel and series–parallel circuits form a part of practically every electronic device. In this chapter we develop the mathematical techniques necessary to solve these circuits: first, those purely parallel circuits, then those with both series and parallel elements. Finally, we study two special types of parallel circuits: the balanced bridge and the loaded voltage divider.

10-1 THE PARALLEL CIRCUIT

Figure 10-1 illustrates a parallel circuit. Note that in a parallel circuit the heads of the resistors are connected together and the tails of the resistors are connected together. Perhaps you have visited a used-car lot where all the cars are parked side by side in a long line, with all the headlights pointing one direction and the tail lights pointing the other direction. Like the resistors, the cars are parked in parallel.

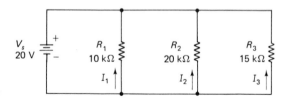

FIGURE 10-1 The parallel circuit.

10-2 VOLTAGE AND CURRENT OF THE PARALLEL CIRCUIT

As can be seen from Fig. 10-1, the voltage across each of the resistors (or branches) is the same in a parallel circuit. This enables us to solve for each of the branch currents using Georg Simon Ohm's law.

EXAMPLE 10-1 Compute the current within each of the resistors of Fig. 10-1.

SOLUTION We can compute each of the currents using Ohm's law.

$$I_1 = \frac{V_s}{R_1} = \frac{20}{10} = 2 \text{ mA}$$

$$I_2 = \frac{V_s}{R_2} = \frac{20}{20} = 1 \text{ mA}$$

$$I_3 = \frac{V_s}{R_3} = \frac{20}{15} = 1.333 \text{ mA}$$

Note that, in each case, the current is found by dividing the applied voltage by the branch resistance.

EXAMPLE 10-2 That champion tropical fish hobbyist, Catfish Hyphessobrycon, has his aquarium pump connected in parallel across the 120-V line with his aquarium heater. If the pump has a resistance of 3600 Ω and the heater, 720 Ω, what is the current through each?

SOLUTION Catfish's pump has current of

$$I = \frac{V}{R} = \frac{120}{3600} = 33.33 \text{ mA}$$

His heater has current of

$$I = \frac{V}{R} = \frac{120}{720} = 166.7 \text{ mA}$$

As noted above, current within a parallel circuit divides. However, the total current is the sum of the branch currents:

$$I_t = I_1 + I_2 + I_3 + \dots$$

where I_t is total current and I_1, I_2, and I_3 are branch currents.

EXAMPLE 10-3 Compute the total current for the circuit of Fig. 10-1.

SOLUTION The total current is the sum of the branch currents we obtained in Example 10-1.

$$I_t = I_1 + I_2 + I_3 = 2 + 1 + 1.333 = 4.333 \text{ mA}$$

EXAMPLE 10-4 Compute the total current used by Catfish's aquarium appliances in Example 10-2.

SOLUTION

$$I_t = 33.33 + 166.7 = 200.0 \text{ mA}$$

PROBLEMS

Problems 10-1 through 10-6 refer to Fig. 10-2. Compute the missing quantities.

	R_1	R_2	R_3	I_1	I_2	I_3	I_t
10-1.	24 kΩ	100 kΩ	51 kΩ				
10-2.	1 kΩ	2 kΩ	1.3 kΩ				
10-3.	5 Ω	10 Ω	7.6 Ω				
10-4.	1 MΩ	100 kΩ	10 kΩ				
10-5.	10 MΩ	4.7 MΩ	1.5 MΩ				
10-6.	5 MΩ	100 kΩ	560 kΩ				

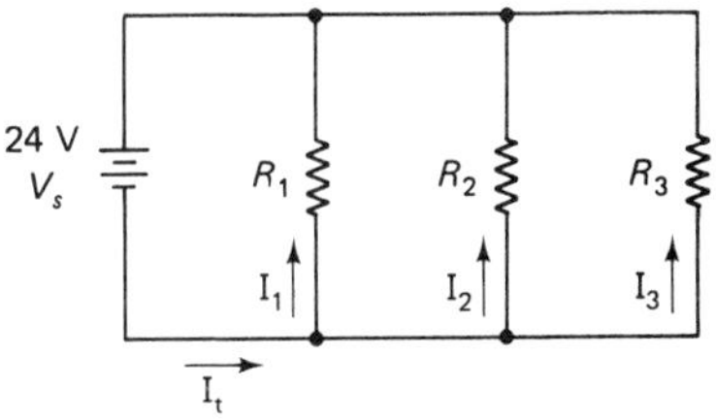

FIGURE 10-2 Problems 10-1 through 10-6.

10-7. Florist Flora Fauna uses lights of 144 Ω and 192 Ω to keep her Australian cacti growing. If the lights are connected in parallel across a 110-V source, what is the current through each lamp? What is the total current?

10-8. When Professor Bea Fuddle designed her flashlight, she connected three bulbs of 5 Ω, 4 Ω, and 3 Ω in parallel with her 4.5-V source. What is the current through each of the lamps? What is the total current?

10-3 RESISTANCE OF A PARALLEL CIRCUIT

Let us develop a general equation for the total resistance of a parallel circuit. We can start by knowing that the sum of the branch currents is equal to the total current:

$$I_t = I_1 + I_2 + I_3 + \ldots$$

But each current is a voltage (the supply voltage) divided by the individual branch resistances. Total current is supply voltage divided by total resistance.

$$\frac{V_s}{R_t} = \frac{V_s}{R_1} + \frac{V_s}{R_2} + \frac{V_s}{R_3} + \ldots$$

Next, divide the whole equation by V_s:

$$\frac{1}{R_t} = \frac{1}{R_1} + \frac{1}{R_2} + \frac{1}{R_3} + \ldots$$

This equation expresses the relationship of total resistance and individual resistors.

EXAMPLE 10-5 Compute the total resistance as seen by the power supply in Fig. 10-3.

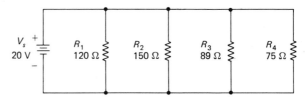

FIGURE 10-3 Example 10-5.

SOLUTION

$$\frac{1}{R_t} = \frac{1}{R_1} + \frac{1}{R_2} + \frac{1}{R_3} + \frac{1}{R_4}$$
$$= \frac{1}{120} + \frac{1}{150} + \frac{1}{89} + \frac{1}{75}$$
$$= 0.008333 + 0.006667 + 0.01124 + 0.01333$$
$$= 0.03957$$
$$R_t = 25.27 \ \Omega$$

Note how easy this is to do on the calculator. Enter the first number, then find its inverse $\boxed{1/X}$, then store it. Next, enter the second value, 150, find its inverse, add it to the stored value, then restore it. Repeat the procedure for the final two values. Finally, recall the sum from memory and invert it, giving the final answer.

EXAMPLE 10-6 When jalopy jockey Jake Jetspeed courts Jenney Jocose in his 1925 Jensen and they are parked watching the stars, he leaves his radio, heater, and parking lights on. If his radio is 1.400 Ω, his heater 0.800 Ω, and his parking lights 0.700 Ω across his 6-V car battery, what is the total resistance as seen by the battery and total current drawn?

SOLUTION We can calculate total resistance and, knowing this, calculate total current.

$$\frac{1}{R_t} = \frac{1}{R_1} + \frac{1}{R_2} + \frac{1}{R_3}$$
$$= \frac{1}{1.4} + \frac{1}{0.8} + \frac{1}{0.7}$$
$$= 3.393$$
$$R_t = 0.2947 \ \Omega = 294.7 \ \text{m}\Omega$$
$$I_t = \frac{V_s}{R_t} = \frac{6}{0.2947} = 20.36 \ \text{A}$$

Two Parallel Resistors

If only two resistors are involved, the general equation can be modified as follows:

$$\frac{1}{R_t} = \frac{1}{R_1} + \frac{1}{R_2} = \frac{R_1 + R_2}{R_1 R_2}$$
$$R_t = \frac{R_1 R_2}{R_1 + R_2}$$

EXAMPLE 10-7 Two resistors, 3.3 kΩ and 4.7 kΩ, are connected in parallel (Fig. 10-4). What is the total resistance?

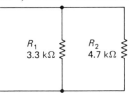

FIGURE 10-4 Example 10-7.

SOLUTION We shall use the short equation:

$$R_t = \frac{R_1 R_2}{R_1 + R_2} = \frac{3.3 \times 4.7}{3.3 + 4.7} = 1.939 \ \text{k}\Omega$$

EXAMPLE 10-8 Two printed circuit cards are connected in parallel across the 5-V supply within the Hopeless One-der computer. If the cards have resistance of 1.869 Ω and 2.316 Ω, respectively, what is the total resistance as seen by the power supply and the total current drawn?

SOLUTION We can use the short equation:

$$R_t = \frac{R_1 R_2}{R_1 + R_2} = \frac{1.869 \times 2.316}{1.869 + 2.316} = 1.034 \ \Omega$$

The total current is

$$I_t = \frac{V_s}{R_t} = \frac{5}{1.034} = 4.834 \ \text{A}$$

Identical Parallel Resistors

When identical-valued resistors are connected in parallel, we can simplify the basic parallel equation as follows. Assume that there are n resistors, each having a value of R ohms.

$$\frac{1}{R_t} = \frac{1}{R} + \frac{1}{R} + \frac{1}{R} + \ldots$$

$$= n\left(\frac{1}{R}\right)$$

$$= \frac{n}{R}$$

$$R_t = \frac{R}{n}$$

EXAMPLE 10-9 Eight 1.3-MΩ resistors are connected in parallel. What is the total resistance?

SOLUTION

$$R_t = \frac{R}{n} = \frac{1.3}{8} = 0.1625 \text{ MΩ}$$
$$= 162.5 \text{ kΩ}$$

PROBLEMS

10-9. Compute the resistance as seen by the power supply of Fig. 10-5.

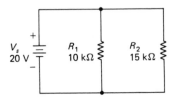

FIGURE 10-5 Problem 10-9.

10-10. If $R_1 = 12$ kΩ and $R_2 = 100$ kΩ, what is the total resistance if they are connected in parallel?

10-11. Resistors of 10 kΩ, 20 kΩ, and 33 kΩ are connected in parallel. What is the total resistance?

10-12. Resistors of 89 kΩ, 10 kΩ, and 51 kΩ are connected in parallel. What is the total resistance?

For each problem, compute the total resistance when the resistors are connected in parallel.

10-13. 2 kΩ, 4.7 kΩ, 3.3 kΩ

10-14. 1 MΩ, 2 MΩ, 2.4 MΩ

10-15. 560 Ω, 510 Ω, 750 Ω, 820 Ω

10-16. 24 Ω, 89 Ω, 100 Ω, 110 Ω

10-17. 2.2 kΩ, 560 Ω, 330 Ω, 39 kΩ

10-18. 1.2 MΩ, 560 kΩ, 270 kΩ, 33 Ω

10-19. 1.8 kΩ, 1.5 kΩ

10-20. 470 Ω, 560 Ω

10-21. 27 Ω, 33 Ω

10-22. 1 MΩ, 1 kΩ

10-23. 1.5 kΩ, 4.3 kΩ

10-24. 13 kΩ, 7.5 kΩ

10-25. Ten 27-kΩ resistors

10-26. Four 3.0-kΩ resistors

10-27. Twenty 8.2-kΩ resistors

10-28. Thirteen 10-MΩ resistors

Power within a parallel circuit is the sum of the power dissipated by each of the elements:

$$P_t = P_1 + P_2 + P_3 + \dots$$

EXAMPLE 10-10 Compute the branch currents and branch power dissipations and the total power and current for the circuit shown in Fig. 10-6.

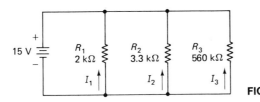

FIGURE 10-6 Example 10-10.

SOLUTION The currents are

$$I_1 = \frac{V_s}{R_1} = \frac{15}{2} = 7.500 \text{ mA}$$

$$I_2 = \frac{V_s}{R_2} = \frac{15}{3.3} = 4.545 \text{ mA}$$

$$I_3 = \frac{V_s}{R_3} = \frac{15}{0.56} = 26.79 \text{ mA}$$

The individual power dissipations can be computed using the voltage and resistance of each branch.

$$P_1 = \frac{V_s^2}{R_1} = \frac{15^2}{2000} = 112.5 \text{ mW}$$

$$P_2 = \frac{V_s^2}{R_2} = \frac{15^2}{3300} = 68.18 \text{ mW}$$

$$P_3 = \frac{V_s^2}{R_3} = \frac{15^2}{560} = 401.8 \text{ mW}$$

The total power is the sum of the individual dissipations:

$$\begin{aligned} P_t &= P_1 + P_2 + P_3 \\ &= 112.5 + 68.18 + 401.8 \\ &= 582.5 \text{ mW} \end{aligned}$$

The total power can also be found by using total current and total voltage:

$$\begin{aligned} P_t &= V_s I_t = V_s(I_1 + I_2 + I_3) \\ &= 15(7.500 + 4.545 + 26.79) \\ &= 582.5 \text{ mW} \end{aligned}$$

PROBLEMS

Problems 10-29 through 10-38 refer to Fig. 10-7. Compute the missing values.

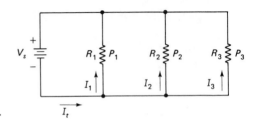

FIGURE 10-7 Problems 10-29 through 10-38.

	R_1	R_2	R_3	V_s	I_1	I_2	I_3	P_1	P_2	P_3	I_t	P_t
10-29.	10 kΩ	12 kΩ	15 kΩ	12 V								
10-30.	1 MΩ	870 kΩ	510 kΩ	100 V								
10-31.	2 kΩ	3 kΩ		10 V		5 mA						
10-32.	33 kΩ	24 kΩ		20 V		1 mA						
10-33.	2 kΩ	1 kΩ		15 V					500 mW			
10-34.	4 kΩ			12 V	2 mA				400 mW			
10-35.	3.3 kΩ			20 V	4 mA						20 mA	
10-36.					20 mA			250 mW	400 mW			1 W
10-37.		10 kΩ		15 V	1 mA	2 mA						
10-38.								20 W	10 W	5 W	4 A	

10-39. Cousin Connie connected three lamps in parallel with her 6-V supply. If the lamps were 1 W, 2 W, and 2.5 W, what is the total resistance of Connie's circuit? The branch currents and total current?

10-40. Math Wizard Wee Winifred has a calculator that has 10 LEDs, each drawing 20 mA, and a circuit board drawing 5 mA connected in parallel with a 4-V supply. What is the total power drawn from the battery, and what are the effective resistances of the individual LEDs and the circuit board?

10-5 THE BALANCED BRIDGE

Figure 10-8 illustrates the balanced Wheatstone bridge. This circuit forms the basis for many measuring devices. In the figure, R_x represents an unknown resistance we are attempting to measure and R_D are calibrated resistance dials. To measure an unknown resistance, the unknown resistor is connected as R_x and the dials manipulated until the meter reads zero. Under these conditions, the bridge is said to be balanced.

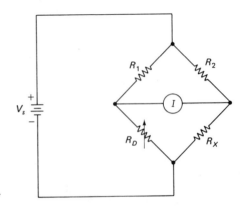

FIGURE 10-8 Balanced Wheatstone bridge.

R_x can then be found by reading the resistance of R_D from the dials and substituting into the following equation:

$$\frac{R_1}{R_D} = \frac{R_2}{R_x} \qquad (10\text{-}1)$$

EXAMPLE 10-11 Using a balanced bridge, R_D is read as 2.563 kΩ. What is the value of the unknown resistance if $R_1 = 10$ kΩ and $R_2 = 10$ kΩ?

SOLUTION

$$\frac{R_1}{R_D} = \frac{R_2}{R_x}$$
$$\frac{10}{2.563} = \frac{10}{R_x}$$
$$R_x = 2.563 \text{ kΩ}$$

In most bridges, R_2 acts as a multiplier and may be used to select the range of resistances that can be measured with the bridge.

EXAMPLE 10-12 Using a balanced bridge, $R_D = 2.563$ kΩ, $R_1 = 10$ kΩ, and $R_2 = 100$ kΩ. Calculate R_x.

SOLUTION

$$\frac{R_1}{R_D} = \frac{R_2}{R_x}$$
$$\frac{10}{2.563} = \frac{100}{R_x}$$
$$R_x = \frac{100 \times 2.563}{10} = 25.63 \text{ kΩ}$$

Compare this result with that of Example 10-11.

Fault Location

The bridge can be used to locate a short in a cable pair by connecting the cable pair to the R_x terminals. Knowing the round-trip resistance of the pair (called the loop resistance) and the resistance of that particular type of wire pair per 1000 ft, the fault can be located.

EXAMPLE 10-13 A loop measurement on a size 18 cable pair reads 250 Ω. How far is the short from the bridge?

SOLUTION An 18-gauge wire has a loop resistance (R_{ES}) of 13.02 Ω/1000 ft. (This represents the resistance of 2000 ft of wire.) The distance to the fault is

$$D = \frac{R_x}{R_{ES}} \times 1000 = \frac{250}{13.02} \times 1000 = 19,201 \text{ ft} \quad \text{or} \quad 3.637 \text{ mi}$$

The Wheatstone bridge is used extensively to locate grounded cable faults. The first step in locating a fault is to have a person in the distant office short the cable pair together. The loop resistance (R_L) between the offices is then measured using the procedure discussed in the preceding paragraph and in Example 10-13. Once this has been done, the bridge is connected in one of two configurations, Murray or Varley. The Murray method uses an adjustable slider, identified as R_1R_2 in Fig. 10-9, which is moved until balance is achieved. With R_L representing the loop resistance of the entire pair, the formula becomes

$$\frac{R_1}{R_2} = \frac{R_L - R_x}{R_x}$$

Solve for R_x:

$$R_x = \frac{R_2}{R_1 + R_2} R_L$$

Thus, by taking a simple loop measurement, R_L is obtained, and by taking a Murray loop, R_x can be found and the distance determined.

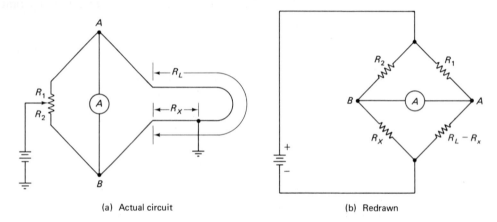

(a) Actual circuit (b) Redrawn

FIGURE 10-9 Murray loop: (a) actual circuit; (b) redrawn.

EXAMPLE 10-14 A grounded wire exists on a pair with a loop measured at 250 Ω and a Murray measured at an R_1 of 650 Ω and an R_2 of 350 Ω. What is the distance to the fault if 18-gauge wire is used?

SOLUTION Solve for R_x:

$$R_x = \frac{R_2}{R_1 + R_2} R_L = \frac{350}{1000} \times 250 = 87.5 \ \Omega$$

Since 1000 ft of size 18 wire has a resistance of 6.51 Ω,

$$D = \frac{R_x}{R_{ES}} \times 1000 = \frac{87.5}{6.51} \times 1000 = 13,440 \ \text{ft} \quad \text{or} \quad 2.55 \ \text{mi}$$

The Varley method uses the circuit shown in Fig. 10-10. Using loop resistances, the equation becomes

$$\frac{R_1}{R_2} = \frac{\dfrac{R_L}{2} + \dfrac{R_y}{2}}{R_3 + \dfrac{R_L - R_y}{2}}$$

Set $R_1 = R_2$ and solve for R_y:

$$1 = \frac{R_L + R_y}{R_L - R_y + 2R_3}$$
$$R_L - R_y + 2R_3 = R_L + R_y$$
$$2R_y = R_L - R_L + 2R_3$$
$$R_y = R_3$$

Thus, the reading, R_3, will tell the operator the distance the fault is from the distant end of the cable pair.

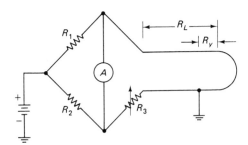

FIGURE 10-10 Varley measurement.

EXAMPLE 10-15 The Varley reads 426 Ω and the loop 2360 Ω for an 18-gauge wire. What is the distance of the fault from the test end?

SOLUTION Since 18-gauge wire has a loop resistance of 13.02 Ω/1000 ft, the distant office is

$$D = \frac{R_x}{R_{ES}} \times 1000 = \frac{2360}{13.02} \times 1000 = 181{,}260 \text{ ft} \quad \text{or} \quad 34.33 \text{ mi}$$

from the test end. The fault is

$$D = \frac{R_3}{R_{ES}} \times 1000 = \frac{426}{13.02} \times 1000 = 32{,}719 \text{ ft} \quad \text{or} \quad 6.197 \text{ mi}$$

from the distant end or $34.33 - 6.20 = 28.13$ mi from the test end.

PROBLEMS

Compute R_x for the following balanced bridge resistances:

	R_1	R_2	R_D
10-41.	5 kΩ	5 kΩ	12 kΩ
10-42.	20 kΩ	20 kΩ	12 kΩ
10-43.	100 kΩ	100 kΩ	12 kΩ
10-44.	50 kΩ	50 kΩ	12 kΩ

10-45.	5 kΩ	50 kΩ	12 kΩ
10-46.	5 kΩ	500 kΩ	12 kΩ
10-47.	50 kΩ	5 kΩ	12 kΩ
10-48.	500 kΩ	5 kΩ	12 kΩ

10-49. A loop measurement on an 18-gauge cable pair reads 397 Ω. How far is the short from the bridge?

10-50. Short L. Shooter measured a 421.7-Ω loop on 18-gauge wire from the Can Ton office to a cable short. How far is the short from the office?

10-51. While grouse hunting, Bead E. Eie leveled his shotgun on a bird perched on Buell City–Donner Junction cable, and fired, resulting in a short within the cable. The Buell technician measured a 401-Ω loop resistance to the short. If the cable carries 18-gauge conductors, how far is the short from Buell City?

10-52. Pierre L. Buckette was excavating a foundation for his house when he severed an 18-gauge telephone cable, resulting in a short. The technician at the central office measured a 169-Ω loop resistance to the short. How far was the short from the central office?

10-53. A grounded wire exists on an 18-gauge pair with a loop measured at 450 Ω. Murray readings result in an R_1 of 670 Ω and an R_2 of 330 Ω. What is the distance to the fault?

10-54. Repeat Prob. 10-53 for a loop of 770 Ω, R_1 of 530 Ω, and R_2 of 470 Ω.

Find the distance to the ground fault for the following problems assuming 18-gauge wire:

	Loop	*Varley*
10-55.	3330 Ω	200 Ω
10-56.	2760 Ω	469 Ω
10-57.	3579 Ω	1013 Ω
10-58.	2695 Ω	231 Ω

10-6 THE LOADED VOLTAGE DIVIDER

In Chapter 7 we studied unloaded voltage dividers. We now expand the subject to include those dividers whose loads are significant. Figure 10-11 shows a divider that

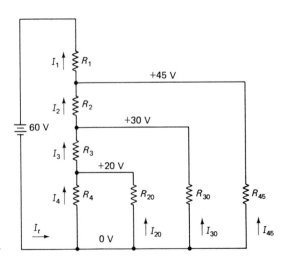

FIGURE 10-11 Loaded voltage divider.

supplies 20, 30, and 45 V to its three separate loads. According to Kirchhoff's current law, the following relationships would be true:

$$I_t = I_1 = I_4 + I_{20} + I_{30} + I_{45}$$
$$I_3 = I_4 + I_{20}$$
$$I_2 = I_3 + I_{30} = I_4 + I_{20} + I_{30}$$

I_4 is referred to as the bleeder current; it could be defined as the current through the string resistor that has the least amount of current flowing through it.

EXAMPLE 10-16 In Fig. 10-11, $I_4 = 5$ mA, $I_{20} = 20$ mA, $I_{30} = 10$ mA, and $I_{45} = 15$ mA. Calculate the string resistors.

SOLUTION Since $I_4 = 5$ mA,

$$R_4 = \frac{V_4}{I_4} = \frac{20}{5} = 4 \text{ k}\Omega$$

$$R_3 = \frac{V_3}{I_3} = \frac{V_3}{I_4 + I_{20}} = \frac{30 - 20}{5 + 20} = 400 \ \Omega$$

$$R_2 = \frac{V_2}{I_2} = \frac{V_2}{I_3 + I_{30}} = \frac{45 - 30}{25 + 10} = 429 \ \Omega$$

$$R_1 = \frac{V_1}{I_1} = \frac{V_1}{I_2 + I_{45}} = \frac{60 - 45}{35 + 15} = 300 \ \Omega$$

Calculate the power:

$$P_1 = V_1 I_1 = 15 \times 50 = 750 \text{ mW}$$
$$P_2 = V_2 I_2 = 15 \times 35 = 525 \text{ mW}$$
$$P_3 = V_3 I_3 = 10 \times 25 = 250 \text{ mW}$$
$$P_4 = V_4 I_4 = 20 \times 5 = 100 \text{ mW}$$

EXAMPLE 10-17 A voltage divider must be designed for the following loads: 20 V at 20 mA, 25 V at 5 mA, and 45 V at 30 mA. Bleeder current should be 10 mA and the supply voltage 75 V. Calculate the divider string.

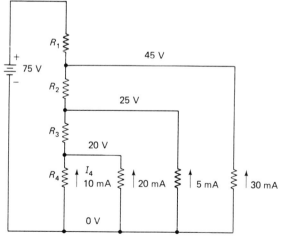

FIGURE 10-12 Example 10-17.

SOLUTION First the schematic is drawn (Fig. 10-12). Then, we calculate the resistances starting at R_4:

$$R_4 = \frac{V_4}{I_4} = \frac{20}{10} = 2 \text{ k}\Omega$$

$$P_4 = V_4 I_4 = 20 \times 10 = 200 \text{ mW}$$

$$R_3 = \frac{V_3}{I_3} = \frac{25-20}{10+20} = \frac{5}{30} = 166.7 \ \Omega$$

$$P_3 = V_3 I_3 = 5 \times 30 = 150 \text{ mW}$$

$$R_2 = \frac{V_2}{I_2} = \frac{45-25}{5+30} = \frac{20}{35} = 571.4 \ \Omega$$

$$P_2 = V_2 I_2 = 20 \times 35 = 700 \text{ mW}$$

$$R_1 = \frac{V_1}{I_1} = \frac{75-45}{30+35} = \frac{30}{65} = 461.5 \ \Omega$$

$$P_1 = V_1 I_1 = 300 \times 65 = 1.95 \text{ W}$$

Changing the Reference Point

As in the unloaded divider, both positive and negative voltages can be supplied if the reference is changed. The only difficulty we will encounter is determining which string resistor has the least amount of current, for this resistor will only have bleeder current flowing through it.

EXAMPLE 10-18 A voltage divider must supply the following: -20 V at 16 mA, -10 V at 3 mA, $+5$ V at 60 mA, and $+15$ V at 20 mA. The bleeder current is 10 mA and the supply voltage is 50 V. Design the divider string.

SOLUTION First, the schematic is drawn (Fig. 10-13). Note that either R_3 or R_4 must contain the bleeder current only. By applying a dose of Kirchhoff's current law

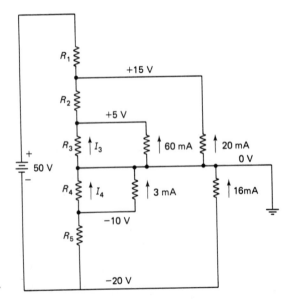

FIGURE 10-13 Example 10-18.

to the 0-V node, we can reveal which resistor has the larger current; then the one with the smaller current must have only bleeder current:

$$I_3 + 60 + 20 = I_4 + 3 + 16$$
$$I_4 - I_3 = 80 - 19 = 61 \text{ mA}$$

This, I_4 is greater than I_3 and I_3 must have only the bleeder current, 10 mA. Therefore, I_4 must be

$$I_4 - I_3 = 61 \text{ mA}$$
$$I_4 = 61 + I_3 = 61 + 10 = 71 \text{ mA}$$

Knowing this, we can solve for the resistors:

$$R_3 = \frac{V_3}{I_3} = \frac{5}{10} = 500 \ \Omega$$
$$P_3 = V_3 I_3 = 5 \times 10 = 50 \text{ mW}$$
$$R_2 = \frac{V_2}{I_2} = \frac{15-5}{10+60} = \frac{10}{70} = 142.9 \ \Omega$$
$$P_2 = V_2 I_2 = 10 \times 70 = 700 \text{ mW}$$
$$R_1 = \frac{V_1}{I_1} = \frac{50 - [15 - (-20)]}{20 + 70} = \frac{15}{90} = 166.7 \ \Omega$$
$$P_1 = V_1 I_1 = 15 \times 90 = 1.350 \text{ W}$$
$$R_4 = \frac{V_4}{I_4} = \frac{10}{71} = 140.8 \ \Omega$$
$$P_4 = V_4 I_4 = 10 \times 71 = 710.0 \text{ mW}$$
$$R_5 = \frac{V_5}{I_5} = \frac{20-10}{71+3} = \frac{10}{74} = 135.1 \ \Omega$$
$$P_5 = V_5 I_5 = 10 \times 74 = 740.0 \text{ mW}$$

EXAMPLE 10-19 A voltage divider supplies +9 V at 100 mA, +5 V at 200 mA, −6V at 120 mA, and −9 V at 250 mA, with a bleeder current of 50 mA and a source of 25 V. Design the divider.

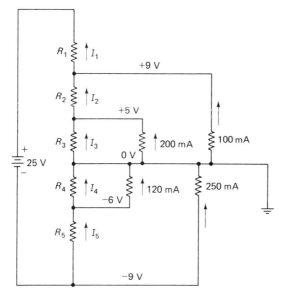

FIGURE 10-14 Example 10-19.

SOLUTION First, the schematic is drawn (Fig. 10-14). Then we determine whether I_3 is greater than I_4:

$$I_3 + 200 + 100 = I_4 + 120 + 250$$
$$I_3 - I_4 = 370 - 300 = 70 \text{ mA}$$

Thus, $I_4 = 50$ mA and

$$I_3 - I_4 = 70 \text{ mA}$$
$$I_3 = 70 + I_4 = 70 + 50 = 120 \text{ mA}$$

Therefore,

$$R_3 = \frac{V_3}{I_3} = \frac{5}{120} = 41.67 \ \Omega$$
$$P_3 = V_3 I_3 = 5 \times 120 = 600 \text{ mW}$$
$$R_2 = \frac{V_2}{I_2} = \frac{9-5}{120+200} = \frac{4}{320} = 12.50 \ \Omega$$
$$P_2 = V_2 I_2 = 4 \times 320 = 1.28 \text{ W}$$
$$R_1 = \frac{V_1}{I_2} = \frac{25 - [9 - (-9)]}{320 + 100} = \frac{7}{420} = 16.67 \ \Omega$$
$$P_1 = V_1 I_1 = 7 \times 0.42 = 2.94 \text{ W}$$
$$R_4 = \frac{V_4}{I_4} = \frac{6}{50} = 120 \ \Omega$$
$$P_4 = V_4 I_4 = 6 \times 50 = 300 \text{ mW}$$
$$R_5 = \frac{V_5}{I_5} = \frac{9-6}{120+50} = \frac{3}{170} = 17.65 \ \Omega$$
$$P_5 = V_5 I_5 = 3 \times 170 = 510 \text{ mW}$$

PROBLEMS

Calculate the resistance and power dissipations for the following voltage dividers:

	Source Voltage	Bleeder	V_1	I_1	V_2	I_2	V_3	I_3
10-59.	100 V	10 mA	20 V	5 mA	50 V	5 mA	60 V	10 mA
10-60.	90 V	8 mA	15 V	5 mA	20 V	5 mA	60 V	15 mA
10-61.	120 V	50 mA	10 V	50 mA	40 V	30 mA	55 V	35 mA
10-62.	130 V	100 mA	10 V	5 mA	20 V	40 mA	80 V	60 mA
10-63.	500 V	1 mA	80 V	100 μA	120 V	80 μA	420 V	600 μA
10-64.	460 V	100 μA	20 V	10 μA	60 V	10 μA	440 V	10 μA
10-65.	60 V	1 mA	−10 V	2 mA	+30 V	100 μA		
10-66.	400 V	50 μA	−80 V	40 μA	+20 V	5 μA	+300 V	100 μA
10-67.	40 V	30 mA	−15 V	10 mA	−5 V	15 mA	+20 V	5 mA
10-68.	80 V	120 mA	−20 V	40 mA	−10 V	60 mA	+40 V	100 mA

11

BATTERIES AND CONDUCTORS

In this chapter we consider the mathematics surrounding the generation of dc and that surrounding its transmission to a load. The battery is considered first, and we analyze the effect of its internal resistance. We then consider conductors: how they are sized, their resistance, and the effects of temperature upon that resistance.

11-1 BATTERY INTERNAL RESISTANCE

Every battery can be considered as a voltage source, V_s, in series with a internal resistance, R_i, as shown in Fig. 11-1. When a load is connected to the terminals, current will flow through the internal resistance, reducing the voltage at the terminals so that it is less than V_s.

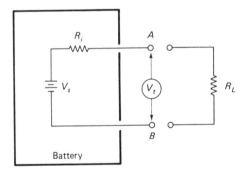

FIGURE 11-1 Equivalent diagram of a battery.

Assuming the load is connected (Fig. 11-1), we can calculate this terminal voltage as follows:

$$I = \frac{V}{R} = \frac{V_s}{R_i + R_L}$$

But the voltage drop across R_i is

$$V_D = IR$$
$$= \left(\frac{V_s}{R_i + R_L} \right) R_i$$

This must be subtracted from V_s to obtain V_t:

$$V_t = V_s - V_D$$
$$= V_s - \frac{V_s R_i}{R_i + R_L}$$
$$= V_s\left(1 - \frac{R_i}{R_i + R_L}\right)$$

EXAMPLE 11-1 A battery with a source voltage of 36 V has an internal resistance of 0.26 Ω. What will the terminal voltage be under a load of (a) 100 Ω? (b) 10 Ω? (c) 1 Ω? (d) 100 mΩ?

SOLUTION

(a) $V_t = V_s\left(1 - \dfrac{R_i}{R_i + R_L}\right)$

$= 36\left(1 - \dfrac{0.26}{0.26 + 100}\right)$

$= 35.9066$ V

(b) $V_t = V_s\left(1 - \dfrac{R_i}{R_i + R_L}\right)$

$= 36\left(1 - \dfrac{0.26}{0.26 + 10}\right)$

$= 35.0877$ V

(c) $V_t = V_s\left(1 - \dfrac{R_i}{R_i + R_L}\right)$

$= 36\left(1 - \dfrac{0.26}{0.26 + 1}\right)$

$= 28.5714$ V

(d) $V_t = V_s\left(1 - \dfrac{R_i}{R_i + R_L}\right)$

$= 36\left(1 - \dfrac{0.26}{0.26 + 0.1}\right)$

$= 10.0000$ V

Note that as R_L gets lower, so does the terminal voltage.

EXAMPLE 11-2 A 6-V battery with an internal resistance of 120 mΩ is connected to a 5-Ω load. What is the current through and the voltage across the load? What is the power dissipated by the battery and the load?

SOLUTION The total current is

$$I = \frac{V_s}{R_{tot}} = \frac{V_s}{R_i + R_L} = \frac{6}{0.120 + 5} = 1.172 \text{ A}$$

The terminal voltage is

$$V_t = I_t R_L = 1.172 \times 5 = 5.859 \text{ V}$$

The power dissipated by the load is

$$P_L = V_L I_L = 5.859 \times 1.172 = 6.866 \text{ W}$$

The power dissipated by the battery is

$$P_i = I^2 R_i = (1.172)^2 (0.120) = 164.8 \text{ mW}$$

PROBLEMS

11-1. A battery with a source voltage of 24 V has an internal resistance of 246 mΩ. What will its terminal voltage be under a **(a)** 40-Ω load? **(b)** 4-Ω load? **(c)** 400-mΩ load?

11-2. Sanderson Supertec invented a flashlight using two 1½-V cells in series. If each cell has a 96-mΩ internal resistance, what will the voltage be across his 20-Ω lamp?

11-3. While attempting to start his 1936 Rawlings automobile, Harrison Elfert III shorted his starter cable across the battery terminals. If the battery no-load voltage was 6 V, its internal resistance 20 mΩ, and that of the cable 15 mΩ, what current was Harrison causing through the cable during the short? What was the terminal voltage? What power was being dissipated by the cable and by the battery?

11-4. Read Prob. 11-3. Finally, Harrison was able to step on his starter. If the cables to the starter had a 20-mΩ resistance and the starter 100 mΩ, how much voltage appeared across the starter? What power was dissipated by the battery, the connecting cables, and the starter?

11-2 MULTIPLE BATTERIES

When batteries are connected in series, their internal resistances can be considered in series (Fig. 11-2) and the voltages in series.

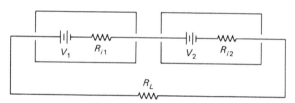

FIGURE 11-2 Batteries in series.

EXAMPLE 11-3 Two 6-V batteries are connected series aiding in series with a 50-Ω load. If the batteries have internal resistance of 1 Ω and 1.8 Ω, what is the current through the load?

SOLUTION Since the batteries are connected series-aiding, they are as shown in Fig. 11-3. Start at point A and form a loop equation:

$$+6 - 1I + 6 - 1.8I - 50I = 0$$
$$52.8I = 12$$
$$I = 227.3 \text{ mA}$$

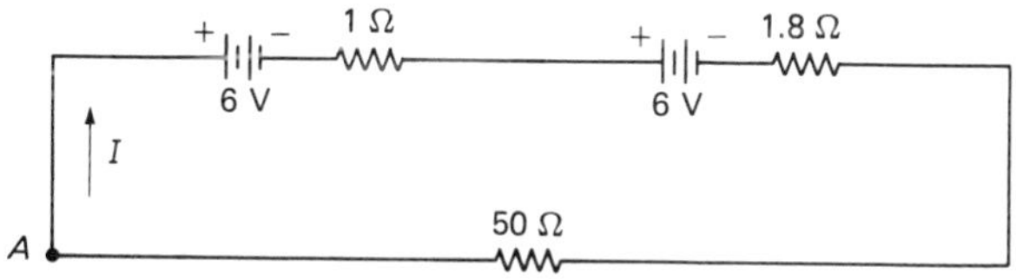

FIGURE 11-3 Example 11-3.

EXAMPLE 11-4 Two batteries are connected series-opposing in series with a 40-Ω load. The first battery is 9 V, having an internal resistance of 60 mΩ, whereas the second unit is 6 V, having an internal resistance of 50 mΩ. Compute the current through and voltage across the load.

SOLUTION Figure 11-4 illustrates the circuit. Form a loop equation starting at *A*:

$$+6 - 0.05\,I - 0.06\,I - 9 - 40\,I = 0$$
$$40.11\,I = -3$$
$$I = 74.79 \text{ mA} \quad \text{in a direction opposite to}$$
$$\text{that shown}$$

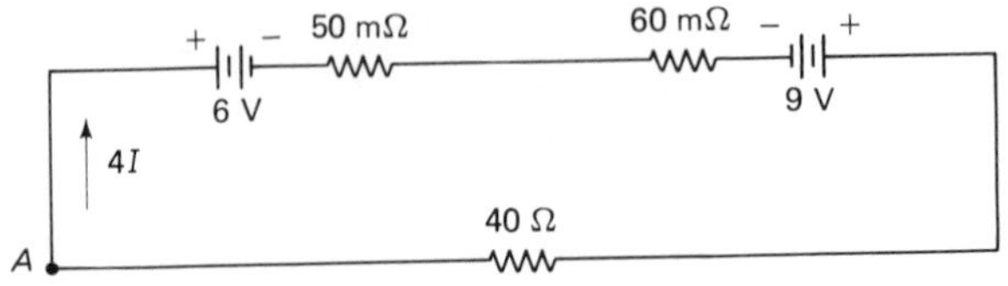

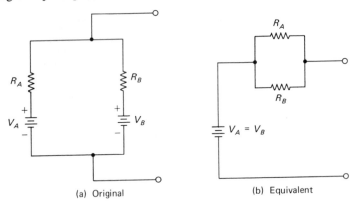

FIGURE 11-4 Example 11-4.

Batteries in Parallel

I hate to pull a quick shuffle on you, but to discuss this subject in detail requires the use of either simultaneous equations (Chapter 13) or network analysis (Chapter 15). Therefore, at this time I shall state the principles involved but actually derive these principles in Chapter 15.

When two batteries of equal source voltages are connected in parallel, their equivalent source voltage can be considered as the same as that of the batteries, and their equivalent internal resistance is that of the two internal resistances in parallel (Fig. 11-5). Using this principle, we can compute load currents and voltages.

(a) Original   (b) Equivalent

FIGURE 11-5 Batteries in parallel.

EXAMPLE 11-5 Two 16-V batteries connected in parallel drive a 3.0-Ω load. If the internal resistance of each battery is 750 mΩ, what is the load current and load voltage?

SOLUTION We can now assume that the circuit is as shown in Fig. 11-6. Total resistance of the circuit is

$$R_T = (R_A \parallel R_B) + R_L$$
$$= (0.750 \parallel 0.750) + 3.0$$
$$= 0.375 + 3.0$$
$$= 3.375 \ \Omega$$

The total current and therefore the load current is

$$I_L = \frac{V_s}{R_T} = \frac{16}{3.375} = 4.741 \text{ A}$$

The load voltage is

$$V_L = I_L R_L = 4.741 \times 3.0 = 14.22 \text{ V}$$

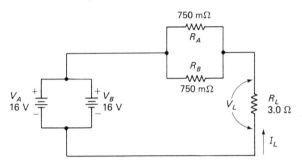

FIGURE 11-6 Example 11-5.

EXAMPLE 11-6 To provide more current to his robot, Herman, Jerry Osenphephyr connected two 24-V batteries in parallel. The larger battery had an internal resistance of 120 mΩ, compared with 190 mΩ for the small battery. If Herman's circuits have a resistance of 5 Ω, what is the voltage across these circuits, and what is the current supplied by each battery?

SOLUTION The actual circuit is as shown in Fig. 11-7(a). However, we can figure load voltages and currents using Fig. 11-7(b).

$$I_L = \frac{V_s}{(R_A \parallel R_B) + R_L}$$
$$= \frac{24}{(0.120 \parallel 0.190) + 5} = \frac{24}{0.07355 + 5}$$
$$= 4.730 \text{ A}$$
$$V_L = I_L R_L = 4.730 \times 5 = 23.65 \text{ V}$$

Referring to Fig. 11-7(a), we now know that the load has 23.65 V across it, whereas each battery supplies 24 V. Thus, the voltage drop across R_A is

$$V_{RA} = V_A - V_L = 24 - 23.65209$$
$$= 0.3479 \text{ V}$$

The current through R_A is

$$I_A = \frac{V_{RA}}{R_A} = \frac{0.3479}{0.120} = 2.899 \text{ A}$$

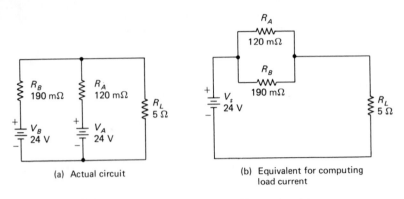

(a) Actual circuit

(b) Equivalent for computing load current

FIGURE 11-7 Batteries with unequal internal resistances.

The current through R_B is

$$I_B = \frac{V_{RB}}{R_B} = \frac{0.3479}{0.190} = 1.831 \text{ A}$$

Note that, as a check, the total current is

$$I_L = I_A + I_B$$
$$= 2.899 + 1.831 = 4.730 \text{ A}$$

PROBLEMS

11-5. Two 12-V batteries are connected series-aiding with a 2.5-Ω load. If the internal resistance of the batteries is 120 mΩ and 150 mΩ, respectively, what is the current through the load?

11-6. Repeat Prob. 11-5 for a 0.25-Ω load.

11-7. The following 3 batteries are connected series-aiding with a 6-Ω load. What is the current, voltage, and power of the load?
    (a) 12 V, internal resistance (R_i) of 120 mΩ **(b)** 6 V, R_i of 50 mΩ **(c)** 24 V, R_i of 80 mΩ

11-8. Repeat Prob. 11-7 using the following batteries:
    (a) 50 V, R_i of 60 mΩ **(b)** 60 V, R_i of 80 mΩ **(c)** 110 V, R_i of 100 mΩ

11-9. Batteries of 64 V $(R_i = 200 \text{ m}\Omega)$ and 50 V $(R_i = 250 \text{ m}\Omega)$ are connected series opposing with a 10-Ω load. Compute total current, voltage across the load, and power dissipated by each battery and by the load.

11-10. Assume that the load in Prob. 11-9 short-circuited. Compute the total current and the power dissipated by each battery.

11-11. Two 6-V batteries with an R_i of 0.5 Ω each are connected in parallel across a 7.5-Ω load. Compute the load current and voltage.

11-12. In his trip around the world in a bathtub, Billy Blubber's bathtub bore bright battle beacons being bedazzled by batteries beset by internal resistance of 510 mΩ each. If two 12-V batteries were used in parallel to drive each 10-Ω beacon, how much power did each battle beacon consume?

11-3 CONDUCTORS

Thus far we have conveniently ignored the wires that connect sources to loads. However, these wires have two very serious limitations: (a) they have resistance, and (b) they can only conduct a finite amount of current. The wires used in electronics come in varied sizes and two common materials, copper and aluminum, with copper being by far the more prevalent. Wire size is specified by an American wire gauge (AWG) number (Table 11-1); the higher the number, the smaller the wire. The table refers to several characteristics of wires that we shall discuss: gauge, diameter, area, resistance, weight, and ampacity.

TABLE 11-1 Copper wire at 20°C.

Gauge	Diameter (mils)	Cross Section (cmils)	Resistance/ 1000 ft (Ω)	Ampacity[a] (A)
0000	460.2	211,600	0.04901	225
000	409.6	167,800	0.06180	175
00	364.8	133,100	0.07793	150
0	324.9	105,500	0.09827	125
1	289.3	83,690	0.1239	100
2	257.6	66,370	0.1563	90
3	229.4	52,640	0.1970	80
4	204.3	41,740	0.2485	70
5	181.9	33,100	0.3133	55
6	162.0	26,250	0.3951	50
7	144.3	20,820	0.4982	
8	128.5	16,510	0.6282	35
9	114.4	13,090	0.7921	
10	101.9	10,380	0.9989	25
11	90.74	8234	1.260	
12	80.81	6530	1.588	20
13	71.96	5178	2.003	
14	64.08	4107	2.525	15
15	57.07	3257	3.184	
16	50.82	2583	4.016	6
17	45.26	2048	5.064	
18	40.30	1624	6.385	3
19	35.89	1288	8.051	
20	31.96	1022	10.15	
21	28.45	810.1	12.80	
22	25.35	642.4	16.14	
23	22.57	509.5	20.36	
24	20.10	404.0	25.67	
25	17.90	320.4	32.37	
26	15.94	254.1	40.81	
27	14.20	201.5	51.47	
28	12.64	159.8	64.90	
29	11.26	126.7	81.83	
30	10.03	100.5	103.2	

[a] Ampacity varies considerably with insulation used.

Wire Gauge

The AWG number represents the size of the wire, the higher numbers representing smaller sizes. Note that the cross-sectional area doubles for roughly every three AWG numbers. Although the table lists many wire sizes, as a practical matter only the even-numbered sizes are readily available. When wires larger than zero are needed,

00, 000, and 0000 sizes are available; wires larger than these are measured in millions of circular mils (Mcmils).

Diameter

The mil is 0.001 in.; thus, a ¼ in.-diameter wire is 250 mils in diameter (about AWG 2).

Area

The cross-sectional area of a wire is measured in a unit called a circular mil (cmil), as shown by the shaded part of Fig. 11-8. The area of a cmil in in² is, therefore,

$$A = \pi r^2$$
$$= \pi \left(\frac{0.001}{2}\right)^2$$
$$= 785.40 \times 10^{-9} \text{ in.}^2$$

The use of the cmil eliminates much of the math involved with square mils. To illustrate, let us find the number of cmils in a wire 0.100 in. in diameter:

$$A = \pi r^2 = \pi \left(\frac{0.100}{2}\right)^2 \text{ in.}^2$$

However, a square mil is $\pi(0.0005)^2$ in.². Thus, the area in cmils is

$$A = \frac{\pi \left(\dfrac{0.100}{2}\right)^2}{\pi \left(\dfrac{0.001}{2}\right)^2}$$
$$= \left(\frac{0.100}{0.001}\right)^2$$
$$= 100^2$$
$$= 10,000 \text{ cmils}$$

This illustrates the advantage of the cmil: π always drops out. Thus, to find the area of a wire in cmils,

$$A = d^2$$

where d is the diameter in mils.

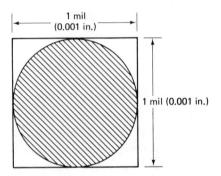

FIGURE 11-8   The circular mil.

EXAMPLE 11-7   Find the cross-sectional area of a wire in cmils if its diameter is ⅛ in.

SOLUTION   Since ⅛ in. is 125 mils,

$$A = d^2 = (125)^2 = 15,625 \text{ cmils}$$

EXAMPLE 11-8   Compute the number of cmils for an AWG 20 wire.

SOLUTION   According to Table 11-1, AWG 20 wire has a diameter of 31.96 mils. The area is

$$A = d^2 = 31.96^2 = 1021 \text{ cmils}$$

This checks closely with the table's value of 1022 cmils.

Resistance

The resistance of wire is measured in $\Omega/1000$ ft. Thus, to compute total resistance, we must first find out how many thousands of feet of wire are under consideration.

EXAMPLE 11-9   Compute the resistance of 1 mi of AWG 16 copper wire.

SOLUTION

$$1 \text{ mi} = 5280 \text{ ft} = 5.28 \text{ thousands of ft}$$

However, 1000 ft of AWG 16 wire has a resistance (R_{ES}) of 4.016 Ω, from the table. Thus,

$$R_1 = R_{ES} \times \text{length} = 4.016 \times 5.28 = 21.20 \ \Omega$$

EXAMPLE 11-10   A certain computer terminal is installed 600 ft from its signal source, from which it receives signals over a pair of AWG 22 wires. What is the resistance of the wire?

SOLUTION   Note that the signal must travel to the terminal and back, a distance of 1200 ft. Therefore,

$$R_t = R_{ES} \times \text{length} = 16.14 \times 1.20 = 19.37 \ \Omega$$

Ampacity

The larger the diameter of the wire, the more current the wire is capable of handling. It is dangerous to exceed the safe current-carrying capacity (ampacity) of a wire, for many fires have been started by allowing an extension cord or other conductor to supply more current than it can handle. Because of their knowledge of electricity, those in electronics should be especially alert to this hazard.

The safe current-carrying capacities of wires are dependent upon adequate ventilation or heat conduction away from the wire. If many wires were tied into a handle, each carrying a large current, heat could build up in the center wires, causing disastrous results. Thus, the environment of the wire must be considered. Wire size is selected by determining actual current, then selecting the minimum even size capable of carrying this current.

EXAMPLE 11-11 A cord must be selected for an electric heater drawing 2000 W at 120 V. What is the minimum safe size of wire that could be used?

SOLUTION We must first know the current:

$$I = \frac{P}{V} = \frac{2000}{120} = 16.67 \text{ A}$$

From the wire tables, size 14 is too small, so we should use a size 12 wire.

EXAMPLE 11-12 A 5-V power supply must provide 40 A to its load. What wire size should be selected?

SOLUTION From the table, an AWG 6 will supply at least 50 A.

Voltage Drop

Since wire has resistance and has current flowing through it, there is a voltage drop across the wire.

EXAMPLE 11-13 A 2-Ω lamp bank draws current from a 120-V source ½ mi away over a pair of AWG 4 wires. What is the actual voltage across the lamp bank?

SOLUTION Since we are using a pair of wires, we have 1 mi of wire. From Table 11-1, AWG 4 wire has a resistance of 0.2485 Ω/1000 ft. Thus, the total wire resistance is

$$
\begin{aligned}
R_W &= R_{ES} \times \text{length} \\
&= 0.2485 \times 5.28 \\
&= 1.312 \ \Omega
\end{aligned}
$$

The total resistance as seen by the source is

$$R_t = R_W + R_L = 1.312 + 2 = 3.312 \ \Omega$$

The total current is

$$I = \frac{V}{R} = \frac{120}{3.312} = 36.23 \text{ A}$$

The voltage across the lamps is

$$V = IR = 36.23 \times 2 = 72.46 \text{ V}$$

Note that this is much lower than the source voltage. The wires are dropping the rest, 47.54 V.

Resistivity

In computing the resistance of a material, it is seen to vary directly with the length of the material and indirectly with the cross-sectional area:

$$\frac{R_1}{R_2} = \frac{L_1/A_1}{L_2/A_2} = \frac{L_1 A_2}{L_2 A_1}$$

This can also be expressed as

$$R = \rho \frac{L_1}{A_1}$$

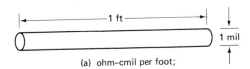

(a) ohm–cmil per foot;

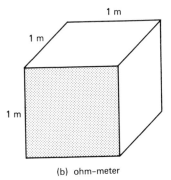

(b) ohm–meter

FIGURE 11-9 Resistivity.

where ρ (Greek letter rho) is a constant of proportionality called resistivity. Thus, the higher the ρ, the greater the resistance a material offers to electrical flow. The English system defines ρ as that resistance of a wire 1 mil in diameter and 1 ft long [Fig. 11-9(a)]. The units of measurement can be found by solving for ρ, then substituting the units known:

$$R = \rho \frac{L_1}{A_1}$$

$$\rho = \frac{RA_1}{L_1} = \frac{\text{ohms} \times \text{cmils}}{\text{ft}}$$

Thus, in the English system, ρ is measured in Ω-cmil/ft. The SI system uses the model shown in Fig. 11-9(b), where it is defined as that resistance of a 1-m³ chunk of material from the shaded face to the opposite face. We can find its units as was done above:

TABLE 11-2 Constants for common metals.

	Resistivity (Ω-cmil/ft)	Temperature Coefficient (°C)
Aluminum	19.3	+0.0039
Carbon	30,000	−0.0005
Copper	10.37	+0.00393
Gold	14.7	+0.0034
Iron	78	+0.005
Lead	125	+0.0043
Platinum	66	+0.003
Silver	9.9	+0.0038
Tungsten	33	+0.0045

$$\rho = \frac{RA_1}{L_1} = \frac{\text{ohms} \times \text{m}^2}{\text{m}}$$
$$= \text{ohm-meters}$$

Thus, the SI unit is ohm-meters. Table 11-2 lists the resistivity for several common materials.

EXAMPLE 11-4 What is the resistivity of the copper shown in Table 11-1?

SOLUTION We shall compute ρ for AWG 0, 10, and 30. Reading the values, for AWG 0:

$$\rho = \frac{RA_1}{L_1} = \frac{0.09827 \times 105{,}500}{1000} = 10.37 \ \Omega\text{-cmil/ft}$$

For AWG 10:

$$\rho = \frac{RA_1}{L_1} = \frac{0.9989 \times 10{,}380}{1000} = 10.37 \ \Omega\text{-cmil/ft}$$

For AWG 30:

$$\rho = \frac{RA_1}{L_1} = \frac{103.2 \times 100.5}{1000} = 10.37 \ \Omega\text{-cmil/ft}$$

Note that ρ is a property of the material used and is not dependent upon the physical shape.

EXAMPLE 11-15 What would be the resistance of a 2-mi length of copper wire of 500,000 cmils?

SOLUTION We found ρ in Example 11-14.

$$R = \rho \frac{L_1}{A_1} = 10.37 \times \frac{2 \times 5280}{500{,}000} = 219.0 \ \text{m}\Omega$$

Effect of Temperature

The values for Table 11-1 were for copper wire at 20°C. The resistance of most metals rises with temperature and can be expressed by the equation

$$R = R_I(1 + \alpha \, \Delta T)$$

in which R_I represents the resistance at a known temperature, α is a constant for the material called the temperature coefficient, and ΔT is the difference in temperature from that at which R_I is known to that at which R is to be determined. Table 11-2 lists temperature coefficients for some common materials.

A common way of expressing the effect of temperature is

$$\frac{R_1}{R_2} = \frac{234.5 + t_1}{234.5 + t_2}$$

where t_1 and t_2 are in °C. This equation only applies to copper, for it assumes that the resistance of copper varies linearly from −234.5°C, where it is zero. This characteristic has been determined by experiment to be true.

EXAMPLE 11-16 Compute the resistance per 1000 ft of AWG 16 wire at 25°C.

Solution From Table 11-1, we know AWG 16 wire has a resistance of 4.016 Ω per 1000 ft.

$$\frac{R_1}{R_2} = \frac{234.5 + t_1}{234.5 + t_2}$$
$$\frac{R_1}{4.016} = \frac{234.5 + 25}{234.5 + 20}$$
$$R_1 = 4.095 \text{ } \Omega$$

PROBLEMS

11-13 Find the cross-sectional area in cmils for wire with the following diameters: **(a)** ¼ in. **(b)** 1 in. **(c)** 1 cm **(d)** 26 mm

11-14. Find the cross-sectional area in cmils for wire with the following diameters: **(a)** ⅜ in. **(b)** 2 in. **(c)** 1.5 cm **(d)** 35 mm

11-15. Compute the resistance of 2 mi of AWG 4 copper wire.

11-16. Compute the resistance of 3 mi of AWG 8 copper wire.

11-17. A service entrance is fed with a 200-ft pair of AWG 0 copper wires. Compute the resistance of the wire.

11-18. A set of speakers is fed with a 100-ft pair of AWG 18 wires. Compute the resistance of the wires.

11-19. A cord must be selected for an electric toaster consuming 1300 W at 120 V. What minimum size of wire should be used?

11-20. A 5-V power supply must provide 45 W to its load. What is the minimum size of wire that can be used?

11-21. Buck Bored had to drill a hole in his 1937 Lemon automobile parked on the rear of his palatial manor. The drill is rated at 2.5 A at 115 V and Buck ran a 500-ft AWG 18 extension cord from his 115-V outlet. What current and voltage was the drill actually receiving?

11-22. Norman Nerd decided to conserve energy by designing and building an electric-powered car. Unfortunately, it required a 3-mi AWG 12 extension cord plugged into a 115-V outlet. If the car was rated at 115 V, 1500 W, what was the actual voltage and current received by Norman's car?

11-23. A 2000-ft length of AWG 14 wire (not copper) has a resistance of 25 Ω. What is its resistivity in Ω-cmil/ft?

11-24. A 250 m length of AWG 16 wire (not copper) has a resistance of 2.65 Ω. What is its resistivity in Ω-cmil/ft?

11-25. What is the resistance of 100 mi of 5-Mcmil copper wire?

11-26. What is the resistance of 250 mi of 7-Mcmil copper wire?

11-27. Compute the resistance of 2 mi of AWG 14 copper wire at 30°C.

11-28. Repeat Prob. 11-27 for 0°C.

11-29. Compute the resistance of 75 mi of 5-Mcmil aluminum wire at 35°C.

11-30. Compute the resistance of 25 mi of AWG 20 gold wire at 40°C.

12

GRAPHS AND EQUATIONS

Perhaps I can paraphrase a prominent Chinese philosopher, "A graph is worth a thousand words." Through it, we electronic bugs can picture that invisible force we all know, love, and avoid contact with. In this chapter we study graphs and how they can be used to represent this alphabet soup we have been discussing thus far. We study the attributes of graphs, then put algebraic equations onto the graphs, and finally, extract equations from graphs.

12-1 INTRODUCTION TO GRAPHS

There are many types of graphs, some of which are shown in Fig. 12-1. The bar graph is useful for comparing one phenomenon with successive phenomena. Using it, we can tell rapidly when a peak or a valley occurred. The circular graph is useful for telling at a glance what parts of the whole are more prominent. The segmented graph is useful for showing trends of data that varies discretely from period to period.

Most of the graphs used in electronics are those shown in Fig. 12-1(d). Here, a voltage varies smoothly with time. This is practically always the case for voltage, current, and power.

Locating Points on a Graph

The general form for a graph is shown in Fig. 12-2 and consists of two directional number lines, called axes. The center point where the two number lines meet is called the origin. It is from here that all measurements are taken. The horizontal axis forms a number line called the X axis or abscissa and represents the number of units traveled from the origin horizontally. For example, traveling right from the origin represents movement in the $+X$ direction, whereas traveling left from the origin represents movement in the $-X$ direction.

The vertical direction forms a second number line called the Y axis or ordinate. Therefore, traveling up the graph represents a number in the $+Y$ direction and traveling down the graph represents the $-Y$ direction.

Points are plotted on the graph using a pair of numbers. By convention, the first number represents the number of units traveled in the horizontal direction, whereas the second number represents the number of units traveled in the vertical direction from the X axis.

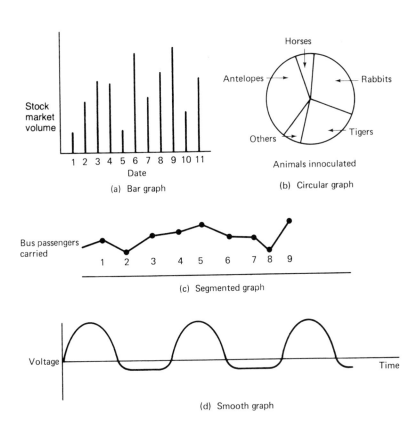

(a) Bar graph

(b) Circular graph

(c) Segmented graph

(d) Smooth graph

FIGURE 12-1 Types of graphs.

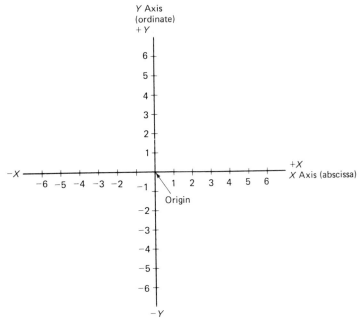

FIGURE 12-2 Coordinate axes.

EXAMPLE 12-1 Plot the point (3, 6) on a graph.

SOLUTION There are two methods that can be used for locating a point: (a) coordinates and (b) following a trail.

(a) To locate (3, 6) by the coordinate method, start at the origin and move 3 units to the right. Draw a vertical line through this point. Next, start at the origin and move up the Y axis 6 units. Draw a horizontal line through this point. The point at which these two drawn lines intersect forms the solution [Fig. 12-3(a)].

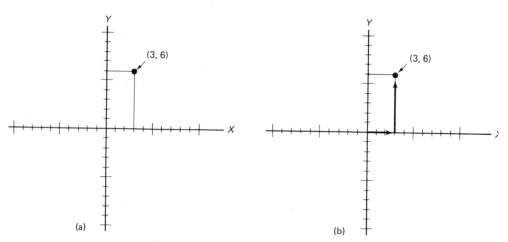

FIGURE 12-3 Example 12-1, locating a point.

(b) To locate (3, 6) by the following a trail method, start at the origin and proceed 3 hops to the right. From this point, travel 6 points up vertically. This locates the point [Fig. 12-3(b)].

Note that either method can be used, for the results are identical.

EXAMPLE 12-2 Locate the following points on a graph: (−3, 2), (5, −3), (−1, −4), (4, 3).

SOLUTION Figure 12-4 illustrates the solution.

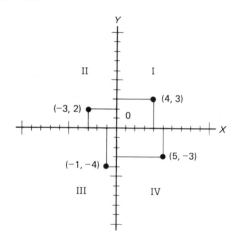

FIGURE 12-4 Example 12-2.

Figure 12-4 illustrates another attribute of a graph. Note that each point is located in a separate area of the graph. These areas are called quadrants. Points in quadrant I will contain X and Y values that are both positive. All points in quadrant II will have negative X values and positive Y values. Those in quadrant III will have negative X values and negative Y values. Those in quadrant IV will have positive X values and negative Y values.

Plotting Experimental Data

Graphs are very often used in electronics for plotting data gathered from readings of voltmeters, ammeters, and other devices. These readings usually form a table and the data must be plotted from the table. The procedure can be summarized as follows:

(a) Select suitable horizontal and vertical scales.
(b) Plot each point on the graph.
(c) Draw a smooth graph through (or near) the points.

EXAMPLE 12-3 The voltage across a certain circuit was varied and the current through the circuit was measured [Fig. 12-5(a)]. The results are shown in Fig. 12-5(b). Plot the graph.

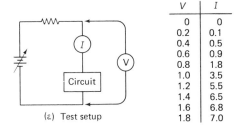

V	I
0	0
0.2	0.1
0.4	0.5
0.6	0.9
0.8	1.8
1.0	3.5
1.2	5.5
1.4	6.5
1.6	6.8
1.8	7.0

(a) Test setup

(b) Test results   **FIGURE 12-5** Voltage and current.

SOLUTION The first step is to select suitable scales. The scales are most convenient when major divisions represent 1, 2, or 5 or decades of 1, 2, or 5.

Scanning down the voltage data we see that the maximum value we wish to plot is 1.8 V. Thus, we would not want to choose a maximum scale voltage of 100 V [Fig. 12-6(a)], or 0.1 V [Fig. 12-6(b)]. In this case, we shall select a maximum scale voltage of 2.0 V [Fig. 12-6(c)]. In a similar manner, a maximum current of 10 A seems appropriate.

The next step is to plot each point on the graph [Fig. 12-7(a)]; putting a dot at each plotted point.

Finally, the graph must be drawn. Many beginners whip out their straightedge and draw a segmented graph as shown in Fig. 12-7(b). But do we actually believe that the voltage curve takes a sharp turn at each plotted point? Of course not. Voltage across most circuits varies smoothly from one point to another. Therefore, the correct procedure is to draw a smooth curve as close to each point as possible. Some points may not end up on the curve, but that is fine, for no measured value is perfect. Therefore, the finished curve is that shown in Fig. 12-7(c).

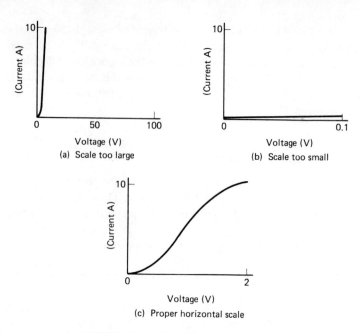

(a) Scale too large

(b) Scale too small

(c) Proper horizontal scale

FIGURE 12-6 Choosing the right scale.

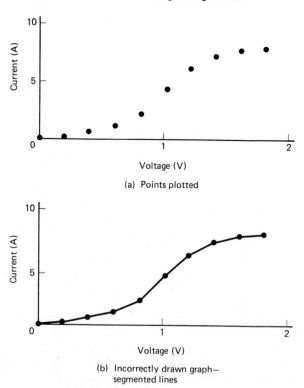

(a) Points plotted

(b) Incorrectly drawn graph—
segmented lines

FIGURE 12-7 Drawing the graph.

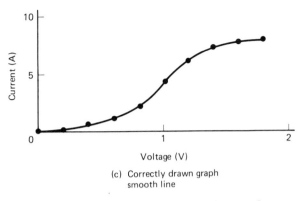

(c) Correctly drawn graph
smooth line

FIGURE 12-7 Drawing the graph *(Continued).*

EXAMPLE 12-4 The output of an amplifier was measured as its input is varied (Fig. 12-8). Plot the curve.

SOLUTION This curve is called a transfer curve, for it expresses how information is transferred from the input to the output. Figure 12-8(c) illustrates the finished graph.

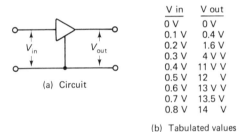

V in	V out
0 V	0 V
0.1 V	0.4 V
0.2 V	1.6 V
0.3 V	4 V V
0.4 V	11 V V
0.5 V	12 V
0.6 V	13 V V
0.7 V	13.5 V
0.8 V	14 V

(a) Circuit

(b) Tabulated values

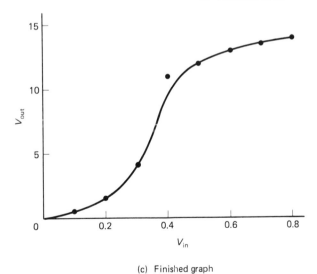

(c) Finished graph

FIGURE 12-8 Example 12-4.

Note that the horizontal and vertical scales are not the same, but were chosen so that each "fills out" the graph.

Nonlinear Axes

The graphs we have thus far discussed use linear scales. That is, the scales start at zero and each unit stepped off along the axis has the same distance as other units along that same axis. However, it is possible to miss some information if linear scales are used with some phenomena. The graph of a diode, for example, cannot be plotted very well because the very low currents are too close to the X axis. Figure 12-9 illustrates a logarithmic (or log, for short) scale in the Y direction and a linear scale in the X direction. The log scale provides uniform distance between decades. The distances from 1 to 10, 10 to 100, 100 to 1000, and 1000 to 10,000 are all equal on a log scale. Either the X or Y or both axes could be plotted on a log scale. Many phenomena in electronics are logarithmic. The ear, for example, hears sound intensity on a logarithmic rather than linear scale. That is, the ear hears an increase from 100 μW to 1 mW about the same as from 1 W to 10 W.

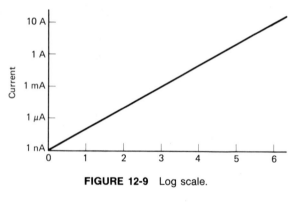

FIGURE 12-9 Log scale.

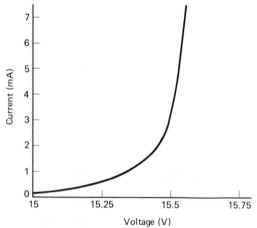

FIGURE 12-10 Truncated scale.

A second type of nonlinear graph is one with a truncated scale. That is, instead of starting at zero, the scale starts at some nonzero value and increases from there. Figure 12-10 illustrates the action of a zener diode. The curve is relatively uninteresting until it starts to curve around 15.25 V, so the scale is chosen to illustrate this characteristic.

PROBLEMS

12-1. Plot the following points on a graph: (3, 6), (4, 2), (−6, 3), (−7, −2).

12-2. Plot the following points on a graph: (1, 3), (−2, 3), (−4, −5), (5, −2).

12-3. Plot the following points on a graph: (0.31, −0.62), (0.8642, 0.3462), (−0.762, −0.3).

12-4. Plot the following points on a graph: (13,623; 18,265), (−25,623; 32,641), (20,001; −13,063).

12-5. Agape Truelove designed his Love Analysis Machine to output a voltage that raises and lowers in response to his emotional state. After years of courting, he finally worked up the courage to kiss Lady Lois L'Amour and recorded the following readings. Plot them on a graph:

Time (s) (X Axis)	Voltage (V) (Y Axis)
0	2.0
1	3.0
2	4.4
3	5.9
4	7.8
5	10.6
6	9.3
7	7.2
8	8.6
9	10.5
10	15.2
11	25.9

12-6. The frequency with which the wings of the Batwing Hummingbird beat was recorded as it traveled over a sunflower. Plot the data:

Time (s) (X Axis)	Frequency (Hz) (Y Axis)
0	400
1	500
2	550
3	550
4	400
5	200
6	180
7	250
8	400
9	600
10	700

12-7. Data from Norbert Navely's newly designed motor were taken to determine its current–torque characteristics. Plot the data:

Torque (kg-m) (X Axis)	Current (A) (Y Axis)
0	3
1	4
2	5
3	6.2
4	7.3
5	8.6
6	10.1
7	11.5
8	13.3
9	14.9

12-8. The input voltage and output voltage of an amplifier were compared, resulting in the following data. Plot the data.

Input Voltage (mV) (X Axis)	Output Voltage (V) (Y Axis)
0	0
1	0.20
2	0.41
3	0.58
4	0.75
5	0.99
6	1.20
7	1.43
8	1.63
9	1.80
10	1.95

12-2 GRAPHING EQUATIONS

In this section we consider how to convert an algebraic equation into a line on a graph. This is a very useful technique to analyze how one variable or another behaves.

Independent and Dependent Variables

Consider the equation

$$Y = 3X + 6$$

The usual way of looking at this equation is to try an X value and see what Y is. Let us try the number 6:

$$Y = 3X + 6$$
$$= 3(6) + 6$$
$$= 24$$

In this analysis, we can see that the value of Y depends upon what X is. Therefore, Y is called the dependent variable and X is referred to as the independent variable. This assignment is quite useful in the electronics laboratory, for here we will set one value, the current for example, and measure the other value, the voltage for example. Thus, the voltage is the dependent variable and the current is the independent variable. Another way of representing this relationship is

$$Y = f(X)$$

This is read, "Y is a function of X." It means that there is an equation involving X as an independent variable and Y is dependent upon X. The following are examples of functions:

$$Y = 3X^2 + 6 \qquad Y = f(X)$$
$$N = 2R^2 - 7 \qquad N = f(R)$$
$$Q = 3P + 4 \qquad Q = f(P)$$

If the function (the expression to the right of the equals sign) involves more than one independent variable, these can be listed within the parentheses, separated by commas. Examples:

$$E = IR^2 + 32N \qquad E = f(I, R, N)$$
$$X_1 = X_2^2 + X_2X_3 + X_3^2 \qquad X_1 = f(X_2, X_3)$$
$$A = \tfrac{1}{2}(b_1 + b_2)h \qquad A = f(b_1, b_2, h)$$

Plotting Equations

The usual method of plotting an equation is to place the independent variable along the X axis and dependent variable along the Y axis. To plot an equation, values are chosen for the independent variable and the dependent value is calculated. Let us apply this to a particular equation.

EXAMPLE 12-5 Plot the equation

$$R = 3V - 4$$

SOLUTION We shall choose values for V and calculate the resultant R, placing each on a table as follows:

V	R	Calculation
-1	-7	$R = 3(-1) - 4 = -7$
-2	-10	$R = 3(-2) - 4 = -10$
0	-4	$R = 3(0) - 4 = -4$
1	-1	$R = 3(1) - 4 = -1$
2	2	$R = 3(2) - 4 = 2$
3	5	$R = 3(3) - 4 = 5$

Note that we chose both negative and positive values. Plotting these values one point at a time results in Fig. 12-11(a). Finally, drawing a line through the points results in the finished graph [Fig. 12-11(b)].

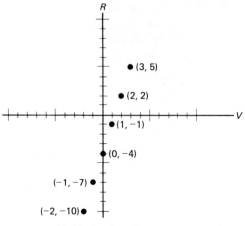

(a) Plotting the points

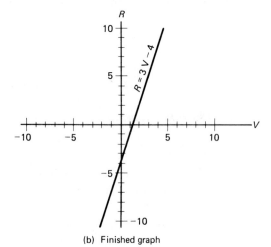

(b) Finished graph

FIGURE 12-11 Example 12-5, plotting an equation.

Note that the equation in Fig. 12-11 plotted as a straight line. This is always true for an equation with two variables raised to the first power. The only difference between one equation of the first power and another of the first power is where the straight line is placed upon the graph.

EXAMPLE 12-6 Plot $V = -\frac{1}{2}R + 6$.

SOLUTION Since the equation contains no higher power of V or R than 1, the equation will plot as a straight line. Therefore, we shall plot only three points: two to make the straight line and the third as a check that we have not made any error. Note that one of the easiest points to calculate is where the independent variable (R) is set to zero.

R	V	Computation
0	6	$V = -\frac{1}{2}(0) + 6 = 6$
-5	8.5	$V = -\frac{1}{2}(-5) + 6 = 8.5$
5	3.5	$V = -\frac{1}{2}(5) + 6 = 3.5$

Plotting the graph results in Fig. 12-12.

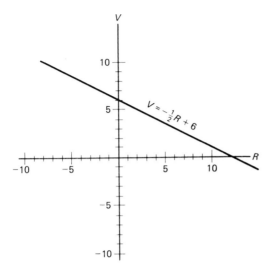

FIGURE 12-12 Plot of $V = -\frac{1}{2}R + 6$.

Higher-Order Equations

The highest exponential value carried by a variable of an equation determines its order. Thus far we have been studying first-order equations. Those with variables having exponents greater than 1 will not plot as a straight line. Therefore, more points will have to be calculated. However, we know what the general shape of these equations will be (Fig. 12-13). Second-order equations will be studied in detail

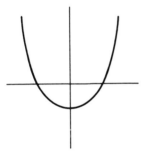

(a) Second order

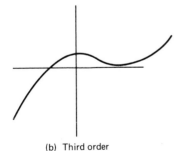

(b) Third order

FIGURE 12-13 Higher-order equations.

in Chapter 18. Third- and higher-order equations are usually not encountered in electronics.

EXAMPLE 12-7   Plot the equation $Y = 2X^2 + X - 4$.

SOLUTION   We have to choose several points.

X	Y	Calculation
−6	62	$Y = 2(-6)^2 + (-6) - 4 = 62$
−4	24	$Y = 2(-4)^2 + (-4) - 4 = 24$
−2	2	$Y = 2(-2)^2 + (-2) - 4 = 2$
0	−4	$Y = 2(0)^2 + (0) - 4 = -4$
2	6	$Y = 2(2)^2 + (2) - 4 = 6$
4	32	$Y = 2(4)^2 + (4) - 4 = 32$
6	74	$Y = 2(6)^2 + (6) - 4 = 74$

The graph is shown plotted in Fig. 12-14.

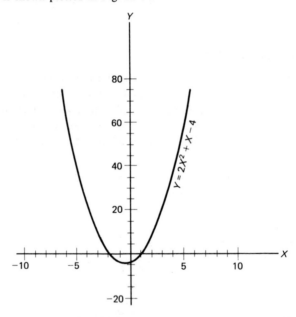

FIGURE 12-14   Example 12-7.

Solving Equations

Graphs can also be a very useful tool for solving equations. Consider the equation

$$3X - 4 = 0 \qquad (12\text{-}1)$$

There is only one value for X at which the equation is true: when X is $4\frac{1}{3}$. This can be shown as follows:

$$3X - 4 = 0$$
$$3X = 4$$
$$X = \frac{4}{3}$$

This is the analytical method of solving an equation. But we can use the graphical method, also, by changing the equation to

$$3X - 4 = Y \qquad (12\text{-}2)$$

When $Y = 0$, the correct value of X has been found. Let us plot Eq. (12-2) (Fig. 12-15). It is a straight line and the value of X at which Y is zero is the solution of Eq. (12-1).

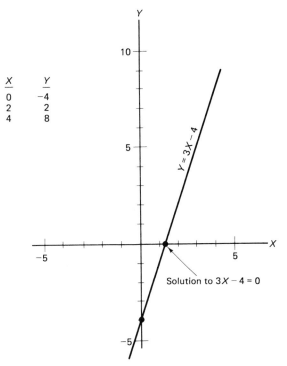

X	Y
0	-4
2	2
4	8

FIGURE 12-15 Graphical solution to the equation $3X - 4 = 0$.

EXAMPLE 12-8 Solve the following equation graphically:

$$X^2 - 2X - 15 = 0$$

SOLUTION The first step is to prepare a table of values for the equation

$$X^2 - 2X - 15 = Y$$

Once this is completed, the solution can be obtained by plotting the data.

X	Y	Computation
−6	33	$(-6)^2 - 2(-6) - 15 = 33$
−4	9	$(-4)^2 - 2(-4) - 15 = 9$
−2	−7	$(-2)^2 - 2(-2) - 15 = -7$
0	−15	$(0)^2 - 2(0) - 15 = -15$
2	−15	$(2)^2 - 2(2) - 15 = -15$
4	−7	$(4)^2 - 2(4) - 15 = -7$
6	9	$(6)^2 - 2(6) - 15 = 9$
8	33	$(8)^2 - 2(8) - 15 = 33$

The solution is shown in Fig. 12-16. Note that there are two values of X that satisfy the equation, −3 and +5. We can verify this by substitution:

$$X^2 - 2X - 15 = Y$$
$$(-3)^2 - 2(-3) - 15 = 0$$
$$(5)^2 - 2(5) - 15 = 0$$

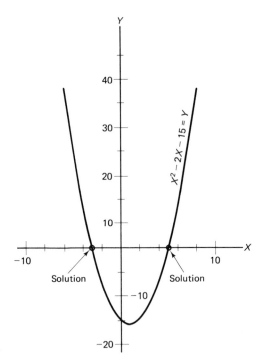

FIGURE 12-16 Example 12-8.

The programmable calculator is of great assistance in plotting these graphs. Merely enter the equation into the calculator. Then, by substituting various values of the independent variable, it will calculate the dependent value. Try it for Example 12-8.

PROBLEMS

12-9. In each of the following equations, state the independent variable(s) and the dependent variable. Also state the equation in function notation [Example: $Z = f(X, Y)$].

(a) $A = 3B + 6$

(b) $R_T = R_1 + R_2 + R_3$

(c) $A = \pi r^2$

(d) $B = \dfrac{f_0}{Q}$

(e) $C_T = \dfrac{1}{\dfrac{1}{C_1} + \dfrac{1}{C_2} + \dfrac{1}{C_3}}$

12-10. In each of the following equations, state the independent variable(s) and the dependent variable. Also state the equation in function notation.

(a) $f_x = mf_1 + nf_2$

(b) $L = 27 \times 10^{-10} \dfrac{D}{p^{5/3}} (1 + r^{-1})^{-5/3}$

(c) $y = \dfrac{y_{22} + y_1}{y_{11}(y_{22} + y_1) - y_{12} y_{21}}$

(d) $N_p = \dfrac{3.49 E_p \times 10^6}{f \Delta_c b_m}$

12-11. Plot the following equations:

(a) $Y = 3X + 2$

(b) $Y = -2X + 3$

(c) $Y = -3X + 4$

(d) $Y = -X + 3$

(e) $Y = 5$

12-12. Plot the following equations:

(a) $R = 3Z + 4$

(b) $R = 4Z + 2$

(c) $R = -Z + 6$

(d) $R = -\tfrac{3}{2} Z + 6$

(e) $R = 0.13Z + 6.25$

12-13. Plot the following equations:

(a) $Y = 3X^2 - 2X - 4$

(b) $Y = X^3 - 2X^2 + X + 3$

12-14. Plot the following equations:

(a) $M = -2N^2 + 3N - 3$

(b) $M = -N^3 + N^2 - N + 1$

12-15. Solve the following equations graphically and check your solution analytically:

(a) $X + 3 = 0$

(b) $3X + 6 = 0$

(c) $2X - 6 = 0$

(d) $3(4R + 6) - 3R = 0$

(e) $\dfrac{R + 6}{4} + 3 = 0$

(f) $\dfrac{R - 2}{3} + R + 6 = 0$

12-16. Solve the following equations graphically and check your solution analytically:

(a) $X - 2 = 0$

(b) $2Y + 3 = 0$

(c) $-2Z + 6 = 0$

(d) $3(2R + 3) - 3R = 0$

(e) $\dfrac{N+1}{3} - \dfrac{N-2}{4} = 0$

(f) $\dfrac{M+3}{6} + \dfrac{M}{2} + \dfrac{M+1}{3} = 0$

12-17. Solve graphically:

    (a) $X^2 - 4 = 0$

    (b) $X^2 + 7X + 12 = 0$

    (c) $X^2 - 6X + 8 = 0$

12-18. Solve graphically:

    (a) $M^2 - 9 = 0$

    (b) $M^2 + 5M + 6 = 0$

    (c) $R^2 + 2R - 24 = 0$

12-3 DEVELOPING EQUATIONS FROM GRAPHS

In the preceding section we had an equation and placed it on a graph. In this section we reverse the process and obtain an equation from a graph. We limit ourselves to straight-line graphs and therefore will obtain first-degree equations.

Let us first analyze a general form for a first-degree equation:

$$Y = mX + b \tag{12-3}$$

Example: $\qquad Y = \tfrac{1}{2}X + 4 \tag{12-4}$

Note the equation consists of four elements:

1. The dependent variable, Y
2. The independent variable, X
3. The numerical constant, m
4. The numerical constant, b

Using Eq. (12-4) as a basis, let us vary the b and see its effect (Fig. 12-17). Note that changing b changes the point at which the plot intersects with the Y axis. Note further that the number b always is the same value as the point at which the plot intersects the Y axis. This point of intersection is called the Y intercept. We can show that b represents the Y intercept by observing that, at this point, X equals zero. Set X to zero in Eqs. (12-3) and (12-4):

$$
\begin{aligned}
Y &= mX + b & Y &= 2X + 4 \\
&= m(0) + b & &= 2(0) + 4 \\
&= b & &= 4
\end{aligned}
$$

Therefore, since b represents the Y intercept, we can develop this part of our equation by reading b directly off the graph.

Whereas b represents the Y intercept, m represents the angle of the plot with respect to the horizontal direction (Fig. 12-18). This angle expressed by m is called the slope of the m. It can be measured directly off the graph by selecting two points along the plot (Fig. 12-19), where they are labeled A and B. Next measure how far it is from A to B in the horizontal direction, ΔX, and how far from A to B in the vertical direction, ΔY. The slope is calculated as $\Delta Y / \Delta X$. It is very important to pay attention to which direction one must move in the horizontal or vertical direction. If, in going from A to B, one must travel in the positive direction, call that number positive. If negative, give that quantity a minus sign.

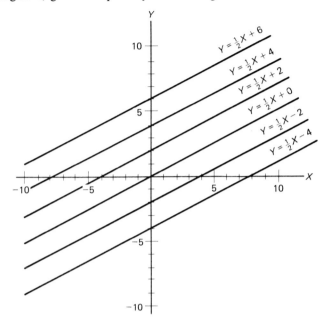

FIGURE 12-17 Effect of varying b in $Y = mX + b$.

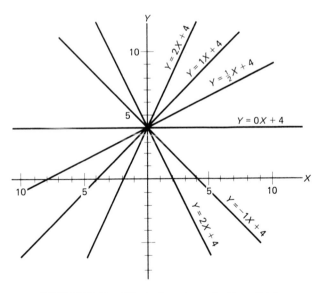

FIGURE 12-18 Effect of varying m in $Y = mX + b$.

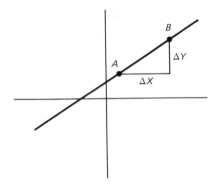

FIGURE 12-19 Measuring the slope.

EXAMPLE 12-9 Find the Y intercept and the slope for the plot shown in Fig. 12-20.

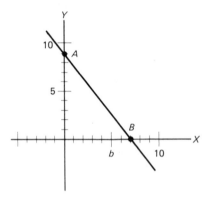

FIGURE 12-20 Example 12-9.

SOLUTION The Y intercept is, by inspection, 9 (point A). The slope can be found by selecting any two points along the plot. However, why not try for the easiest way out and select the Y intercept as one of the points? In this case, the X intercept (B) serves as the second point. We are now ready to calculate m:

$$m = \frac{\Delta Y}{\Delta X}$$

When going from point A to B, we must travel nine units in the negative direction vertically. Thus, $\Delta Y = -9$. At the same time, to go from A to B we must travel seven units in the positive horizontal direction. Therefore, $\Delta X = +7$. Calculate m:

$$m = \frac{\Delta Y}{\Delta X} = \frac{-9}{+7} \quad \text{or} \quad -1.286$$

Having picked both m and b off the graph, we can develop an equation by supplying these to Eq. (12-3).

EXAMPLE 12-10 Develop an equation for the plot shown in Fig. 12-20.

SOLUTION From Example 12-9, we know that m is $-\frac{9}{7}$ and b is 9. Therefore,

$$Y = mX + b$$
$$= -\frac{9}{7}X + 9$$
$$7Y = -9X + 63$$

EXAMPLE 12-11 Develop an equation for the plot shown in Fig. 12-21.

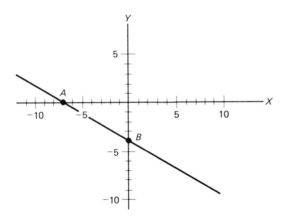

FIGURE 12-21 Example 12-11.

SOLUTION By inspection, the Y intercept is -4 (point B). Traveling from A to B, m is

$$m = \frac{\Delta Y}{\Delta X} = \frac{-4}{+7}$$

Therefore,

$$Y = mX + b$$
$$= -\tfrac{4}{7} X - 4$$
$$7Y = -4X - 28$$

PROBLEMS

12-19. Find the Y intercept, slope, and equation for each of the plots shown in Fig. 12-22.

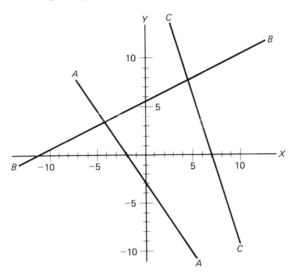

FIGURE 12-22 Problem 12-19.

12-20. Find the Y intercept, slope, and equation for each of the plots shown in Fig. 12-23.

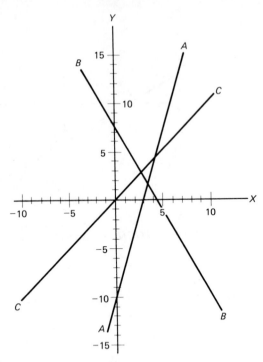

FIGURE 12-23 Problem 12-20.

12-21. A straight line runs through points (15, 33) and (−52, −20). Draw a graph for this line and develop its equation.

12-22. A straight line runs through points (−0.315, 0.264) and (0.131, 0.725). Draw a graph for this line and develop its equation.

13

SIMULTANEOUS EQUATIONS

Do you enjoy a good riddle? Then you should thoroughly enjoy this chapter, for simultaneous equations are precisely that: riddles. The techniques employed in this chapter open a brand new world to the solution of electronic problems, one that allows us to use the methods discussed in Chapter 15. In this chapter we first define simultaneous equations, then examine four techniques for their solution. In Chapter 14 we examine a rather unique method for solution.

13-1 INTRODUCTION

Just what is a simultaneous equation? It is a series of equations defining a series of variables. One such example could be as follows:

The sum of two voltages is 25 V and their difference is 5 V. What is the value of each voltage?

This problem has two different variables, the voltages. We shall call them V_1 and V_2. The problem also has two sets of conditions from which two equations can be derived. One equation is the sum of the voltages and the second equation is the difference of these same voltages. We can express these as follows:

$$\text{(A)} \quad V_1 + V_2 = 25 \tag{13-1}$$
$$\text{(B)} \quad V_1 - V_2 = 5 \tag{13-2}$$

Thus, we have two equations expressing the relationship between two variables. There is a general principle here as follows:

In order to obtain a unique solution, there must be the same number of equations as there are variables.

Thus, if we have four variables, *A, B, C,* and *D,* we must have four equations. Furthermore, these equations must be unique. That is, no one of the equations could be derived from the others. In our example, if we multiply the top equation by 2, we have

$$\text{(A)} \quad 2V_1 + 2V_2 = 50$$

This equation is not unique to the other two, for it is derived from one of these two. We shall use letters with a circle to identify which of the equations we are manipulating.

13-2 SOLUTION BY GRAPH

Two variable simultaneous equations can be solved quite easily by graphical methods. We have merely to plot each equation and the point at which they intersect is the solution, for it is only at this point that the X and Y values will be correct for both equations.

EXAMPLE 13-1 Solve Eqs. (13-1) and (13-2) by graph.

SOLUTION First, each is plotted (Fig. 13-1). Note that there is only one point on the entire graph at which they intersect, $V_1 = 15$, $V_2 = 10$. Thus, this is the solution. We can verify this solution by substituting these values back into the equations:

$$\text{\textcircled{A}} \quad V_1 + V_2 = 25 \qquad 15 + 10 = 25$$
$$\text{\textcircled{B}} \quad V_1 - V_2 = 5 \qquad 15 - 10 = 5$$

Note that these solutions simultaneously satisfy Eqs. (13-1) and (13-2).

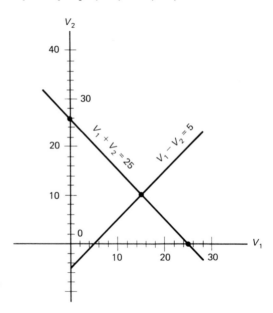

FIGURE 13-1 Example 3-1, simultaneous equations.

EXAMPLE 13-2 Solve by graphing:

$$2I_1 - 3I_2 = 6$$
$$I_1 + 6I_2 = -4$$

SOLUTION The two plots are shown in Fig. 13-2. The graphs must be interpreted, yielding a solution of 1.60 for I_1 and -0.93 for I_2. Substitute into the two equations:

$$\text{\textcircled{A}} \qquad 2I_1 - 3I_2 = 6$$
$$2(1.60) - 3(-0.93) = 6$$
$$5.99 = 6$$
$$\text{\textcircled{B}} \qquad I_1 + 6I_2 = -4$$
$$(1.60) + 6(-0.93) = -4$$
$$-3.98 = -4$$

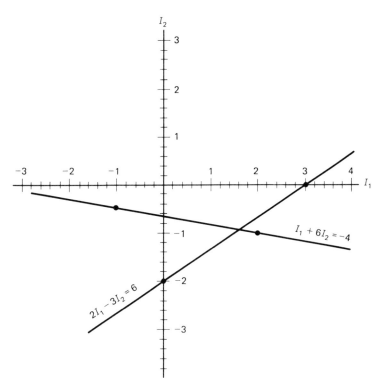

FIGURE 13-2 Example 13-2.

These solutions are as close as we can get (two places) using the graphical method.

In the foregoing example, there was a solution. However, there are two types of problems for which a solution does not exist. The first has already been mentioned: that which occurs when one equation can be derived from another. Consider the following two equations:

$$\text{(A)} \quad 2X + Y = 10$$
$$\text{(B)} \quad 6X + 3Y = 30$$

Note that Ⓑ is exactly three times that of Ⓐ. Plotting each results in the single line shown in Fig. 13-3, where there an infinite number of solutions.

A second type of problem occurs when the two lines are exactly parallel (Fig. 13-4). In this case, there is no value of X and Y that will simultaneously satisfy both equations. It can also be seen that the two equations cannot be true, for either $2X + Y = 5$ or $2X + Y = 7$.

PROBLEMS

13-1. The sum of two currents is 5 A and their difference is 3 A. Compute the currents by graphing the equations.

13-2. The sum of two resistances is 10 kΩ and their difference is 0 kΩ. Compute the resistances by graphs.

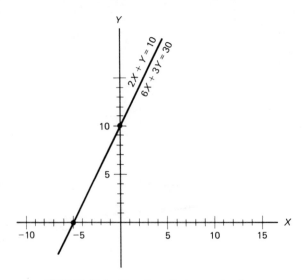

FIGURE 13-3 Equations that are not unique.

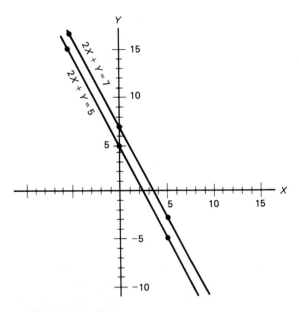

FIGURE 13-4 Simultaneous equations for which no solution exists.

13-3. Solve by graphing:

$$3X + 4Y = 2$$
$$2X - 3Y = 6$$

13-4. Solve by graphing:

$$X + 6Y = 12$$
$$2X + 3Y = 4$$

13-5. Mortimer Mindmiss had two types of coins in his pocket but could not remember how much each was worth. He found that when he took three of the first type and five of the second, the bank gave him a $1 bill. Further, he found that when he took seven of the first type and five of the second, the bank gave him a $2 bill. How much was each coin worth? Solve by graph.

13-6. Hennifer Holly purchased two types of integrated circuits (ICs) for a total of $6.20. One type cost $0.20 each and the other type $0.30 each. If she received a total of 24 ICs, how many of each type did she receive?

13-3 SOLUTION BY ADDITION

Although the graphical method allows us to visualize simultaneous equations, it has the dual disadvantage of being inaccurate and time consuming. The method used within this section is useful for simple problems that have integer coefficients. A solution by addition requires the following steps:

(a) Put the equations in standard form, such as

$$3X + 2Y = 50$$
$$2X - 4Y = 70$$

Note that the variables are beneath each other and the constant is on the right of the equation.

(b) Multiply one of the equations by a constant such that one of the variables will sum to zero in the next step.

(c) Add the two equations.

(d) Solve for one of the variables.

(e) Substitute this solution in one of the equations and solve for the second variable.

(f) Check the solution using the second equation.

EXAMPLE 13-3 Solve by addition:

$$ⓐ \quad 3X - 50 = -2Y$$
$$ⓑ \quad 70 = -4Y + 2X$$

SOLUTION

Step (a)

$$ⓐ \quad 3X + 2Y = 50$$
$$ⓑ \quad 2X - 4Y = 70$$

Step (b) If we multiply ⓐ by 2, we can add $4Y$ to $-4Y$, causing this variable to drop out:

$$ⓐ \quad 6X + 4Y = 100$$
$$ⓑ \quad 2X - 4Y = 70$$

Step (c) Add the two equations:

$$ⓐ \quad 6X + 4Y = 100$$
$$ⓑ \quad \underline{2X - 4Y = 70}$$
$$ⓐ+ⓑ \quad 8X = 170$$

Step (d)   Solve for X:

$$8X = 170$$
$$X = 21.25$$

Step (e)   Let us substitute this solution into Ⓐ and solve for Y:

$$6(21.25) + 4Y = 100$$
$$Y = -6.875$$

Step (f)   Substituting into Ⓑ to check our solution:

$$2(21.25) - 4(-6.875) = 70$$
$$70 = 70$$

The solutions check:

$$X = 21.25$$
$$Y = -6.875$$

EXAMPLE 13-4   Solve by addition:

$$\text{Ⓐ} \quad 4V_1 = 20 - 7V_2$$
$$\text{Ⓑ} \quad V_2 + V_1 = 20$$

SOLUTION

Step (a)

$$\text{Ⓐ} \quad 4V_1 + 7V_2 = 20$$
$$\text{Ⓑ} \quad V_1 + V_2 = 20$$

Step (b)   We shall multiply Ⓑ by -4:

$$\text{Ⓑ} \quad -4V_1 - 4V_2 = -80$$

Step (c)

$$
\begin{array}{rl}
\text{Ⓐ} & 4V_1 + 7V_2 = 20 \\
\text{Ⓑ} & -4V_1 - 4V_2 = -80 \\
\hline
\text{Ⓐ} + \text{Ⓑ} & 3V_2 = -60
\end{array}
$$

Step (d)

$$V_2 = -20$$

Step (e)   Substitute into Ⓑ:

$$\text{Ⓑ} \quad -20 + V_1 = 20$$
$$V_1 = 40$$

Step (f)   Substitute into Ⓐ

$$\text{Ⓐ} \quad 4V_1 = 20 - 7V_2$$
$$4(40) = 20 - 7(-20)$$
$$160 = 160$$

The answer is

$$V_1 = 40$$
$$V_2 = -20$$

PROBLEMS

Solve by addition:

13-7. $X + 2Y = 3$
$\quad\quad 2X - \ Y = 1$

13-8. $4R_1 + 3R_2 = 58$
$\quad\quad 2R_1 + 4R_2 = 44$

13-9. $4A + 2B = 18$
$\quad\quad 3A - \ B = \ 1$

13-10. $9P - 5Q = 11$
$\quad\quad\ 3P + 4Q = 32$

13-11. $2Z = 3X - 24$
$\quad\quad -2X + 24 = -4Z$

13-12. $4X = 3Y - 5$
$\quad\quad 2X + 5 = 2Y$

13-13. $16 = 2M + 3N$
$\quad\quad -3 - M = 4N$

13-14. $50L_1 + 6L_2 = 25$
$\quad\quad\ 36L_2 - 7L_1 = 33$

13-4 SOLUTION BY SUBSTITUTION

The technique of substitution can be summarized as:

(a) Solve for one of the variables in equation Ⓐ.
(b) Substitute the expression for this variable into Ⓑ.
(c) Substitute back into one of the equations to find the second variable.
(d) Check the solution using the other equation.

EXAMPLE 13-5 Solve by substitution:

$$Ⓐ \quad 2X + 3Y = 4$$
$$Ⓑ \quad 5X - 4Y = 3$$

SOLUTION

Step (a) We shall solve for X in Ⓐ:

$$Ⓐ \quad 2X + 3Y = 4$$
$$X = \frac{4 - 3Y}{2}$$

Step (b) We can now substitute the expression on the right of the equals sign in step (a) for the X variable in Ⓑ.

$$Ⓑ \quad 5X - 4Y = 3$$
$$Ⓑ \quad 5\left(\frac{4 - 3Y}{2}\right) - 4Y = 3$$
$$Ⓑ \quad 5(4 - 3Y) - 8Y = 6$$
$$Ⓑ \quad 20 - 15Y - 8Y = 6$$
$$Ⓑ \quad -23Y = -14$$
$$Ⓑ \quad Y = 0.6087$$

Step (c) We shall substitute into Ⓐ:

$$2X + 3(0.6087) = 4$$
$$X = 1.087$$

Step (d) Substitute into Ⓑ:

$$5(1.087) - 4(0.6087) = 3$$
$$3.000 = 3$$

Thus, the solution is

$$X = 1.087$$
$$Y = 0.6087$$

EXAMPLE 13-6 Three times the sum of two currents is 1.087 mA, whereas twice their difference is 290 μA. Find the currents.

SOLUTION We must first set up the equations. From the sum statement we have

(A) $\quad 3(I_1 + I_2) = 1.087$

From the difference statement

(B) $\quad 2(I_1 - I_2) = 0.290$

Step (a) Solve for I_1 in (A):

(A) $\quad 3(I_1 + I_2) = 1.087$
(A) $\quad 3I_1 + 3I_2 = 1.087$
(A) $\quad I_1 = \dfrac{1.087 - 3I_2}{3}$

Step (b)

(B) $\quad 2(I_1 - I_2) = 0.290$
(B) $\quad 2I_1 - 2I_2 = 0.290$
(A)(B) $\quad 2\left(\dfrac{1.087 - 3I_2}{3}\right) - 2I_2 = 0.290$
(A)(B) $\quad 2.174 - 6I_2 - 6I_2 = 0.870$
(A)(B) $\quad 12I_2 = 1.304$
(A)(B) $\quad I_2 = 0.1087 \text{ mA}$

Step (c) Substitute into (A):

(A) $\quad 3(I_1 + I_2) = 1.087$
(A) $\quad 3I_1 + 3I_2 = 1.087$
(A)(B) $\quad 3I_1 + 3(0.1087) = 1.087$
(A)(B) $\quad I_1 = 0.2537 \text{ mA}$

Step (d) Substitute into (B):

(B) $\quad 2(I_1 - I_2) = 0.290$
(B) $\quad 2(0.2537 - 0.1087) = 0.290$
(B) $\quad 0.2900 = 0.290$

Thus, the answers are

$$I_1 = 253.7 \text{ μA}$$
$$I_2 = 108.7 \text{ μA}$$

PROBLEMS

Solve by substitution:

13-15. $\quad 2W + X = 30$
$\quad\quad\quad 3W + 2X = 40$

13-16. $\quad 9M + 10Q = 27$
$\quad\quad\quad 11M - 3Q = 40$

13-17. $8M = 3N + 25$
$7N = 3M + 6$

13-18. $3L_1 + L_2 = 32$
$3L_1 - 2L_2 = 5$

13-19. $25M_1 + 6M_2 = 180$
$20M_1 + 9M_2 = 250$

13-20. $15X_1 - 10X_2 = 12$
$5X_1 + 6X_2 = 25$

13-5 SOLUTION BY EQUALITY

The method of equality is very similar to that of solution by substitution. The steps can be summarized as:

(a) Solve for one of the variables in Ⓐ.
(b) Solve for that same variable in Ⓑ.
(c) Set these two expressions equal to one another and solve for the variable.
(d) Substitute back into one of the equations to find the second variable.
(e) Check the solution using the other equation.

EXAMPLE 13-7 Solve by equality:

Ⓐ $4S + 12T = 7$
Ⓑ $7S + 9T = 6$

SOLUTION

Step (a)   We shall solve for S in Ⓐ:

Ⓐ $4S + 12T = 7$

Ⓐ $\quad S = \dfrac{7 - 12T}{4}$

Step (b)   We shall solve for S in Ⓑ:

Ⓑ $7S + 9T = 6$

Ⓑ $\quad S = \dfrac{6 - 9T}{7}$

Step (c)   We can now set these equal to each other:

$$\frac{7 - 12T}{4} = \frac{6 - 9T}{7}$$
$$49 - 84T = 24 - 36T$$
$$48T = 25$$
$$T = 0.5208$$

Step (d)   We can now substitute T into Ⓐ and solve for S:

$$4S + 12(0.5208) = 7$$
$$S = 0.1875$$

Step (e)   Checking with Ⓑ:

$$7(0.1875) + 9(0.5208) = 6$$
$$6.000 = 6$$

Thus, the solution is

$$S = 0.1875$$
$$T = 0.5208$$

EXAMPLE 13-8 Solve by equality:

$$\text{Ⓐ} \quad 26D + 41E = 32$$
$$\text{Ⓑ} \quad 12D - 16E = 3$$

SOLUTION

Step (a) Solve for D in Ⓐ:

$$\text{Ⓐ} \quad D = \frac{32 - 41E}{26}$$

Step (b) Solve for D in Ⓑ:

$$\text{Ⓑ} \quad D = \frac{16E + 3}{12}$$

Step (c) Set these equal to each other:

$$\frac{32 - 41E}{26} = \frac{16E + 3}{12}$$
$$384 - 492E = 416E + 78$$
$$908E = 306$$
$$E = 0.3370$$

Step (d) Substitute into Ⓐ:

$$26D + 41(0.3370) = 32$$
$$D = 0.6993$$

Step (e) Check with Ⓑ:

$$12(0.6993) - 16(0.3370) = 3$$
$$3.000 = 3$$

Thus, the solution is

$$D = 0.6993$$
$$E = 0.3370$$

PROBLEMS

Solve by equality:

13-21. $4X + 6Y = 20$
$3X - 2Y = 12$

13-22. $13A - 9B = -10$
$15A + 2B = 13$

13-23. $26Y_1 + 3Y_2 = 50$
$4Y_1 - 10Y_2 = -30$

13-24. $1.03A_5 = 6A_6 + 7.21$
$3.92A_6 - 4.16A_5 = -50.31$

13-25. $26.3N - 6.21M = -3.69$
$8.31 = 2.65M + 7.27N$

13-26. $126R_1 = 312R_2 - 615$
$341R_2 = 12R_1 - 633$

13-6 THREE VARIABLES

A three-variable problem requires three equations, which we shall label Ⓐ, Ⓑ, and Ⓒ. Figure 13-5 illustrates the general approach used to solve for the three unknowns. Equations Ⓐ and Ⓑ are first combined by one of the foregoing techniques to eliminate

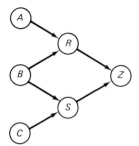

FIGURE 13-5 Three variable problems.

one of the variables. The resulting equation we shall call Ⓡ. Next, Ⓑ and Ⓒ are combined to eliminate the same variable, resulting in Ⓢ. We now have two equations in two unknowns, and these can be solved as shown previously for two of the variables. Once these two have been found, the third can be found by substitution.

EXAMPLE 13-9 Solve the following simultaneous equations:

$$\begin{array}{ll} Ⓐ & 2X - Y + Z = -4 \\ Ⓑ & X - 2Y - 3Z = -19 \\ Ⓒ & 3X + 2Y + 2Z = 3 \end{array}$$

SOLUTION We must first combine Ⓐ and Ⓑ to eliminate a variable. Let us eliminate X. Multiply Ⓑ by -2 and add to Ⓐ:

$$\begin{array}{ll} Ⓑ & -2X + 4Y + 6Z = 38 \\ Ⓐ & \underline{2X - Y + Z = -4} \\ Ⓡ & 3Y + 7Z = 34 \end{array}$$

Combine Ⓑ and Ⓒ by multiplying Ⓑ by -3 to eliminate X:

$$\begin{array}{ll} Ⓑ & -3X + 6Y + 9Z = 57 \\ Ⓒ & \underline{3X + 2Y + 2Z = 3} \\ Ⓢ & 8Y + 11Z = 60 \end{array}$$

We shall solve Ⓡ and Ⓢ by equality:

$$\begin{array}{ll} Ⓡ & 3Y + 7Z = 34 \\ Ⓡ & Y = \dfrac{34 - 7Z}{3} \\ Ⓢ & 8Y + 11Z = 60 \\ Ⓢ & Y = \dfrac{60 - 11Z}{8} \\ ⓇⓈ & \dfrac{60 - 11Z}{8} = \dfrac{34 - 7Z}{3} \\ ⓇⓈ & 180 - 33Z = 272 - 56Z \\ ⓇⓈ & 23Z = 92 \\ ⓇⓈ & Z = 4 \end{array}$$

Substitute into Ⓡ to obtain Y:

$$\begin{array}{ll} Ⓡ & 3Y + 7(4) = 34 \\ & Y = 2 \end{array}$$

Substitute Y and Z into Ⓐ to obtain X:

$$2X - (2) + (4) = -4$$
$$X = -3$$

We can check the answer by substituting into Ⓒ:

$$3(-3) + 2(2) + 2(4) = 3$$
$$3 = 3$$

The answers are, therefore,

$$X = -3$$
$$Y = 2$$
$$Z = 4$$

EXAMPLE 13-10 Solve the following simultaneous equations:

Ⓐ $\quad 2D = 3E - 4F + 14$
Ⓑ $\quad 6D - 5E - 2F = 60$
Ⓒ $\quad 5D - 4E = -7F + 23$

SOLUTION Multiply Ⓑ by 2 and combine with Ⓐ to eliminate F:

Ⓑ $\quad 12D - 10E - 4F = 120$
Ⓐ $\quad \underline{2D - 3E + 4F = 14}$
Ⓡ $\quad 14D - 13E = 134$

Combine Ⓑ and Ⓒ by equality to eliminate F:

Ⓑ $\quad F = \dfrac{60 - 6D + 5E}{-2}$

Ⓒ $\quad F = \dfrac{23 - 5D + 4E}{7}$

Ⓢ $\quad \dfrac{23 - 5D + 4E}{7} = \dfrac{60 - 6D + 5E}{-2}$

Ⓢ $\quad -46 + 10D - 8E = 420 - 42D + 35E$

Ⓢ $\quad 52D - 43E = 466$

Combine Ⓡ and Ⓢ by equality to find D:

Ⓡ $\quad D = \dfrac{134 + 13E}{14}$

Ⓢ $\quad D = \dfrac{466 + 43E}{52}$

ⓇⓈ $\quad \dfrac{466 + 43E}{52} = \dfrac{134 + 13E}{14}$

$$6524 + 602E = 6968 + 676E$$
$$74E = -444$$
$$E = -6$$

Substitute into Ⓡ:

Ⓡ $\quad D = \dfrac{134 + 13(-6)}{14} = 4$

Substitute D and E into Ⓐ:

$$2(4) = 3(-6) - 4F + 14$$
$$F = -3$$

Check the solution by substituting into Ⓒ:

$$5(4) - 4(-6) = -7(-3) + 23$$
$$44 = 44$$

Thus, the solution is

$$D = 4$$
$$E = -6$$
$$F = -3$$

PROBLEMS

Solve the following simultaneous equations:

13.27. $\begin{aligned} W + 2X - Y &= 9 \\ 2W - 3X + 2Y &= -7 \\ W + 3X + Y &= 10 \end{aligned}$

13-28. $\begin{aligned} 3R_1 + R_2 - 3R_3 &= -12 \\ R_1 + 3R_2 - 9R_3 &= -4 \\ 2R_1 - 2R_2 - R_3 &= -22 \end{aligned}$

13-29. $\begin{aligned} 6C_1 - 2C_2 + 3C_3 &= 7 \\ 4C_1 &= -2 + 6C_3 + 5C_2 \\ 27 - 2C_3 + C_1 - 8C_2 &= 0 \end{aligned}$

13-30. $\begin{aligned} -3M_1 &= 2M_2 - 2M_3 - 11 \\ 2M_2 - 6M_3 &= 3M_1 - 55 \\ -7M_3 - 6M_2 &= 5M_1 - 130 \end{aligned}$

13-31. $\begin{aligned} 4P - 2Q &= 24 \\ 3P + 2Q - 5R &= 4 \\ 5Q - 4R &= -20 \end{aligned}$

13-32. $\begin{aligned} 3L_1 - 2L_2 - L_3 &= 3 \\ 4L_1 + 3L_2 &= 6L_3 - 19 \\ 2L_2 - 6L_3 &= -19 \end{aligned}$

13-7 LITERAL EQUATIONS

The preceding work considered numerical coefficients. However, the methods discussed apply to any type of coefficients.

EXAMPLE 13-11 Solve the following simultaneous equations for V_1 and V_2:

$$\text{Ⓐ} \quad AV_1 + BV_2 = C$$
$$\text{Ⓑ} \quad RV_1 + SV_2 = T$$

SOLUTION The methods of substitution and equality are most useful for this type of problem. Use the latter to solve for V_1:

$$\text{Ⓐ} \quad AV_1 + BV_2 = C$$
$$\text{Ⓐ} \quad V_2 = \frac{C - AV_1}{B}$$
$$\text{Ⓑ} \quad RV_1 + SV_2 = T$$
$$\text{Ⓑ} \quad V_2 = \frac{T - RV_1}{S}$$
$$\text{Ⓐ Ⓑ} \quad \frac{C - AV_1}{B} = \frac{T - RV_1}{S}$$
$$\text{Ⓐ Ⓑ} \quad CS - ASV_1 = TB - BRV_1$$
$$\text{Ⓐ Ⓑ} \quad BRV_1 - ASV_1 = TB - CS$$

$$\text{Ⓐ Ⓑ} \quad V_1(BR - AS) = TB - CS$$

$$\text{Ⓐ Ⓑ} \quad V_1 = \frac{TB - CS}{BR - AS}$$

Substitute into Ⓐ to solve for V_2:

$$\text{Ⓐ} \quad AV_1 + BV_2 = C$$

$$\text{Ⓐ Ⓑ} \quad A\left(\frac{TB - CS}{BR - AS}\right) + BV_2 = C$$

$$A(TB - CS) + BV_2(BR - AS) = C(BR - AS)$$

$$BV_2(BR - AS) = C(BR - AS) - A(TB - CS)$$

$$V_2 = \frac{C(BR - AS) - A(TB - CS)}{B(BR - AS)}$$

$$V_2 = \frac{BCR - ACS - ABT + ACS}{B^2R - ABS}$$

$$V_2 = \frac{CR - AT}{BR - AS}$$

PROBLEMS

13-33. Solve for X and Y:

$$AX + BY = C$$
$$DX + EY = F$$

13-34. Solve for R and S:

$$(A + 1)R - 3S = B$$
$$CR - (B + C)S = A + B$$

14

DETERMINANTS

Although the methods shown in Chapter 13 are valuable for simultaneous equations with integer coefficients, the method shown in this chapter is appropriate for any set of simultaneous equations. The advent of the hand-held calculator has now made the determinant method even more attractive. In fact, the author is checking his work in this chapter by means of a program he developed for his calculator using determinants. In this chapter we first define the determinant matrix and how it is evaluated. Next, we examine a method of solving simultaneous equations called Cramer's rule. Finally, we examine a second determinant method called pivotal reduction. It is this method that is frequently programmed into calculators and computers.

14-1 THE DETERMINANT MATRIX

The determinant method is a cookbook method of solving simultaneous equations whereby the coefficients of the variables are set down in an array similar to tic-tac-toe. These numbers are then manipulated in a regular fashion and voilà—the answer appears.

Two Variables

The two variable determinant matrix has the form of

$$\begin{vmatrix} a & b \\ c & d \end{vmatrix}$$

where a, b, c, and d are numbers. This matrix has a value of

$$\begin{vmatrix} a & b \\ c & d \end{vmatrix} = ad - bc$$

Note that the arrows show what numbers are to be multiplied. Those products formed in the positive direction are given positive signs. Those products formed in the negative direction are given a negative sign.

EXAMPLE 14-1 Evaluate:

$$\begin{vmatrix} 3 & -4 \\ 2 & 6 \end{vmatrix}$$

SOLUTION

$$\begin{vmatrix} 3 & -4 \\ 2 & 6 \end{vmatrix} = [(3)(6)] - [(-4)(2)] = 26$$

Note that the signs of the products are determined by whether we are on a northwest-southeast (+) diagonal or a northeast-southwest (−) diagonal.

EXAMPLE 14-2 Evaluate:

$$\begin{vmatrix} -14 & 3 \\ -6 & 2 \end{vmatrix}$$

SOLUTION

$$\begin{vmatrix} -14 & 3 \\ -6 & 2 \end{vmatrix} = [(-14)(2)] - [(3)(-6)] = -10$$

Three Variables

Three variable determinant matrices are evaluated in precisely the same manner as two variables. However, to make the visualization easier, we have copied the first two columns.

$$= [aei + bfg + cdh] - [ceg + afh + bdi]$$

Note that there are three product terms in the positive direction and three in the negative direction.

EXAMPLE 14-3 Evaluate:

$$\begin{vmatrix} 3 & 6 & 7 \\ 2 & 4 & -2 \\ -1 & -5 & 1 \end{vmatrix}$$

SOLUTION Repeat the first two columns and draw arrows to indicate products:

$$= [(3)(4)(1) + (6)(-2)(-1) + (7)(2)(-5)] - [(7)(4)(-1) + (3)(-2)(-5) + (6)(2)(1)]$$
$$= [12 + 12 - 70] - [-28 + 30 + 12]$$
$$= -46 - 14$$
$$= -60$$

EXAMPLE 14-4   Evaluate:

$$\begin{vmatrix} 17 & 0 & 3 \\ 6 & -1 & -4 \\ -2 & 7 & -16 \end{vmatrix}$$

SOLUTION

$$[(17)(-1)(-16) + (0)(-4)(-2) + (3)(6)(7)] - [(3)(-1)(-2) + (-4)(7)(17)$$
$$+ (-16)(0)(6)]$$
$$= [272 + 0 + 126] - [6 - 476 + 0]$$
$$= 398 + 470$$
$$= 868$$

More Than Three Variables

This method we have been using can only be applied to two or three variables. Those determinants having four or more variables must be solved using the techniques of Section 14-2 or 14-3.

PROBLEMS

Evaluate the following determinants:

14-1. $\begin{vmatrix} 2 & 5 \\ 1 & 6 \end{vmatrix}$

14-2. $\begin{vmatrix} 4 & -2 \\ 6 & 1 \end{vmatrix}$

14-3. $\begin{vmatrix} -6 & -4 \\ 2 & -1 \end{vmatrix}$

14-4. $\begin{vmatrix} 8 & -4 \\ 4 & -2 \end{vmatrix}$

14-5. $\begin{vmatrix} 17.3 & -3.15 \\ 18.2 & 6.27 \end{vmatrix}$

14-6. $\begin{vmatrix} -17.6 & -17.3 \\ -31.2 & -2.6 \end{vmatrix}$

14-7. $\begin{vmatrix} 1 & 4 & 7 \\ 2 & 5 & 8 \\ 3 & 6 & 9 \end{vmatrix}$

14-8. $\begin{vmatrix} 4 & -6 & 2 \\ 3 & -4 & -6 \\ 5 & 10 & 8 \end{vmatrix}$

14-9. $\begin{vmatrix} -6.2 & -7.3 & 3.17 \\ 2.1 & 7.3 & 7.4 \\ -3.6 & -4.9 & -2.8 \end{vmatrix}$

14-10. $\begin{vmatrix} -8.3 & -17 & -35 \\ 3.6 & 4.6 & 5.6 \\ 6.6 & 7.6 & -8.6 \end{vmatrix}$

14-11. $\begin{vmatrix} 173 & -216 & 14.3 \\ 431 & 872 & 165 \\ -17 & 0 & 888 \end{vmatrix}$

14-12. $\begin{vmatrix} 444 & 0.0157 & -3.6 \\ 876 & -0.00132 & 8.6 \\ 777 & 0.163 & -10 \end{vmatrix}$

14-2 CRAMER'S RULE FOR EQUATIONS

In this section we study the solution of simultaneous equations using a technique called Cramer's rule.

Two Variables

To solve a two-variable problem, the equations must first be put in proper form such that the variables and their coefficients are on the left side of the equals sign and the constants on the right. Additionally, the X terms and Y terms must be lined up. For example, the following are in proper form:

$$2X - 3Y = -13$$
$$4X + 5Y = 7$$

Note that the variables and their coefficients are lined up beneath each other and that the constants are on the right of the equals sign.

The next step is to form the denominator determinant by placing all the variable coefficients in a 2 x 2 format. Note that the signs are retained:

$$\begin{vmatrix} 2 & -3 \\ 4 & 5 \end{vmatrix}$$

To solve for X, form a numerator determinant by substituting the constants on the right for the X coefficients in the denominator determinant; divide this by the denominator determinant:

$$X = \frac{\begin{vmatrix} -13 & -3 \\ 7 & 5 \end{vmatrix}}{\begin{vmatrix} 2 & -3 \\ 4 & 5 \end{vmatrix}}$$

We are now ready to evaluate the determinants according to the equation

$$\begin{vmatrix} a & c \\ b & d \end{vmatrix} = ad - bc$$

Thus, solve for X:

$$X = \frac{\begin{vmatrix} -13 & -3 \\ 7 & 5 \end{vmatrix}}{\begin{vmatrix} 2 & -3 \\ 4 & 5 \end{vmatrix}} = \frac{(-13 \times 5) - (-3 \times 7)}{(2 \times 5) - (-3 \times 4)} = \frac{-44}{22} = -2 \qquad X = -2$$

In a similar manner, we can solve for Y by forming its numerator determinant and dividing by the denominator determinant. The numerator is formed by substituting the constants on the right of the equation for the Y coefficients:

$$Y = \frac{\begin{vmatrix} 2 & -13 \\ 4 & 7 \end{vmatrix}}{\begin{vmatrix} 2 & -3 \\ 4 & 5 \end{vmatrix}} = \frac{(2 \times 7) - (-13 \times 4)}{(2 \times 5) - (-3 \times 4)} = \frac{66}{22} = 3$$

EXAMPLE 14-5 Solve the equations:

$$4V_1 = 10V_2 - 16$$
$$7V_2 - 5V_1 - 12 = 0$$

SOLUTION Put them into standard form:

$$4V_1 - 10V_2 = -16$$
$$-5V_1 + 7V_2 = 12$$

Solve for V_1:

$$V_1 = \frac{\begin{vmatrix} -16 & -10 \\ 12 & 7 \end{vmatrix}}{\begin{vmatrix} 4 & -10 \\ -5 & 7 \end{vmatrix}} = \frac{(-16 \times 7) - (-10 \times 12)}{(4 \times 7) - [-10 \times (-5)]} = \frac{8}{-22} = -0.3636 \text{ V}$$

Solve for V_2:

$$V_2 = \frac{\begin{vmatrix} 4 & -16 \\ -5 & 12 \end{vmatrix}}{-22} = \frac{(4 \times 12) - [-5 \times (-16)]}{-22} = \frac{-32}{-22} = 1.455 \text{ V}$$

EXAMPLE 14-6   Solve the equations:

$$13.6A - 36.5 = 2.65B$$
$$17.21 - 13.78B = 2.05A$$

SOLUTION   Put into standard form:

$$13.6A - 2.65B = 36.5$$
$$2.05A + 13.78B = 17.21$$

Solve for A:

$$A = \frac{\begin{vmatrix} 36.5 & -2.65 \\ 17.21 & 13.78 \end{vmatrix}}{\begin{vmatrix} 13.6 & -2.65 \\ 2.05 & 13.78 \end{vmatrix}} = \frac{(36.5 \times 13.78) - (-2.65 \times 17.21)}{(13.6 \times 13.78) - (-2.65 \times 2.05)}$$

$$= \frac{502.97 + 45.61}{187.41 + 5.433} = \frac{548.58}{192.84} = 2.845$$

Solve for B:

$$B = \frac{\begin{vmatrix} 13.6 & 36.5 \\ 2.05 & 17.21 \end{vmatrix}}{192.84} = \frac{(13.6 \times 17.21) - (36.5 \times 2.05)}{129.84}$$

$$= \frac{234.06 - 74.825}{192.84} = \frac{159.23}{192.84} = 0.8257$$

Three Variables

Three-variable equations are solved the same way as the two-variable:

(a)   Put the equations in standard form.
(b)   Form a 3 x 3 denominator determinant from the coefficients of the variables.
(c)   Form numerator determinants by substituting the constants on the right of the equation for the coefficients of the variable under consideration.

EXAMPLE 14-7   Solve the following simultaneous equations:

$$3X + 1 - Z = 0$$
$$2X = 14 + 4Y - Z$$
$$-3X - Y = -1$$

SOLUTION Rearrange in standard form:

$$3X+0Y-\ Z=-1$$
$$2X-4Y+\ Z=\ 14$$
$$-3X-1Y+0Z=-1$$

Form determinants:

$$X=\frac{\begin{vmatrix} -1 & 0 & -1 \\ 14 & -4 & 1 \\ -1 & -1 & 0 \end{vmatrix}}{\begin{vmatrix} 3 & 0 & -1 \\ 2 & -4 & 1 \\ -3 & -1 & 0 \end{vmatrix}} \qquad Y=\frac{\begin{vmatrix} 3 & -1 & -1 \\ 2 & 14 & 1 \\ -3 & -1 & 0 \end{vmatrix}}{\begin{vmatrix} 3 & 0 & -1 \\ 2 & -4 & 1 \\ -3 & -1 & 0 \end{vmatrix}} \qquad Z=\frac{\begin{vmatrix} 3 & 0 & -1 \\ 2 & -4 & 14 \\ -3 & -1 & -1 \end{vmatrix}}{\begin{vmatrix} 3 & 0 & -1 \\ 2 & -4 & 1 \\ -3 & -1 & 0 \end{vmatrix}}$$

Solve for each of the variables:

$$X=\frac{\begin{vmatrix} -1 & 0 & -1 \\ 14 & -4 & 1 \\ -1 & -1 & 0 \end{vmatrix}\begin{matrix} -1 & 0 \\ 14 & -4 \\ -1 & -1 \end{matrix} \begin{matrix} +(-1)(-4)(0) + (0)(1)(-1) \\ +(-1)(14)(-1) - (-1)(-4)(-1) \\ -(-1)(1)(-1) - (0)(14)(0) \end{matrix}}{\begin{vmatrix} 3 & 0 & -1 \\ 2 & -4 & 1 \\ -3 & -1 & 0 \end{vmatrix}\begin{matrix} 3 & 0 \\ 2 & -4 \\ -3 & -1 \end{matrix} \begin{matrix} +(3)(-4)(0) + (0)(1)(-3) \\ +(-1)(2)(-1) - (-1)(-4)(-3) \\ -(3)(1)(-1) - (0)(2)(0) \end{matrix}}$$

$$=\frac{17}{17}=1$$

$$Y=\frac{\begin{vmatrix} 3 & -1 & -1 \\ 2 & 14 & 1 \\ -3 & -1 & 0 \end{vmatrix}\begin{matrix} 3 & -1 \\ 2 & 14 \\ -3 & -1 \end{matrix} \begin{matrix} +(3)(14)(0) + (-1)(1)(-3) \\ +(-1)(2)(-1) - (-3)(14)(-1) \\ -(-1)(1)(3) - (0)(2)(-1) \end{matrix}}{17}$$

$$=\frac{-34}{17}=-2$$

$$Z=\frac{\begin{vmatrix} 3 & 0 & -1 \\ 2 & -4 & 14 \\ -3 & -1 & -1 \end{vmatrix}\begin{matrix} 3 & 0 \\ 2 & -4 \\ -3 & -1 \end{matrix} \begin{matrix} +(3)(-4)(-1) + (0)(14)(-3) \\ +(-1)(2)(-1) - (-3)(-4)(-1) \\ -(-1)(14)(3) - (-1)(2)(0) \end{matrix}}{17}$$

$$=\frac{68}{17}=4$$

EXAMPLE 14-8 Solve the following equations:

$$3I_1 = 2I_2 + 6I_3$$
$$4I_1 - I_3 = 6$$
$$7I_1 + 6I_2 - I_3 - 7 = 0$$

SOLUTION Rearrange in standard form:

$$3I_1 - 2I_2 - 6I_3 = 0$$
$$4I_1 + 0I_2 - 1I_3 = 6$$
$$7I_1 + 6I_2 - 1I_3 = 7$$

Solve for each of the variables:

$$I_1 = \frac{\begin{vmatrix} 0 & -2 & -6 \\ 6 & 0 & -1 \\ 7 & 6 & -1 \\ 3 & -2 & -6 \\ 4 & 0 & -1 \\ 7 & 6 & -1 \end{vmatrix}}{\begin{vmatrix} 3 & -2 & -6 \\ 4 & 0 & -1 \\ 7 & 6 & -1 \end{vmatrix}} \quad \begin{array}{l} +(0)(0)(-1) + (-2)(-1)(7) \\ +(-6)(6)(6) - (7)(0)(-6) \\ -(6)(-1)(0) - (-1)(6)(-2) \\ +(3)(0)(-1) + (-2)(-1)(7) \\ +(-6)(4)(6) - (7)(0)(-6) \\ -(6)(-1)(3) - (-1)(4)(-2) \end{array} = \frac{-214}{-120} = 1.783 \text{ A}$$

$$I_2 = \frac{\begin{vmatrix} 3 & 0 & -6 \\ 4 & 6 & -1 \\ 7 & 7 & -1 \end{vmatrix}}{-120} \quad \begin{array}{l} +(3)(6)(-1) + (0)(-1)(7) \\ +(-6)(4)(7) - (7)(6)(-6) \\ -(7)(-1)(3) - (-1)(4)(0) \end{array} = \frac{87}{-120} = 0.7250 \text{ A}$$

$$I_3 = \frac{\begin{vmatrix} 3 & -2 & 0 \\ 4 & 0 & 6 \\ 7 & 6 & 7 \end{vmatrix}}{-120} \quad \begin{array}{l} +(3)(0)(7) + (-2)(6)(7) \\ +(0)(4)(6) - (7)(0)(0) \\ -(6)(6)(3) - (7)(4)(-2) \end{array} = \frac{-136}{-120} = 1.133 \text{ A}$$

Minors

We cannot use the diagonal multiplication technique on problems of greater than three variables. Therefore, we must study a technique whereby larger problems can be broken down into smaller problems. The minor of a determinant can be defined as the determinant remaining after one row and one column have been deleted. Assume, for example, the following fourth-order determinant:

$$\begin{vmatrix} 1 & 2 & 3 & 4 \\ 5 & 6 & 7 & 8 \\ 9 & 10 & 11 & 12 \\ 13 & 14 & 15 & 16 \end{vmatrix}$$

The following are some examples of minors of this determinant. Note that each minor is associated with the number at the intersection of the row and column deleted:

The minor of the number 1:

$$\begin{vmatrix} 6 & 7 & 8 \\ 10 & 11 & 12 \\ 14 & 15 & 16 \end{vmatrix} \quad \text{(First row and first column deleted)}$$

The minor of the number 2:

$$\begin{vmatrix} 5 & 7 & 8 \\ 9 & 11 & 12 \\ 13 & 15 & 16 \end{vmatrix} \quad \text{(First row and second column deleted)}$$

The minor of the number 11:

$$\begin{vmatrix} 1 & 2 & 4 \\ 5 & 6 & 8 \\ 13 & 14 & 16 \end{vmatrix} \quad \text{(Third row and third column deleted)}$$

Note that these minors also contain 2 × 2 minors. For example, some minors of the last determinant would be:

$$\begin{vmatrix} 6 & 8 \\ 14 & 16 \end{vmatrix}$$

$$\begin{vmatrix} 5 & 8 \\ 13 & 16 \end{vmatrix}$$

$$\begin{vmatrix} 1 & 4 \\ 5 & 8 \end{vmatrix}$$

Evaluating Higher-Order Determinants

A fourth-order determinant can be evaluated as a sum of products. Each product consists of a minor and its associated number (that is, the number at the intersection of the row and column deleted). Each product will have a sign given it in accordance with the position of the associated number as follows:

$$\begin{vmatrix} + & - & + & - \\ - & + & - & + \\ + & - & + & - \\ - & + & - & + \end{vmatrix}$$

The factors are formed by finding all the minors of a particular row or a particular column. These summed factors are called cofactors.

EXAMPLE 14-9 Evaluate the following determinant:

$$\begin{vmatrix} 1 & 2 & 3 & 4 \\ 5 & 6 & 7 & 8 \\ 9 & 10 & 11 & 12 \\ 13 & 14 & 15 & 16 \end{vmatrix}$$

SOLUTION We shall arbitrarily follow row 1, finding all its minors:

$$= +(1)\begin{vmatrix} 6 & 7 & 8 \\ 10 & 11 & 12 \\ 14 & 15 & 16 \end{vmatrix} - (2)\begin{vmatrix} 5 & 7 & 8 \\ 9 & 11 & 12 \\ 13 & 15 & 16 \end{vmatrix} + (3)\begin{vmatrix} 5 & 6 & 8 \\ 9 & 10 & 12 \\ 13 & 14 & 16 \end{vmatrix} - (4)\begin{vmatrix} 5 & 6 & 7 \\ 9 & 10 & 11 \\ 13 & 14 & 15 \end{vmatrix}$$

This expression is the value of the original 4 × 4 matrix. Evaluate these:

$$
\begin{aligned}
&= +(1)\{[(6)(11)(16) + (7)(12)(14) + (8)(10)(15)] \\
&\quad - [(8)(11)(14) + (7)(10)(16) + (6)(12)(15)]\} \\
&\quad - (2)\{[(5)(11)(16) + (7)(12)(13) + (8)(9)(15)] \\
&\quad - [(8)(11)(13) + (7)(9)(16) + (5)(12)(15)]\} \\
&\quad + (3)\{[(5)(10)(16) + (6)(12)(13) + (8)(9)(14)] \\
&\quad - [(8)(10)(13) + (6)(9)(16) + (5)(12)(14)]\} \\
&\quad - (4)\{[(5)(10)(15) + (6)(11)(13) + (7)(9)(14)] \\
&\quad - [(7)(10)(13) + (6)(9)(15) + (5)(11)(14)] \\
&= (1)(3432 - 3432) - (2)(3052 - 3052) \\
&\quad + (3)(2744 - 2744) - (4)(2490 - 2490) \\
&= (1)(0) - (2)(0) + (3)(0) - (4)(0) \\
&= 0
\end{aligned}
$$

Thus, the value of this particular determinant is 0.

EXAMPLE 14-10 Evaluate the following determinant:

$$\begin{vmatrix} 2 & -1 & 3 & 6 \\ 4 & -7 & -8 & 9 \\ -10 & 11 & 5 & 14 \\ 18 & 16 & -13 & 12 \end{vmatrix}$$

SOLUTION This time we shall arbitrarily follow the second column, finding each of the minors. Note that the determinant can be evaluated by following any row or any column.

$$= -(-1)\begin{vmatrix} 4 & -8 & 9 \\ -10 & 5 & 14 \\ 18 & -13 & 12 \end{vmatrix} + (-7)\begin{vmatrix} 2 & 3 & 6 \\ -10 & 5 & 14 \\ 18 & -13 & 12 \end{vmatrix}$$

$$-(11)\begin{vmatrix} 2 & 3 & 6 \\ 4 & -8 & 9 \\ 18 & -13 & 12 \end{vmatrix} + (16)\begin{vmatrix} 2 & 3 & 6 \\ 4 & -8 & 9 \\ -10 & 5 & 14 \end{vmatrix}$$

$$= -(-1)[(-606) - (1042)] + (-7)[(1656) - (-184)]$$
$$- (11)[(-18) - (-954)] + (16)[(-374) - (738)]$$
$$= -1648 - 12{,}880 - 10{,}296 - 17{,}792$$
$$= -42{,}616$$

Higher-Order Equations

We are now in a position to solve higher-order simultaneous equations using Cramer's rule. The denominator and numerator matrices are formed, then divided, yielding the results.

EXAMPLE 14-11 Solve the following simultaneous equations:

Ⓐ $3A + 6B - 2C + D = 0$
Ⓑ $5B + 3C + 2D = 11$
Ⓒ $4A - B + 3C = 27$
Ⓓ $A - 4B - 6C - 2D = -18$

SOLUTION We must first form the denominator matrix:

$$\begin{vmatrix} 3 & 6 & -2 & 1 \\ 0 & 5 & 3 & 2 \\ 4 & -1 & 3 & 0 \\ 1 & -4 & -6 & -2 \end{vmatrix}$$

We shall evaluate this using the first column:

$$d = 3\begin{vmatrix} 5 & 3 & 2 \\ -1 & 3 & 0 \\ -4 & -6 & -2 \end{vmatrix} - 0 + 4\begin{vmatrix} 6 & -2 & 1 \\ 5 & 3 & 2 \\ -4 & -6 & -2 \end{vmatrix} - 1\begin{vmatrix} 6 & -2 & 1 \\ 5 & 3 & 2 \\ -1 & 3 & 0 \end{vmatrix}$$

$$= 3(0) - 0 + 4(14) - 1(-14)$$
$$= 70$$

Looking over the 4 × 4 matrix, solving for B and C will allow us to substitute into equation Ⓑ and find D. Then, we can find A. Forming the numerator for the B variable:

$$N_B = \begin{vmatrix} 3 & 0 & -2 & 1 \\ 0 & 11 & 3 & 2 \\ 4 & 27 & 3 & 0 \\ 1 & -18 & -6 & -2 \end{vmatrix}$$

$$= 3\begin{vmatrix} 11 & 3 & 2 \\ 27 & 3 & 0 \\ -18 & -6 & -2 \end{vmatrix} - 0 + 4\begin{vmatrix} 0 & -2 & 1 \\ 11 & 3 & 2 \\ -18 & -6 & -2 \end{vmatrix}$$

$$-1\begin{vmatrix} 0 & -2 & 1 \\ 11 & 3 & 2 \\ 27 & 3 & 0 \end{vmatrix}$$

$$= 3(-120) - 0 + 4(16) - 1(-156)$$
$$= -140$$

Solve for B:

$$B = \frac{N_B}{d} = \frac{-140}{70} = -2$$

Find the numerator for the C determinant:

$$N_C = \begin{vmatrix} 3 & 6 & 0 & 1 \\ 0 & 5 & 11 & 2 \\ 4 & -1 & 27 & 0 \\ 1 & -4 & -18 & -2 \end{vmatrix}$$

$$= 3\begin{vmatrix} 5 & 11 & 2 \\ -1 & 27 & 0 \\ -4 & -18 & -2 \end{vmatrix} - 0 + 4\begin{vmatrix} 6 & 0 & 1 \\ 5 & 11 & 2 \\ -4 & -18 & -2 \end{vmatrix} - 1\begin{vmatrix} 6 & 0 & 1 \\ 5 & 11 & 2 \\ -1 & 27 & 0 \end{vmatrix}$$

$$= 3(-40) - 0 + 4(38) - 1(-178)$$
$$= 210$$

Solve for C:

$$C = \frac{N_C}{d} = \frac{210}{70} = 3$$

Substitute into Ⓑ:

$$5B + 3C + 2D = 11$$
$$5(-2) + 3(3) + 2D = 11$$
$$D = 6$$

Substitute into Ⓐ:

$$3A + 6B - 2C + D = 0$$
$$3A + 6(-2) - 2(3) + (6) = 0$$
$$A = 4$$

Check with Ⓓ:

$$A - 4B - 6C - 2D = -18$$
$$4 - 4(-2) - 6(3) - 2(6) = -18$$
$$-18 = -18$$

Thus, the solutions are

$$A = 4 \quad B = -2 \quad C = 3 \quad D = 6$$

It should be noted that equations of any order can be solved in this manner. A fifth order must be broken down to a sum of fourth-order cofactors, for example. Also, the alternating-sign scheme is used for all orders.

PROBLEMS

Solve the following sets of simultaneous equations:

14-13. $3T + 7U = 48$
$6T - 2U = 12$

14-14. $12M - 7N = -71$
$4M + 5N = 13$

14-15. $6.3X - 17.3Y = 13.62$
$15.2X + 2.35Y = 6.872$

14-16. $39.6A - 12.3B = -3.68$
$27.2A + 16.4B = 100$

14-17. $13.6X_1 = 3X_2 - 12.39$
$18.31X_2 = -192.6 + 3.1X_1$

14-18. $372 = 27.6L_2 - 17.3L_1$
$32L_1 = 27L_2 - 5.16$

14-19. $84.3R_1 = 185.6R_2$
$2.31R_1 - 853 = 3.92R_2$

14-20. $13.6 \times 10^{-6}R_1 + 16.31 \times 10^{-3}R_2$
$= 3.15 \times 10^{-4}$
$1.315 \times 10^{-6}R_2 - 78 \times 10^{-3}R_1$
$= -4.13 \times 10^{-5}$

14-21. $A + B - C = -3$
$2A - B + 2C = 12$
$3A + 2B + C = 5$

14-22. $3X_1 + 2X_2 - X_3 = 22$
$X_1 + 6X_2 + 2X_3 = 32$
$4X_1 - 5X_2 - 3X_3 = 4$

14-23. $2V_1 = 2.295 + 3V_2 - 6V_3$
$4V_2 = 15.675 + 3V_1 + 2V_3$
$6V_3 = -2.565 + 3V_2 + 2V_1$

14-24. $4T_1 - 3T_2 + 5T_3 = 21.99$
$3T_1 = T_3 - T_2 - 24.85$
$2T_1 + 2T_2 = 5T_3 - 67.84$

14-25. $8X_{L1} + 7X_{L2} - 3X_{L3} = 510.9$
$3.15X_{L1} = 4X_{L2} - 2.16X_{L3} - 90.23$
$44.1X_{L2} = 3X_{L1} - 62.15X_{L3} + 2005$

14-26. $2.169R_1 - 3.865R_2 = 84.3R_3 - 36,410$
$84.6R_1 = 2.16R_3 - 4.314R_2 - 535$
$43.9R_1 - 26.3R_3 = 4.16R_2 - 11,200$

14-27. $2A + B + 6D = 20$
$4A + 3B + 4C = -2$
$6A + 5B - 2C - 3D = -15$
$3B + 4C - 2D = -24$

14-28. $3A + B - C + D = 7$
$4A + 3B + 2C + 6D = -2$
$5A - 2B - 2C - D = 8$
$2A + 7B - 3C + 2D = 21$

14-29. $4.61R + 3.26S - 6.43T + 6.43U = -58.05$
$3.61R - 8.31S + 6.15T = 38.76$
$5.39R = 3.88T + 4.65U - 15.73$
$43T = 36.3S + 4.31R - 6.67U + 268$

14-30. $3A + B - 2C + D + 3E = 25$
$6A + 3B - 4D + 2E = 3$
$2B + 2C - 3D - 4E = -21$
$5A - 2B + 3C - 5D = -33$
$4A - 3C + 6D + 2E = 52$

14-3 DETERMINANT MANIPULATIONS

Determinants can be manipulated by several methods to make calculations easier. We study some of these methods in this section.

Determinants With Zero Value

A determinant with any row or any column containing all 0's has a value of zero.

EXAMPLE 14-12 Evaluate:

$$\begin{vmatrix} 1 & 2 & 4 \\ 0 & 0 & 0 \\ 5 & 6 & -2 \end{vmatrix}$$

SOLUTION

$$\begin{aligned} \text{Value} &= [(1)(0)(-2) + (2)(0)(5) + (4)(0)(6)] \\ &\quad - [(4)(0)(5) + (2)(0)(-2) + (1)(0)(6)] \\ &= 0 \end{aligned}$$

A determinant where all the elements of one row are proportional to the corresponding elements in another row has a value of zero. Similarly, a determinant where all the elements of one column are proportional to the corresponding elements of another column has a value of zero.

EXAMPLE 14-13 Evaluate:

$$\begin{vmatrix} 1 & 2 & 4 \\ 2 & 3 & 8 \\ 3 & 5 & 12 \end{vmatrix}$$

SOLUTION Note that all the values in the third column are exactly four times those in the first column.

$$\begin{aligned} \text{Value} &= [(1)(3)(12) + (2)(8)(3) + (4)(2)(5)] \\ &\quad - [(4)(3)(3) + (2)(2)(12) + (1)(8)(5)] \\ &= 124 - 124 \\ &= 0 \end{aligned}$$

Interchanging Rows or Columns

Interchanging any two rows or interchanging any two columns multiplies the value of the determinant by -1.

EXAMPLE 14-14 Compute the value of the following determinant, then interchange columns 2 and 3 and recompute:

$$\begin{vmatrix} 2 & 4 & -6 \\ 3 & 1 & 2 \\ -5 & 7 & 8 \end{vmatrix}$$

SOLUTION

$$\begin{aligned} \text{Value} &= [(2)(1)(8) + (4)(2)(-5) + (-6)(3)(7)] \\ &\quad - [(-6)(1)(-5) + (4)(3)(8) + (2)(2)(7)] \\ &= (-150) - (154) \\ &= -304 \end{aligned}$$

Interchange the second and third columns:

$$\begin{vmatrix} 2 & -6 & 4 \\ 3 & 2 & 1 \\ -5 & 8 & 7 \end{vmatrix}$$

$$\begin{aligned} \text{Value} &= [(2)(2)(7) + (-6)(1)(-5) + (4)(3)(8)] \\ &\quad - [(4)(2)(-5) + (-6)(3)(7) + (2)(1)(8)] \\ &= (154) - (-150) \\ &= 304 \end{aligned}$$

Note the change in sign.

Factoring

Factoring a constant from any row or any column does not change the value of the expression.

EXAMPLE 14-15 Prove that:

$$\begin{vmatrix} 2 & 3 & -2 \\ 4 & 1 & -4 \\ 6 & 7 & 3 \end{vmatrix} = 2 \begin{vmatrix} 1 & 3 & -2 \\ 2 & 1 & -4 \\ 3 & 7 & 3 \end{vmatrix}$$

SOLUTION Note that a 2 has been factored from column 1. Evaluate the left side:

$$\begin{aligned} \text{Value} &= [(2)(1)(3) + (3)(-4)(6) + (-2)(4)(7)] \\ &\quad - [(-2)(1)(6) + (3)(4)(3) + (2)(-4)(7)] \\ &= (-122) - (-32) \\ &= -90 \end{aligned}$$

The value of the right side is

$$\begin{aligned} \text{Value} &= 2\{[(1)(1)(3) + (3)(-4)(3) + (-2)(2)(7)] \\ &\quad - [(-2)(1)(3) + (3)(2)(3) + (1)(-4)(7)]\} \\ &= 2[(-61) - (-16)] \\ &= 2(-45) \\ &= -90 \end{aligned}$$

It should also be noted that the reverse is true. A determinant is multiplied by a factor when any one of its rows or columns has been multiplied by that factor.

The Zero Quadrant

When all elements of a determinant below the diagonal or above the diagonal are zero, the value of the determinant is the product of the diagonal elements.

EXAMPLE 14-16 Evaluate:

$$\begin{vmatrix} 3 & 1 & 2 \\ 0 & 2 & 4 \\ 0 & 0 & -6 \end{vmatrix}$$

SOLUTION Note that all the elements beneath the diagonal are zero. We should, therefore, expect the answer to be $(3)(2)(-6)$ or -36. However, let us perform the evaluation "long hand" to make sure.

$$\text{Value} = [(3)(2)(-6) + (1)(4)(0) + (2)(0)(0)] - [(2)(2)(0) + (1)(0)(-6) + (3)(4)(0)]$$
$$= -36$$

Note that all the products are zero except the diagonal product.

Reduction Theorem

In the preceding paragraph we noted that if all the elements below or above the diagonal were zero, the value of the determinant is the product of the diagonal elements. The reduction theorem provides a method for making any element within the determinant a zero. Thus, it can be used on any determinant to make all the elements below the diagonal zero.

The value of a determinant remains unchanged when the following procedure is performed:

(a) Choose a row or column. We shall call this the source.
(b) Choose another row or column. We shall call this the destination. The source and destination must both be columns or both be rows.
(c) Multiply each element in the source by a constant, add it to corresponding element of the destination, and substitute this result into the destination.

This procedure can be expressed as

$$nS + D_I = D_F \tag{14-1}$$
$$(n)(\text{source}) + (\text{destination, initial}) \rightarrow (\text{destination, final})$$

By choosing n carefully, one of the elements within the destination can be made zero.

We illustrate the process in the following example.

EXAMPLE 14-17 Make the 6 within the following determinant a zero.

$$\begin{vmatrix} -2 & 3 & 2 \\ 4 & -2 & 6 \\ 8 & 4 & 2 \end{vmatrix}$$

SOLUTION We shall choose row 1 as the source and row 2 as the destination. If we were to choose a factor of -3, then element 3 of row 2 would be

$$nS + D = (-3)(2) + (6)$$
$$= 0$$

where n is the factor, S the source, and D the destination. Thus, we shall choose an n of -3. Apply this to each element of the first and second rows:

Element 1:

$$nS + D = (-3)(-2) + 4 = 10$$

Element 2:

$$nS + D = (-3)(3) + (-2) = -11$$

Element 3:

$$nS + D = (-3)(2) + 6 = 0$$

Thus, our new determinant becomes

$$D_{NEW} = \begin{vmatrix} -2 & 3 & 2 \\ 10 & -11 & 0 \\ 8 & 4 & 2 \end{vmatrix}$$

Evaluate both the new and the old:

$$D_{OLD} = [(-2)(-2)(2) + (3)(6)(8) + (2)(4)(4)] - [(2)(-2)(8) + (3)(4)(2) \\ + (-2)(6)(4)] \\ = 240$$
$$D_{NEW} = [(-2)(-11)(2) + (3)(0)(8) + (2)(10)(4)] - [(2)(-11)(8) + (3)(10)(2) \\ + (-2)(0)(4)] \\ = 240$$

Note that applying this procedure does not change the value of the determinant.

PROBLEMS

Evaluate the following determinants using the principles of Section 14-3:

14-31. (a) $\begin{vmatrix} 0 & 0 & 0 \\ 1 & 3 & -2 \\ 3 & -1 & 5 \end{vmatrix}$ (b) $\begin{vmatrix} 2 & 4 & 6 \\ 1 & 2 & 3 \\ 4 & -2 & 6 \end{vmatrix}$

(c) $\begin{vmatrix} 1 & 3 & 2 \\ 0 & 4 & 3 \\ 0 & 0 & 5 \end{vmatrix}$ (d) $\begin{vmatrix} 8 & 0 & 2 \\ 6 & 0 & 7 \\ 7 & 0 & 4 \end{vmatrix}$

14-32. (a) $\begin{vmatrix} 6 & 4 & -5 \\ 0 & -1 & 2 \\ 0 & 0 & 4 \end{vmatrix}$ (b) $\begin{vmatrix} 1 & -3 & 6 \\ 0 & 0 & 0 \\ 2 & 1 & -5 \end{vmatrix}$

(c) $\begin{vmatrix} 3 & 4 & -1 \\ 0 & -2 & 4 \\ 0 & 0 & -3 \end{vmatrix}$ (d) $\begin{vmatrix} 2 & 3 & 4 \\ 0 & -1 & 5 \\ 0 & 0 & -3 \end{vmatrix}$

14-33. $\begin{vmatrix} 3 & 2 & 1 & 5 & 6 \\ 0 & -6 & -4 & 2 & 5 \\ 0 & 0 & 1 & 3 & 4 \\ 0 & 0 & 0 & 2 & -1 \\ 0 & 0 & 0 & 0 & 1 \end{vmatrix}$

14-34. $\begin{vmatrix} 2 & -1 & 3 & 4 & 6 & 5 \\ 0 & 2 & 1 & -5 & -4 & 3 \\ 0 & 0 & -1 & 2 & -2 & 1 \\ 0 & 0 & 0 & -3 & 1 & -6 \\ 0 & 0 & 0 & 0 & -4 & 1 \\ 0 & 0 & 0 & 0 & 0 & -2 \end{vmatrix}$

Factor as far as possible, then evaluate:

14-35. (a) $\begin{vmatrix} 2 & 4 & 6 \\ 3 & 6 & -12 \\ 0 & -4 & -2 \end{vmatrix}$ (b) $\begin{vmatrix} 5 & -10 & 5 \\ -1 & 4 & -3 \\ -2 & 6 & 4 \end{vmatrix}$

14-36. (a) $\begin{vmatrix} 1 & 7 & 3 \\ 10 & -14 & -6 \\ 6 & 21 & 9 \end{vmatrix}$ (b) $\begin{vmatrix} 1 & 9 & 5 \\ 8 & 12 & -20 \\ 14 & 6 & 10 \end{vmatrix}$

Make element 1 of row 3 a 0, then evaluate:

14-37. (a) $\begin{vmatrix} 1 & 2 & 4 \\ 6 & 1 & -3 \\ -4 & 3 & 7 \end{vmatrix}$ (b) $\begin{vmatrix} -2 & 4 & -3 \\ -1 & 2 & 6 \\ 10 & 5 & 7 \end{vmatrix}$

14-38. (a) $\begin{vmatrix} 4 & 2 & 6 \\ 3 & -2 & 5 \\ 6 & 4 & 5 \end{vmatrix}$ (b) $\begin{vmatrix} 3 & 1 & 7 \\ 2 & 1 & 7 \\ 8 & 4 & 6 \end{vmatrix}$

14-4 PIVOTAL REDUCTION METHOD

The method for solution of simultaneous equations we examine in this section is one that is used extensively to solve multivariable problems. Once the method has been programmed into a computer or calculator, solutions to 12-variable problems are as easy as those to two-variable problems.

Matrix Format

Before we can use this method, we must become a bit more formal in setting up the matrix. Let us define a series of equations as

$$M_{VAR} = M_{CON}$$

where M_{VAR} is the $n \times n$ square denominator determinant we have used previously and M_{CON} is a $1 \times n$ matrix consisting of all the constants to the right of the equal sign. Thus, the following series of equations would be represented by the matrix shown:

Equation	Matrix Format				
$2W + X - 2Y = 18$	$\begin{vmatrix} 2 & 1 & -2 \\ 2 & 3 & 1 \\ 4 & -5 & 6 \end{vmatrix}$		$=$	$\begin{vmatrix} 18 \\ 11 \\ -42 \end{vmatrix}$	

$2W + X - 2Y = 18$
$2W + 3X + Y = 11$
$4W - 5X + 6Y = -42$

This is merely a shorthand definition. The method of pivotal reduction seeks to make all the elements below the diagonal of the $n \times n$ matrix zeros. It uses the reduction theorem to accomplish this end, applying it to entire rows on both sides of the equals sign. Let us apply this method to the aforementioned matrix. Our goal is to make the 2, −5, and 4 all zeros. We shall start by using row 1 as source, and row 2 as destination.

The first step in the process is to find the proper n, called the pivotal term, that will make the 2 a 0. Let us use Eq. (14-1).

$$D_F = nS + D_I$$

but D_F must be 0. Therefore,

$$0 = nS + D_I$$

$$n = -\frac{D_I}{S} \tag{14-2}$$

The pivotal term, n, can be found according to Eq. (14-2). Apply this to rows 1 and 2:

$$n = -\frac{D_I}{S} = -\frac{2}{2} = -1$$

We can now apply the reduction theorem to each element of row 2, using row 1 as a source:

$$D_{F1} = nS + D_I$$
$$D_{F1} = (-1)(2) + 2 = 0$$
$$D_{F2} = (-1)(1) + 3 = 2$$
$$D_{F3} = (-1)(-2) + 1 = 3$$
$$D_{F4} = (-1)(18) + (11) = -7$$

Our matrix now becomes

$$\begin{vmatrix} 2 & 1 & -2 \\ 0 & 2 & 3 \\ 4 & -5 & 6 \end{vmatrix} = \begin{vmatrix} 18 \\ -7 \\ -42 \end{vmatrix}$$

Use row 1 as a source and row 3 as a destination:

$$n = -\frac{D_I}{S} = -\frac{4}{2} = -2$$
$$D_{F1} = nS + D_I$$
$$D_{F1} = (-2)(2) + 4 = 0$$
$$D_{F2} = (-2)(1) + (-5) = -7$$
$$D_{F3} = (-2)(-2) + 6 = 10$$
$$D_{F4} = (-2)(18) + (-42) = -78$$

Our matrix now becomes

$$\begin{vmatrix} 2 & 1 & -2 \\ 0 & 2 & 3 \\ 0 & -7 & 10 \end{vmatrix} = \begin{vmatrix} 18 \\ -7 \\ -78 \end{vmatrix}$$

We now move over one column and must make the -7 a zero. Use row 2 as a source and row 3 as a destination:

$$n = \frac{D_I}{S} = \frac{-7}{2} = 3.5$$

Note that it is not necessary to apply the process to column 1, for these are now both zero:

$$D_{F1} = nS + D_I$$
$$D_{F1} = (3.5)(0) + (0) = 0$$
$$D_{F2} = (3.5)(2) + (-7) = 0$$
$$D_{F3} = (3.5)(3) + (10) = 20.5$$
$$D_{F4} = (3.5)(-7) + (-78) = -102.5$$

Our matrix now becomes

$$\begin{vmatrix} 2 & 1 & -2 \\ 0 & 2 & 3 \\ 0 & 0 & 20.5 \end{vmatrix} = \begin{vmatrix} 18 \\ -7 \\ -102.5 \end{vmatrix}$$

To illustrate what we have accomplished, let us convert this matrix form back into equation form:

$$\begin{array}{ll} Ⓐ & 2W + X - 2Y = \quad 18 \\ Ⓑ & \qquad 2X + 3Y = \quad -7 \\ Ⓒ & \qquad\qquad 20.5Y = -102.5 \end{array}$$

Note that Ⓒ has only one variable, Y. Therefore, we can easily solve for Y. But Ⓑ has only the variables X and Y, and if we know Y, we can solve for X. Similarly, knowing X and Y, we can solve for W in Eq. Ⓐ. This entire process is known as back substitution. Solve for Y:

$$\begin{array}{ll} Ⓒ & 20.5Y = -102.5 \\ & \quad Y = \quad -5 \end{array}$$

Solve for X:

$$\begin{array}{ll} Ⓑ & \quad 2X + 3Y = -7 \\ Ⓑ & \quad 2X + 3(-5) = -7 \\ Ⓑ & \qquad\qquad X = \quad 4 \end{array}$$

Solve for W:

$$\begin{array}{ll} Ⓐ & \quad 2W + X - 2Y = 20 \\ Ⓐ & \quad 2W + (4) - 2(-5) = 20 \\ Ⓐ & \qquad\qquad W = 3 \end{array}$$

EXAMPLE 14-18 Solve the following simultaneous equations:

$$\begin{array}{rcrcrcrcrcr} 2V &+& 2W &+& X &-& Y &-& 2Z &=& -4 \\ 4V &-& 4W &-& 2X &+& 3Y &+& 3Z &=& 37 \\ 3V &+& 3W &+& X &-& 2Y &+& Z &=& -26 \\ 5V &-& 2W &+& 2X &+& 2Y &-& 3Z &=& 61 \\ -2V &+& 3W &+& 3X &+& 3Y &+& 3Z &=& 20 \end{array}$$

SOLUTION The matrix is

$$\begin{vmatrix} 2 & 2 & 1 & -1 & -2 \\ 4 & -4 & -2 & 3 & 3 \\ 3 & 3 & 1 & -2 & 1 \\ 5 & -2 & 2 & 2 & -3 \\ -2 & 3 & 3 & 3 & 3 \end{vmatrix} = \begin{vmatrix} -4 \\ 37 \\ -26 \\ 61 \\ 20 \end{vmatrix}$$

Reducing rows 2, 3, 4, and 5 using pivotal terms of -2, $-\frac{3}{2}$, $-\frac{5}{2}$, and 1, respectively, we have a matrix of

Pivotal Term

$$\begin{vmatrix} 2 & 2 & 1 & -1 & -2 \\ 0 & -8 & -4 & 5 & 7 \\ 0 & 0 & -0.5 & -0.5 & 4 \\ 0 & -7 & -0.5 & 4.5 & 2 \\ 0 & 5 & 4 & 2 & 1 \end{vmatrix} = \begin{vmatrix} -4 \\ 45 \\ -20 \\ 71 \\ 16 \end{vmatrix} \qquad \begin{matrix} -2 \\ -1.5 \\ -2.5 \\ 1 \end{matrix}$$

Moving to column 2, we will reduce rows 4 and 5, for row 3 is already reduced. Row 2 is the source. The matrix now becomes

Pivotal Term

$$\begin{vmatrix} 2 & 2 & 1 & -1 & -2 \\ 0 & -8 & -4 & 5 & 7 \\ 0 & 0 & -0.5 & -0.5 & 4 \\ 0 & 0 & 3 & 0.125 & 4.125 \\ 0 & 0 & 1.5 & 5.125 & 5.375 \end{vmatrix} = \begin{vmatrix} -4 \\ 45 \\ -20 \\ 31.625 \\ 44.125 \end{vmatrix} \qquad \begin{matrix} -0.875 \\ 0.625 \end{matrix}$$

Moving to column 3, we shall use row 3 as our source. Reducing the 3 and the 1.5, our matrix becomes

Pivotal Term

$$\begin{vmatrix} 2 & 2 & 1 & -1 & -2 \\ 0 & -8 & -4 & 4 & 5 \\ 0 & 0 & -0.5 & -0.5 & 4 \\ 0 & 0 & 0 & -2.875 & 19.875 \\ 0 & 0 & 0 & 3.625 & 17.375 \end{vmatrix} = \begin{vmatrix} -4 \\ 41 \\ -20 \\ -88.375 \\ -15.875 \end{vmatrix} \qquad \begin{matrix} 6 \\ 3 \end{matrix}$$

Now move to column 4 and use row 4 as the source:

Pivotal Term

$$\begin{vmatrix} 2 & 2 & 1 & -1 & -2 \\ 0 & -8 & -4 & 4 & 5 \\ 0 & 0 & -0.5 & -0.5 & 4 \\ 0 & 0 & 0 & -2.875 & 19.875 \\ 0 & 0 & 0 & 0 & 42.435 \end{vmatrix} = \begin{vmatrix} -4 \\ 41 \\ -20 \\ -88.375 \\ -127.304 \end{vmatrix} \qquad 1.2609$$

Back-substitute:

Ⓔ $42.435Z = -127.304$
$Z = -3.00$
Ⓓ $-2.875Y + 19.875Z = -88.375$
Ⓓ $-2.875Y + 19.875(-3) = -88.375$
$Y = 10$

$$\begin{array}{ll} \text{ⓒ} & -0.5X - 0.5Y + 4Z = -20 \\ \text{ⓒ} & -0.5X - 0.5(10) + 4(-3) = -20 \\ & X = 6 \\ \text{ⓑ} & -8W - 4X + 4Y + 5Z = 41 \\ \text{ⓑ} & -8W - 4(6) + 4(10) + 5(-3) = 41 \\ & W = -5 \\ \text{ⓐ} & 2V + 2W + X - Y - 2Z = -4 \\ \text{ⓐ} & 2V + 2(-5) + (6) - (10) - 2(-3) = -4 \\ & V = 2 \end{array}$$

Thus, the answers are

$$V = 2 \qquad W = -5 \qquad X = 6 \qquad Y = 10 \qquad Z = -3$$

PROBLEMS

Solve using pivotal reduction:

14-39. $2X + 3Y = 13$
$4X - Y = 5$

14-40. $6M - 3N = -6$
$3M + 4N = -25$

14-41. $3.6L_1 + 2.6L_2 = 12.9$
$2.3L_1 - 1.6L_2 = 3.35$

14-42. $3R_1 + 7R_2 = 48,200$
$5R_1 + 4R_2 = 44,300$

14-43. $3I_1 + 2I_2 + I_3 = 22$
$6I_1 - I_2 + 3I_3 = 15$
$-9I_1 + 3I_2 - 5I_3 = -14$

14-44. $V_1 + V_2 = V_3 + 15$
$2V_1 + 3V_2 = -4V_3 + 50$
$-3V_1 = 2V_2 - V_3 - 65$

14-45. $2R_1 - R_2 + 3R_3 = 183$
$4R_1 - 6R_2 - 5R_3 = -1275$
$3R_1 + 2R_2 + 2R_3 = 1092$

14-46. $5I_1 + 2I_2 - I_3 = 4.45$
$-2I_1 + I_2 - 3I_3 = -1.9$
$I_1 + 5I_2 + 4I_3 = 4.45$

14-47. $3.6L_1 + 2.3L_2 - 3.95L_3 = -10.7$
$2.6L_1 - 4.85L_2 + 6.31L_3 = 41.4$
$8.23L_1 + 4.31L_2 + 6.22L_3 = 100$

14-48. $1.04X_1 + 3.69X_2 - 20X_3 = -88$
$3.15X_1 - 4.3X_2 + 6X_3 = 15$
$3.06X_1 + 2.6X_2 - 3.15X_3 = 15.63$

14-49. $A + 2B + C - D = 4$
$2A + 3B - C + 2D = 13$
$3A + B - 3C + D = 0$
$2A - B + 2C + 3D = 18$

14-50. $P + 2Q + 3R + 4S = 70$
$2P + 2Q - R + S = 61$
$3P - 3Q - 2R - 3S = -9$
$2P + 5Q + 6R - 7S = -32$

14-51. $1.31L_1 - 3L_2 + 6.2L_3 + 7.3L_4 = 20$
$6.3L_1 + 5.2L_2 - 3.1L_3 + 8.7L_4 = 48$
$3.3L_1 + 6.6L_2 + 5.5L_3 - 7.7L_4 = 40$
$1.6L_1 - 2.8L_2 + 8.7L_3 + 12L_4 = -10$

14-52. $3A + 4B + 6C + D - 6E = 22$
$3A + 2B - 3C - 7D + 5E = -72$
$4A - B - 2C - 3D - 4E = -24$
$2A + 3B + 2C - 3D + 4E = -48$
$-3A + 2B - 6C + 3D - 2E = 86$

15

NETWORK ANALYSIS

In this chapter we examine six methods of analyzing circuits. These methods will allow us not only to solve some problems easier, but to solve new ones not capable of being solved by our previous methods. These new methods of solution are:

1. The loop method—used when a circuit has multiple supplies.
2. The nodal method—used in circuits having many parallel paths
3. Thévenin's theorem—a method whereby an entire circuit can be reduced to a single voltage supply in series with a resistance.
4. Norton's theorem—a method whereby an entire circuit can be reduced to a single current supply in parallel with a resistance.
5. Wye–delta—a "cookbook" method of converting one three-resistor circuit to another three-resistor circuit.

15-1 THE LOOP METHOD

This method of circuit analysis is based upon Kirchhoff's voltage law and is especially useful when:

1. There are few parallel branches.
2. There are multiple supplies.

Figure 15-1(a) illustrates such a circuit.

The first step in solving a loop problem is to assign algebraic current designators to all branches, such as I_1, I_2, and I_3. Remember, it is unimportant which direction of current flow we assume, for if the sign of the answer is positive, our assumed direction will have been correct; if negative, our assumed direction will have been incorrect. In assigning currents we should assign a minimum number of unknowns by combining them whenever possible. Note that, instead of assigning the bottom branch I_3, a combination of I_1 and I_2 was used [Fig. 15-1(b)]. Note further that it is easier to assign currents by looking at the nodes rather than at the resistances and power supplies.

Next, we shall follow Kirchhoff's voltage law around a complete path [Fig. 15-1(c)]. There are three that we might choose: the top closed path, the bottom closed path, or all around the outside of the circuit; each path is referred to as a mesh. We must choose the same number of meshes as there are unknowns within the circuit; since there are two unknowns, I_1 and I_2, we must choose two meshes. We shall arbitrarily choose M_1 and M_2.

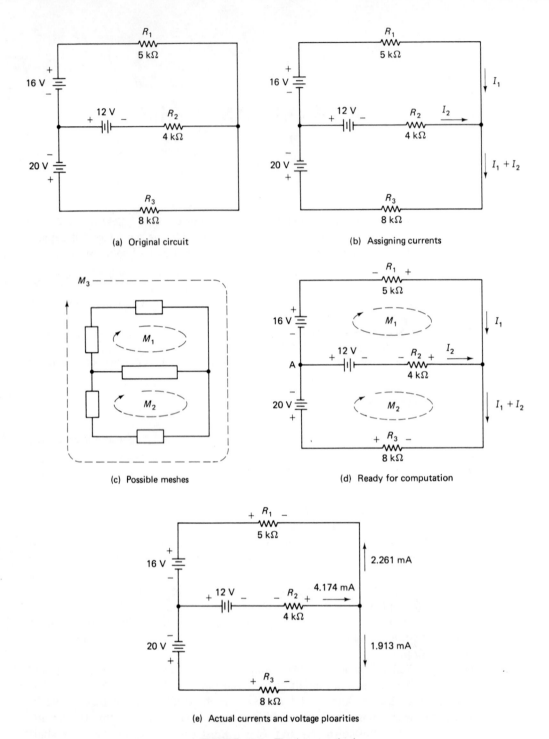

(a) Original circuit

(b) Assigning currents

(c) Possible meshes

(d) Ready for computation

(e) Actual currents and voltage ploarities

FIGURE 15-1 The loop method.

We have used Kirchhoff's voltage law before around a closed path. However, we now have an additional dimension, for there may be times that the direction of the mesh opposes the direction of the current. When this happens, we shall call a resistor voltage drop positive instead of negative. One way to remember this is that if we were to affix polarity signs to the resistors in accordance with the assumed current flows [Fig. 15-1(d)], we would encounter a plus sign whenever we travel through a resistor in a direction that is opposite to current flow. Thus, our signs will be totally determined by the sign we first encounter when we travel the meshy route.

Let us now develop the equation for mesh M_1 starting at point A:

$$\text{Ⓐ} \quad -16 - 5I_1 + 4I_2 - 12 = 0$$

Note that resistor voltage drops are equal to the resistance multiplied by the current. Thus, the voltage across R_1 is

$$V_{R1} = IR_1 = I_1 R_1 = 5I_1$$

Note further that the voltage drop across R_2 is taken as positive because our mesh direction was opposite the assumed current direction, I_2. To prevent mixing up simultaneous equations, they will be labeled with a circle, as Ⓐ for mesh M_1.

Follow M_2 from point A:

$$\text{Ⓑ} \quad +12 - 4I_2 - 8(I_1 + I_2) + 20 = 0$$

Note that the current through R_3 is $I_1 + I_2$. Rearrange these equations and reduce Ⓑ:

$$\text{Ⓐ} \quad 5I_1 - 4I_2 = -28$$
$$\text{Ⓑ} \quad 2I_1 + 3I_2 = \quad 8$$

Solve for I_1 and I_2:

$$I_1 = \frac{\begin{vmatrix} -28 & -4 \\ 8 & 3 \end{vmatrix}}{\begin{vmatrix} 5 & -4 \\ 2 & 3 \end{vmatrix}} = \frac{-52}{23} = -2.261 \text{ mA}$$

$$I_2 = \frac{\begin{vmatrix} 5 & -28 \\ 2 & 8 \end{vmatrix}}{23} = \frac{96}{23} = 4.174 \text{ mA}$$

The current through R_3 is

$$I_{R3} = I_1 + I_2 = -2.261 + 4.174 = 1.913 \text{ mA}$$

Since the sign of I_1 was negative, actual current flow is opposite to that we assumed [Fig. 15-1(e)]; the others were assumed correctly. Note that with the currents computed, all the voltage drops can be obtained, the actual polarities being determined by the actual current directions.

EXAMPLE 15-1 Compute the voltage across R_1, R_2, and R_3 of Fig. 15-2(a).

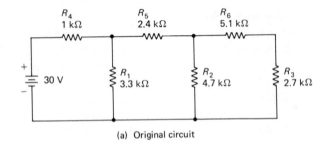

(a) Original circuit

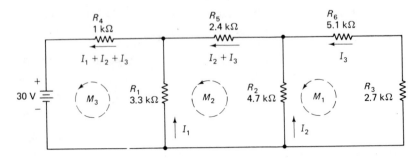

(b) Ready for computation

FIGURE 15-2 Example 15-1.

SOLUTION The circuit is shown ready for mathematical analysis in Fig. 15-2(b). Since there are three unknowns, three meshes must be used. First we analyze M_1:

$$\text{Ⓐ} \quad 4.7I_2 - 2.7I_3 - 5.1I_3 = 0$$
$$\text{Ⓐ} \quad 0I_1 + 4.7I_2 - 7.8I_3 = 0$$

Then, analyze M_2:

$$\text{Ⓑ} \quad -4.7I_2 - 2.4(I_2 + I_3) + 3.3I_1 = 0$$
$$\text{Ⓑ} \quad 3.3I_1 - 7.1I_2 - 2.4I_3 = 0$$

Next, analyze M_3:

$$\text{Ⓒ} \quad 30 - 3.3I_1 - 1(I_1 + I_2 + I_3) = 0$$
$$\text{Ⓒ} \quad 4.3I_1 + 1I_2 + 1I_3 = 30$$

Solve for the currents:

$$I_1 = \frac{\begin{vmatrix} 0 & 4.7 & -7.8 \\ 0 & -7.1 & -2.4 \\ 30 & 1 & 1 \end{vmatrix}}{\begin{vmatrix} 0 & 4.7 & -7.8 \\ 3.3 & -7.1 & -2.4 \\ 4.3 & 1 & 1 \end{vmatrix}} = \frac{-1999.8}{-327.89} = 6.099 \text{ mA}$$

$$I_2 = \frac{\begin{vmatrix} 0 & 0 & -7.8 \\ 3.3 & 0 & -2.4 \\ 4.3 & 30 & 1 \end{vmatrix}}{-327.89} = \frac{-772.2}{-327.89} = 2.355 \text{ mA}$$

$$I_3 = \frac{\begin{vmatrix} 0 & 4.7 & 0 \\ 3.3 & -7.1 & 0 \\ 4.3 & 1 & 30 \end{vmatrix}}{-327.89} = \frac{-465.3}{-327.89} = 1.419 \text{ mA}$$

Given this, the voltages would be

$$V_{R1} = I_1 R_1 = 6.099 \times 3.3 = 20.13 \text{ V}$$
$$V_{R2} = I_2 R_2 = 2.355 \times 4.7 = 11.07 \text{ V}$$
$$V_{R3} = I_3 R_3 = 1.419 \times 2.7 = 3.831 \text{ V}$$

The Distribution Problem

One problem that is often encountered is that of distribution of both positive and negative voltages over three wires to a load [Fig. 15-3(a)]. In this figure, R_1 and R_2 represent the load and R_3, R_4, and R_5 the resistance of the wire connecting the load to the supply.

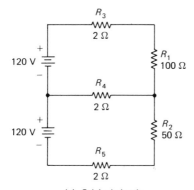

(a) Original circuit

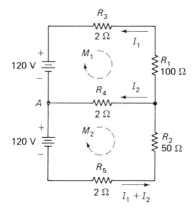

(b) Ready for computation

FIGURE 15-3 Example 15-2, voltage distribution.

EXAMPLE 15-2 Compute the voltages across the load R_1 and R_2 of Fig. 15-3(a).

SOLUTION Figure 15-3(b) shows the assumed currents, polarities, and mesh directions. Start at point A; from mesh M_1,

$$\text{Ⓐ} \quad -120 + 2I_1 + 100I_1 - 2I_2 = 0$$
$$102I_1 - 2I_2 = 120$$

From mesh M_2,

$$\text{Ⓑ} \quad 2I_2 + 50(I_1 + I_2) + 2(I_1 + I_2) - 120 = 0$$
$$\text{Ⓑ} \quad 52I_1 + 54I_2 = 120$$

Solve for the branch currents:

$$I_1 = \frac{\begin{vmatrix} 120 & -2 \\ 120 & 54 \end{vmatrix}}{\begin{vmatrix} 102 & -2 \\ 52 & 54 \end{vmatrix}} = \frac{6720}{5612} = 1.197 \text{ A}$$

$$I_2 = \frac{\begin{vmatrix} 102 & 120 \\ 52 & 120 \end{vmatrix}}{5612} = \frac{6000}{5612} = 1.069 \text{ A}$$

$$I_1 + I_2 = 1.197 + 1.069 = 2.266 \text{ A}$$

Thus, the load voltage drops are

$$V_{R1} = I_1 R_1 = 1.197 \times 100 = 119.7 \text{ V}$$
$$V_{R2} = (I_1 + I_2)(R_2) = 2.266 \times 50 = 113.3 \text{ V}$$

Note that all currents and voltages were assumed correctly.

Let us now assume that R_1 is much lower and observe V_{R1} and V_{R2}.

EXAMPLE 15-3 Assume that R_1 is 4 Ω and repeat Example 15-2.

SOLUTION The redrawn circuit is shown in Fig. 15-4. Analyze mesh M_1:

$$\text{Ⓐ} \quad -120 + 2I_1 + 4I_1 - 2I_2 = 0$$
$$\text{Ⓐ} \quad 6I_1 - 2I_2 = 120$$

Analyze mesh M_2:

$$\text{Ⓑ} \quad 2I_2 + 50(I_1 + I_2) + 2(I_1 + I_2) - 120 = 0$$
$$\text{Ⓑ} \quad 52I_1 + 54I_2 = 120$$

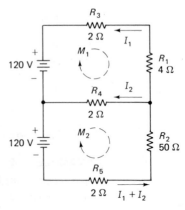

FIGURE 15-4 Example 15-3, unbalanced loads.

Solve for currents:

$$I_1 = \frac{\begin{vmatrix} 120 & -2 \\ 120 & 54 \end{vmatrix}}{\begin{vmatrix} 6 & -2 \\ 52 & 54 \end{vmatrix}} = \frac{6720}{428} = 15.701 \text{ A}$$

$$I_2 = \frac{\begin{vmatrix} 6 & 120 \\ 52 & 120 \end{vmatrix}}{428} = \frac{-5520}{428} = -12.897 \text{ A}$$

$$I_1 + I_2 = 15.701 - 12.897 = 2.804 \text{ A}$$

Solve for V_{R1} and V_{R2}:

$$V_{R1} = I_1 R_1 = 15.701 \times 4 = 62.80 \text{ V}$$
$$V_{R2} = (I_1 + I_2)(R_2) = 2.804 \times 50 = 140.2 \text{ V}$$

Note that because of the excessive current drawn by R_1, the voltage across R_2 far exceeds 120 V.

PROBLEMS

Using loop analysis, compute the following:

15-1. Compute I_1 and I_2 of Fig. 15-5.

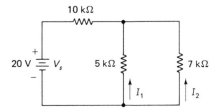

FIGURE 15-5 Problems 15-1 and 15-2.

15-2. Recompute I_1 and I_2 of Fig. 15-5 if V_s is changed to 10 V.
15-3. Compute I_1 and I_2 of Fig. 15-6.
15-4. Recompute I_1 and I_2 of Fig. 15-6 if V_1 is changed to 40 V.

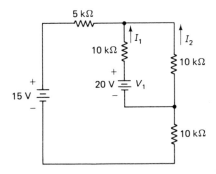

FIGURE 15-6 Problems 15-3 and 15-4.

Referring to Fig. 15-7, compute I_1, I_2, and I_3:

	V_1	V_2	R_1	R_2	R_3	R_4	R_5
15-5.	10 V	10 V	10 kΩ	10 kΩ	10 kΩ	10 kΩ	10 kΩ
15-6.	10 V	20 V	5 kΩ	3 kΩ	6 kΩ	8 kΩ	20 kΩ
15-7.	100 V	100 V	5 Ω	5 Ω	5 Ω	5 Ω	5 kΩ
15-8.	100 V	−100 V	5 kΩ	5 kΩ	5 kΩ	5 kΩ	5 kΩ
15-9.	100 V	−100 V	300 Ω	200 Ω	400 Ω	100 Ω	50 Ω
15-10.	50 V	−100 V	10 Ω	20 Ω	30 Ω	100 Ω	200 Ω

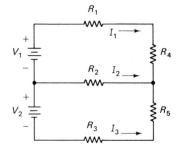

FIGURE 15-7 Problems 15-5 through 15-10.

15-2 THE NODAL METHOD

Based upon Kirchhoff's current law, the method of nodes is especially useful in analyzing circuits with many parallel branches. Although it can be used with circuits having multiple voltage supplies, the loop method is usually easier to use for this type of problem.

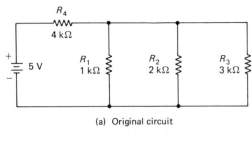

(a) Original circuit

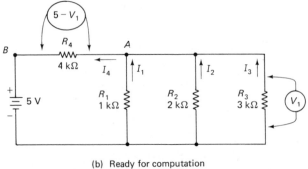

(b) Ready for computation

FIGURE 15-8 Nodal analysis.

Let us apply the nodal method to Fig. 15-8(a). The first step is to assign algebraic labels to all voltages across all resistors using a minimum number of labels [Fig. 15-8(b)]. Note that each resistor has a label, and that instead of labeling the voltage across R_4 as V_2, we can label it in terms of V_1 and the supply voltage. This keeps the number of variables to one instead of two.

Next, we need to select as many nodes as we have unknown voltages, in this case, one. The nodes should be selected so that each is completely surrounded by resistors; thus, node A is a good choice, but node B is not, for it is directly connected to the voltage supply.

The third step is to recognize Kirchhoff's current law within each node: the currents flowing into the node must equal the currents flowing out of the node. Finally, each current is set equal to a voltage divided by a resistance:

$$I_1 + I_2 + I_3 = I_4$$
$$\frac{V_1}{R_1} + \frac{V_1}{R_2} + \frac{V_1}{R_3} = \frac{5 - V_1}{R_4}$$

Substitute the known resistance values:

$$\frac{V_1}{1} + \frac{V_1}{2} + \frac{V_1}{3} = \frac{5 - V_1}{4}$$
$$12V_1 + 6V_1 + 4V_1 = 15 - 3V_1$$
$$25V_1 = 15$$
$$V_1 = 0.600 \text{ V}$$

Thus,

$$V_{R4} = 5 - V_1 = 5 - 0.6 = 4.40 \text{ V}$$

Knowing the voltages, each of the branch currents can be found.

EXAMPLE 15-4 Compute the current through the 4-kΩ and 8-kΩ resistors of Fig. 15-9(a).

SOLUTION Current directions and voltages are first assigned [Fig. 15-9(b)]. Since two variables are required, V_1 and V_2, two nodes must be chosen. Nodes A and B are good choices for they are surrounded by resistors. Analyze node A:

Ⓐ $\qquad I_1 = I_2 + I_3 + I_4$

Ⓐ $\qquad \dfrac{20 - V_1 - V_2}{5} = \dfrac{V_1 + V_2}{5} + \dfrac{V_1}{10} + \dfrac{V_1}{8}$

Ⓐ $\quad 160 - 8V_1 - 8V_2 = 8V_1 + 8V_2 + 4V_1 + 5V_1$

Ⓐ $\qquad 25V_1 + 16V_2 = 160$

Analyze node B:

Ⓑ $\qquad I_3 + I_4 = I_5$

Ⓑ $\qquad \dfrac{V_1}{10} + \dfrac{V_1}{8} = \dfrac{V_2}{4}$

Ⓑ $\qquad 4V_1 + 5V_1 = 10V_2$

Ⓑ $\qquad 9V_1 - 10V_2 = 0$

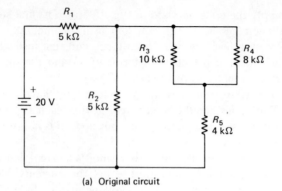

(a) Original circuit

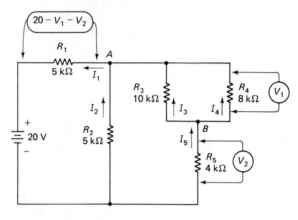

(b) Ready for computation

FIGURE 15-9 Example 15-4.

Solve for V_1:

$$V_1 = \frac{\begin{vmatrix} 160 & 16 \\ 0 & -10 \end{vmatrix}}{\begin{vmatrix} 25 & 16 \\ 9 & -10 \end{vmatrix}} = \frac{-1600}{-394} = 4.061 \text{ V}$$

$$V_2 = \frac{\begin{vmatrix} 25 & 160 \\ 9 & 0 \end{vmatrix}}{-394} = \frac{-1440}{-394} = 3.655 \text{ V}$$

Solve for I_4 and I_5:

$$I_4 = \frac{V_1}{R} = \frac{4.061}{8} = 507.6 \text{ }\mu\text{A}$$

$$I_5 = \frac{V_2}{R} = \frac{3.655}{4} = 913.8 \text{ }\mu\text{A}$$

The Unbalanced Bridge

In Chapter 10 we studied the balanced bridge, where no current flowed through the center branch of the circuit. We shall now consider an unbalanced bridge [Fig. 15-10(a)]. Note that

$$\frac{R_1}{R_3} \neq \frac{R_2}{R_4}$$

Therefore, current will flow through R_5. Using the nodal method, let us analyze such a circuit.

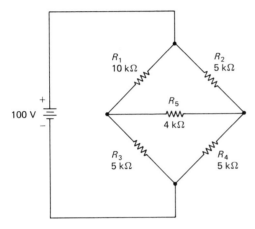

(a) Original circuit

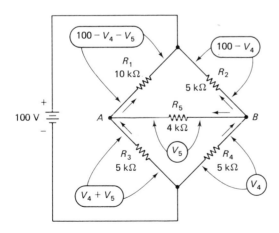

(b) Ready for computation

FIGURE 15-10 Example 15-5, the unbalanced bridge.

EXAMPLE 15-5 Compute currents through all the resistors in Fig. 15-10(a).

SOLUTION First, current directions must be assigned, then voltage designations [Fig. 15-10(b)]. Note that, by assigning current through R_5 as shown, node A is at a higher voltage than node B. Thus, the voltage across R_3 is $V_4 + V_5$. Note further that the problem is easier to work if the V's are made additive (that is, if V_{R3} were assigned V_3 and V_{R5} were assigned V_5, then V_{R4} would be $V_3 - V_5$). The nodes are then selected for analysis; with only two variables, V_4 and V_5, only two are needed. We shall select nodes A and B, since they are both surrounded by resistors. Analyze node A:

$$\text{(A)} \qquad I_{R3} + I_{R5} = I_{R1}$$
$$\text{(A)} \qquad \frac{V_4 + V_5}{5} + \frac{V_5}{4} = \frac{100 - V_4 - V_5}{10}$$
$$\text{(A)} \quad 4V_4 + 4V_5 + 5V_5 = 200 - 2V_4 - 2V_5$$
$$\text{(A)} \qquad 6V_4 + 11V_5 = 200$$

Analyze node B:

$$\text{(B)} \qquad I_{R5} + I_{R2} = I_{R4}$$
$$\text{(B)} \qquad \frac{V_5}{4} + \frac{100 - V_4}{5} = \frac{V_4}{5}$$
$$\text{(B)} \quad 5V_5 + 400 - 4V_4 = 4V_4$$
$$\text{(B)} \qquad 8V_4 - 5V_5 = 400$$

Solve for V_4:

$$V_4 = \frac{\begin{vmatrix} 200 & 11 \\ 400 & -5 \end{vmatrix}}{\begin{vmatrix} 6 & 11 \\ 8 & -5 \end{vmatrix}} = \frac{-5400}{-118} = 45.763 \text{ V}$$

$$V_5 = \frac{\begin{vmatrix} 6 & 200 \\ 8 & 400 \end{vmatrix}}{-118} = \frac{+800}{-118} = -6.780 \text{ V}$$

Note that the direction of current flow through R_5 was assumed incorrectly, as indicated by the minus sign of V_5. Solve for the resistor currents:

$$I_1 = \frac{V}{R_1} = \frac{100 - V_4 - V_5}{R_1} = \frac{100 - 45.763 + 6.780}{10} = 6.102 \text{ mA}$$

$$I_2 = \frac{V}{R_2} = \frac{100 - V_4}{R_2} = \frac{100 - 45.763}{5} = 10.85 \text{ mA}$$

$$I_3 = \frac{V}{R_3} = \frac{V_4 + V_5}{R_3} = \frac{45.763 - 6.780}{5} = 7.797 \text{ mA}$$

$$I_4 = \frac{V_4}{R_4} = \frac{45.763}{5} = 9.153 \text{ mA}$$

$$I_5 = \frac{V_5}{R_5} = \frac{-6.780}{4} = -1.695 \text{ mA}$$

Thus, I_5 flows from point A to point B.

EXAMPLE 15-6 Compute the value of R_4 in Fig. 15-11(a).

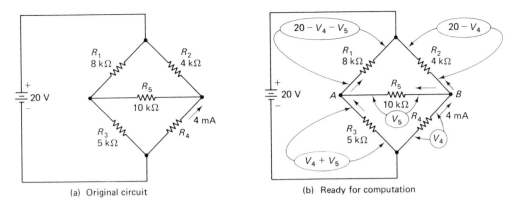

(a) Original circuit (b) Ready for computation

FIGURE 15-11 Example 15-6.

SOLUTION The current directions and voltage designators are first assigned [Fig. 15-11(b)]. Analyze node A:

$$\text{(A)} \quad I_{R3} + I_{R5} = I_{R1}$$
$$\text{(A)} \quad \frac{V_4 + V_5}{5} + \frac{V_5}{10} = \frac{20 - V_4 - V_5}{8}$$
$$\text{(A)} \quad 8V_4 + 8V_5 + 4V_5 = 100 - 5V_4 - 5V_5$$
$$\text{(A)} \quad 13V_4 + 17V_5 = 100$$

Analyze node B (note that I_{R4} was given):

$$\text{(B)} \quad I_{R4} = I_{R5} + I_{R2}$$
$$\text{(B)} \quad 4 = \frac{V_5}{10} + \frac{20 - V_4}{4}$$
$$\text{(B)} \quad 80 = 2V_5 + 100 - 5V_4$$
$$\text{(B)} \quad 5V_4 - 2V_5 = 20$$

Solve for V_4:

$$V_4 = \frac{\begin{vmatrix} 100 & 17 \\ 20 & -2 \end{vmatrix}}{\begin{vmatrix} 13 & 17 \\ 5 & -2 \end{vmatrix}} = \frac{-540}{-111} = 4.865 \text{ V}$$

Therefore, $R_4 = V_4/I_4 = 4.865/4 = 1.216 \text{ k}\Omega$

PROBLEMS

Using nodal analysis, complete the following:

15-11. Compute V_1 in Fig. 15-12 if $V_s = 30$ V, $R_1 = 3$ kΩ, $R_2 = 4$ kΩ, $R_3 = 3$ kΩ, and $R_4 = 2$ kΩ.

15-12. Compute V_s in Fig. 15-12 if $V_1 = 20$ V, $R_1 = 500$ Ω, $R_2 = 200$ Ω, $R_3 = 400$ Ω, and $R_4 = 600$ Ω.

15-13. Referring to Fig. 15-13, compute V_1 if $R_1 = R_2 = R_5 = 3$ kΩ, $R_3 = 4$ kΩ, $R_4 = 5$ kΩ, $R_6 = 10$ kΩ, and $V_S = 30$ V.

15-14. Referring to Fig. 15-13, compute V_1 if $R_2 = R_4 = R_6 = 10$ kΩ, $R_1 = 8$ kΩ, $R_3 = 6$ kΩ, $R_5 = 4$ kΩ, and $V_s = 20$ V.

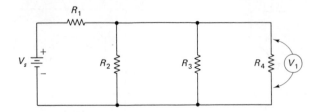

FIGURE 15-12 Problems 15-11 and 15-12.

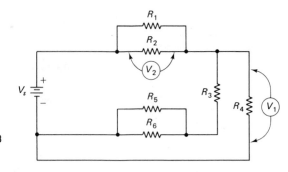

FIGURE 15-13 Problems 15-13 through 15-16.

15-15. Referring to Fig. 15-13, compute V_s if $V_2 = 20$ V, $R_1 = 5$ kΩ, $R_2 = 3$ kΩ, $R_3 = 5$ kΩ, $R_4 = 4$ kΩ, $R_5 = 10$ kΩ, and $R_6 = 8$ kΩ.

15-16. Referring to Fig. 15-13, compute V_s if $V_2 = 30$ V, $R_1 = 300$ Ω, $R_2 = 200$ Ω, $R_3 = R_4 = R_5 = 500$ Ω, and $R_6 = 100$ Ω.

15-17. Referring to Fig. 15-14, compute the voltage across R_3 if $V_s = 100$ V, $R_1 = R_2 = R_3 = 10$ kΩ, $R_4 = 20$ kΩ, and $R_5 = 10$ kΩ.

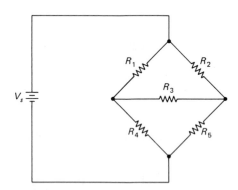

FIGURE 15-14 Problems 15-17 through 15-20.

15-18. Referring to Fig. 15-14, compute the voltage across R_3 if $V_s = 80$ V, $R_1 = 3$ kΩ, $R_2 = 4$ kΩ, $R_3 = 5$ kΩ, $R_4 = 4$ kΩ, and $R_5 = 6$ kΩ.

15-19. Referring to Fig. 15-14, compute the voltage across R_4 if $V_s = 20$ V, $R_1 = R_2 = 10$ kΩ, $R_3 = 5$ kΩ, $R_4 = 6$ kΩ, and $R_5 = 4$ kΩ.

15-20. Referring to Fig. 15-14, compute V_s if the current through R_1 is 1 mA, $R_1 = R_2 = 10$ kΩ, $R_3 = 2$ kΩ, $R_4 = 10$ kΩ, and $R_5 = 5$ kΩ.

Thévenin's theorem is both simple to state and simple to implement. It states:

Any two-terminal network can be replaced with a thevenized voltage supply in series with a thevenized resistance.

Thus, pick any two terminals of any network—it can have as many resistors and as many supplies as you want—and the network can be replaced with a simple voltage supply in series with a single resistor. For example, the very complex network shown in Fig. 15-15(a) as seen by terminals A and B can be replaced by a V_{th} and an R_{th}, as shown in Fig. 15-15(b). Equipment connected to terminals A and B of the original circuit could not tell the difference between it and the equivalent circuit.

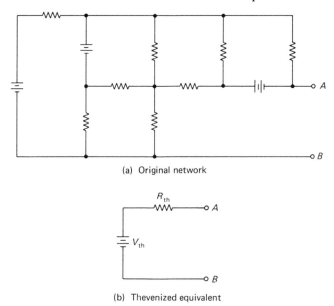

(a) Original network

(b) Thevenized equivalent

FIGURE 15-15 Thévenin's theorem.

Not only is the theorem simple to state, it is simple to implement. However, it is easier to picture this implementation as happening on the test bench than on paper. To use the theorem:

1. Remove the load.
2. To find V_{th}, measure the voltage between terminals A and B.
3. To find R_{th}:
   (a) Remove all voltage and current supplies, leaving only their internal resistance (if known).
   (b) Short the former terminals of each voltage supply and open the terminals of each current supply.
   (c) Measure the resistance between A and B; this is R_{th}.

EXAMPLE 15-7 Find the current through R_L of Fig. 15-16(a) using Thévenin's theorem.

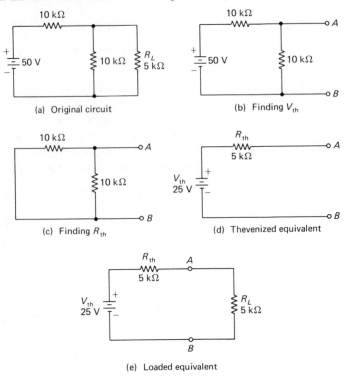

FIGURE 15-16 Example 15-7, Thévenin's theorem.

SOLUTION First, we shall select terminals A and B, then discard the load, R_L [Fig. 15-16(b)]. To find V_{th}, we must now measure between A and B, obviously 25 V in this circuit. To find R_{th}, remove the 50-V supply and short its terminals [Fig. 15-16(c)]. Measure R_{th} between A and B, finding 5 kΩ; thus, we have the equivalent circuit shown in Fig. 15-16(d). Since this circuit will react the same as the original circuit to a load, we can now reconnect the load and find the current through it [Fig. 15-16(e)], resulting in

$$I = \frac{V}{R} = \frac{25}{5+5} = 2.5 \text{ mA}$$

Thus, the load has 2.5 mA when connected to the original circuit or when connected to the thevenized circuit.

Thévenin's theorem is very useful for finding the current through the center branch of an unbalanced bridge.

EXAMPLE 15-8 Find the current through R_5 of Fig. 15-17(a).

SOLUTION We must first select terminals, then discard the load, R_5 [Fig. 15-17(b)]. To find V_{th}, find the difference in voltage between point A and point B; point A is 50 V; point B is

$$I = \frac{V}{R} = \frac{100}{20 + 10} = \frac{100}{30} = 3.333 \text{ mA}$$
$$V_{R4} = IR_4 = 3.333 \times 10 = 33.33 \text{ V}$$

Thus,

$$V_{\text{th}} = V_A - V_B = 50 - 33.33 = 16.67 \text{ V}$$

To find R_{th}, remove V_s and short its former terminals [Fig. 15-17(c)], resulting in C and D being the same electrical point. We can now compute the resistance, R_{th}:

$$R_{\text{th}} = [R_1 \| R_3] + [R_2 \| R_4]$$
$$= 6.667 + 5 = 11.67 \text{ k}\Omega$$

The thevenized circuit is shown in Fig. 15-17(d). By connecting the load back to terminals A and B, we can compute current through R_5:

$$I = \frac{V}{R} = \frac{16.67}{11.67 + 10} = 769.2 \text{ μA}$$

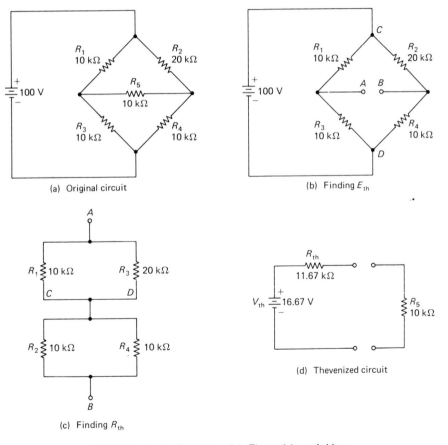

(a) Original circuit

(b) Finding E_{th}

(c) Finding R_{th}

(d) Thevenized circuit

FIGURE 15-17 Example 15-8, Thevenizing a bridge.

Batteries in Parallel

In Chapter 11 I promised to show that the internal resistances of two batteries in parallel can be considered to be themselves in parallel (Fig. 15-18). Let us apply Thévenin's theorem to it. The thevenized resistance can be seen to be

$$R_{th} = R_1 \| R_2$$

If the two batteries have equal voltages, they are series bucking and no current flows. Thus,

$$V_{th} = V_1 = V_2$$

However, if one battery, V_1, is greater than V_2, we have

$$V_1 - IR_1 - IR_2 - V_2 = 0$$
$$I = \frac{V_1 - V_2}{R_1 + R_2}$$
$$V_{AB} = V_1 - IR_1$$
$$= V_1 - (V_1 - V_2)\frac{R_1}{R_1 + R_2}$$

Note that if $R_1 = R_2$ and $V_1 = V_2$:

$$V_{AB} = V_1 - (V_1 - V_2)\frac{R_2}{R_1 + R_2}$$

$$= V_1 - (V_1 - V_1)\frac{R_1}{R_1 + R_1}$$

$$= V_1$$

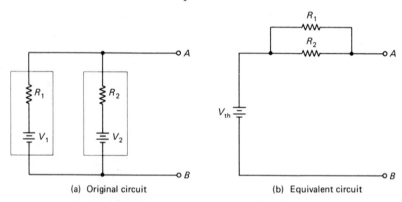

(a) Original circuit   (b) Equivalent circuit

FIGURE 15-18 Batteries in parallel.

PROBLEMS

15-21. Compute R_{th} and V_{th} of Fig. 15-19 looking into terminals A and B.

15-22. Recompute Prob. 15-21 assuming that R_3 is 5 kΩ.

15-23. Compute Prob. 15-13 using Thévenin's theorem.

15-24. Compute Prob. 15-14 using Thévenin's theorem.

15-25. Two 6-V batteries having internal resistances of 3 Ω are connected in parallel. What is V_{th} and R_{th} looking into the battery terminals?

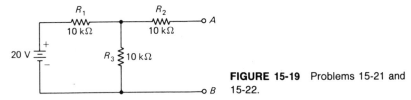

FIGURE 15-19 Problems 15-21 and 15-22.

15-26. Sam Screwloose connected a 6-V battery with an internal resistance of 1 Ω in parallel with a 12-V battery with a 0.5 Ω internal resistance. What current would flow through a 10-Ω load? (Use Thévenin's theorem.)

15-4 NORTON'S THEOREM

Norton's theorem is closely related to Thévenin's theorem and states:

Any two terminal network can be replaced with a nortonized current supply in parallel with a nortonized resistance.

Note that the primary differences between Norton and Thévenin's theorems are that:

1. Norton uses a constant current supply, whereas Thévenin uses a constant voltage supply.
2. Norton has a resistor in parallel with his supply, whereas Thévenin has a resistor in series with his.

Finding the Norton Equivalent

As in Thévenin's theorem, Norton's theorem is both simple to state and simple to implement. The theorem itself states that any two-terminal network can be replaced with a constant current supply, I_n, in parallel with a resistance, R_n. To find I_n and R_n (Fig. 15-20):

1. Remove the load.
2. To find I_n, place an ammeter between terminals A and B and read the meter. Note that the theoretical resistance of an ammeter is zero ohms.
3. To find R_n:
   (a) Remove all voltage and current supplies, leaving only their internal resistance (if known).

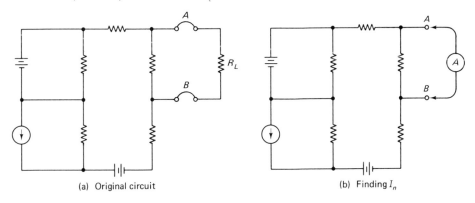

(a) Original circuit   (b) Finding I_n

FIGURE 15-20 Implementing Norton's theorem.

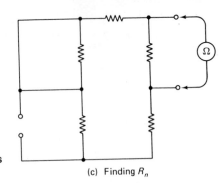

FIGURE 15-20 Implementing Norton's theorem *(Continued).*

(c) Finding R_n

(b) Short the former terminals of each voltage supply and open the terminals of each current supply.

(c) Measure the resistance, R_n, between terminals A and B.

Note that R_n and R_{th} are identical and are found in the same manner.

EXAMPLE 15-9 Find the Norton equivalent circuit for Fig. 15-21(a).

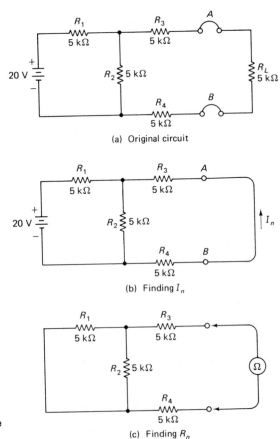

(a) Original circuit

(b) Finding I_n

(c) Finding R_n

FIGURE 15-21 Example 15-9, the Norton equivalent.

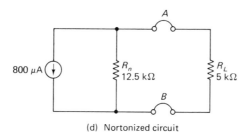

FIGURE 15-21 Example 15-9, the
Norton equivalent *(Continued).*

(d) Nortonized circuit

SOLUTION After discarding the load resistor, terminals *A* and *B* are shorted together and the current measured through the short. Measuring would be simpler, but since we cannot do that here, we shall calculate the voltage across R_2 using nodes, assigning it V_2. Then

$$\frac{20 - V_2}{R_1} = \frac{V_2}{R_2} + \frac{V_2}{R_3 + R_4}$$

$$\frac{20 - V_2}{5} = \frac{V_2}{5} + \frac{V_2}{10}$$

$$40 - 2V_2 = 2V_2 + V_2$$

$$V_2 = 8 \text{ V}$$

Since V_2 is 8 V, the current through R_3 (and thus the short circuit) would be

$$I_n = \frac{V_2}{R_3 + R_4} = \frac{8}{5 + 5} = 800 \ \mu\text{A}$$

To find R_n [Fig. 15-21(c)],

$$R_n = 5 + (5\|5) + 5 = 12.5 \text{ k}\Omega$$

The final Norton equivalent is shown in Fig. 15-21(d).

PROBLEMS

15-27. Find the Norton equivalent of Fig. 15-22.

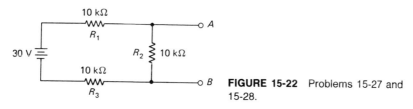

FIGURE 15-22 Problems 15-27 and
15-28.

15-28. Repeat Prob. 15-27 assuming that $R_1 = 0 \ \Omega$.

15-29. What is the Norton equivalent of a 12-V battery with an internal resistance of 0.1 Ω?

15-30. What is the Norton equivalent of a 10-kΩ resistor?

15-5 THE SUPERPOSITION METHOD

Where current is to be computed in a circuit having multiple supplies, the superposition method can be used. The principle of the method is:

The current along a particular path is the algebraic sum of the currents produced by the individual supplies along the path.

Thus, connecting only one supply at a time, current is found along a path. The total current is then the sum of all these computed currents.

EXAMPLE 15-10 Find I_1, I_2, and I_3 of Fig. 15-23(a).

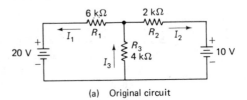

(a)   Original circuit

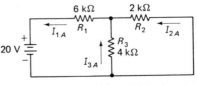

(b)   Effect of 20 V supply

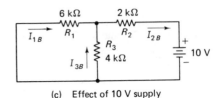

FIGURE 15-23 Example 15-10, superposition method.

(c)   Effect of 10 V supply

SOLUTION In order to use the superposition method, all supplies except one must be removed from the circuit [Fig. 15-23(b)].[1] The currents are then computed in this figure:

$$R_t = R_1 + R_2 \| R_3$$
$$= 6 + \frac{4 \times 2}{4 + 2} = 7.333 \text{ k}\Omega$$
$$I_{1A} = \frac{V}{R} = \frac{20}{7.333} = 2.727 \text{ mA}$$

Because of circuit values, two-thirds of the current will flow through R_2 and one-third through R_3.

$$I_{2A} = \frac{2}{3} \times 2.727 = 1.818 \text{ mA}$$
$$I_{3A} = \frac{1}{3} \times 2.727 = 0.909 \text{ mA}$$

---

[1] Leave the internal resistance of the supplies in the circuit if they are known.

Next, the 20-V supply is removed and the effect of the 10-V supply determined [Fig. 15-23(c)]:

$$R_t = R_2 + R_1 \| R_3 = 2 + 4 \| 6 = 4.400 \text{ k}\Omega$$
$$I_{2B} = \frac{V}{R_t} = \frac{10}{4.400} = 2.273 \text{ mA}$$
$$V_{R3} = 10 - V_{R2} = 10 - IR_2 = 10 - (2.273 \times 2) = 5.454 \text{ V}$$
$$I_{1B} = \frac{V}{R_3} = \frac{5.454}{6} = 0.909 \text{ mA}$$
$$I_{3B} = \frac{V}{R_3} = \frac{5.454}{4} = 1.364 \text{ mA}$$

The total current in each branch is the algebraic sum of each of the individual currents computed above. Note that current direction is important.

$$I_1 = I_{1A} - I_{1B} = 2.727 - 0.909 = 1.818 \text{ mA}$$
$$I_2 = I_{2B} - I_{2A} = 2.273 - 1.818 = 0.455 \text{ mA}$$
$$I_3 = I_{3B} + I_{3A} = 1.364 + 0.909 = 2.273 \text{ mA}$$

EXAMPLE 15-11 A distribution system feeds loads as shown in Fig. 15-24(a). Compute the voltage across the loads if the wires used are 14 gauge with a R_{ES} of 2.58 Ω at 25°C and the distance to the load is 1000 ft.

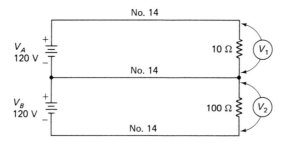

(a) Original circuit

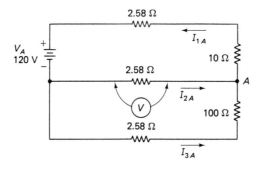

(b) Effect of V_A

FIGURE 15-24 Example 15-11, distribution system.

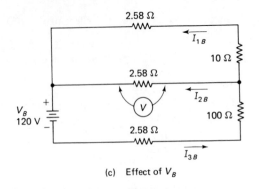

(c) Effect of V_B

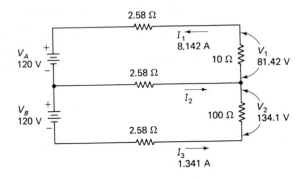

(d) Final computed results

FIGURE 15-24 Example 15-11, distribution system *(Continued).*

SOLUTION A 1000-ft length of 14-gauge wire has a resistance of 2.58 Ω. Compute the effect of V_A [Fig. 15-24(b)], using node A:

$$I_{2A} + I_{3A} = I_{1A}$$

$$\frac{V}{2.58} + \frac{V}{100 + 2.58} = \frac{120 - V}{10 + 2.58}$$

$$V = 20.00 \text{ V}$$

Thus,

$$I_{2A} = \frac{V}{2.58} = \frac{20}{2.58} = 7.752 \text{ A}$$

$$I_{3A} = \frac{V}{2.58 + 100} = \frac{20}{102.58} = 0.195 \text{ A}$$

$$I_{1A} = I_{2A} + I_{3A} = 7.752 + 0.195 = 7.947 \text{ A}$$

Repeat for the effect of V_B [Fig. 15-24(c)]:

$$I_{3B} = I_{2B} + I_{1B}$$
$$\frac{120 - V}{100 + 2.58} = \frac{V}{2.58} + \frac{V}{10 + 2.58}$$
$$V = 2.453 \text{ V}$$
$$I_{2B} = \frac{V}{2.58} = \frac{2.453}{2.58} = 0.951 \text{ A}$$
$$I_{1B} = \frac{V}{2.58 + 10} = \frac{2.453}{12.58} = 0.195 \text{ A}$$
$$I_{3B} = I_{1B} + I_{2B} = 0.195 + 0.951 = 1.146 \text{ A}$$

Compute the total I_1 and I_3:

$$I_1 = I_{1A} + I_{1B} = 7.947 + 0.195 = 8.142 \text{ A}$$
$$I_3 = I_{3A} + I_{3B} = 0.195 + 1.146 = 1.341 \text{ A}$$

Thus, the load voltages are

$$V_1 = I_1 R = 8.142 \times 10 = 81.42 \text{ V}$$
$$V_2 = I_3 R = 1.341 \times 100 = 134.1 \text{ V} \quad \text{(note that this exceeds 120 V)}$$

PROBLEMS

Use superposition to solve the following:

15-31. A distribution system similar to Fig. 15-24(a) feeds two 100-Ω loads over 1 mi of AWG 16 wire from 120-V sources. What are the voltages across the loads?

15-32. A distribution system similar to Fig. 15-24(a) feeds two 5-Ω loads over 3000 ft of AWG 10 wire from 120-V sources. What are the voltages across the loads?

15-33. Three 12-V batteries are connected in parallel with a 100-Ω load. If the batteries have internal resistances of 0.1 Ω, 0.2 Ω, and 0.4 Ω, what is the load current?

15-34. Compute I in Fig. 15-25.

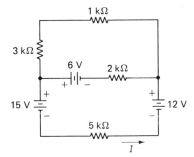

FIGURE 15-25 Problem 15-34.

15-6 *WYE–DETLA TRANSFORMATIONS*

The wye–delta transformation is a set of cookbook formulas for converting a wye network [Fig. 15-26(a)] to a delta [Fig. 15-26(b)], or vice versa. The formulas for these conversions are shown in Fig. 15-26(c).

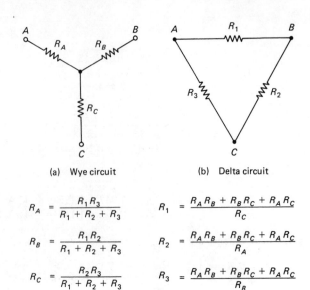

(a) Wye circuit   (b) Delta circuit

$$R_A = \frac{R_1 R_3}{R_1 + R_2 + R_3}$$ $$R_1 = \frac{R_A R_B + R_B R_C + R_A R_C}{R_C}$$

$$R_B = \frac{R_1 R_2}{R_1 + R_2 + R_3}$$ $$R_2 = \frac{R_A R_B + R_B R_C + R_A R_C}{R_A}$$

$$R_C = \frac{R_2 R_3}{R_1 + R_2 + R_3}$$ $$R_3 = \frac{R_A R_B + R_B R_C + R_A R_C}{R_B}$$

(c) Transformation formulas

FIGURE 15-26 Wye-delta transformation.

EXAMPLE 15-12 Find the total resistance of the unbalanced bridge shown in Fig. 15-27.

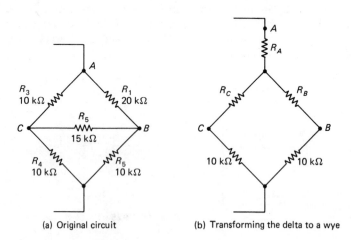

(a) Original circuit   (b) Transforming the delta to a wye

FIGURE 15-27 Example 15-12, wye-delta transformation.

SOLUTION The top three resistors can be visualized as a delta circuit, labeled as shown, and converted to the wye shown in Fig. 15-27(b). Be very careful to define the nodes and the resistors precisely as shown in Fig. 15-26(a) and (b):

$$R_A = \frac{R_1 R_3}{R_1 + R_2 + R_3} = \frac{20 \times 10}{20 + 15 + 10} = \frac{200}{45} = 4.444 \text{ k}\Omega$$

$$R_B = \frac{R_1 R_2}{R_1 + R_2 + R_3} = \frac{20 \times 15}{45} = 6.667 \text{ k}\Omega$$

$$R_C = \frac{R_2 R_3}{R_1 + R_2 + R_3} = \frac{15 \times 10}{45} = 3.333 \text{ k}\Omega$$

Total resistance of the network can now be easily obtained:

$$\begin{aligned} R_t &= R_A + [(R_C + 10)\|(R_B + 10)] \\ &= 4.444 + [(3.333 + 10)\|(6.667 + 10)] \\ &= 11.85 \text{ k}\Omega \end{aligned}$$

EXAMPLE 15-13 Compute the total resistance between points X and W of Fig. 15-28(a).

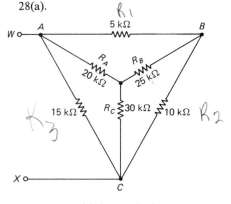

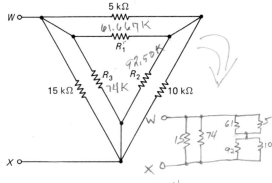

(a) Original circuit

(b) Transformed circuit

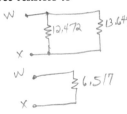

FIGURE 15-28 Example 15-13.

SOLUTION Label the circuit per Fig. 15-26 and convert the center three resistors to a delta arrangement as shown in Fig. 15-28(b):

$$\begin{aligned} R_1 &= \frac{R_A R_B + R_B R_C + R_A R_C}{R_C} \\ &= \frac{(20 \times 25) + (25 \times 30) + (20 \times 30)}{30} = \frac{1850}{30} \\ &= 61.667 \text{ k}\Omega \end{aligned}$$

$$R_2 = \frac{1850}{R_A} = \frac{1850}{20} = 92.50 \text{ k}\Omega$$

$$R_3 = \frac{1850}{R_B} = \frac{1850}{25} = 74.00 \text{ k}\Omega$$

The circuit now becomes a simple series–parallel network:

$$\begin{aligned} R_t &= (15\|R_3)\|[(R_1\|5) + (R_2\|10)] \\ &= 12.472\|[4.625 + 9.024] \\ &= 12.472\|13.65 = 6.517 \text{ k}\Omega \end{aligned}$$

PROBLEMS

Use wye–delta transformation:

15-35. Compute the resistance between A and B of Fig. 15-29 if $R_1 = R_3 = R_5 = 10 \ k\Omega$, $R_2 = 5 \ k\Omega$, and $R_4 = 6 \ k\Omega$.

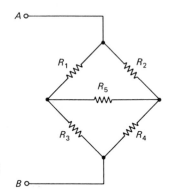

FIGURE 15-29 Problems 15-35 and 15-36.

15-36. Compute the resistance between A and B of Fig. 15-29 if $R_1 = 5 \ k\Omega$, $R_2 = 4 \ k\Omega$, $R_3 = 3 \ k\Omega$, $R_4 = 6 \ k\Omega$, and $R_5 = 10 \ k\Omega$.

15-37. Compute the resistance between A and B of Fig. 15-30 if $R_1 = 10 \ k\Omega$, $R_2 = 5 \ k\Omega$, $R_3 = 6 \ k\Omega$, $R_4 = 5 \ k\Omega$, $R_5 = 5 \ k\Omega$, and $R_6 = 1 \ k\Omega$.

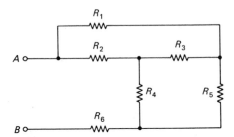

FIGURE 15-30 Problems 15-37 and 15-38.

15-38. Compute the resistance between A and B of Fig. 15-30 if $R_1 = 50 \ \Omega$, $R_2 = 100 \ \Omega$, $R_3 = 60 \ \Omega$, $R_4 = 70 \ \Omega$, $R_5 = 80 \ \Omega$, and $R_6 = 90 \ \Omega$.

16

DIRECT-CURRENT METERS

In this chapter we peer into the mysterious cases housing the measuring devices we use, to see what makes them work. We analyze both voltmeters and ammeters, comparing several circuits and learning how to compute the values of the components used. There are two different display types used for meters: digital and analog. The digital types display the actual numbers read by the voltmeter, whereas the analog types have a needle pointing to a value on a scale. Thus, they require interpretation. Since the usual computations are those required of analog meters that have no amplifiers, we concentrate on these.

16-1 THE VOLTMETER

Two specifications are usually supplied for a basic meter movement: its full-scale current, I_{FS}, and its resistance R_m. Using these two specifications, we can design a voltmeter for any scale. Since voltmeters must be placed in parallel with the circuit they are to measure, they should ideally be of infinite resistance. Thus, we should expect to put resistances in series with the basic meter movement to make a voltmeter.

EXAMPLE 16-1 A 50-μA meter movement having a resistance of 2 kΩ is to be used as a 0–50-V voltmeter. Design the circuit.

SOLUTION The circuit shown in Fig. 16-1 must be used. We have only to calculate R. However, at 50 V, we know that the meter must be drawing its full-scale current, I_{FS}, of 50 μA. Therefore, the total resistance of the circuit must be

$$R_T = \frac{V_{FS}}{I_{FS}} = \frac{50}{50 \times 10^{-6}} = 1 \text{ M}\Omega$$

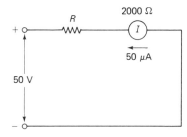

FIGURE 16-1 Example 16-1, the voltmeter.

But the meter has 2 kΩ of this, so the resistance of R is

$$R = R_T - R_m = 1,000,000 - 2000$$
$$= 998,000 \ \Omega$$
$$= 998 \ k\Omega$$

EXAMPLE 16-2 Using the same movement as in Example 16-1, design a meter for 500 V full scale (FS).

SOLUTION

$$R_T = \frac{V_{FS}}{I_{FS}} \ \frac{500}{50 \times 10^{-6}} = 10 \ M\Omega$$
$$R = R_T - R_m = 10,000,000 - 2000 = 9.998 \ M\Omega$$

Note that as the voltage range increases, the resistance of the meter increases.

Ohms/Volt Rating

From Examples 16-1 and 16-2 we can derive a specification that is very frequently used to describe voltmeters—its ohms/volt rating—also called its sensitivity. There are two methods by which this rating can be calculated:

1. Knowing the total resistance of the meter and its voltage range, ohms/volt rating equals total resistance of the meter divided by the full-scale voltage on that range:

$$\frac{\Omega}{V} = \frac{R_t}{V_{FS}}$$

2. It is the inverse of I_{FS}:

$$\frac{\Omega}{V} = \frac{1}{I_{FS}}$$

Note that Ω/V is constant for a particular voltmeter, regardless of its voltage range. This ohms per volt (Ω/V) rating simplifies calculations considerably.

EXAMPLE 16-3 Using the meter movement from Examples 16-1 and 16-2, design a meter for a V_{FS} of 10 V.

SOLUTION The previous equation we used was

$$R_T = \frac{V_{FS}}{I_{FS}}$$

However, knowing that

$$\frac{1}{I_{FS}} = \frac{\Omega}{V}$$
$$R_T = V_{FS} \left(\frac{\Omega}{V} \right)$$

Thus, let us find the sensitivity of the meter movement:

$$\frac{\Omega}{V} = \frac{1}{I_{FS}} = \frac{1}{50 \times 10^{-6}} = 20 \ k\Omega/V$$

Calculate R_T:

$$R_T = V_{FS}\left(\frac{\Omega}{V}\right) = 10 \times 20 = 200 \text{ k}\Omega$$
$$R = R_T - R_m = 200 - 2 = 198 \text{ k}\Omega$$

EXAMPLE 16-4 Design a 15-V voltmeter from a movement having a sensitivity of 1500 Ω/V and a resistance of 1500 Ω.

SOLUTION

$$R_T = V_{FS}\left(\frac{\Omega}{V}\right) = 15 \times 1.5 = 22.5 \text{ k}\Omega$$
$$R = R_T - R_m = 22.5 - 1.5 = 21 \text{ k}\Omega$$

Multirange Voltmeters

A single meter can be designed for a variety of voltage ranges by adding a selector switch and properly sized resistors. Figure 16-2(a) shows one possible circuit.

EXAMPLE 16-5 Calculate the resistors necessary for the multirange voltmeter in Fig. 16-2(a).

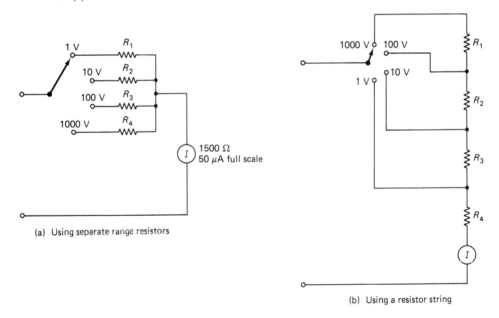

(a) Using separate range resistors

(b) Using a resistor string

FIGURE 16-2 Multi-range voltmeter.

SOLUTION Compute the sensitivity of the movement:

$$\frac{\Omega}{V} = \frac{1}{V_{FS}} = \frac{1}{50 \times 10^{-6}} = 20 \text{ k}\Omega/V$$

For the 1-V scale,

$$R_1 = V_{FS} \times \frac{\Omega}{V} - R_m$$
$$= (1 \times 20{,}000) - 1500 = 18.50 \text{ k}\Omega$$

For the 10-V scale,

$$R_2 = V_{FS} \times \frac{\Omega}{V} - R_m$$
$$= (10 \times 20{,}000) - 1500 = 198.5 \text{ k}\Omega$$

For the 100-V scale,

$$R_3 = V_{FS} \times \frac{\Omega}{V} - R_m$$
$$= (100 \times 20{,}000) - 1500 = 1{,}998{,}500 \text{ }\Omega$$

For the 1000-V scale,

$$R_4 = V_{FS} \times \frac{\Omega}{V} - R_m$$
$$= (1000 \times 20{,}000) - 1500$$
$$= 19{,}998{,}500 \text{ }\Omega$$

Note the peculiar values of resistances computed in this example. These values would be very difficult to obtain in 1%-tolerance resistors, resulting in a rather high cost. By revising the basic circuit a bit [Fig. 16-2(b)], more conventional values can be used.

EXAMPLE 16-6 Compute the resistor string values shown in Fig. 16-2(b).

SOLUTION Although "brute-force" Ohm's law techniques can be used, the sensitivity rating makes life much simpler. For the 1-V scale,

$$R_4 = V_{FS} \times \frac{\Omega}{V} - R_m$$
$$= (1 \times 20{,}000) - 1500 = 18{,}500 \text{ }\Omega$$

Since the difference between the 10-V and 1-V scales is 9 V:

$$R_3 = V_{DIFF} \times \frac{\Omega}{V}$$
$$= 9 \times 20{,}000 = 180 \text{ k}\Omega$$

Similarly,

$$R_2 = V_{DIFF} \times \frac{\Omega}{V}$$
$$= (100 - 10) \times 20{,}000$$
$$= 1.8 \text{ M}\Omega$$
$$R_1 = V_{DIFF} \times \frac{\Omega}{V}$$
$$= (1000 - 100) \times 20{,}000$$
$$= 18.0 \text{ M}\Omega$$

Note that each of the preceding values, except R_4, is a commonly available resistor, resulting in a much lower cost.

EXAMPLE 16-7 A 1800-Ω 100-μA meter movement is to be used as a voltmeter with ranges of 1.5, 5, 15, 50, 150, and 500 V. Design the divider string.

SOLUTION The schematic is shown in Fig. 16-3.

$$\frac{\Omega}{V} = \frac{1}{I_{\text{FS}}} = \frac{1}{100 \times 10^{-6}} = 10 \text{ k}\Omega/\text{V}$$

$$R_6 = V_{\text{FS}} \times \frac{\Omega}{V} - R_m$$
$$= (1.5 \times 10{,}000) - 1800$$
$$= 13.20 \text{ k}\Omega$$

$$R_5 = V_{\text{DIFF}} \times \frac{\Omega}{V}$$
$$= (5 - 1.5) \times (10{,}000)$$
$$= 35.0 \text{ k}\Omega$$

$$R_4 = V_{\text{DIFF}} \times \frac{\Omega}{V}$$
$$= (15 - 5) \times (10{,}000)$$
$$= 100 \text{ k}\Omega$$

$$R_3 = V_{\text{DIFF}} \times \frac{\Omega}{V}$$
$$= (50 - 15) \times (10{,}000)$$
$$= 350 \text{ k}\Omega$$

$$R_2 = V_{\text{DIFF}} \times \frac{\Omega}{V}$$
$$= (150 - 50) \times (10{,}000)$$
$$= 1.00 \text{ M}\Omega$$

$$R_1 = V_{\text{DIFF}} \times \frac{\Omega}{V}$$
$$= (500 - 150) \times (10{,}000)$$
$$= 3.50 \text{ M}\Omega$$

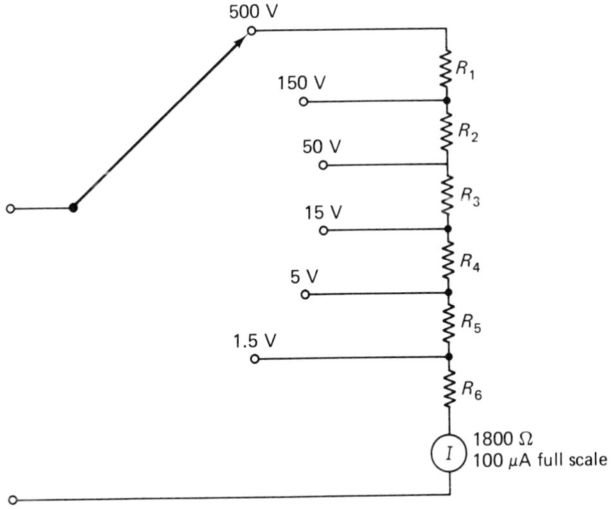

FIGURE 16-3 Example 16-7.

PROBLEMS

16-1. Using a 100-μA 2-kΩ meter movement, design a voltmeter having a full-scale voltage of **(a)** 50 V **(b)** 100 V **(c)** 200 V **(d)** 500 V.

16-2. Using a 1-mA 100-Ω meter movement, design a voltmeter having a full-scale voltage of **(a)** 10 V **(b)** 25 V **(c)** 50 V **(d)** 100 V.

16-3. Using a 100-μA 2-kΩ meter movement, design a multirange voltmeter having scales of 1 V, 2 V, 5 V, 10 V, 20 V, 50 V, and 100 V. Use a divider string.

16-4. Repeat Prob. 16-3 using a 20-μA 5-kΩ meter movement.

16-5. Using a 1-mA 150-Ω meter movement, design a multirange voltmeter having scales of 10 V, 50 V, 100 V, 500 V, 1 kV, 5 kV, and 10 kV. Use a divider string.

16-6. Repeat Prob. 16-5 using a 200-μA 1500-Ω meter movement.

16-2 THE AMMETER

An ammeter is placed in series with the circuit it is to measure and, ideally, should have zero ohms resistance. We can construct such a device from a basic meter movement by placing a resistor in parallel with the meter.

EXAMPLE 16-8 Design a 1-mA FS ammeter from a meter movement having a FS current of 50 μA and a resistance of 1500 Ω.

SOLUTION We have two requirements that must be met simultaneously:

1. The maximum current through the movement is to be 50 μA.
2. With this current of 50 μA through the meter movement, the total current through the entire circuit of the meter must be 1 mA.

These two conditions can only be met by using a parallel circuit as that shown in Fig. 16-4. The resistor placed in parallel with the meter movement is called a shunt resistor,

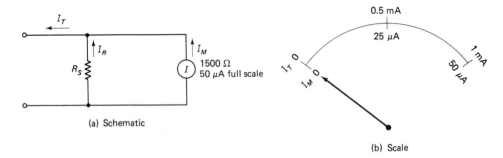

(a) Schematic

(b) Scale

FIGURE 16-4 Example 16-8, the ammeter.

for it shunts most of the current around the sensitive meter movement. Its value can be calculated as follows. By Kirchhoff's law,

$$I_t = I_R + I_m$$

Therefore,

$$I_R = I_t - I_m$$

but

$$V = I_R R_S = (I_t - I_m) R_S$$
$$V = I_m R_m$$

Equate these two:

$$I_m R_m = (I_t - I_m) R_S$$

Solve for R_S:

$$R_S = \frac{I_m R_m}{I_t - I_m} \qquad (16\text{-}1)$$

Substitute the values given:

$$
R_S = \frac{I_m R_m}{I_t - I_m}
$$
$$
= \frac{50 \times 10^{-6} \times 1500}{1 \times 10^{-3} - 50 \times 10^{-6}}
$$
$$
= 78.95 \ \Omega
$$

Note that, with this resistance in parallel with the 1500-Ω meter resistance, the total ammeter resistance is

$$
R_A = R_S \| R_m
$$
$$
= \frac{1500 \times 78.95}{1500 + 78.95}
$$
$$
= 75.00 \ \Omega
$$

EXAMPLE 16-9 Design a 1-A FS ammeter using the meter movement in Example 16-8.

SOLUTION From Eq. (16-1),

$$
R_S = \frac{I_m R_m}{I_t - I_m}
$$
$$
= \frac{50 \times 10^{-6} \times 1500}{1 - (50 \times 10^{-6})}
$$
$$
= 75.00 \ \text{m}\Omega
$$

Multirange Ammeters

The circuit of Fig. 16-5 could be used for a multiple-range ammeter. However, should one of the switch positions become dirty, preventing a good contact, all the current would pass through the meter movement, possibly destroying it. For this reason, the universal, or Ayrton shunt is usually used [Fig. 16-6(a)]. Note that the meter is always shunted, and dirt between the switch contacts would lessen meter current rather than increase it. Using Ohm's law, we can calculate each of the resistances. However, let us develop a general-purpose formula for examining the circuit. From Fig. 16-6(b), we can calculate current in the divider string, I_s:

$$I_s = I_t - I_m$$

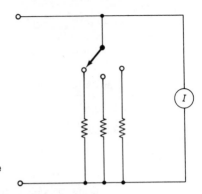

FIGURE 16-5 Possible multirange ammeter.

Knowing this, we can compute the total resistance of the string, R_s:

$$I_m R_m = I_s R_s$$

$$R_s = \frac{I_m R_m}{I_s}$$

for we know I_m, R_m, and R_s. Now we shall move the selector switch to the next higher range [Fig. 16-6(c)]. Note that the voltage between points A and B can be computed two ways:

$$V_A = I_m(R_1 + R_m)$$
$$V_A = I_s R_p$$

Equate these two:

$$I_m(R_1 + R_m) = I_s R_p$$

Substitute $R_s - R_p$ for R_1 and $I_t - I_m$ for I_s and solve for R_p:

$$I_m(R_s - R_p + R_m) = (I_t - I_m)R_p$$
$$R_p = \frac{I_m(R_t + R_m)}{I_t} \qquad (16\text{-}2)$$

Note that the numerator is independent of the range.

EXAMPLE 16-10 Compute the Ayrton shunt resistors shown in Fig. 16-6(a).

SOLUTION Going to the most sensitive range, 50 μA will flow through the total shunt and 50 μA through the meter movement [Fig. 16-6(b)]. Thus,

$$R_s = R_m = 1200 \ \Omega$$

Moving to the 1-mA range [Fig. 16-6(c)], from Eq. (16-2):

$$
\begin{aligned}
R_p &= \frac{I_m(R_t + R_m)}{I_t} \\
&= \frac{50 \times 10^{-6}(1200 + 1200)}{1 \times 10^{-3}} \\
&= \frac{0.120}{1 \times 10^{-3}} \\
&= 120 \ \Omega
\end{aligned}
$$

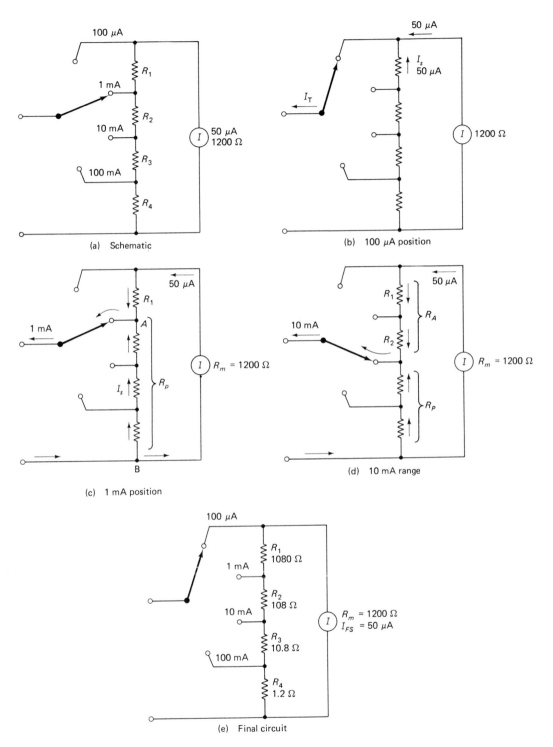

(a) Schematic

(b) 100 μA position

(c) 1 mA position

(d) 10 mA range

(e) Final circuit

FIGURE 16-6 Universal or Ayrton shunt.

Thus,

$$R_1 = R_s - R_p$$
$$= 1200 - 120 = 1080 \ \Omega$$

Move to the 10-mA range [Fig. 16-6(d)]:

$$R_p = \frac{I_m(R_s + R_m)}{I_t} = \frac{0.120}{0.01} = 12.0 \ \Omega$$
$$R_A = R_s - R_p = 1200 - 12.0 = 1188 \ \Omega$$

But

$$R_A = R_1 + R_2$$
$$R_2 = R_A - R_1 = 1188 - 1080 = 108 \ \Omega$$

Move to the 100-mA range:

$$R_4 = \frac{I_m(R_s + R_m)}{I_t} = \frac{0.12}{0.1} = 1.200 \ \Omega$$
$$R_3 = R_s - R_1 - R_2 - R_4$$
$$= 1200 - 1080 - 108 - 1.2$$
$$= 10.80 \ \Omega$$

The finished schematic is shown in Fig. 16-6(e).

EXAMPLE 16-11 Compute the values for an Ayrton shunt with scales of 100 μA, 500 μA, 1.0 mA, and 5 mA using a 50-μA movement with a resistance of 2000 Ω.

SOLUTION The schematic is shown in Fig. 16-7. $R_s = 2000 \ \Omega$ since on the 100-μA scale, 50 μA must flow through the meter and 50 μA through the shunt. Move to the 500-μA position:

$$R_p = \frac{I_m(R_s + R_m)}{I_t} = \frac{50 \times 10^{-6}(2000 + 2000)}{500 \times 10^{-6}}$$
$$= \frac{0.200}{500 \times 10^{-6}} = 400 \ \Omega$$
$$R_1 = R_s - R_p = 2000 - 400 = 1600 \ \Omega$$

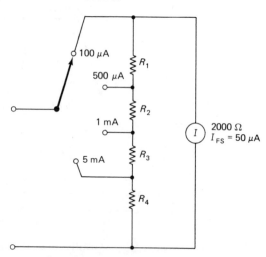

FIGURE 16-7 Example 16-11.

In the 1-mA position:

$$R_p = \frac{I_m(R_s + R_m)}{I_t} = \frac{0.200}{0.001} = 200.0 \ \Omega$$
$$R_2 = R_s - R_1 - R_p = 2000 - 1600 - 200 = 200 \ \Omega$$

In the 5-mA position:

$$R_p = \frac{I_m(R_s + R_m)}{I_t} = \frac{0.200}{0.005} = 40.00 \ \Omega$$
$$R_4 = R_p = 40.00 \ \Omega$$
$$R_3 = R_s - R_1 - R_2 - R_4 = 2000 - 2600 - 200 - 40 = 160 \ \Omega$$

PROBLEMS

16-7. Using a 1-mA 60-Ω meter movement, design an ammeter for **(a)** 5 mA **(b)** 10 mA **(c)** 100 mA **(d)** 500 mA **(e)** 1 A.

16-8. Repeat Prob. 16-7 using a 20-μA 2-kΩ meter movement.

16-9. Design a multirange ammeter using a 500-μA 200-Ω meter movement, providing scales of 1 mA, 5 mA, 10 mA, 50 mA, and 100 mA. Use an Ayrton shunt.

16-10. Repeat Prob. 16-9 using a 50-μA 2500-Ω meter movement.

16-11. Design a multirange ammeter using a 1-mA 75-Ω movement, providing scales of 2 mA, 5 mA, 10 mA, 100 mA, and 1 A. Use an Ayrton shunt.

16-12. Repeat Prob. 16-11 using a 50-μA 1-kΩ meter movement.

16-3 THE OHMMETER

The ohmmeter is constructed from a basic meter movement, a battery, and scaling resistors (Fig. 16-8). With leads A and B open, the meter reads 0 mA (infinite ohms on the resistance scale, indicated by the "∞" symbol). When leads A and B are shorted together, resistor R_A is adjusted to read full scale, in this case 1 mA (zero ohms on the resistance scale). If a 10-kΩ resistor were now placed between leads A and B, the meter would read half-scale, since there would be twice as much resistance in the circuit as when the leads were shorted. The resistance scale would be marked "10 kΩ" as this point. Thus, the resistance scale is nonlinear, reading from right to left.

As the battery within the meter ages, its terminal voltage is reduced, providing erroneous resistance readings. Thus, every time the meter is used, the two leads should be shorted and R_A adjusted for the meter to read full scale. In this manner the effects of the aging, feeble battery are minimized.

Multiple resistance ranges can be provided by selecting various values of R_1 [Fig. 16-8(c)]. Note that the resistance selected represents the half-point on the scale. That is, if the 10-kΩ resistor is selected and a 10-kΩ resistor were being measured, the meter, when properly calibrated, would read half-scale.

Many ohmmeters have two batteries: a high-voltage battery (about 6 V) and a low-voltage battery (1.5 V). The high-voltage battery is switched into the circuit on the high-resistance scale to obtain enough EMF to provide the 1 mA of current for a meter reading. (A 6-V battery is high voltage???)

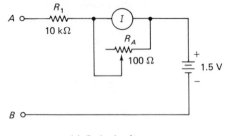

(a) Basic circuit

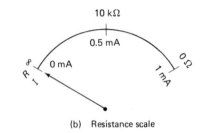

(b) Resistance scale

FIGURE 16-8 The ohmmeter.   (c) Multiple range ohmmeter

16-4 THE VOLT-OHM-MILLIAMMETER

The volt-ohm-milliammeter (VOM) is one of the most common types of meters found in the electronic industry. It is a combination voltmeter, ohmmeter, and ammeter requiring no external power supply and containing a variety of ranges. Figure 16-9 illustrates the device. The function to be performed is selected using the large, central selector switch, with the test leads connected between the "+" and "−" jacks.

16-5 EFFECTS OF METERS UPON A CIRCUIT

A famous physicist once observed: "No system can be measured without disturbing that system." This is certainly true within electronics: whenever a meter is used to

FIGURE 16-9 The volt-ohm-milliammeter.

measure an electrical property of a circuit, it disturbs that circuit. It changes the original voltage and current from that which existed prior to attaching the meter.

This phenomenon is quite apparent in the use of an ohmmeter; prior to measurement, no current flows through the resistor under test. While making the measurement, however, current does flow. Although this current flow usually has no effect upon a resistor, there are some devices that are greatly affected. An ordinary incandescent lamp has different resistances when it is cold and when it is hot. The very act of using an ohmmeter will cause current flow through the lamp, heating the filament and changing its resistance. Thus, the reading will not represent the resistance when the lamp is cold.

Voltmeter Loading

A voltmeter also affects the circuit it is measuring. The important question we must ask, however, is whether its effect is negligible. Consider the circuit shown in Fig. 16-10(a). Here, the voltage between *A* and *B* is seen to be 12 V. However, when a 20-kΩ/V voltmeter is added and its 15-V range selected, we note that its resistance is

$$R_m = \frac{\Omega}{V} \times V_{\text{FS}}$$
$$= 20,000 \times 15 = 300 \text{ k}\Omega$$

When this 300-kΩ meter is placed in parallel with R_2, the resultant resistance between A and B is

$$R_{AB} = R_2 \| R_m$$
$$= \frac{R_2 R_m}{R_2 + R_m} = \frac{(10^6) \times (0.3 \times 10^6)}{(10^6) + (0.3 \times 10^6)} = 230.77 \text{ k}\Omega$$

Thus, we now have a voltage across R_2 of

$$V_{AB} = \frac{R_{AB}}{R_{AB} + R_1} \times V_S = \frac{230.77}{230.77 + 1000} \times 24 = 4.500 \text{ V}$$

Whereas we had 12 V across AB without the meter, we have 4.5 V with it in the circuit.

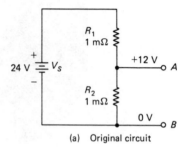

(a)   Original circuit

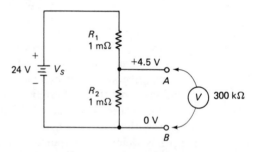

FIGURE 16-10 Voltmeter loading.   (b) Connecting a voltmeter

Although we can always compute the effect a voltmeter loading as was done above, it is much more important to recognize when such a problem may exist. Consider the circuit shown in Fig. 16-11. Here, the voltage between A and B without the voltmeter is 12.000 V. However, with the voltmeter it is 11.803 V, a 1.6% error. If we are using a 5% meter, this error is obviously negligible. How, then, can we approximate our percent loading error? Read on.

The error of a meter can be approximated by considering the ratio of the thevenized resistance of the circuit being measured to the resistance of the meter. Thus, if we

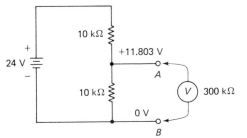

FIGURE 16-11 Negligible voltmeter loading.

were to look into terminals A and B of Fig. 16-11 and short the battery (mentally, of course), the thevenized resistance of the circuit would be

$$R_{th} = 10 \text{ k}\Omega \| 10 \text{ k}\Omega = 5 \text{ k}\Omega$$

The ratio of this to meter resistance is

$$\text{ratio} = \frac{R_{th}}{R_m} = \frac{5}{300} = 0.01667 \quad \text{or} \quad 1.667\%$$

Thus, our error when using this meter is about 1.7%.

EXAMPLE 16-12 Approximate the measurement error of the circuit shown in Fig. 16-12.

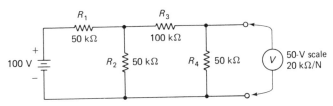

FIGURE 16-12 Example 16-12.

SOLUTION We shall use approximations. Mentally shorting out the battery yields

$$R_{eq} = R_1 \| R_2 = 50 \text{ k}\Omega \| 50 \text{ k}\Omega = 25 \text{ k}\Omega$$

This is in series with a 100-kΩ resistor, yielding 125 kΩ. This resultant resistance is in parallel with R_4, producing

$$R_{th} = R_{eq} \| R_4 = 125 \| 50 = 35.71 \text{ k}\Omega$$

The resistance of the meter is

$$R_m = \frac{\Omega}{V} \times V_{FS} = 20{,}000 \times 50 = 1 \text{ M}\Omega$$

The ratio of R_{th} and R_m is

$$\text{ratio} = \frac{R_{th}}{R_m} = \frac{35.71}{1000} = 0.0357 \quad \text{or} \quad 3.57\%$$

Thus, the meter reading would be approximately 4% in error, owing to the loading effect of the voltmeter.

Electronic Voltmeters

Electronic voltmeters, both digital and scale types, differ in their resistance characteristics from the voltmeters discussed above. These devices contain a power supply and amplifiers, thus extracting very little energy from the circuit they are measuring. Most have a constant resistance of 11 MΩ regardless of the scale used. Thus, when figuring their loading effect, R_m should be considered as 11 MΩ.

Ammeter Insertion Loss

Like the voltmeter, the ammeter also affects the circuit it measures. Since the ammeter has a very low resistance, its effect can be neglected in all but circuits of very low resistance. Consider the circuit shown in Fig. 16-13(a). According to Ohm's law,

$$I = \frac{V}{R} = \frac{0.1}{0.1} = 1 \text{ A}$$

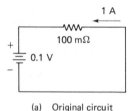

(a) Original circuit

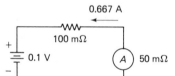

FIGURE 16-13 Ammeter insertion loss.   (b) Ammeter inserted

If, however, an ammeter with a resistance of 50 mΩ were inserted in the circuit, the current would be

$$I = \frac{V}{R} = \frac{0.1}{0.100 + 0.050} = 0.667 \text{ A}$$

This represents a 33.3% error and is called an insertion loss.

It is quite easy to estimate the error caused by an ammeter. Knowing the ammeter resistance (this can be approximated by knowing most ammeters have a voltage drop of from 50 to 100 mV) and the thevenized circuit resistance, we obtain

$$\% \text{ error} \approx \frac{R_m}{R_{th}} \times 100$$

Thus, in Fig. 16-13,

$$\% \text{ error } \frac{R_m}{R_{th}} \times 100 = \frac{50}{100} \times 100 = 50\%$$

This should tell us that the error is substantial and cannot be discounted.

EXAMPLE 16-13 Approximate the insertion error of Fig. 16-14:

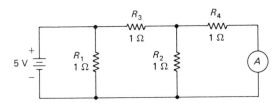

FIGURE 16-14 Example 16-13.

SOLUTION The thevenized resistance is

$$R_{th} = (R_2 \| R_3) + R_4$$

Note that when the 5-V battery is shorted, it shorts R_1.

$$R_{th} = 1.500 \ \Omega$$

A rough guess of the current would be about 3 A. Thus, assuming that we use the 5-A scale of the ammeter and it has a 50-mV drop,

$$R_m = \frac{V}{1} = \frac{50 \text{ mV}}{3 \text{ A}} = 16.67 \text{ m}\Omega$$

$$\% \text{ error} = \frac{R_m}{R_{th}} \times 100 = \frac{16.67}{1500} \times 100$$

$$= 1.11\%$$

PROBLEMS

16-13. Approximate the measurement error of the circuit shown in Fig. 16-15 for a 20-kΩ/V meter on the 10-V scale.

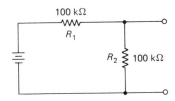

FIGURE 16-15 Problem 16-13.

16-14. Repeat Prob. 16-13 for $R_2 = 10$ kΩ.

16-15. Approximate the measurement error of the circuit shown in Fig. 16-16 for a 20-kΩ/V meter on the 100-V scale.

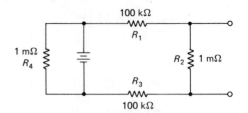

FIGURE 16-16 Problem 16-15.

16-16. Repeat Prob. 16-15 for $R_2 = 1$ kΩ.

16-17. Approximate the measurement error of the circuit shown in Fig. 16-17.

16-18. Repeat Prob. 16-17 assuming that R is 100 mΩ.

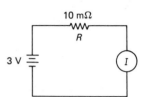

FIGURE 16-17 Problem 16-17.

PART II

ALTERNATING-CURRENT MATHEMATICS

In this part of the book, we study the mathematics required to solve alternating current (ac) circuit problems. The ac circuit is one in which electrons flow first one direction, then the opposite direction along a wire, alternating between directions periodically. This type of circuit forms the basis for radio waveforms and the 60-Hz power used in our homes. The primary mathematical tool we use in ac mathematics is trigonometry, the study of the triangle.

17

EXPONENTS AND RADICALS

Although we have studied exponents before in Chapters 3 and 4, we now expand the subject, concentrating on terms raised to the half power. We will find that our study of the triangle will use these terms extensively, for they are not only interesting subjects to put under our mathematical microscope, but they are useful in making life easier for us.

17-1 LAWS OF EXPONENTS

There are five operations we can consider for use with exponents: addition, subtraction, multiplication, division, and exponentiation.

Addition and Subtraction

Terms with bases that differ in their literal or their exponent cannot be added or subtracted, whereas terms with identical bases and exponents can be added or subtracted. In all cases, the exponent remains unchanged. Thus, the following are examples that can be added:

$$3^2 + 3^2 = 2(3^2)$$
$$a^2 + a^2 = 2a^2$$
$$3a^2 + 2a^2 = 5a^2$$
$$4a^5 + 9a^5 = 13a^5$$

Similarly, the following are examples that can be subtracted:

$$5(3^2) - 3(3^2) = 2(3^2)$$
$$5a^3 - 3a^3 = 2a^3$$
$$7a^2 - 2a^2 = 5a^3$$
$$-3a^3 - 4a^3 = -7a^3$$

However, because the bases or the exponents differ, the following cannot be added or subtracted:

$$a^3 + b^3 \qquad \text{differing bases}$$
$$a^3 - b^3 \qquad \text{differing bases}$$
$$3a^2 - 2a^3 \qquad \text{differing exponents}$$
$$2a^3 - 2a^2 \qquad \text{differing exponents}$$

Multiplication

In the multiplication process, literals with identical bases can be multiplied by adding their exponents. Examples include:

$$(a^3)(a^2) = a^{3+2} = a^5$$
$$(3a^2)(a^3) = 3a^{2+3} = 3a^5$$
$$(3a)(a^3) = 3a^{1+3} = 3a^4$$

EXAMPLE 17-1 Compute the following:

(a) $(X^2)(X^n)$ (b) $(X^{-2})(X^{-6})$ (c) $(X^{2N+3})(X^{3N-6})$ (d) $(3X^{2N})(4X^{3N})$

SOLUTION In each case the exponents are added.

(a) $(X^2)(X^n) = X^{2+n}$
(b) $(X^{-2})(X^{-6}) = X^{-2-6} = X^{-8}$
(c) $(X^{2N+3})(X^{3N-6}) = X^{2N+3+3N-6} = X^{5N-3}$
(d) $(3X^{2N})(4X^{3N}) = 12X^{3N+2N} = 12X^{5N}$

Note that, in each case, the exponents were added. It makes no difference what the form of the exponent is, they are added.

Division

Literals with identical bases can be divided by subtracting the exponent of the denominator from that of the numerator. Examples include:

$$\frac{a^3}{a^2} = a^{3-2} = a^1 = a$$

$$\frac{3a^2b^3}{4a^3b^2} = \frac{3}{4}a^{2-3}b^{3-2} = \frac{3}{4}a^{-1}b^1 = \frac{3}{4}a^{-1}b$$

$$\frac{6a^2b^3c^5}{3a^6b^3c^3} = \frac{6}{3}a^{2-6}b^{3-3}c^{5-3} = 2a^{-4}c^2$$

Note that in Example 17-1, a zero exponent was computed for b. This is to be expected, since dividing any number by itself results in a zero exponent. Therefore, any number (except zero) raised to the zero power is 1:

$$\frac{a^n}{a^n} = a^{n-n} = a^0 = 1$$

Note further that a negative exponent in the numerator is the same as a positive exponent in the denominator:

$$\frac{1}{a^n} = \frac{a^0}{a^n} = a^{0-n} = a^{-n}$$

Thus, $a^{-n} = 1/a^n$.

Similarly, a positive exponent in the numerator is the same as a negative exponent in the denominator:

$$a^n = \frac{1}{a^{-n}}$$

Thus, taking a literal from the numerator to the denominator or from the denominator to the numerator can be done by changing the sign of the exponent.

EXAMPLE 17-2 Perform the following division and express the result using positive exponents:

$$\frac{5a^5b^6c^{-2}d^3}{3a^3b^{10}c^{-6}d^3}$$

SOLUTION

$$\frac{5a^5b^6c^{-2}d^3}{3a^3b^{10}c^{-6}d^3} = \frac{5}{3}\,a^{5-3}b^{6-10}c^{-2-(-6)}d^{3-3} = \frac{5}{3}\,a^2b^{-4}c^4$$

Note that the d literals divide, resulting in 1. Further, by putting the b term into the denominator, we can change the sign of its exponent:

$$= \frac{5a^2c^4}{3b^4}$$

EXAMPLE 17-3 Compute the following:

(a) $\dfrac{a^m}{a^n}$ (b) $\dfrac{a^{3R}}{a^{3S}}$ (c) $\dfrac{X^{WN+2}}{X^{2\,WN-6}}$

SOLUTION In each case the exponents are subtracted.

(a) $\dfrac{a^m}{a^n} = a^{m-n}$

(a) $\dfrac{a^{3R}}{a^{3S}} = a^{3R-3S}$

(c) $\dfrac{X^{WN+2}}{X^{2\,WN-6}} = X^{WN+2-(2\,WN-6)} = X^{-WN+8}$

Note that, no matter the form of the exponent, if the bases are identical, the exponents subtract.

Exponentiation

Terms with exponents can also be raised to a power. When this occurs, the exponents are multiplied.

$$(a^m)^n = a^{mn}$$

We can illustrate this by the following:

$$(a^3)^4 = (a^3)(a^3)(a^3)(a^3)$$
$$= a^{3+3+3+3} = a^{12}$$

It should be emphasized that all literals or constants within the parentheses are affected:

$$(3a^3b^4)^2 = 3^2a^{3\times2}b^{4\times2} = 9a^6b^8$$

Note that each literal and constant was raised to the power.

EXAMPLE 17-4 Compute the following:

(a) $(3X^2)^4$ (b) $(9M^{-2}N^3)^3$ (c) $(6m^{N+2})^3$ (d) $(3A^{4.1N})^{3.2N}$

SOLUTION

(a) $(3X^2)^4 = 3^4 X^{2\times4} = 81X^8$

(b) $(9M^{-2}N^3)^3 = 9^3 M^{(-2\times3)}N^{3\times3}$
$$= 729M^{-6}N^9$$

(c) $(6m^{N+2})^3 = 6^3 m^{(N+2)3} = 216m^{3N+6}$

(d) $(3A^{4.1N})^{3.2N} = 3^{3.2N}A^{3.2\times4.1N^2} = 3^{3.2N}A^{13.12N^2}$

Note that it does not matter whether the exponents are positive, negative, literal, or constant: raising a power to a power is done by multiplying exponents.

EXAMPLE 17-5 Perform the indicated operation:

$$\left(\frac{3A^{2+3N}B^{3-4N}}{A^{6+2N}B^{7-2N}}\right)^{3N+6}$$

SOLUTION We must first perform the division, yielding:

$$(3A^{2+3N-(6+2N)}B^{3-4N-(7-2N)})^{3N+6}$$
$$= (3A^{-4+N}B^{-4-2N})^{3N+6}$$

We can now raise to a power:

$$= 3^{3N+6}A^{(N-4)(3N+6)}B^{(-4-2N)(3N+6)}$$

The exponents can be left in this form or multiplied, yielding

$$= 3^{3N+6}A^{3N^2-6N-24}B^{-6N^2-24N-24}$$

PROBLEMS

Multiply:

17-1. (a) $(X^2)(X^4)$ (b) $(3X^2)(X^6)$

17-2. (a) $(4m^2)(3m^3)$ (b) $(5L_1^2)(6L_2^4)$

17-3. (a) $(5R^N)(6R^M)$ (b) $(5A^P)(3A^2)$

17-4. (a) $(2L_2^2)(6L_1^N)$ (b) $(6F^{3L})(7F^{4L})$

17-5. (a) $(6L^{3N}M^{2N+1})(3L^{4N}M^{3-6N})$

17-6. (a) $(32A^{3A+B}B^{2A-6})(-6A^{3A+2C}B^{2A+3B+C}C^{2A})$

17-7. (a) $(16Z_1^{3.5A+2.15}Z_2^{4.3A-6.7})(-3Z_1^{2A-6.5}Z_2^{6.7A+6})$

17-8. (a) $(23B^{3Q+5R+6}C^{7Q-3P})(21A^{2A+6}B^{2Q+6}C^{5P+2R})$

Divide, expressing the result in nonfractional forms:

17-9. (a) $X^2 \div X^4$ (b) $3A^3 \div 2A^2$

17-10. (a) $3X^3 \div 3X^6$ (b) $2N^2 \div 2MN^3$

17-11. (a) $3R^N \div R^2$ (b) $2A^{4N} \div A^3$

17-12. (a) $46L_1^{3N+2} \div 23L_1$ (b) $16B^{N+2} \div 4AB^3$

17-13. $(4A^{2N+3}B^{3N-2}) \div (6A^{2N-3}B^{6N+7})$

17-14. $(-6L_1^{3P+2Q}L_2^{P-2}) \div (3L_1^{3P-5}Q L_2^{2P-Q+1})$

17-15. $(25R_1^{3.55-6.9}R_2^{8.25}) \div (5R_1^{2.15+6}R_2^{3.55+6})$

17-16. $(23.6S^{2P+3Q+1}T^{3P-2Q}) \div (-2.16S^{-3.1P+6.2}T^{3.2P+Q})$

Perform the exponentiation:

17-17. (a) $(3A)^2$ (b) $(2A^3C^2)^4$

17-18. (a) $(2A^3C^4)^3$ (b) $(4L_1^2L_2^3)^4$

17-19. (a) $(3A^{-2}B^3)^3$ (b) $(4L_1^{-2}L_2^{-4})^{-5}$

17-20. (a) $(P^{12}Q^{-3}R^{-4})^{-3}$ (b) $(A_1^2B_1^{-3}C_1^{-4})^{-2}$

17-21. (a) $(L^{P+2}M^{P-2})^3$ (b) $(R_1^{3R+T}R_2^{2R-T})^4$

17-22. (a) $(Z_1^{2S+3}Z_2^{3S-2})^{3S}$ (b) $(Q_1^{5N+3}Q_2^{3N-2})^{2N}$

17-23. $(4R_1^{3A+B}R_2^{2A-3B})^{A+B}$

17-24. $(17S_1^{3A-2B}S_2^{A+B})^{2A+1}$

Perform the indicated operation:

17-25. $\left(\dfrac{3A^2B^3}{A^{-4}B^6}\right)^2$

17-26. $\left(\dfrac{2B_1^3B_2^4}{B_1^4B_2^6}\right)^3$

17-27. $\left(\dfrac{3L_1^{2N+6}L_2^{2N-3}}{2L_1^3NL_2^{12N+6}}\right)^3$

17-28. $\left[\dfrac{(4A^NB^{M+1})(5A^2B^{N+3})}{3A^{3N}B^{5N}}\right]^3$

17-2 FRACTIONAL EXPONENTS

The system of fractional exponents is a convenient way of representing some quantities. But just what is the meaning of a fractional exponent? Let us try it on a number:

$$49^{1/2} = ?$$

If we were to raise this quantity to the second power, by the rules of exponents we would have

$$(49^{1/2})^2 = 49^{1/2 \times 2} = 49$$

Thus, the original quantity must be such that when multiplied by itself (squared, as was done above) it is equal to 49. Obviously, the answer is 7. Therefore,

$$49^{1/2} = 7$$

In a similar manner:

$$25^{1/2} = 5$$
$$36^{1/2} = 6$$
$$100^{1/2} = 10$$

Therefore, the fractional power, $\frac{1}{2}$, represents the square root of the number: that number that, when multiplied by itself, is equal to the original number. Consider the following:

$$64^{1/3} = ?$$

By raising this quantity to the third power (cubing it) we find:

$$(64^{1/3})^3 = 64^{1/3 \times 3} = 64^1 = 64$$

Thus, the quantity before cubing must represent a number that when cubed will equal 64. The number 4 satisfies this requirement:

$$64^{1/3} = 4$$
$$64 = 4^3$$

Therefore, the fractional exponent represents the root of a number: square root ($\frac{1}{2}$), cube root ($\frac{1}{3}$), fourth root ($\frac{1}{4}$), and so on. Note, however, that there is nothing different in how it is handled. The fractional power obeys the same exponential rules as any power.

As an exercise, try the fractional root on your calculator. First, raise 6 to the seventh power. Then, having obtained the result, 279,936, raise this number to the $\frac{1}{7}$ power. Note that the 6 reappears.

EXAMPLE 17-6 Raise the following numbers to the roots indicated, then raise the result to the inverse of the original root:

(a) 3^6 (b) 4.9^3 (c) $5^{1/3}$ (d) $16^{1/8}$

SOLUTION

(a) $3^6 = 729$ $729^{1/6} = 3$
(b) $4.9^3 = 117.649$ $117.649^{1/3} = 4.9$
(c) $5^{1/3} = 1.709975947$ $1.709975947^3 = 5$
(d) $16^{1/8} = 1.414213562$ $1.414213562^8 = 16$

There will be some slight roundoff errors.

PROBLEMS

Raise the following numbers to the roots indicated, then raise the result to the inverse of the original root.

17-29. (a) 4^3 (b) 5^2 (c) 3^3
17-30. (a) 3^5 (b) 4^4 (c) 6^3
17-31. (a) $3^{1/2}$ (b) $5^{1/3}$ (c) $25^{1/6}$
17-32. (a) $35^{1/4}$ (b) $86^{1/2}$ (c) $123^{1/5}$
17-33. (a) $6^{2.1}$ (b) $10^{1.23}$ (c) $21^{1.68}$
17-34. (a) $7^{3.3}$ (b) $8^{3.05}$ (c) $13^{2.69}$

17-3 THE RADICAL SIGN

The radical sign, $\sqrt{}$, means "root of." When no number appears with the root, the square root is inferred. Because roots can be represented as fractional powers, the following are true:

$$\sqrt{5} = 5^{1/2}$$
$$\sqrt{36} = 36^{1/2}$$
$$\sqrt[3]{5} = 5^{1/3}$$
$$\sqrt[7]{6} = 6^{1/7}$$

EXAMPLE 17-7   For the following numbers, find the square root ($\sqrt{}$) on your calculator. Then, raise these numbers to the ½ power and compare the results.

   (a) 35   (b) 69   (c) 17.16

SOLUTION

   (a) $\sqrt{35} = 5.916079783$
       $35^{1/2} = 5.916079783$
   (b) $\sqrt{69} = 8.306623863$
       $69^{1/2} = 8.306623863$
   (c) $\sqrt{17.16} = 4.142463035$
       $17.16^{1/2} = 4.142463035$

     Note that in each case, $\sqrt{n} = n^{1/2}$.

It should also be noted that there are two square roots of a number: a positive quantity and a negative quantity. In Example 17-7(a), to be precise:

$$\sqrt{35} = 5.916079783 \qquad \text{or} \qquad -5.916079783$$

The negative quantity, when multiplied by itself, also results in 35. We indicate this by the notation $\pm$, which reads "plus or minus."

$$\sqrt{35} = \pm 5.916079783$$

Although most of our work will use the positive quantity, the negative quantity is also a solution.

Rational and Irrational Numbers

At this point, let us define two types of numbers. A rational number is one that can be found by dividing one integer by another. The following are examples:

$$3.156 \qquad 3156 \div 1000$$
$$46.32 \qquad 4632 \div 100$$

On the other hand, irrational numbers cannot be so expressed. Square roots and cube roots are examples:

$$\sqrt{2} = 1.414213562 \text{ (approximately)}$$
$$\sqrt[5]{36} = 2.047672511 \text{ (approximately)}$$
$$\pi = 3.141592654 \text{ (approximately)}$$

In the text that follows, we refer to these two types of numbers.

Literals

The radical is used very frequently to represent square roots of literals. In this case, it can be thought of as raising the literal to a fractional power. For example:

$$\sqrt{L^2 M^4} = (L^2 M^4)^{1/2} = L^{2 \times 1/2} M^{4 \times 1/2} = LM^2$$

Many times, odd powers can appear beneath the radical sign. The result should be expressed as the square root of the literal (or its fractional power) multiplied by the whole literal:

$$\sqrt{T^5} = (T^5)^{1/2} = T^{21/2} = (T^2)(T^{1/2})$$
$$= T^2\sqrt{T}$$

The final result is said to be reduced.

EXAMPLE 17-8   Reduce the following:

(a) $\sqrt{N^3 P^4}$ (b) $\sqrt{Q^{23} Z^{16}}$

SOLUTION

(a) $\sqrt{N^3 P^4} = (N^3 P^4)^{1/2} = N^{3 \times 1/2} P^{4 \times 1/2} = NP^2\sqrt{N}$

Although we have gone a long way around the barn, you can see that the pattern is to leave the single, "odd" power under the radical sign. Another way of looking at this is:

$$\sqrt{N^3 P^4} = \sqrt{N \times N^2 \times P^4}$$
$$= \sqrt{N^2 P^4}\sqrt{N}$$
$$= NP^2\sqrt{N}$$

(b) $\sqrt{Q^{23} Z^{16}} = \sqrt{Q^{22} Z^{16}}\sqrt{Q}$
$$= Q^{11} Z^8 \sqrt{Q}$$

In a similar manner, the cube root of a literal would leave any excess after division by 3 underneath the radical.

EXAMPLE 17-9   Reduce the following:

$$\sqrt[3]{L^3 M^7 N^{11}}$$

SOLUTION

$$\sqrt[3]{L^3 M^7 N^{11}} = \sqrt[3]{L^3 M^6 N^9 M^1 N^2}$$
$$= \sqrt[3]{L^3 M^6 N^9} \sqrt[3]{MN^2} = LM^2 N^3 \sqrt[3]{MN^2}$$

Note that the cube root requires multiplying the exponents by $\frac{1}{3}$.

EXAMPLE 17-10   Reduce the following:

$$\sqrt[5]{Z^{12} T^{23}}$$

SOLUTION   We must now leave any powers in excess of those divisible by 5 under the radical:

$$\sqrt[5]{Z^{12} T^{23}} = \sqrt[5]{Z^{10} T^{20} Z^2 T^3}$$
$$= \sqrt[5]{Z^{10} T^{20}} \sqrt[5]{Z^2 T^3}$$
$$= Z^2 T^4 \sqrt[5]{Z^2 T^3}$$

Numerical Constants

When a number appears under the radical sign, we can find the power indicated. Thus,

$$\sqrt{3} = 1.732050808$$

However, this has the disadvantage of being inexact, for it is accurate to only 10 places, in this case. A better way to express the result is to leave it as the square root of 3. In this manner, the reader is restricted in precision to whatever device he or she chooses. There are, however, many occasions where some reduction can be done before the final result is expressed. Consider the $\sqrt{32}$. By factoring, we can express it as

$$\sqrt{32} = \sqrt{16 \times 2} = \sqrt{16}\,\sqrt{2} = 4\sqrt{2}$$

This form is preferred over the initial form. Thus, we have reduced the constant.

EXAMPLE 17-11 Reduce the following:

(a) $\sqrt{72}$ (b) $\sqrt{1800}$ (c) $\sqrt[3]{960}$

SOLUTION

(a) We can determine if 72 contains a perfect square by factoring it:

$$72 = 2 \cdot 2 \cdot 2 \cdot 3 \cdot 3$$

Therefore, grouping an even number of 2's and an even number of 3's:

$$\sqrt{72} = \sqrt{2 \cdot 2 \cdot 3 \cdot 3}\,\sqrt{2}$$
$$= 6\sqrt{2}$$

(b) $\sqrt{1800} = \sqrt{2 \cdot 2 \cdot 2 \cdot 3 \cdot 3 \cdot 5 \cdot 5}$
$\qquad\quad = \sqrt{2 \cdot 2 \cdot 3 \cdot 3 \cdot 5 \cdot 5}\,\sqrt{2}$
$\qquad\quad = 30\sqrt{2}$

(c) In this case we must look for groups of three factors:

$$\sqrt[3]{960} = \sqrt[3]{2 \cdot 2 \cdot 2 \cdot 2 \cdot 2 \cdot 2 \cdot 3 \cdot 5}$$
$$= \sqrt[3]{2 \cdot 2 \cdot 2 \cdot 2 \cdot 2 \cdot 2}\,\sqrt[3]{3 \cdot 5}$$
$$= 2 \cdot 2\,\sqrt[3]{15}$$
$$= 4\,\sqrt[3]{15}$$

Complete Monomials

Let us now apply the preceding principles to those radicals having both literals and numbers:

EXAMPLE 17-12 Reduce:

(a) $\sqrt{5P^5Q^6R^{-3}}$ (b) $\sqrt{1350L_1^5L_2^{-3}L_3^7}$

(c) $\sqrt[3]{48Z_1^{10}Z_2^{16}}$ (d) $\sqrt[5]{11,664\,V_1^{16}V_2^{20}V_3^{-13}}$

SOLUTION

(a) $\sqrt{5P^5Q^6R^{-3}} = \sqrt{P^4Q^6R^{-2}}\,\sqrt{5PR^{-1}} = P^2Q^3R^{-1}\,\sqrt{5PR^{-1}}$

Note that the 5 could not be factored and, therefore, must be left under the radical.

(b) $\sqrt{1350L_1^5L_2^{-3}L_3^7} = \sqrt{2 \cdot 3 \cdot 3 \cdot 3 \cdot 5 \cdot 5L_1^5L_2^{-3}L_3^7}$

$\qquad\qquad\qquad = \sqrt{3 \cdot 3 \cdot 5 \cdot 5 \cdot L_1^4L_2^{-2}L_3^6}\,\sqrt{2 \cdot 3 \cdot L_1L_2^{-1}L_3}$

$\qquad\qquad\qquad = 15L_1^2L_2^{-1}L_3^3\,\sqrt{6L_1L_2^{-1}L_3}$

(c) $\sqrt[3]{48Z_1^{10}Z_2^{16}}$ $= \sqrt[3]{2 \cdot 2 \cdot 2 \cdot 2 \cdot 3Z_1^{10}Z_2^{16}}$

$= \sqrt[3]{2 \cdot 2 \cdot 2Z_1^9Z_2^{15}} \sqrt[3]{2 \cdot 3Z_1Z_2}$

$= 2Z_1^3Z_2^5 \sqrt[3]{6Z_1Z_2}$

(d) $\sqrt[5]{11,664\,V_1^{16}V_2^{20}V_3^{-13}}$

$= \sqrt[5]{2 \cdot 2 \cdot 2 \cdot 2 \cdot 3 \cdot 3 \cdot 3 \cdot 3 \cdot 3 \cdot 3 \cdot V_1^{16}V_2^{20}V_3^{-13}}$

$= \sqrt[5]{3 \cdot 3 \cdot 3 \cdot 3 \cdot 3\,V_1^{15}V_2^{20}V_3^{-10}} \sqrt[5]{2 \cdot 2 \cdot 2 \cdot 2 \cdot 3 \cdot V_1V_3^{-3}}$

$= 3V_1^3V_2^4V_3^{-2} \sqrt[5]{48V_1V_3^{-3}}$

PROBLEMS

Find the square root of the following numbers. Compare this with raising the numbers to the ½ power.

17-35. (a) 26 (b) 35 (c) 24

17-36. (a) 88 (b) 92 (c) 71

17-37. (a) 35.9 (b) 17.21 (c) 81.31

17-38. (a) 66.6 (b) 84.2 (c) 1029

Reduce the following:

17-39. (a) $\sqrt{W^2N^3}$ (b) $\sqrt{W^4P^5}$ (c) $\sqrt{Z_1^{10}Z_2^{11}}$

17-40. (a) $\sqrt{A^3B^4}$ (b) $\sqrt{M^5N^6}$ (c) $\sqrt{R_1^2R_2^{11}}$

17-41. (a) $\sqrt{A^4B^6C^{-9}}$ (b) $\sqrt[3]{R_1^3R_2^{-5}R_3^{-6}}$ (c) $\sqrt[4]{M_1^4M_2^{22}M_3^{-31}}$

17-42. (a) $\sqrt{Z_1^4Z_2^{-30}Z_3^{-15}}$ (b) $\sqrt[3]{A_1^{17}A_2^{-15}A_3^{30}}$ (c) $\sqrt[5]{C_1^{45}C_2^{-32}C_3^{48}}$

17-43. (a) $\sqrt{32}$ (b) $\sqrt{74}$ (c) $\sqrt{60}$

17-44. (a) $\sqrt{64}$ (b) $\sqrt{63}$ (c) $\sqrt{48}$

17-45. (a) $\sqrt{1200}$ (b) $\sqrt[3]{405}$ (c) $\sqrt[4]{768}$

17-46. (a) $\sqrt{216}$ (b) $\sqrt[3]{2592}$ (c) $\sqrt[5]{46,656}$

17-47. (a) $\sqrt{10A_1^2A_2^3}$ (b) $\sqrt{20Z_1^2Z_2^{-6}}$

17-48. (a) $\sqrt{12L_1^2L_2^{-2}}$ (b) $\sqrt{24M_1^{-3}M_2^{-6}}$

17-49. (a) $\sqrt{36A^2B^6}$ (b) $\sqrt{50L^3M^{-5}}$

17-50. (a) $\sqrt{30W^5X^{-4}}$ (b) $\sqrt{35M^3N^{-5}}$

17-51. (a) $\sqrt{225L_1^{16}L_2^{13}}$ (b) $\sqrt{625R_1^{16}R_2^{-15}}$

17-52. (a) $\sqrt{1014A^{15}B^{-26}}$ (b) $\sqrt{720F^{22}G^{23}H^{-24}}$

17-53. (a) $\sqrt[3]{768Z_1^{15}Z_2^{-22}}$ (b) $\sqrt[4]{720J_1^{16}J_2^{-28}J_3^{32}}$

17-54. (a) $\sqrt[5]{93,312A^{46}B^{72}}$ (b) $\sqrt[6]{24,576M^{84}N^{63}}$

17-4 ADDITION AND SUBTRACTION

In the processes of addition and subtraction, treat the entire radical the same way a literal would be treated. That is, if the radicals are identical their coefficients may be added or subtracted. For example:

$$2\sqrt{2} + 4\sqrt{2} = 6\sqrt{2}$$
$$7\sqrt[3]{63} + 14\sqrt[3]{63} = 21\sqrt[3]{63}$$
$$16\sqrt{A} + 12\sqrt{A} = 28\sqrt{A}$$
$$\sqrt{33} - 5\sqrt{33} = -4\sqrt{33}$$
$$12\sqrt{A} - 10\sqrt{A} = 2\sqrt{A}$$
$$6\sqrt{3B} - 8\sqrt{3B} = -2\sqrt{3B}$$

However, when the radicals are not identical, they cannot be added or subtracted. The following are examples:

$$2\sqrt{2} + 3\sqrt{3}$$
$$2\sqrt{2} + 2\sqrt[3]{2}$$
$$12\sqrt[3]{3} - 11\sqrt{3}$$
$$3\sqrt{2A} - 5\sqrt{A}$$

This restriction points up the necessity of reducing radicals. At first glance, the following cannot be added:

$$\sqrt{18A^4B} + 3\sqrt{12A^4B}$$

However, by reducing, we obtain

$$= \sqrt{3 \cdot 3A^4}\sqrt{2B} + 3\sqrt{2 \cdot 2A^4}\sqrt{3B}$$
$$= 3A^2\sqrt{2B} + (3)(2A^2)\sqrt{3B}$$
$$= 3A^2\sqrt{2B} + 6A^2\sqrt{3B}$$
$$= 9A^2\sqrt{2B}$$

One error commonly made by students is to try to combine differing radicals. Note the following:

$$\sqrt{4} + \sqrt{9} \neq \sqrt{4+9}$$
$$2 + 3 \neq 3.606$$

Remember: Treat the radical as if it is a literal when adding or subtracting.

EXAMPLE 17-13 Perform the following operations:

(a) $2\sqrt{2} + \sqrt{18}$ (b) $5\sqrt{3} - 2\sqrt{12}$
(c) $5\sqrt{3B} + 7\sqrt{12B}$ (d) $7\sqrt{3A^2B} + \sqrt{48A^4B}$

SOLUTION

(a) $2\sqrt{2} + \sqrt{18} = 2\sqrt{2} + \sqrt{2 \cdot 3 \cdot 3} = 2\sqrt{2} + 3\sqrt{2}$
     $= 5\sqrt{2}$
(b) $5\sqrt{3} - 2\sqrt{12} = 5\sqrt{3} - 2\sqrt{2 \cdot 2 \cdot 3} = 5\sqrt{3} - 4\sqrt{3}$
     $= \sqrt{3}$
(c) $5\sqrt{3B} + 7\sqrt{12B} = 5\sqrt{3B} + 7\sqrt{2 \cdot 2 \cdot 3B}$
     $= 5\sqrt{3B} + 14\sqrt{3B} = 19\sqrt{3B}$
(d) $7\sqrt{3A^2B} + \sqrt{48A^4B} = 7A\sqrt{3B} + \sqrt{16 \cdot 3A^4B}$
     $= 7A\sqrt{3B} + 4A^2\sqrt{3B}$
     These cannot be combined because of the differing powers of A.

Radicals occur many times as a quantity added to a rational number:

$$3 + \sqrt{3} \qquad 8 + \sqrt{5}$$

Note that the radical cannot be added to the rational number. Again, treat the radical as a literal.

EXAMPLE 17-14 Add the following:

(a) $3 + \sqrt{3}$ and $5 + 6\sqrt{3}$
(b) $6 + \sqrt{7}$ and $7 + \sqrt{6}$

SOLUTION

(a) The integers can be added and, because they have identical radicals, the radicals can be added:

$$3 + \sqrt{3}$$
$$5 + 6\sqrt{3}$$
$$\overline{8 + 7\sqrt{3}}$$

(b) In this case, the radicals cannot be added, for they differ:

$$6 + \sqrt{7}$$
$$7 \quad\quad + \sqrt{6}$$
$$\overline{13 + \sqrt{7} + \sqrt{6}}$$

EXAMPLE 17-15 Subtract the following:

(a) $3 + \sqrt{3}$ from $6 + 5\sqrt{3}$
(b) $7 + \sqrt{8}$ from $7 - 6\sqrt{2}$
(c) $6 + \sqrt{10}$ from $7 + \sqrt{5}$

SOLUTION

(a) $$6 + 5\sqrt{3}$$
$$\underline{-(3 + \sqrt{3})}$$
$$3 + 4\sqrt{3}$$

(b) $$7 + \sqrt{8} = 7 + 2\sqrt{2}$$
$$7 - 6\sqrt{2}$$
$$\underline{-(7 + 2\sqrt{2})}$$
$$-8\sqrt{2}$$

(c) $$7 + \sqrt{5}$$
$$\underline{-(6 + \sqrt{10})}$$
$$1 + \sqrt{5} - \sqrt{10}$$

PROBLEMS

Add:

17-55. (a) $4\sqrt{2}, 6\sqrt{2}$ (b) $7\sqrt{3}, 8\sqrt{3}$
17-56. (a) $15\sqrt{10}, 23\sqrt{10}$ (b) $11\sqrt{13}, 16\sqrt{13}$
17-57. (a) $6\sqrt{2}, -7\sqrt{2}$ (b) $8\sqrt{3}, 9\sqrt{3}$
17-58. (a) $5\sqrt{5}, -7\sqrt{5}$ (b) $-6\sqrt{5}, -7\sqrt{15}$
17-59. (a) $3\sqrt{2}, \sqrt{8}$ (b) $5\sqrt{3}, \sqrt{27}$
17-60. (a) $3\sqrt{45}, 6\sqrt{20}$ (b) $3\sqrt{24}, -6\sqrt{54}$
17-61. (a) $2A\sqrt{12}, 3\sqrt{27A^2}$ (b) $5T_1\sqrt{8T_2^2}, 6T_2\sqrt{18T_1^2}$
17-62. (a) $3\sqrt{54M^2}, -2\sqrt{24M^2}$ (b) $-3L_1\sqrt{162L_2^2}, -5L_2\sqrt{50L_1^2}$
17-63. $3 + 4\sqrt{2}, 5 - 6\sqrt{2}$
17-64. $7 - \sqrt{3}, 5 - 6\sqrt{3}$
17-65. $8 + 5\sqrt{8}, 6 + 10\sqrt{18}$
17-66. $20 + 6\sqrt{10}, 5 + 7\sqrt{20}$

Subtract:

17-67. $5\sqrt{2}$ from $7\sqrt{2}$
17-68. $7\sqrt{2}$ from $3\sqrt{3}$
17-69. $3\sqrt{8}$ from $7\sqrt{32}$
17-70. $15\sqrt{75}$ from $6\sqrt{48}$
17-71. $7L_1\sqrt{20L_2^2}$ from $6\,L_2\sqrt{45L_1^2}$
17-72. $12M\sqrt{32Z_1}$ from $3\sqrt{18M^2Z_1}$
17-73. $8+3\sqrt{2}$ from $6+4\sqrt{8}$
17-74. $6-7\sqrt{5}$ from $-7+3\sqrt{15}$

17-5 MULTIPLICATION

Two radicals may be multiplied by each other if the powers are the same: both square root, both cube root, for example. If the powers are not the same, they cannot be combined.

$$\sqrt{2}\,\sqrt{3}=\sqrt{2\cdot 3}=\sqrt{6}$$
$$1.414\times 1.732 = 2.449$$
$$\sqrt[3]{2}\,\sqrt[3]{3}=\sqrt[3]{2\cdot 3}=\sqrt[3]{6}$$
$$1.260\times 1.442 = 1.817$$
$$\sqrt{2}\,\sqrt[3]{3}\text{ cannot be combined.}$$

Note that, when dealing with identical powers, the radicals may freely be factored into individual radicals multiplied by one another.

To multiply expressions containing radicals, treat the radicals as literals during the process. Finally, combine radicals of similar powers and reduce to correct form.

EXAMPLE 17-16 Multiply:

(a) $3\sqrt{2}$ by $7\sqrt{8}$
(b) $1-4\sqrt{6}$ by $7+6\sqrt{3}$
(c) $7+3\sqrt{24}$ by $6+7\sqrt{6}$

SOLUTION

(a) Treat the 3 and 7 as coefficients, multiplying them:

$$(3\sqrt{2})(7\sqrt{8})=21\sqrt{2}\,\sqrt{8}=21\sqrt{16}$$
$$=21\times 4 = 84$$

(b)
$$\begin{array}{r} 1-4\sqrt{6} \\ 7+6\sqrt{3} \\ \hline +6\sqrt{3}-24\sqrt{6}\,\sqrt{3} \\ 7-28\sqrt{6} \\ \hline 7-28\sqrt{6}+6\sqrt{3}-24\sqrt{18} \\ =7-28\sqrt{6}+6\sqrt{3}-72\sqrt{2} \end{array}$$

Note that radicals can be added only if they are identical.

(c) Reduce: $7 + 3\sqrt{24} = 7 + 6\sqrt{6}$

Multiply:

$$\begin{array}{r} 7 + 6\sqrt{6} \\ 6 + 7\sqrt{6} \\ \hline 49\sqrt{6} + 42\sqrt{36} \\ 42 + 36\sqrt{6} \\ \hline 42 + 85\sqrt{6} + 42 \times 6 \\ 294 + 85\sqrt{6} \end{array}$$

PROBLEMS

Multiply the following:

17-75. (a) $2\sqrt{3}, 4\sqrt{6}$ (b) $6\sqrt{5}, -20\sqrt{15}$
17-76. (a) $7\sqrt{6}, -8\sqrt{10}$ (b) $3\sqrt{14}, -7\sqrt{21}$
17-77. $3 + 7\sqrt{2}, 7 - 6\sqrt{2}$
17-78. $8 + 6\sqrt{3}, 17 + 7\sqrt{3}$
17-79. $6 + 4\sqrt{12}, 7 + 3\sqrt{27}$
17-80. $6 - 7\sqrt{3}, 7 + 8\sqrt{6}$
17-81. $25 + 40\sqrt{2}, 30 - 16\sqrt{18}$
17-82. $-6 + 3\sqrt{40}, -10 - 5\sqrt{30}$

17-6 DIVISION

Division of radicals uses multiplication, not division. How is that for an opener? We can divide by multiplying both numerator and denominator by a factor chosen such that the radical in the denominator disappears. Let us start by removing the radical from a simple fraction. This procedure is called rationalizing.

EXAMPLE 17-17 Rationalize the following:

(a) $\dfrac{1}{\sqrt{3}}$ (b) $\dfrac{3\sqrt{6}}{2\sqrt{5}}$

SOLUTION In each case we shall multiply by the radical in the denominator:

(a) $\dfrac{1}{\sqrt{3}} = \dfrac{1}{\sqrt{3}} \cdot \dfrac{\sqrt{3}}{\sqrt{3}} = \dfrac{\sqrt{3}}{\sqrt{9}} = \dfrac{\sqrt{3}}{3}$

(b) $\dfrac{3\sqrt{6}}{2\sqrt{5}} = \dfrac{3\sqrt{6}}{2\sqrt{5}} \dfrac{\sqrt{5}}{\sqrt{5}} = \dfrac{3\sqrt{30}}{2 \times 5} = \dfrac{3\sqrt{30}}{10}$

Where an expression in the denominator consists of both a rational term and an irrational term, the denominator and numerator must be multiplied by the *conjugate* of the denominator. The conjugate of an expression is that expression with the opposite sign given to the irrational term. The following are examples:

Term	Conjugate
$3 + 6\sqrt{2}$	$3 - 6\sqrt{2}$
$5 - 7\sqrt{6}$	$5 + 7\sqrt{6}$
$8 + 33\sqrt{10}$	$8 - 33\sqrt{10}$

EXAMPLE 17-18 Divide:

(a) $\dfrac{1}{3 - 2\sqrt{3}}$ (b) $\dfrac{6 + 7\sqrt{5}}{7 - 3\sqrt{5}}$

SOLUTION

(a) We must multiply both the numerator and the denominator by the conjugate of $3 - 2\sqrt{3}$, $3 + 2\sqrt{3}$.

$$\frac{1}{3 - 2\sqrt{3}} \cdot \frac{3 + 2\sqrt{3}}{3 + 2\sqrt{3}} = \frac{3 + 2\sqrt{3}}{9 + 6\sqrt{3} - 6\sqrt{3} - 4\sqrt{3}\sqrt{3}}$$

$$= \frac{3 + 2\sqrt{3}}{9 - 12}$$

$$= \frac{3 + 2\sqrt{3}}{-3}$$

Note that, by multiplying by the conjugate, the irrational term drops out of the denominator. This process comes from the factoring model:

$$(a + b)(a - b) = a^2 - b^2$$

(b) $\dfrac{6 + 7\sqrt{5}}{7 - 3\sqrt{5}} = \dfrac{6 + 7\sqrt{5}}{7 - 3\sqrt{5}} \cdot \dfrac{7 + 3\sqrt{5}}{7 + 3\sqrt{5}}$

$$= \frac{42 + 18\sqrt{5} + 49\sqrt{5} + 21\sqrt{5}\sqrt{5}}{49 - 9(5)}$$

$$= \frac{42 + 105 + 67\sqrt{5}}{4}$$

$$= \frac{147 + 67\sqrt{5}}{4}$$

PROBLEMS

Rationalize:

17-83. (a) $\dfrac{1}{\sqrt{2}}$ (b) $\dfrac{1}{3\sqrt{5}}$ (c) $\dfrac{1}{17\sqrt{7}}$

17-84. (a) $\dfrac{1}{\sqrt{3}}$ (b) $\dfrac{1}{6\sqrt{6}}$ (c) $\dfrac{1}{32\sqrt{13}}$

17-85. (a) $\dfrac{1}{3\sqrt{36}}$ (b) $\dfrac{1}{7\sqrt{48}}$ (c) $\dfrac{1}{6\sqrt{12}}$

17-86. (a) $\dfrac{1}{6\sqrt{20}}$ (b) $\dfrac{1}{5\sqrt{30}}$ (c) $\dfrac{1}{3\sqrt{96}}$

Divide:

17-87. (a) $\dfrac{1}{3 - 6\sqrt{5}}$ (b) $\dfrac{1}{8 + 2\sqrt{3}}$

17-88. (a) $\dfrac{1}{7 - 6\sqrt{5}}$ (b) $\dfrac{1}{3 + 4\sqrt{7}}$

17-89. (a) $\dfrac{3 - 2\sqrt{5}}{3 + 6\sqrt{5}}$ (b) $\dfrac{8 + 2\sqrt{3}}{6 - 7\sqrt{3}}$

17-90. (a) $\dfrac{5 + 7\sqrt{6}}{2 - 16\sqrt{6}}$ (b) $\dfrac{8 + 7\sqrt{6}}{7 + 3\sqrt{6}}$

17-91. (a) $\dfrac{6 + 3\sqrt{5}}{7 - 6\sqrt{3}}$ (b) $\dfrac{8 + 7\sqrt{3}}{7 + 6\sqrt{5}}$

17-92. (a) $\dfrac{3 + 4\sqrt{10}}{5 - 3\sqrt{11}}$ (b) $\dfrac{13 - 10\sqrt{5}}{11 - 5\sqrt{7}}$

18

QUADRATIC EQUATIONS

So far we have been carefully avoiding equations that have squared variables. In this chapter we face the subject head on, defeating the squared terms with such weapons as graphing, factoring, and a cookbook method.

18-1 INTRODUCTION

A quadratic equation is one that has the variable raised to the second power. Examples include:

$$X^2 + 3X + 1 = 0$$
$$X^2 + 3X = 0$$
$$X^2 - 2X + 1 = 0$$
$$X^2 = 5$$

There are three methods of solving these equations:

1. Solution by graph
2. Solution by factoring
3. The quadratic formula

The graphical method allows us to visualize the curve and its solution, giving us some insights into the behavior of these equations. The factoring method is the first usually attempted on any quadratic equation, because it is the easiest to perform. The formula method always yields the correct solution, but it is a bit more difficult than the factoring method.

18-2 GRAPHING

When we studied graphs in Chapter 12, we limited ourselves to first-order equations (no squared terms). We now consider quadratic equations, a general form of which is

$$X^2 + aX + b = Y$$

where a and b are constants, X is the independent variable, and Y is the dependent variable. These types of equations take four forms: circle, ellipse, parabola, and hyperbola. These four forms can be visualized as planes intersecting a cone as shown in

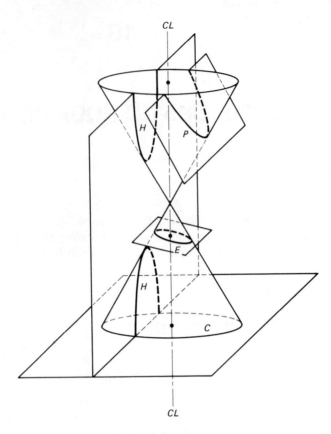

FIGURE 18-1 Conics.

Fig. 18-1. Depending upon the angle and placement of the plane, the resulting intersection will be a circle, ellipse, parabola, or hyperbola.

The Circle

A circle with its center at the origin (Fig. 18-2), has the algebraic form

$$X^2 + Y^2 = r^2$$

In this formula, r represents the radius of the circle. The circle, C of Fig. 18-1, is formed from a cone by intersecting the cone with a plane that is perpendicular to the centerline (CL on the figure).

The Ellipse

An ellipse with its center at the origin (Fig. 18-3), has the algebraic form

$$aX^2 + bY^2 = c$$

If a and b are equal, the figure is a circle rather than an ellipse. The ellipse E of Fig. 18-1 can be formed by intersecting a cone at an angle between that of the circle-generating plane and the slope of the side of the cone.

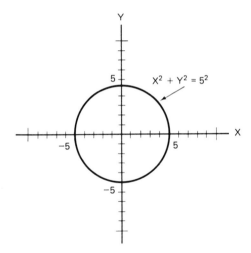

FIGURE 18-2 Circle.

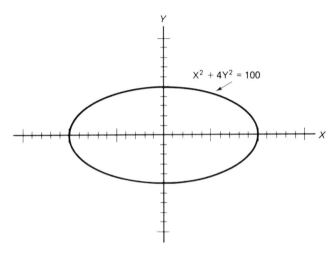

FIGURE 18-3 Ellipse.

The Hyperbola

An hyperbola with its center at the origin (Fig. 18-4), has the equation

$$aX^2 - bY^2 = c$$

Note that it has two portions, called branches, shown as H in Fig. 18-1. It can be formed by intersecting a cone with a plane that is parallel to the centerline.

The Parabola

The parabola (Fig. 18-5), has the equation

$$Y = aX^2 + bX + c$$

Shown as P in Fig. 18-1, it can be generated from a cone by intersection with a plane that is parallel to the slope of the cone. Most of the equations we shall encounter will be parabolas.

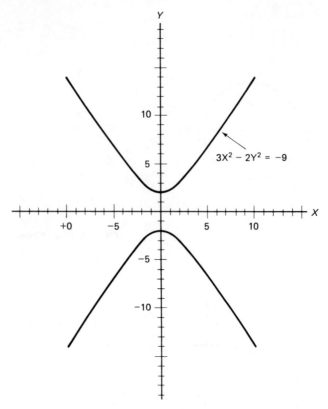

FIGURE 18-4 The hyperbola.

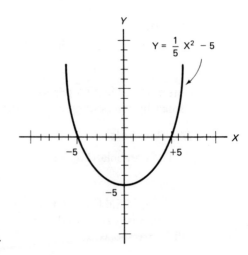

FIGURE 18-5 The parabola.

Solving Equations

Solution by graphical means is done by forming a table of X and Y values and graphing the result. The points at which Y equals zero are the solutions. In general, quadratics will have two solutions.

EXAMPLE 18-1 Solve the equation

$$3X^2 - 2X - 6 = 0$$

SOLUTION We must first graph the equation

$$Y = 3X^2 - 2X - 6$$

by selecting values of X and computing Y.

X	Y
-3	27
-2	10
-1	-1
0	-6
1	-5
2	2
3	15

Since we are expecting a parabola, we should select X values such that we see Y rising then dipping or dipping then rising. Furthermore, since we are expecting two solutions, we should find two places where Y goes from negative to positive or positive to negative. Next, we can graph the equation (Fig. 18-6). Note that the solutions lie where Y goes from positive to negative or negative to positive. To be perfectly honest, I saw that the solution was between -1 and -2, so I selected several intermediate points to give a more accurate graph.

The two solutions are 1.79 and -1.12.

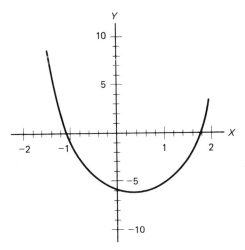

FIGURE 18-6 Example 18-1.

EXAMPLE 18-2   Solve for X:

$$-X^2 - 2X + 7 = 0$$

SOLUTION   Set up a table for $-X^2 - 2X + 7 = Y$:

X	Y
-5	-8
-4	-1
-3	4
-2	7
-1	8
0	7
1	4
2	-1
3	-8

The graph is shown in Fig. 18-7. The two solutions are -3.8 and $+1.8$.

PROBLEMS

Solve the following problems graphically:

18-1. $X^2 + 7X + 3 = 0$
18-2. $X^2 - 3X + 2 = 0$
18-3. $X^2 + 54X + 72 = 0$
18-4. $2X^2 - 30X + 5 = 0$
18-5. $100X^2 - 3X - 1 = 0$
18-6. $X^2 - 12,000X + 30,000,000 = 0$

18-3   SOLUTION BY FACTORING

If a quadratic equation can be factored, its solution becomes very easy. Consider the following equation:

$$X^2 - 2X - 15 = 0$$

Note that it can be factored into

$$(X + 3)(X - 5) = 0$$

Look that expression over carefully. If the quantity $(X + 3)$ were zero, the equation would be valid, for:

$$(X + 3)(X - 5) = 0$$
$$(0)(X - 5) = 0$$
$$0 = 0$$

Furthermore, if the quantity $(X - 5)$ were zero, the equation would also be valid, for

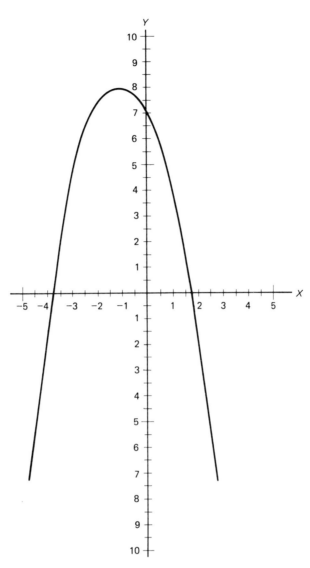

FIGURE 18-7 Example 18-2.

$$(X+3)(X-5)=0$$
$$(X+3)(0)=0$$
$$0=0$$

Therefore, we have two conditions under which the equation is true:

$$X+3=0 \quad \text{and} \quad X-5=0$$

Solving these, we have the two required solutions:

$$X+3=0 \qquad X-5=0$$
$$X=-3 \qquad X=5$$

We can prove that these two are the correct solution:

$$X^2 - 2X - 15 = 0$$
$$(-3)^2 - 2(-3) - 15 = 0$$
$$9 + 6 - 15 = 0$$
$$X^2 - 2X - 15 = 0$$
$$(5)^2 - 2(5) - 15 = 0$$
$$25 - 10 - 15 = 0$$

From the foregoing it can be seen that the solution to a factorable equation can be found by setting each of the factors equal to zero.

EXAMPLE 18-3 Solve for X:

$$2X^2 - 9X - 18 = 0$$

SOLUTION We can factor this equation into

$$(2X + 3)(X - 6) = 0$$

Set each factor equal to zero:

$$2X + 3 = 0 \qquad X - 6 = 0$$
$$X = -\frac{3}{2} \qquad X = 6$$

EXAMPLE 18-4 Solve for X:

$$X^2 + 3X = 0$$

SOLUTION Factor: $X(X + 3) = 0$. Set these factors to zero:

$$X = 0$$

and

$$X + 3 = 0$$
$$X = -3$$

The solutions are 0 and -3.

PROBLEMS

Solve for the unknown by factoring:

18-7. $X^2 - 6X + 8 = 0$

18-8. $X^2 - 5X + 6 = 0$

18-9. $Y^2 - 2Y - 8 = 0$

18-10. $Z^2 - 16 = 0$

18-11. $2N^2 - 50 = 0$

18-12. $9R_1^2 - 64 = 0$

18-13. $R_2^2 - R_2 - 56 = 0$

18-14. $2X^2 + 7X + 3 = 0$

18-15. $6P^2 - 5P - 6 = 0$

18-16. $3Q^2 - 10Q + 8 = 0$

18-17. $4B^2 + 4B - 3 = 0$

18-18. $35X^2 + 59X + 14 = 0$

18-4 SOLUTION BY QUADRATIC EQUATION

The quadratic equation is usually memorized and values plugged into it to obtain the solution. However, let us examine its origins. The formula starts by assuming that the following equation must be solved for X:

$$AX^2 + BX + C = 0 \qquad \text{(18-1)}$$

If we could make the first term a perfect square, the last term a perfect square, and the middle term twice the product of the square roots of the first and last term, we would have an equation that could be factored according to the form

$$(ax + b)^2 = n$$

By taking the root of both sides, we could solve for X. But we can do this for Eq. (18-1) by the following:

Move C:

$$AX^2 + BX = -C$$

Multiply by $4A$:

$$4A^2X^2 + 4ABX = -4AC$$

Add B^2:

$$4A^2X^2 + 4ABX + B^2 = B^2 - 4AC$$

Factor:

$$(2AX + B)^2 = B^2 - 4AC$$

Take the square root:

$$2AX + B = \pm\sqrt{B^2 - 4AC}$$

Solve for X:

$$X = \frac{-B \pm \sqrt{B^2 - 4AC}}{2A} \qquad \text{(18-2)}$$

Thus, we have a cookbook formula for solving for X.

EXAMPLE 18-5 Solve for X:

$$6X^2 - X - 40 = 0$$

SOLUTION Plug into Eq. (18-2):

$$X = \frac{-B \pm \sqrt{B^2 - 4AC}}{2A}$$
$$= \frac{-(-1) \pm \sqrt{(-1)^2 - 4(6)(-40)}}{2(6)}$$
$$= \frac{1 \pm \sqrt{961}}{12}$$
$$= \frac{1 \pm 31}{12}$$

Thus, one solution is

$$X = \frac{1 + 31}{12} = \frac{32}{12} = 2\frac{2}{3}$$

The second solution is

$$X = \frac{1 - 31}{12} = \frac{-30}{12} = 2\frac{1}{2}$$

This problem is also factorable and we can verify that these are the correct solutions by

$$6X - X - 40 = 0$$
$$(2X + 5)(3X - 8) = 0$$
$$2X + 5 = 0 \qquad 3X - 8 = 0$$
$$X = -\frac{5}{2} = -2\frac{1}{2} \qquad X = \frac{8}{3} = 2\frac{2}{3}$$

The answers obtained from the factoring method are identical to those obtained by Eq. (18-2).

The major advantage to using Eq. (18-2) is that it can be used for any quadratic equation, factorable or not.

EXAMPLE 18-6 Solve for R_1:

$$3.6R_1^2 + 2.9R_1 - 6.73 = 0$$

SOLUTION

$$R_1 = \frac{-B \pm \sqrt{B^2 - 4AC}}{2A}$$
$$= \frac{-(2.9) \pm \sqrt{2.9^2 - 4(3.6)(-6.73)}}{2(3.6)}$$
$$= \frac{-2.9 \pm \sqrt{105.322}}{7.2}$$

The two solutions are

$$R_1 = \frac{-2.9 + \sqrt{105.322}}{7.2} = 1.0226$$
$$R_1 = \frac{-2.9 - \sqrt{105.322}}{7.2} = -1.8281$$

Imaginary Roots

There are some types of problems that result in some mighty peculiar results. For example, let us solve the following:

$$X^2 - 6X + 25 = 0$$
$$X = \frac{-B \pm \sqrt{B^2 - 4AC}}{2A}$$
$$= \frac{6 \pm \sqrt{36 - 4(1)(25)}}{2(1)}$$
$$= \frac{6 \pm \sqrt{-64}}{2}$$

Note that this requires us to find the square root of -64. This occurs because the parabola does not cross the X axis (Fig. 18-8). However, our mathematical friends have supplied us with an "escape" by defining $\sqrt{-1}$ to be the constant j. (Refer to Chapter 21 for an in-depth discussion.) Making this assumption, we can solve the equation:

$$X = \frac{6 \pm \sqrt{-64}}{2}$$
$$= \frac{6 \pm \sqrt{-1}\sqrt{64}}{2}$$
$$= \frac{6 \pm j8}{2}$$
$$= 3 \pm j4$$

This solution is correct, for we can substitute it back into the original equation. Substitute $3 + j4$:

$$X^2 - 6X + 25 = 0$$
$$(3 + j4)^2 - 6(3 + j4) + 25 = 0$$
$$9 + j24 + j^2 16 - 18 - j24 + 25 = 0$$
$$9 - 18 + 25 + j^2 16 = 0$$
$$16 + j^2 16 = 0$$

Recognize that j is $\sqrt{-1}$:

$$16 + (\sqrt{-1})(\sqrt{-1})(16) = 0$$
$$16 + (-1)(16) = 0$$
$$16 - 16 = 0$$

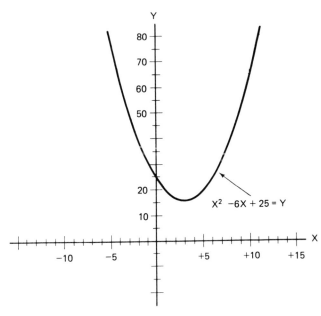

FIGURE 18-8 Equation with imaginary roots.

Thus, by a little sleight of hand, we can solve problems having imaginary roots.

PROBLEMS

Solve by the quadratic formula:

18-19. $X^2 - 13X + 42 = 0$

18-20. $X^2 + 5X - 24 = 0$

18-21. $R_1^2 + 3R_1 - 4 = 0$

18-22. $N^2 + 8N + 2 = 0$

18-23. $3A^2 + 25A + 4 = 0$

18-24. $4.13 L_1^2 + 32L_1 - 6.39 = 0$

18-25. $3A^2 - 2A + 3 = 0$

18-26. $8.2N^2 - 4.51N + 84.9 = 0$

19

TRIANGLES

There is a remarkable parallel between the forces of ac electricity and the study of the right triangle. This parallel enables us to more fully understand circuit concepts, while providing a system of mathematics that we can use to compute circuit values. Although the word "trigonometry" strikes fear in the heart of all who hear it, it comes from three Greek words: "tri" meaning "three," "gon," meaning "angle" (thus, triangle), and "metron," meaning "measure." Thus, it is the study of triangles: the lengths of their sides and the extent of their angles. In this chapter we study triangles in order that we can more fully understand the phenomenon of ac electricity.

19-1 ANGLES

Since a triangle is formed from three (tri) angles, it seems only reasonable that we first consider angles. An angle is formed any time two lines intersect (Fig. 19-1). The angle consists of two lines, sides, intersecting at a point called the vertex. Angles are frequently designated by indicating the three points used to form the triangle. Thus, angle B in the triangle of Fig. 19-1 can be referred to as $\underline{/ABC}$ or $\underline{/CBA}$, using the symbol "$\underline{/}$" to represent "angle." The term "$\underline{/ABC}$" is then pronounced, "angle aye bee cee."

An angle can also be represented by placing a curved line near the vertex and labeling the angle, usually with a Greek letter.[1] Thus, the angle in Fig. 19-1(b) would be called $\underline{/\alpha}$, "angle alpha."

Types of Angles

There are three categories of the extent of an angle: acute, right, and obtuse (Fig. 19-2). A right angle is one in which the two lines meet in a square or perpendicular arrangement. The top edge of this book forms a right angle with its left side. The top edge also forms a right angle with its right side. A square used by a carpenter is formed from two pieces of steel meeting at a right angle. The right angle is very often represented by placing a small square at its vertex [Fig. 19-2(a)].

Any angle that is less than a right angle is called an acute angle [Fig. 19-2(b)]. Any angle that is greater than a right angle is called an obtuse angle [Fig. 19-2(c)].

Two angles which, when added, result in a right angle are called complementary

[1] Appendix II lists the Greek letters.

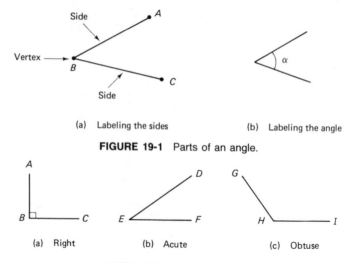

(a) Labeling the sides   (b) Labeling the angle

FIGURE 19-1 Parts of an angle.

(a) Right   (b) Acute   (c) Obtuse

FIGURE 19-2 Types of angles.

angles. Angles α and β of Fig. 19-3 are complementary angles. Angles *ABC* added to $\underline{\vert CBD}$ result in right $\underline{\vert ABD}$.

Two angles which, when added, result in a straight line are said to be supplementary. Angles γ and δ of Fig. 19-4 are supplementary. Angles *EFG*, when added to $\underline{\vert GFH}$ results in $\underline{\vert EFH}$, a straight line. They are, therefore, supplementary.

Plotting Angles

When angles are placed on a graph, the angle is measured with respect to the horizontal, positive *X* axis (Fig. 19-5). The angle can be thought of as a ray,[2] *OA*, that has been rotated clockwise from ray *OB* by the angle α. An angle generated by rotating a ray in the counterclockwise direction is a positive angle, whereas an

FIGURE 19-3 Complementary angles.

FIGURE 19-4 Supplementary angles.

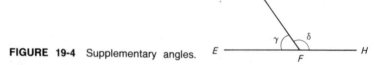

[2] A ray is a line that originates at a point, the origin, and extends in only one direction from that point.

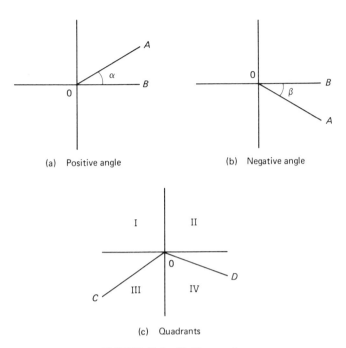

(a) Positive angle (b) Negative angle

(c) Quadrants

FIGURE 19-5 Plotting angles.

angle rotated from *OB* in the clockwise direction is a negative angle. Angle α of Fig. 19-5(a) is a positive angle, whereas /β of Fig. 19-5(b) is a negative angle.

The graph shown is divided into four sections, called quadrants [Fig. 19-5(c)]. The quadrants are labeled as **I, II, III,** and **IV.** Thus, ray *OC* lies in the third quadrant, whereas ray *OD* lies in the fourth quadrant.

19-2 UNITS OF ANGULAR MEASUREMENT

Since the word trigonometry refers to measurement ("-metry"), we should examine methods whereby angles are measured. There are three systems of angular measurement: sexagesimal, radians, and grads. These are summarized in Table 19-1.

Sexagesimal System

The sexagesimal system defines one full circle as 360 degrees (Fig. 19-6). Thus, the right angle shown in Fig. 19-7(a) represents one fourth of a full circle or

$$\frac{1}{4} \times 360 = 90°$$

The sexagesimal system subdivides the degree into 60 minutes and subdivides each minute into 60 seconds. The following conversion fractions can, therefore, be used:

$$\frac{1 \text{ deg}}{60 \text{ min}} \quad \text{and} \quad \frac{1 \text{ min}}{60 \text{ sec}}$$

EXAMPLE 19-1 Convert to degrees, minutes, and seconds:

(a) 867 min   (b) 87,423 sec

SOLUTION

(a) $867 \text{ min} = 867 \text{ min} \times \dfrac{1 \text{ deg}}{60 \text{ min}} = 14.45°$

We know that we have 14 deg. Convert the fraction to min:

$$0.45 \text{ deg} = 0.45 \text{ deg} \times \dfrac{60 \text{ min}}{1 \text{ deg}} = 27.0 \text{ min}$$

Therefore, 867 min = 14 deg 27 min.

TABLE 19-1   Units of angular measurement.

	Sexagesimal	Radians	Grads
Units per circle	360	2π	400
Symbol	deg, °	rad, ^{r}	grad, ^{g}
Subunits	Minutes (min)	None	None
	Seconds (sec)		
	1° = 60 min		
	1′ = 60 sec		

(b) $87,423 \text{ sec} = 87,423 \text{ sec} \times \dfrac{1 \text{ min}}{60 \text{ sec}} \times \dfrac{1 \text{ deg}}{60 \text{ min}}$

$= 24.28416667°$

We know we have 24 whole degrees. Convert the fraction to min:

$$0.2841667 \text{ deg} = 0.2841667 \text{ deg} \times \dfrac{60 \text{ min}}{1 \text{ deg}}$$

$$= 17.05 \text{ min}$$

Convert the fraction of min to sec:

$$0.05 \text{ min} = 0.05 \text{ min} \times \dfrac{60 \text{ sec}}{1 \text{ min}} = 3 \text{ sec}$$

Therefore, 87,423 sec = 24 deg 17 min 3 sec.

EXAMPLE 19-2   Convert to degrees and decimal fractions of degrees:

(a) 35 deg 2 min   (b) 27 deg 7 min 35 sec

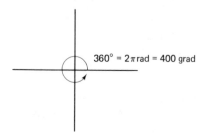

360° = 2π rad = 400 grad

FIGURE 19-6   One full circle.

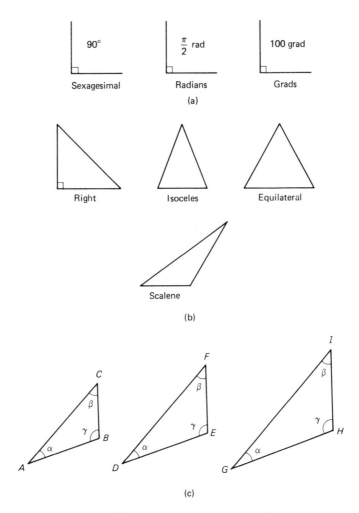

FIGURE 19-7 Angular measurement.

SOLUTION

(a) We already have 35.00°. Convert the min:

$$2 \text{ min} = 2 \text{ min} \times \frac{1 \text{ deg}}{60 \text{ min}} = 0.03333°$$

Add:

$$\text{total} = 35.00 + 0.03333 = 35.03°$$

(b) We have 27 deg. Convert the min to deg:

$$7 \text{ min} = 7 \text{ min} \times \frac{1 \text{ deg}}{60 \text{ min}} = 0.1166667°$$

Convert the sec to deg:

$$35 \text{ sec} = 35 \text{ sec} \times \frac{1 \text{ min}}{60 \text{ sec}} \times \frac{1 \text{ deg}}{60 \text{ min}} = 0.009722°$$

Sum:

$$\text{total} = 27.00 + 0.1166667 + 0.009722$$
$$= 27.1264°$$

Note that the original problem was accurate to six places (xx deg xx min xx sec), so we expressed the answer to six places.

Radian System

The radian system defines one full circle as 2π radians. Therefore, one full circle expressed to 10 places is

$$2\pi = 2 \times 3.1415926535 = 6.283185307 \text{ rad}$$

Since a right angle represents one-fourth circle, it is

$$\frac{2\pi \text{ rad}}{\text{circle}} \times \frac{1}{4} \text{ circle} = \frac{1}{2}\pi \text{ rad}$$

We can convert from degrees to radians or radians to degrees by using the following conversion fractions:

$$\frac{1 \text{ circle}}{360 \text{ deg}} \quad \text{and} \quad \frac{1 \text{ circle}}{2\pi \text{ rad}}$$

EXAMPLE 19-3 Convert to radians:

    (a) 26°   (b) −34°   (c) −276°

SOLUTION

Use the foregoing conversion fractions:

    (a) $26 \text{ deg} = 26 \text{ deg} \times \dfrac{1 \text{ circle}}{360 \text{ deg}} \times \dfrac{2\pi \text{ rad}}{1 \text{ circle}}$

$$= 0.1444\pi \text{ rad} = 0.4538 \text{ rad}$$

    (b) $-34 \text{ deg} = -34 \text{ deg} \times \dfrac{1 \text{ circle}}{360 \text{ deg}} \times \dfrac{2\pi \text{ rad}}{1 \text{ circle}}$

$$= -0.18889\pi \text{ rad} = -0.5934 \text{ rad}$$

    (c) $-276 \text{ deg} = -276 \text{ deg} \times \dfrac{1 \text{ circle}}{360 \text{ deg}} \times \dfrac{2\pi \text{ rad}}{1 \text{ circle}}$

$$= 1.5333\pi \text{ rad} = -4.817 \text{ rad}$$

EXAMPLE 19-4 Convert to degrees:

    (a) 0.39 rad   (b) −1.23 rad   (c) −4.33 rad

SOLUTION

    (a) $0.39 \text{ rad} = 0.39 \text{ rad} \times \dfrac{1 \text{ circle}}{2\pi \text{ rad}} \times \dfrac{360 \text{ deg}}{1 \text{ circle}}$

$$= \frac{70.2}{\pi} \text{ deg} = 22.35°$$

(b) $-1.23 \text{ rad} = -1.23 \text{ rad} \times \dfrac{1 \text{ circle}}{2\pi \text{ rad}} \times \dfrac{360 \text{ deg}}{1 \text{ circle}}$

$\qquad = -\dfrac{221.4}{\pi} \text{ deg}$

$\qquad = -70.47°$

(c) $-4.33 \text{ rad} = -4.33 \text{ rad} \times \dfrac{1 \text{ circle}}{2\pi \text{ rad}} \times \dfrac{360 \text{ deg}}{1 \text{ circle}}$

$\qquad = \dfrac{-779.4}{\pi} \text{ deg}$

$\qquad = -248.1°$

Grad System

The grad system defines one full circle as 400 grads, subdividing only in decimal fractional units. We can use the conversion fraction:

$$\dfrac{1 \text{ circle}}{400 \text{ grads}}$$

EXAMPLE 19-5   Convert to grads:

(a) 27°   (b) −3.1 rad

SOLUTION

(a) $27 \text{ deg} = 27 \text{ deg} \times \dfrac{1 \text{ circle}}{360 \text{ deg}} \times \dfrac{400 \text{ grads}}{1 \text{ circle}} = 30 \text{ grads}$

(b) $-3.1 \text{ rad} = -3.1 \text{ rad} \times \dfrac{1 \text{ circle}}{2\pi \text{ rad}} \times \dfrac{400 \text{ grads}}{1 \text{ circle}} = -197.4 \text{ grads}$

EXAMPLE 19-6   Convert 130.3 grads to (a) deg, min, sec and (b) rad.

SOLUTION

(a) $130.3 \text{ grads} = 130.3 \text{ grads} \times \dfrac{1 \text{ circle}}{400 \text{ grads}} \times \dfrac{360 \text{ deg}}{1 \text{ circle}} = 117.27 \text{ deg}$

Convert the fraction:

$$0.27 \text{ deg} = 0.27 \text{ deg} \times \dfrac{60 \text{ min}}{1 \text{ deg}} = 16.2 \text{ min}$$

Convert the fraction to sec:

$$0.2 \text{ min} = 0.2 \text{ min} \times \dfrac{60 \text{ sec}}{1 \text{ min}} = 12 \text{ sec}$$

Therefore, 130.3 grads = 117 deg 16 min 12 sec

(b) $130.3 \text{ grads} = 130.3 \text{ grads} \times \dfrac{1 \text{ circle}}{400 \text{ grads}} \times \dfrac{2\pi \text{ rad}}{1 \text{ circle}}$

$\qquad = 2.047 \text{ rad}$

Calculator Conversion

Although many calculators have a button for converting from one angular system of measurement to another, the following procedure can be used on any calculator having an angular mode switch. It is valid only for angles between 0 and 180 deg, however. We shall change an angle from one system (the source) to another (the destination).

(a) Put the calculator in the source mode.
(b) Key in the source number.
(c) Find the cos of this source.
(d) Put the calculator into the destination mode.
(e) Find the cos⁻¹ (also called the arc cos).
(f) The number displayed on the calculator is the answer in the destination mode.

PROBLEMS

Convert into degrees, minutes, and seconds:

19-1. (a) 400 min (b) 70,163 sec (c) −43,830 sec
19-2. (a) 430 min (b) −100,630.3 sec (c) −23,630.31 sec
19-3. (a) 235 min (b) 1,830,642 sec (c) 796,013 sec
19-4. (a) 186 min (b) −12,820,146 sec (c) 349,623 sec

Convert into degrees and decimal fractions of degrees:

19-5. (a) 25 deg 32 min (b) 63 deg 44 min
19-6. (a) 72 deg 48 min (b) −82 deg 22 min
19-7. 14 deg 36 min 27 sec
19-8. −13 deg 44 min 23 sec
19-9. 75 deg 15 min 2 sec
19-10. 58 deg 3 min 14 sec
19-11. 123 deg 13 sec
19-12. −138 deg 44 sec

Convert to radians:

19-13. (a) 27 deg (b) 77 deg (c) −33.69 deg
19-14. (a) −45 deg (b) 99 deg (c) 145.632 deg
19-15. 63 deg 43 min 26 sec
19-16. 143 deg 58 min 51 sec
19-17. 38 deg 6 min 13 sec
19-18. 340 deg 40 min 33 sec

Convert to degrees:

19-19. (a) 0.63 rad (b) −0.07 rad (c) 0.31 rad
19-20. (a) 1.31 rad (b) −2.31 rad (c) 1.06 rad
19-21. (a) 0.88 rad (b) −0.73 rad (c) 4.31 rad
19-22. (a) −4.86 rad (b) 1.88 rad (c) 5.93 rad

Convert to grads:

19-23. (a) 3.59 deg (b) 46.3 deg (c) 75.9 deg
19-24. (a) 83.6 deg (b) −135 deg (c) −331 deg

19-3 TRIANGLES

A triangle is a plane figure containing three straight sides. There are several classifications of triangles [Fig. 19-7(b)]:

1. A right triangle has one right angle.
2. An isoceles triangle has two sides of equal length.
3. An equilateral triangle has all three sides of equal length.
4. A scalene triangle has three sides, each differing in length.

The right triangle is, by far, the most important in the study of electronics.

Similar Triangles

Triangles are said to be similar when their angles are identical. In Fig. 19-7(c), all the triangles have identical angles of α, β, and γ. Therefore, the triangles are similar. For such triangles, the sides are proportional. That is,

$$\frac{AC}{DF} = \frac{AB}{DE} = \frac{BC}{EF} \quad \text{and} \quad \frac{AB}{GH} = \frac{BC}{HI} = \frac{AC}{GI}$$

EXAMPLE 19-7 In Fig. 19-7(c), side AB is 12 m and DE is 16 m. If side AC is 15 m, what is the length of DF?

SOLUTION

$$\frac{AC}{DF} = \frac{AB}{DE}$$
$$\frac{15}{DF} = \frac{12}{16}$$
$$DF = 20 \text{ m}$$

EXAMPLE 19-8 In Fig. 19-7(c), DE is 50 mm, GH is 93 mm, and EF is 40 mm. How long is HI?

SOLUTION

$$\frac{DE}{GH} = \frac{EF}{HI}$$
$$\frac{50}{93} = \frac{40}{HI}$$
$$HI = 74.4 \text{ mm}$$

One very important characteristic of a triangle is that the sum of its interior angles is always 180° (Fig. 19-8). Therefore, if we know any two of the angles, we can compute the third.

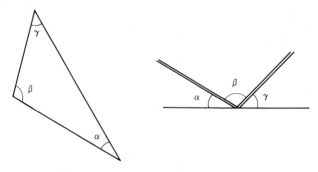

FIGURE 19-8 The sum of interior angles of a triangle equals 180°.

EXAMPLE 19-9 In Fig. 19-8, $\underline{/\alpha}$ is 30° and $\underline{/\beta}$ is 110°. Compute $\underline{/\gamma}$.

SOLUTION The sum of all the angles is 180°.

$$\alpha + \beta + \gamma = 180°$$
$$30 + 110 + \gamma = 180°$$
$$\gamma = 40°$$

The Right Triangle

The right triangle is used extensively in ac electronics to represent phase relationships. Figure 19-9 illustrates this important triangle, labeling its sides in lowercase letters, its vertices in uppercase letters, and its angles in Greek letters. Side c of the figure is called the hypotenuse, after a Greek word meaning to stretch. It will always be the longest of the three sides and it will always be opposite (across from) the right angle. Note that $\underline{/\gamma}$ is 90°, a right angle. Further, since the sum of all three interior angles is 180° and one of these is 90°, the sum of the two acute angles is 90 deg.

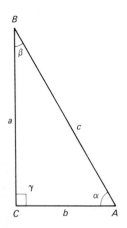

FIGURE 19-9 The right triangle.

EXAMPLE 19-10 The guy wire to a transmission tower makes an angle of 55° with the ground. What is the angle between the guy wire and the tower (Fig. 19-10)?

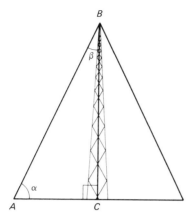

FIGURE 19-10 Example 19-10.

SOLUTION The sum of the acute angles is 90°.

$$\alpha + \beta = 90°$$
$$55 + \beta = 90°$$
$$\beta = 35°$$

Note that α and β will always be acute angles, less than 90°.

The right triangle has a second relationship that is of importance. Referring to Fig. 19-9, we have

$$a^2 + b^2 = c^2 \qquad (19-1)$$

Thus, the square of the hypotenuse, c, is equal to the sum of the squares of the other two sides. This relationship is called the Pythagorean theorem after its discoverer, the Greek philosopher Pythagoras. It should be stressed that this relationship is only true for right triangles. Using Eq. (19-1), we can compute any side, given the other two.

EXAMPLE 19-11 In Fig. 19-10, if the tower is 40 m high and the wire is guyed 20 m from the base, what is the length of the guy wire?

SOLUTION We first have to identify the hypotenuse. It is (a) the longest side and (b) the side opposite the right triangle. In the figure, this is side AB, which is represented as c in Eq. (19-1). We shall call side AC, b and side BC, a.

$$a^2 + b^2 = c^2$$
$$40^2 + 20^2 = c^2$$
$$c = 44.72 \text{ m}$$

EXAMPLE 19-12 Figure 19-11 illustrates an end roof truss. If 2 ft must be added for overhang, what is the total length of the top element of the truss, AC? Also, what is the length of support elements d, e, f, and g if they are spaced uniformly along the ceiling joist BD?

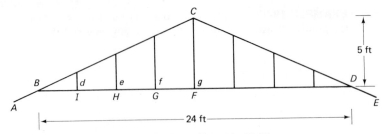

FIGURE 19-11 Example 19-12.

SOLUTION We shall first compute length BC using the Pythagorean theorem:

$$a^2 + b^2 = c^2$$
$$(BF)^2 + (CF)^2 = (BC)^2$$
$$\left(\frac{24}{2}\right)^2 + (5)^2 = BC^2$$

$$BC = \sqrt{169} = 13.00 \text{ ft}$$

Adding the 2 ft for overhang, length AC is 15.00 ft. We can compute the support elements using the principles of similar triangles. Element g is given as 5 ft. Since these elements are equally spaced,

$$BI = IH = HG = GF = 3 \text{ ft}$$

Compute f:

$$\frac{f}{g} = \frac{BG}{BF}$$
$$\frac{f}{5} = \frac{9}{12}$$
$$f = 3.75 \text{ ft}$$

Compute e:

$$\frac{e}{g} = \frac{BH}{BF}$$
$$\frac{e}{5} = \frac{6}{12}$$
$$e = 2.50 \text{ ft}$$

Compute d:

$$\frac{d}{g} = \frac{BI}{BF}$$
$$\frac{d}{5} = \frac{3}{12}$$
$$d = 1.25 \text{ ft}$$

PROBLEMS

Problems 19-27 through 19-32 refer to Fig. 19-12. Compute the missing items.

	AB	BC	CD	DE	AE	BD
19-27.	5 m	5 m	5 m		4 m	
19-28.	3 m	4 m	4.5 m			2.9 m

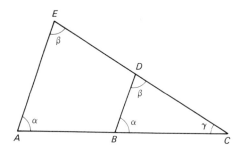

FIGURE 19-12 Problems 19-27 through 19-32 (not to scale).

	AB	BC	CD	DE	AE	BD
19-29.		8 cm		10 cm	10 cm	9 cm
19-30.	1.2 in.		1.6 in.	1.3 in.	1.1 in.	

19-31. If α is 80° and β is 70°, compute γ.

19-32. If α is 75° and γ is 35°, compute β.

19-33. A guy wire to a power pole forms an angle of 70° with the ground. What is the angle between the guy wire and the pole?

19-34. A ladder leaning against a house forms an angle of 60° with the ground. What is its angle to the house?

19-35. Dilapidated Moonstruck Manor must be braced to prevent it from collapsing (Fig. 19-13). How long must the brace be if *AB* is 12 ft and *BC* is 16 ft?

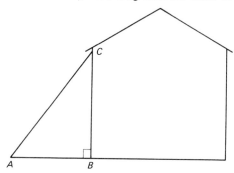

FIGURE 19-13 Problem 19-35.

19-36. When Wrongway Willard flew his Sopwith Camel east from Seattle to Duluth, he forgot to take into the account a wind blowing from north to south, and ended up in Kansas City (Fig. 19-14). If Seattle and Duluth are 1420 mi apart and Duluth is 550 mi from Kansas City, how far had Wrongway Willard traveled?

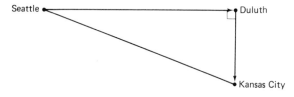

FIGURE 19-14 Problem 19-36.

19-37. When Dippie Dennis takes the shortcut home from school, he proceeds from one corner of Farmer Folie's field to the opposite corner. The field is rectangular, and one side is 430 ft long. If Dippie travels 520 ft across the field, how long is the other side of the field?

19-38. Resistance, reactance, and impedance are related as shown in Fig. 19-15. If the resistance is 4.3 kΩ and the reactance is 7.2 kΩ, what is the value of impedance expressed in kΩ?

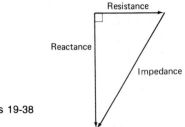

FIGURE 19-15 Problems 19-38 through 19-40.

19-39. Read Prob. 19-38. If the resistance is 930 Ω and the impedance 1.3 kΩ, how much is the reactance?

19-40. Read Prob. 19-38. If the reactance is 4.9 mΩ and the impedance 7.3 mΩ, how much is the resistance?

20

TRIGONOMETRY

In this chapter we apply the principles of Chapter 19 to the solution of right-triangle problems. We shall compute each of the sides and each of the angles of a triangle. This work is used extensively in alternating current to relate the electrical forces within a circuit.

20-1 TRIGONOMETRIC FUNCTIONS

In Chapter 19 we noted that similar triangles have sides that are proportional. This principle also applies to right triangles (Fig. 20-1). All three are similar. Therefore, the following ratios apply:

$$\frac{o_1}{h_1} = \frac{o_2}{h_2} = \frac{o_3}{h_3}$$

$$\frac{o_1}{h_1} = \frac{2}{2.5} = 0.80$$

$$\frac{o_2}{h_2} = \frac{1.6}{2.0} = 0.80$$

$$\frac{o_3}{h_3} = \frac{1}{1.25} = 0.80$$

Therefore, for any triangle similar to those in Fig. 20-1, the ratios of the sides are identical. But similar triangles have identical angles. Therefore, all right triangles having the angle α shown in Fig. 20-1 will have a ratio of 0.80 for the o side divided by the h side. This angle is 53.13°.

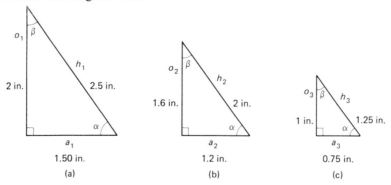

FIGURE 20-1 Similar right triangles.

Let us examine α a bit more carefully. This angle is formed from two sides: an a side and an h side. The h side we have already defined as being the hypotenuse. The hypotenuse will always be the longest side of a right triangle and it will always be the side opposite the right angle. The a side is called the adjacent side. It is the side that, with the hypotenuse, forms one of the acute angles. The third element of this trilogy is the opposite side. This is defined as that side which is opposite the acute angle we are examining. In summary, whenever we examine a particular acute angle within a right triangle, that angle is formed by an intersection of the hypotenuse and the adjacent sides. The third side is the opposite side, that which is directly across from the angle we are examining.

Note that these definitions are relative. In Fig. 20-2, if we are examining the angle α, side e is the hypotenuse, side f the adjacent, and side d the opposite side.

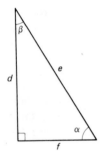

FIGURE 20-2 Opposite and adjacent sides are relative.

If we are examining angle β, side e is still the hypotenuse, but side d is the adjacent, and side f the opposite side.

EXAMPLE 20-1 A triangle similar to that of Fig. 20-1 has an angle α of 53.13°. If its opposite side is 253 km, what is the length of its adjacent and hypotenuse sides?

SOLUTION We have already observed that in triangles with an α of 53.13°, the opposite side divided by the hypotenuse is 0.8. Therefore,

$$\frac{\text{opposite}}{\text{hypotenuse}} = 0.8$$

But we were given its opposite side:

$$\frac{253}{\text{hypotenuse}} = 0.8$$

$$\text{hypotenuse} = \frac{253}{0.8} = 316.25 \text{ km}$$

We can use similar reasoning to find the length of the adjacent side. From Fig. 20-1:

$$\frac{o_1}{a_1} = \frac{2}{1.50} = 1.333$$

But the sides of similar triangles have identical ratios. Therefore, for this problem,

$$\frac{\text{opposite}}{\text{adjacent}} = 1.333$$

But we were given the opposite side:

$$\frac{253}{\text{adjacent}} = 1.333$$

$$\text{adjacent} = \frac{253}{1.333} = 189.75 \text{ km}$$

Sine, Cosine, and Tangent

In the foregoing discussion, we observed that similar triangles have identical side ratios. Therefore, for any particular angle, 53.13° for example, there is a fixed ratio for any two sides. It follows for any other angle, 27° for example, another ratio would exist and this ratio would be identical for all triangles having an acute angle of 27°. These ratios have been given the names sine, cosine, and tangent. The ratios are defined as follows:

$$\text{sine } \alpha = \frac{\text{opposite}}{\text{hypotenuse}}$$

$$\text{cosine } \alpha = \frac{\text{adjacent}}{\text{hypotenuse}}$$

$$\text{tangent } \alpha = \frac{\text{opposite}}{\text{adjacent}}$$

EXAMPLE 20-2 Find the sine, cosine, and tangent of angle α of Fig. 20-1(a).

SOLUTION The sine (sin) is defined as the opposite divided by the hypotenuse:

$$\sin 53.13° = \frac{o_1}{h_1} = \frac{2}{2.5} = 0.8$$

The cosine (cos) is defined as the adjacent divided by the hypotenuse:

$$\cos 53.13° = \frac{a_1}{h_1} = \frac{1.5}{2.5} = 0.6$$

The tangent (tan) is defined as the opposite divided by the adjacent:

$$\tan 53.13° = \frac{o_1}{a_1} = \frac{2}{1.50} = 1.333$$

Note that these ratios are identical for the other two triangles:

$$\sin 53.13° = \frac{o_2}{h_2} = \frac{1.6}{2} = 0.8$$

$$= \frac{o_3}{h_3} = \frac{1}{1.25} = 0.8$$

$$\cos 53.13° = \frac{a_2}{h_2} = \frac{1.2}{2} = 0.6$$

$$= \frac{a_3}{h_3} = \frac{0.75}{1.25} = 0.6$$

$$\tan 53.13° = \frac{o_2}{a_2} = \frac{1.6}{1.2} = 1.333$$

$$= \frac{o_3}{a_3} = \frac{1}{0.75} = 1.333$$

If the angle is known, these ratios can easily be found on the calculator. Enter the angle, then press the $\boxed{\sin}$, $\boxed{\cos}$, or $\boxed{\tan}$ button.

EXAMPLE 20-3   Find the sin, cos, and tan of 53.13°.

SOLUTION

$$\sin 53.13° = 0.799998928$$
$$\cos 53.13° = 0.600001429$$
$$\tan 53.13° = 1.333328371$$

EXAMPLE 20-4   Find the sin, cos, and tan for the following angles:

(a) 33°   (b) 87.18°   (c) 1.21 rad

SOLUTION

(a) $\sin 33° = 0.544639035$
  $\cos 33° = 0.838670568$
  $\tan 33° = 0.649407593$

(b) $\sin 87.18° = 0.998789025$
  $\cos 87.18° = 0.049198416$
  $\tan 87.18° = 20.30124356$

(c) For radians, put the calculator into the radian mode, key in the angle, then press the $\boxed{\sin}$, $\boxed{\cos}$, or $\boxed{\tan}$ key.

$$\sin 1.21 \text{ rad} = 0.935616002$$
$$\cos 1.21 \text{ rad} = 0.353019401$$
$$\tan 1.21 \text{ rad} = 2.650324595$$

These ratios should be thoroughly memorized. The following acrostics may help:

Scot Oliver Had A Handful Of Ants

S (ine)	Oliver Had	$\text{Sin} = \dfrac{O}{H}$
C (osine)	A Handful	$\text{Cos} = \dfrac{A}{H}$
o		
T (an)	Of Ants	$\text{Tan} = \dfrac{O}{A}$

PROBLEMS

20-1. A right triangle similar to that of Fig. 20-1 has an angle of 53.13°. If its opposite side is 431 cm, what is the length of its adjacent and hypotenuse sides?

20-2. A right triangle similar to that of Fig. 20-1 has an angle of 53.13° and a hypotenuse of 23.5 m. What is the length of its opposite and adjacent sides?

20-3. A right triangle has an angle whose opposite side is 30, its adjacent side is 35, and its hypotenuse is 46.10. What are its sin, cos, and tan?

20-4. A right triangle has an angle whose opposite side is 43, its adjacent side is 25, and its hypotenuse is 49.74. What are its sin, cos, and tan?

Given below are the sides of a right triangle. In each case compute the sin, cos, and tan of the angle.

	Opposite	Adjacent	Hypotenuse
20-5.	439	860	965.6
20-6.	0.30	1.21	1.247
20-7.	131	27	133.8
20-8.	33.6	42.8	54.4
20-9.	26.13	33.34	42.36
20-10.	8888	9999	13,378
20-11.	832.1	126.3	841.6
20-12.	13,000	0.03	13,000
20-13.	0.0015	863,472	863,472
20-14.	43.6	43.6	61.7

20-15. Find the sin, cos, and tan for the following angles:
    (a) 30°   **(e)** 127.6°
    (b) 27.3°   **(f)** 831.9°
    (c) 43.1°   **(g)** 0.896 rad
    (d) 88.2°   **(h)** 0.021 rad

20-16. Find the sin, cos, and tan for the following angles:
    (a) 5°   **(e)** −267°
    (b) 83.6°   **(f)** 3.15 rad
    (c) 45°   **(g)** −0.89 rad
    (d) 830°   **(h)** 45 rad

20-2 FINDING UNKNOWN SIDES

The main use for these ratios is to calculate unknown values within triangles. Notice that each ratio has three variables. For example, the sine has the angle, the opposite side, and the hypotenuse. Knowing any two of these will enable us to obtain the third.

EXAMPLE 20-5 Compute sides X and Y and angle β of Fig. 20-3.

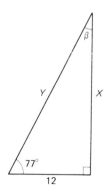

FIGURE 20-3 Example 20-5.

SOLUTION We know the angle and its adjacent side. Which of the three ratios includes these two? The cos and the tan. Using the cos:

$$\cos \alpha = \frac{a}{h}$$

$$\cos 77° = \frac{12}{Y}$$

$$Y = \frac{12}{\cos 77°} = 53.34$$

We must now compute the opposite side, X. We can use the sin or tan.

$$\tan \alpha = \frac{o}{a}$$

$$\tan 77° = \frac{X}{12}$$

$$X = 12 \tan 77° = 51.98$$

We can easily compute β:

$$\alpha + \beta = 90°$$
$$77 + \beta = 90$$
$$\beta = 13°$$

In Example 20-5, we had to calculate 12 divided by the cos 77°. This was accomplished on the calculator as follows:

(a)   Algebraic notation
1.   Enter the 12 $\boxed{1}\ \boxed{2}$
2.   Divide $\boxed{\div}$
3.   Enter 77 $\boxed{7}\ \boxed{7}$
4.   Find the cosine $\boxed{\cos}$
5.   Press the equals sign $\boxed{=}$

(b)   RPN notation:
1.   Enter 12 $\boxed{1}\ \boxed{2}\ \boxed{\text{Enter}}$
2.   Enter 77 $\boxed{7}\ \boxed{7}$
3.   Find the cosine $\boxed{\cos}$
4.   Divide $\boxed{\div}$

Note that there is no need to copy the sin, cos, or tan on paper: let the calculator remember them.

EXAMPLE 20-6   Compute sides M and N and $\underline{/\beta}$ of Fig. 20-4.

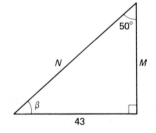

FIGURE 20-4   Example 20-6.

SOLUTION First, let us reflect upon what measurements were given. The angle is known and its opposite side. Let us compute the adjacent side, M. Which of the three ratios includes the opposite side, adjacent side, and the angle? The tangent does.

$$\tan \alpha = \frac{o}{a}$$

$$\tan 50° = \frac{43}{M}$$

$$M = \frac{43}{\tan 50°} = 36.08$$

Compute N:

$$\sin \alpha = \frac{o}{h}$$

$$\sin 50° = \frac{43}{N}$$

$$N = \frac{43}{\sin 50°} = 56.13$$

Compute β:

$$\beta = 90 - \alpha = 90 - 50 = 40°$$

EXAMPLE 20-7 Compute R, S, and β of Fig. 20-5.

FIGURE 20-5 Example 20-7.

SOLUTION

$$\sin \alpha = \frac{o}{h}$$

$$\sin 10° = \frac{R}{20}$$

$$R = 20 \sin 10° = 3.473$$

$$\cos \alpha = \frac{a}{h}$$

$$\cos 10° = \frac{S}{20}$$

$$S = 20 \cos 10° = 19.70$$

$$= 90 - \alpha = 90 - 10 = 80°$$

EXAMPLE 20-8 Sam Surveyor surveyed a sycamore tree by sighting from a point 20 ft from the base of the tree to the top of the sycamore. He measured 71.3°. How tall is the sycamore?

SOLUTION Figure 20-6 shows that this is a trigonometry problem. We are given the angle and the adjacent side and must find the opposite side.

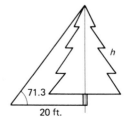

FIGURE 20-6 Example 20-8.

$$\tan \alpha = \frac{o}{a}$$

$$\tan 71.3° = \frac{h}{20}$$

$$h = 20 \tan 71.3° = 59.09 \text{ ft}$$

PROBLEMS

Solve for the unknown quantities in Fig. 20-7.

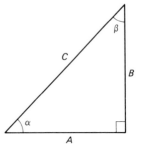

FIGURE 20-7 Problems 20-17 through 20-28.

	α	β	A	B	C
20-17.	10°		12		
20-18.	44°		37		
20-19.	72.16°		38.21		
20-20.	0.86 rad		72.13		
20-21.		84°		32	
20-22.		76°		0.8	
20-23.		3°		8.2	
20-24.		10°		33.6	
20-25.	47°				33.6
20-26.	52°				6.51
20-27.	83°				7.72
20-28.	0.21 rad				0.056

20-29. When Shorty Collins stands 80 ft from the old Endicott building, he has to raise his sights 75° to see the top of the structure. If Shorty is 5 ft 3 in. tall, how tall is the Endicott Building?

20-30. When Simantha Sweetheart swoons seeing Steve Standhi, she stands 60 cm from Steve, staring sensuously at Steve's eyes. If she has to lift her eyes 20°, how much higher are Steve's eyes from the ground than Simantha's?

20-3 FINDING UNKNOWN ANGLES

Each of the trigonometric ratios has three elements: the angle and two of the sides. The examples we have studied so far have had a side as one of the unknowns. However, many times two of the sides are known and we must compute the angle. Thus, we compute the ratio and must determine the angle itself. For this reason, we use inverse functions. Assume we know that tan 45° = 1.0. We can express the inverse function of this in the three following notations, called the arc tangent or inverse tangent:

$$\tan^{-1} 1.0 = 45°$$
$$\text{arc tan } 1.0 = 45°$$
$$\text{inv tan } 1.0 = 45°$$

Each of these statements is saying "the angle whose tangent is 1.0 equals 45°." Try it on your calculator. Compute the tan 45°, yielding 1.0, then convert this 1.0 back into the angle by finding the inverse tan. In a similar manner, the inverse sine can be expressed as follows:

$$\text{If} \qquad \sin 30° = 0.5$$
$$\text{then} \qquad \sin^{-1} 0.5 = 30°$$
$$\text{arc sin } 0.5 = 30°$$
$$\text{inv sin } 0.5 = 30°$$

The latter three statements can be read "the arc sine of 0.5 is 30°" or "the inverse sine of 0.5 is 30°" or "the angle whose sine is 0.5 equals 30°." Perform these functions on your calculator. In a similar manner:

$$\text{If} \qquad \cos 45° = 0.7071$$
$$\text{then} \qquad \cos^{-1} 0.7071 = 45°$$
$$\text{arc cos } 0.7071 = 45°$$
$$\text{inv sin } 0.7071 = 45°$$

This reads "the arc cosine of 0.7071 is 45°" or "the inverse cosine of 0.7071 is 45°" or "the angle whose cosine is 0.7071 equals 45°." Perform this function on your calculator.

The inverse functions are useful for finding an angle, given two of the sides.

EXAMPLE 20-9 Compute α, β, and X of Fig. 20-8.

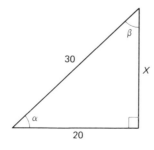

FIGURE 20-8 Example 20-9.

SOLUTION We must first select an unknown angle. We shall select α, noting that we are given its adjacent side and hypotenuse. What function uses the adjacent side, hypotenuse, and the angle? The cosine.

$$\cos \alpha = \frac{a}{h}$$

$$\cos \alpha = \frac{20}{30} = 0.6667$$

Pressing the $\boxed{\cos^{-1}}$, or $\boxed{\text{arc}}$ $\boxed{\cos}$, or $\boxed{\text{inv}}$ $\boxed{\cos}$ yields the angle:

$$= \cos^{-1} 0.6667 = 48.19°$$

We can now compute:

$$\beta = 90 - \alpha = 90 - 48.19 = 41.81°$$

Compute X:

$$\sin \alpha = \frac{o}{h}$$

$$\sin 48.19° = \frac{X}{30}$$

$$X = 30 \sin 48.19° = 22.36$$

EXAMPLE 20-10 In constructing my house, I wanted to assure that it was square. If it is 26 ft on one side and 52 ft on the other, what should a tape read when stretched corner to corner?

SOLUTION We can use the trigonometric functions or the Pythagorean theorem. We shall use the former:

$$\tan \alpha = \frac{o}{a}$$

$$\tan \alpha = \frac{26}{52} = 0.5$$

$$\alpha = \tan^{-1} 0.5 = 26.57°$$

$$\sin \alpha = \frac{o}{h}$$

$$\sin 26.57° = \frac{26}{h}$$

$$h = \frac{26}{\sin 26.57} = 58.14 \text{ ft}$$

EXAMPLE 20-11 A guy wire 40 ft in length supports a 28-ft antenna. What angle does the wire make with the ground, and how far is the ground anchor from the antenna?

SOLUTION Consider the angle first:

$$\sin \alpha = \frac{o}{h}$$

$$\sin \alpha = \frac{28}{40}$$

$$\alpha = \sin^{-1} \frac{28}{40} = 44.43°$$

The distance from anchor to antenna is

$$\cos \alpha = \frac{a}{h}$$

$$\cos 44.43° = \frac{a}{40}$$

$$a = 40 \cos 44.43° = 28.57 \text{ ft}$$

EXAMPLE 20-12 Resistance *(R)*, impedance *(Z)*, and reactance *(X)* are related as shown in Fig. 20-9. If resistance is 4 kΩ, and impedance 5.3 kΩ, compute the reactance and phase angle *(θ)*.

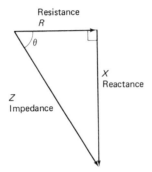

FIGURE 20-9 Example 20-12.

SOLUTION

$$\cos \theta = \frac{a}{h}$$

$$\cos \theta = \frac{4}{5.3}$$

$$\theta = \cos^{-1} \frac{4}{5.3} = 41.00°$$

$$\sin \theta = \frac{o}{h}$$

$$\sin 41.00° = \frac{X}{5.3}$$

$$X = 5.3 \sin 41.00° = 3.477 \text{ kΩ}$$

PROBLEMS

Compute the unknown quantities of Fig. 20-10.

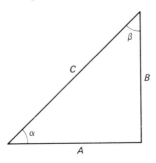

FIGURE 20-10 Problems 20-31 through 20-42.

	α	β	A	B	C
20-31.			30	30	
20-32.			46	80	
20-33.			2.36	4.86	
20-34.			0.89	0.21	
20-35.			783		962
20-36.			46.1		73.2
20-37.			0.15		1.03
20-38.			77.7		141.2
20-39.				1365	8412
20-40.				87.31	152.73
20-41.				1.86	2.15
20-42.				0.015	0.0836

20-43. In a correctly designed staircase (in the United States), the sum of the run (horizontal dimension) plus the rise (vertical dimension) for each step is about 17 in. If the run is 10 in., what is the rise of each step and the angle the entire staircase would make with the horizontal? If there are 10 steps, how long is the staircase, measured along the hypotenuse?

20-44. A keyboard must be designed with a sloping face (Fig. 20-11). How long is the face, and what is angle α [Fig. 20-11(b)]?

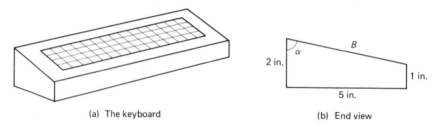

(a) The keyboard   (b) End view

FIGURE 20-11 Problem 20-44.

20-45. If the resistance is 265 Ω and the impedance 400 Ω, compute the reactance and the phase angle (Fig. 20-9).

20-46. If the resistance is 24 kΩ and the reactance is 35 kΩ, compute the impedance and the phase angle (Fig. 20-9).

20-4 FUNCTIONS OF LARGE ANGLES

The trigonometric quantities are valid for all angles, both negative and positive. The easiest way to understand the principles involved is to examine a graph of a large angle (Fig. 20-12). Here we wish to find the sine, cosine, and tangent for 150°. The angle is first plotted, then a triangle formed using the line just plotted as the hypotenuse. It is important that the triangle is formed with the X axis. Next, we label each of the sides with reference to the angle formed with the X axis, α in the figure. Note that the opposite side is positive, for it extends above the X axis. However, the adjacent side is negative, for it extends to the left from the origin. The hypotenuse is always considered to be a positive quantity.

Let us assume that

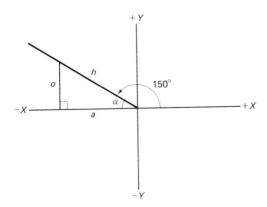

FIGURE 20-12 Functions of 150°.

$$o = 3.560$$
$$a = 6.1661$$
$$h = 7.120$$

We can compute each of the functions as follows:

$$\sin 150° = \frac{o}{h} = \frac{3.560}{7.120} = 0.5000$$

$$\sin 30° = \frac{o}{h} = \frac{3.560}{7.120} = 0.5000$$

Note that the sine of 150° is identical to the sine of 30°. Try it on your calculator. Similarly, the cosine is:

$$\cos 150° = \frac{a}{h} = \frac{-6.1661}{7.120} = -0.8660$$

$$\cos 30° = \frac{a}{h} = \frac{6.1661}{7.120} = 0.8660$$

Note that, because the adjacent side is negative, the cosine 150° is negative. Thus, in this case:

$$\cos 150° = -\cos(180° - 150°) = -\cos 30°$$

We can compute the tangent as follows:

$$\tan 150° = \frac{o}{a} = \frac{3.560}{-6.1661} = -0.5774$$

$$\tan 30° = \frac{o}{a} = \frac{3.560}{6.1661} = 0.5774$$

Again, the tangent of 150° is identical to the tangent of 30° except for the sign. We can summarize the entire process as follows:

> *The magnitudes of functions of large angles are always equal to the functions of the small angles they make with the X axis. However, the signs of the results are dependent upon the signs of the opposite and adjacent sides.*

The calculator always gives the correct answer, both in magnitude and in sign. Figure 20-13 illustrates functions of the third and fourth quadrants. For the third quadrant:

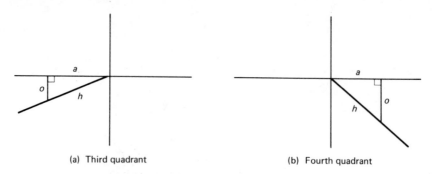

(a) Third quadrant   (b) Fourth quadrant

FIGURE 20-13 Third and fourth quadrant functions.

$$\sin = \frac{o}{h} = \frac{-}{+} = \text{negative}$$

$$\cos = \frac{a}{h} = \frac{-}{+} = \text{negative}$$

$$\tan = \frac{o}{a} = \frac{-}{-} = \text{positive}$$

For the fourth quadrant:

$$\sin = \frac{o}{h} = \frac{-}{+} = \text{negative}$$

$$\cos = \frac{a}{h} = \frac{+}{+} = \text{positive}$$

$$\tan = \frac{o}{a} = \frac{-}{+} = \text{negative}$$

EXAMPLE 20-13 Find the sine, cosine, and tangent of each of the following angles by finding the function of the angle the hypotenuse makes with the X axis:

(a) 127°   (b) 213°   (c) −62°   (d) 395°

SOLUTION

(a) The angle is $180° - 127° = 53°$ (Fig. [20-14(a)]).

$$\sin 53° = 0.7986$$
$$\cos 53° = 0.6018$$
$$\tan 53° = 1.3270$$

From the previous discussion, the sine is positive, the cosine is negative, and the tangent is negative. Thus:

$$\sin 127° = 0.7986$$
$$\cos 127° = -0.6018$$
$$\tan 127° = -1.3270$$

Check the results on your calculator.

(b) This angle, 213°, falls into the third quadrant [Fig. 20-14(b)]. The angle is $213 - 180 = 33°$.

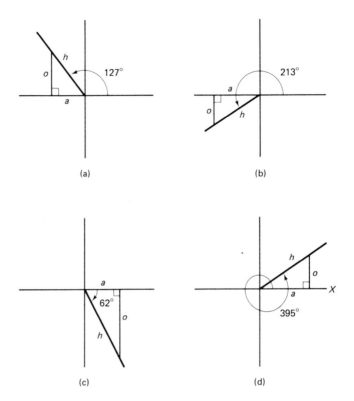

FIGURE 20-14   Example 20-13.

$$\sin 33° = 0.5446$$
$$\cos 33° = 0.8387$$
$$\tan 33° = 0.6494$$

But the sine and cosine are negative and tangent positive:

$$\sin 213° = -0.5446$$
$$\cos 213° = -0.8387$$
$$\tan 213° = 0.6494$$

Check the results with your calculator.

(c)   This angle, $-62°$, falls into the fourth quadrant [Fig. 20-14(c)]. Note that positive angles start with the positive X axis and sweep counterclockwise, whereas negative angles sweep clockwise. The angle is $62°$.

$$\sin 62° = 0.8829$$
$$\cos 62° = 0.4695$$
$$\tan 62° = 1.8807$$

Thus,

$$\sin -62° = -0.8829$$
$$\cos -62° = 0.4695$$
$$\tan -62° = -1.8807$$

Confirm these answers on your calculator.

(d)   The angle, 395°, is shown in Fig. 20-14(d). The angle with the X axis is

$$X = 395 - 360 = 35°$$

Since the angle falls in the first quadrant, all quantities are positive.

$$\sin 395° = \sin 35° = 0.5736$$
$$\cos 395° = \cos 35° = 0.8192$$
$$\tan 395° = \tan 35° = 0.7002$$

Ranges of the Functions

The functions vary in magnitude as shown in Fig. 20-15. The sine and cosine have values between +1 and −1, for the opposite and adjacent sides can never exceed the hypotenuse. However, the tangent can have values approaching negative infinity and positive infinity. This can be shown by examining Fig. 20-16(a). Here the adjacent

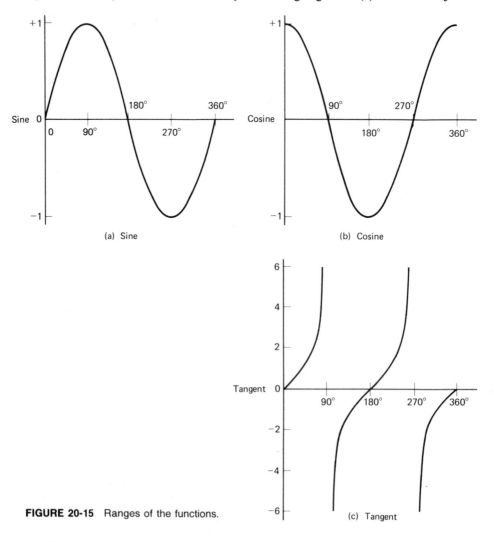

(a) Sine

(b) Cosine

(c) Tangent

FIGURE 20-15   Ranges of the functions.

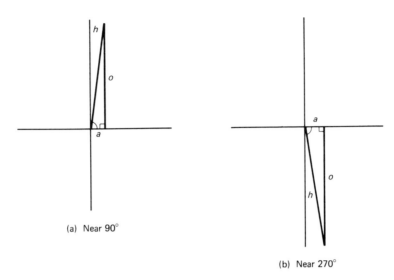

(a) Near 90°

(b) Near 270°

FIGURE 20-16 Critical tangent values.

approaches zero from the positive direction and the opposite has a finite value. Thus, the result is a very large positive number. This problem also occurs at 270°. Verify this phenomenon on your calculator.

PROBLEMS

Find the sin, cos, and tan of each of the following angles by finding the function of the angle that the hypotenuse makes with the X axis.

20-47. **(a)** 146° **(b)** 131° **(c)** 193°
20-48. **(a)** 173° **(b)** 243° **(c)** 265°
20-49. **(a)** −30° **(b)** 383° **(c)** −126°
20-50. **(a)** −109° **(b)** 3 rad **(c)** −2 rad
20-51. What is the range of the sin?
20-52. What is the range of the cos?
20-53. What is the range of the tan?
20-54. Using the definitions of sin, cos, and tan, show that

$$\tan \alpha = \frac{\sin \alpha}{\cos \alpha}$$

20-5 ADVANCED CONCEPTS

In this section we consider some interesting characteristics, definitions, and equations that make trigonometric manipulations a little easier.

Law of Sines

The law of sines applies to any triangle whether it has a right angle or not. Briefly stated, the sin of an angle is proportional to the side opposite that angle. Referring to Fig. 20-17, we have

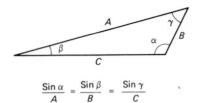

$$\frac{Sin\ \alpha}{A} = \frac{Sin\ \beta}{B} = \frac{Sin\ \gamma}{C}$$

FIGURE 20-17 Law of sines.

$$\frac{\sin \alpha}{A} = \frac{\sin \beta}{B} = \frac{\sin \gamma}{C}$$

Thus, if we know two sides and an angle, we can compute all the remaining quantities.

EXAMPLE 20-14 If A is 12 m, B is 8.6 m, and β is 25°, compute C, α, and γ of Fig. 20-17.

SOLUTION

$$\frac{\sin \alpha}{A} = \frac{\sin \beta}{B}$$

$$\frac{\sin \alpha}{12} = \frac{\sin 25°}{8.6}$$

$$\alpha = 36.14°$$

Since $\alpha + \beta + \gamma = 180$;

$$\gamma = 180 - \alpha - \beta$$
$$= 180 - 36.14 - 25$$
$$= 118.86°$$

Compute C:

$$\frac{\sin \beta}{B} = \frac{\sin \gamma}{C}$$

$$\frac{\sin 25°}{8.6} = \frac{\sin 118.86}{C}$$

$$C = 17.82°$$

Note that we could not use Pythagorean's theorem to find C, for it only applies to right triangles.

We can also use the law of sines if we know two of the angles and one of the sides.

EXAMPLE 20-15 If α is 106°, β is 33°, and A is 26.3 cm, compute γ, B, and C of Fig. 20-16.

SOLUTION We can easily compute the third angle.

$$\gamma = 180 - \alpha - \beta$$
$$= 180 - 106 - 33$$
$$= 41°$$

We can compute the other two sides:

$$\frac{\sin \alpha}{A} = \frac{\sin \beta}{B}$$

$$\frac{\sin 106°}{26.3} = \frac{\sin 33°}{B}$$

$$B = 14.90 \text{ cm}$$

$$\frac{\sin \alpha}{A} = \frac{\sin \gamma}{C}$$

$$\frac{\sin 106°}{26.3} = \frac{\sin 41°}{C}$$

$$C = 17.95 \text{ cm}$$

PROBLEMS

Compute the missing quantities:

	α	β	γ	A	B	C
20-55.	26°			10	20	
20-56.		44°			36.9	17.21
20-57.			103°		44.6	122.6
20-58.	88°			0.61		0.43
20-59.		77°		3.68	4.77	
20-60.			13°		43	136
20-61.	33°		41°		77	
20-62.	83°	15°		16		
20-63.	109°		9°			16
20-64.		33°	116°		13.6	

Law of Cosines

The law of cosines also applies to any triangle, whether or not it has a right angle. Use Fig. 20-18 as a reference:

$$A^2 = B^2 + C^2 - 2BC \cos \alpha$$

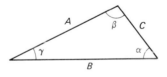

$$A^2 = B^2 + C^2 - 2BC \cos \alpha$$

FIGURE 20-18 Law of cosines.

Notice that it assumes we know two sides and their shared angle. It computes the side opposite that shared angle. The law can be applied to any part of the triangle. Apply it to sides B and C:

$$B^2 = A^2 + C^2 - 2AC \cos \beta$$
$$C^2 = A^2 + B^2 - 2AB \cos \gamma$$

EXAMPLE 20-16 In Fig. 20-18, if B is 20 m, C is 30 m, and α is 27°, compute side C.

SOLUTION

$$A^2 = B^2 + C^2 - 2BC \cos \alpha$$
$$= 20^2 + 30^2 - (2 \times 20 \times 30) \times \cos 27°$$
$$A^2 = 230.79$$
$$A = \sqrt{230.79} = 15.19 \text{ m}$$

EXAMPLE 20-17 In Fig. 20-18, compute C if A is 0.63, B is 0.83, and γ is 44°.

SOLUTION

$$C^2 = A^2 + B^2 - 2AB \cos \gamma$$
$$= 0.63^2 + 0.83^2 - (2 \times 0.63 \times 0.83) \cos 44°$$
$$= 0.3335$$
$$C = \sqrt{0.3335} = 0.5775$$

PROBLEMS

Compute C in Fig. 20-18:

20-65. $A = 48$, $B = 72$, $\gamma = 43.6°$
20-66. $A = 7.31$, $B = 6.87$, $\gamma = 77°$
20-67. $A = 596$, $B = 1263$, $\gamma = 32.6°$
20-68. $A = 773$, $B = 431$, $\gamma = 100°$

Compute B in Fig. 20-18:

20-69. $A = 4.31$, $C = 5.26$, $\beta = 77°$
20-70. $A = 735$, $C = 627$, $\beta = 25.9°$
20-71. $A = 13,746$, $C = 10,093$, $\beta = 1.3$ r
20-72. $A = 0.031$, $C = 0.36$, $\beta = 0.06$ r

Secant, Cosecant, and Cotangent

The most commonly used trigonometric functions are the sin, cos, and tan. However, three others exist and are most easily defined using these three.

$$\text{secant } \alpha = \frac{\text{hypotenuse}}{\text{adjacent}} = \frac{1}{\cos \alpha}$$
$$\text{cosecant } \alpha = \frac{\text{hypotenuse}}{\text{opposite}} = \frac{1}{\sin \alpha}$$
$$\text{cotangent } \alpha = \frac{\text{adjacent}}{\text{opposite}} = \frac{1}{\tan \alpha}$$

It should be emphasized that these are of more value to the mathematician than to the technician.

EXAMPLE 20-18 Find the secant, cosecant, and cotangent of the following angles:

(a) 27°   (b) −2 rad

SOLUTION To find the secant, find the cos and find its inverse $(1/x)$. In a similar manner, find the cosecant by finding the sin and then finding its inverse. The cotangent is the inverse of the tan.

(a) $\text{secant } 27° = \dfrac{1}{\cos 27°} = 1.122$

　　　　$\text{cosecant } 27° = \dfrac{1}{\sin 27°} = 2.203$

　　　　$\text{cotangent } 27° = \dfrac{1}{\tan 27°} = 1.963$

(b) $\text{secant } -2 \text{ rad} = \dfrac{1}{\cos -2 \text{ rad}} = -2.403$

　　　　$\text{cosecant } -2 \text{ rad} = \dfrac{1}{\sin -2 \text{ rad}} = -1.100$

　　　　$\text{cotangent } -2 \text{ rad} = \dfrac{1}{\tan -2 \text{ rad}} = 0.4577$

PROBLEMS

Compute the secant, cosecant, and cotangent for the following angles.

20-73. (a) 26°   (b) 123°
20-74. (a) −43°   (b) 196°
20-75. (a) 369°   (b) 4.18 rad
20-76. (a) 700°   (b) −3.69 rad

Trigonometric Identities

There are a great number of equations that relate the trigonometric functions to one another. Some of the common ones are shown in Table 20-1. The fundamental relationships shown in the table are, by far, the most useful.

EXAMPLE 20-19 Prove that

$$\sin 3X = 3 \sin X - 4 \sin^3 X$$

SOLUTION From Table 20-1:

$$\sin (X + Y) = \sin X \cos Y + \cos X \sin Y$$

Substitute $2X$ for Y:

$$\sin (X + 2X) = \sin X \cos 2X + \cos X \sin 2X$$

But $\cos 2X = \cos^2 X - \sin^2 X$ and $\sin 2X = 2 \sin X \cos X$.

$$\sin 3X = \sin X(\cos^2 X - \sin^2 X) + \cos X(2 \sin X \cos X)$$
$$= \sin X \cos^2 X - \sin^3 X + 2 \sin X \cos^2 X$$

TABLE 20-1 Trigonometric identities.

Fundamental

$$\sin^2 X + \cos^2 X = 1 \qquad \tan X = \frac{\sin X}{\cos X}$$

$$\sin X = \frac{1}{\operatorname{cosec} X} \qquad \cos X = \frac{1}{\sec X}$$

$$\tan X = \frac{1}{\operatorname{cotan} X}$$

Sum of angles

$$\sin (X \pm Y) = \sin X \cos Y \pm \cos X \sin Y$$
$$\cos (X \pm Y) = \cos X \cos Y \mp \sin X \sin Y$$
$$\tan (X \pm Y) = \frac{\tan X \pm \tan Y}{1 + \tan X \tan Y}$$

Double angles

$$\sin 2X = 2 \sin X \cos X$$
$$\cos 2X = \cos^2 X - \sin^2 X$$
$$\tan 2X = \frac{2 \tan X}{1 - \tan^2 X}$$

But $\sin^2 X + \cos^2 X = 1$. Thus, $\cos^2 X = 1 - \sin^2 X$.

$$\sin 3X = \sin X(1 - \sin^2 X) - \sin^3 X + 2 \sin X(1 - \sin^2 X)$$
$$= \sin X - \sin^3 X - \sin^3 X + 2 \sin X - 2 \sin^3 X$$
$$= 3 \sin X - 4 \sin^3 X$$

PROBLEMS

20-77. Prove that $\sin 4X = 8 \cos^3 X \sin X - 4 \cos X \sin X$.

20-78. Prove that

$$\tan 3X = \frac{3 \tan X - \tan^3 X}{1 - 3 \tan^2 X}$$

20-79. Prove that $\cos^2 X - \sin^2 Y = \cos(X + Y) \cos(X - Y)$.

20-80. Prove that

$$\tan \frac{1}{2}X = \frac{1 - \cos X}{\sin X}$$

21

INTRODUCTION TO ALTERNATING CURRENT

In Chapter 20, we discussed briefly the graph of $Y = \sin X$. In this chapter, we see that this curve, called a sine wave, is the fundamental waveform used in analysis of alternating current. We first investigate how this waveform is generated, then examine each element within the sine wave. Finally, we discuss different methods of expression of sine-wave values.

21-1   GENERATING ALTERNATING CURRENT

Alternating current can be generated electronically or electromechanically. We shall examine the method used by a generator such as that at Grand Coulee Dam by which ac is produced. This device generates alternating current by rotating a loop of wire in a magnetic field (Fig. 21-1). The faster the wire cuts magnetic lines of force, the more voltage is generated. Thus, when the loop is horizontal [Fig. 21-2(a)], a high voltage is generated. As the loop rotates to B, less voltage is generated. Finally, at point C, no voltage is generated, for no lines of force are being cut. Figure 21-2(f) illustrates the complete sine wave that the generator produces in one revolution. This sine wave will continue to be produced as long as the rotation is continued. Note that, for half of the waveform, the voltage is positive and for the second half

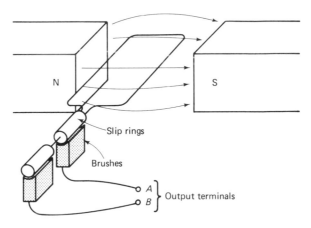

FIGURE 21-1   ac generator.

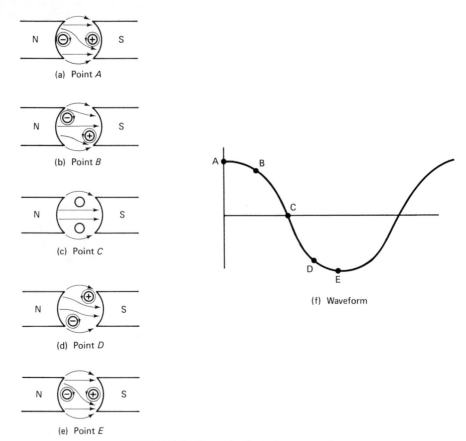

(a) Point A

(b) Point B

(c) Point C

(d) Point D

(e) Point E

(f) Waveform

FIGURE 21-2 Generator instantaneous voltage.

negative. If this voltage were applied to a resistor, current would flow one direction, then another. Thus, it would alternate and is called alternating current.

21-2 THE SINE WAVE

The voltage produced by the generator has the equation

$$v = V \sin \theta$$

where v is the voltage at a particular point, V the maximum voltage attained by the generator, and θ the angle through the cycle.

Figure 21-3(a) illustrates this waveform. If we assume that the generator starts at point C of Fig. 21-2, at 0° through the cycle, 0 V is produced. When the loop has rotated 90°, V volts is produced. At 180°, 0 V is again produced; at 270°, $-V$ volts, and at 360°, 0 V again. But what about an angle 45° through the cycle? How much voltage is produced at this point? We can use the sine-wave equation:

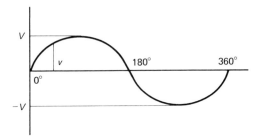

(a) The sine wave

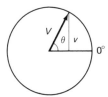

(b) Rotating vector

FIGURE 21-3 The sine wave.

$$v = V \sin \theta$$
$$= V \sin 45°$$
$$= 0.7071 \ V$$

Thus, if the maximum voltage reached is 100 V ($V = 100$), at 45° the voltage is 70.71 V.

EXAMPLE 21-1 What is the voltage when the loop has rotated 136° through its cycle and the maximum voltage is 100 V?

SOLUTION

$$v = V \sin \theta$$
$$= 100 \sin 136°$$
$$= 69.47 \ V$$

The sine wave can also be thought of as a vector (like the spoke of a wheel) rotating at a constant angular velocity or rate of rotation. [Fig. 21-3(b)]. Here, V represents the length of the vector and v the vertical height from the horizontal axis. As the vector is rotated, v alternates between positive and negative, producing the sine wave shown in Fig. 21-3(a).

EXAMPLE 21-2 What is the voltage at a point 48° through the cycle if the maximum is 60 V?

SOLUTION

$$v = V \sin \theta$$
$$= 60 \sin 48°$$
$$= 44.59 \ V$$

EXAMPLE 21-3 At a point 78° through the cycle, the voltage is 44 V. What is the maximum voltage, V?

SOLUTION

$$V = \frac{v}{\sin \theta} = \frac{44}{\sin 78°} = 44.98 \text{ V}$$

The complete sine-wave equation used in electronics is

$$v = V \sin (2\pi ft + \theta)$$

where v is the instantaneous voltage, V the maximum value (called amplitude), f the frequency, t is time, and θ is the phase angle. Note that the equation has three basic parts:

$$\begin{aligned} v \quad & \text{instantaneous voltage} \\ V \quad & \text{maximum amplitude} \\ (2\pi ft + \theta) \quad & \text{the angle of rotation} \end{aligned}$$

We shall pick apart $(2\pi ft + \theta)$, then return to the entire equation.

Frequency

Frequency is the measure of how many complete cycles occur within 1 s and is expressed in hertz (Hz). Consider Fig. 21-4. Here, instead of plotting the sine wave against an angle, it is plotted relative to time. Thus, five cycles occur within a time of 1 s. Therefore, the frequency of this waveform is said to be 5 Hz.

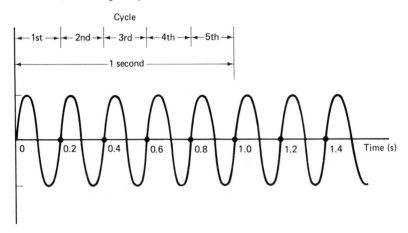

FIGURE 21-4 Frequency.

In electronics, signals occur all the way from Hz to GHz. Commercial FM stations generate frequencies ranging from 88.1 to 107.9 MHz.

EXAMPLE 21-4 A waveform traverses 26 cycles in 31 ms. What is its frequency?

SOLUTION

$$f = \frac{\text{cycles}}{\text{sec}} = \frac{26}{0.031} = 838.7 \text{ Hz}$$

EXAMPLE 21-5 How many cycles will a 2.3-MHz waveform traverse in 1.496 ms?

SOLUTION

$$f = \frac{\text{cycles}}{\text{sec}}$$
$$\text{cycles} = f \times \text{sec}$$
$$= 2.3 \times 10^6 \times 1.496 \times 10^{-3}$$
$$= 3440.8 \text{ cycles}$$

Period

Period is the inverse of frequency:

$$P = \frac{1}{f} = \frac{\text{sec}}{\text{cycle}}$$

It is therefore the measure of how many seconds it takes to complete one cycle of the waveform.

EXAMPLE 21-6 A certain waveform traverses 1826.3 cycles in 1 min. What is the period and the frequency of the waveform?

SOLUTION

$$P = \frac{\text{sec}}{\text{cyc}} = \frac{60}{1826.3} = 32.85 \text{ ms}$$

Therefore, it takes 32.85 ms to complete one cycle. The frequency is

$$f = \frac{1}{P} = \frac{1}{32.85 \times 10^{-3}} = 30.44 \text{ Hz}$$

Thus, 30.44 cycles occur every second.

Phase Angle

The phase angle is a measurement of the angular difference between two waveforms of identical frequency. Consider Fig. 21-5. In this diagram, waveform A is said to have a phase angle of 20° with reference to waveform B. Furthermore, A is said to lead B by 20°. One way to picture this is to move your eye from left to right. The first waveform to cross the X axis is said to lead the second waveform. The second waveform is said to lag the first. Thus, the following are true concerning Fig. 21-5:

1. A leads B by 20°
2. B lags A by 20°
3. C lags A by 35°
4. C lags B by 15°
5. A leads C by 35°
6. B leads C by 15°

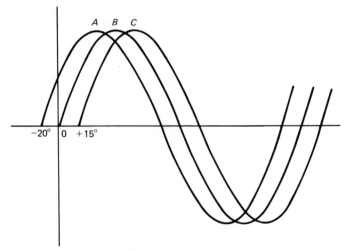

FIGURE 21-5 Phase angle.

Study each of these to verify the definitions of lead and lag.

It should be noted that phase angle can be expressed in radians, degrees, or grads.

Angular Velocity

The general equation for voltage that we are studying is

$$v = V \sin (2\pi ft + \theta)$$

The term $2\pi f$ is called angular velocity and represents the number of radians per second through which our waveform travels. It is usually indicated by the symbol ω (Greek lowercase omega). We can examine its units of measure by the following:

$$\omega = 2\pi f = 2\pi \, \frac{\text{radians}}{\text{cycle}} \times f \, \frac{\text{cycles}}{\text{sec}}$$
$$= 2\pi f \, \text{rad/sec}$$

The angular velocity also corresponds to a vector rotating at a constant rate—the spoke on a bicycle wheel, for example. Note that if we multiply the angular velocity by time, we have an angle, expressed in radians:

$$\omega t = 2\pi ft = 2\pi f \, \frac{\text{rad}}{\text{sec}} \times t \, \text{sec}$$
$$= 2\pi ft \, \text{rad}$$

Thus, the term $2\pi ft$ is an angle expressed in radians. Note this has to be, for we are finding the sin of an angle: $(2\pi ft + \theta)$.

The Sine-Wave Equation

Now, having discussed each part of the equation in detail, let us put it together. The basic waveform for voltage at any time t is represented as

$$v = V \sin (2\pi ft + \theta)$$

where $v =$ instantaneous amplitude at time t, volts
$V =$ maximum amplitude, volts
$f =$ frequency, hertz
$t =$ time, seconds
$\theta =$ phase angle, radians or degrees
$2\pi f =$ angular velocity, radians/second
$2\pi ft =$ angle at time t, radians

EXAMPLE 21-7   The equation for a certain voltage waveform is

$$v = 230 \sin (459t + 26°)$$

What is the waveform's (a) maximum amplitude, (b) frequency, (c) period, (d) phase angle, (e) angular velocity, and (f) graph?

SOLUTION

(a)   The waveform's maximum amplitude is 230.
(b)   The waveform's frequency can be found:

$$2\pi f = 459$$
$$f = \frac{459}{2\pi} = 73.05 \text{ Hz}$$

(c)   The waveform's period is

$$P = \frac{1}{f} = \frac{1}{73.05} = 13.69 \text{ ms}$$

(d)   The phase angle is 26°.
(e)   The angular velocity is 459 rad/s.
(f)   A graph of the waveform is shown in Fig. 21-6.

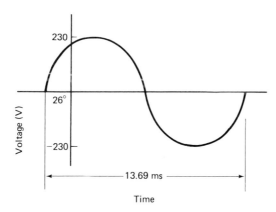

FIGURE 21-6   Example 21-7.

EXAMPLE 21-8   The waveform for a certain current is

$$i = 20 \sin (6380t - 44°)$$

What is the waveform's (a) maximum amplitude, (b) frequency, (c) period, (d) phase angle, (e) angular velocity, and (f) graph?

SOLUTION

(a)   The maximum amplitude is 20 A.

(b)   The frequency is

$$2\pi f = 6380$$
$$f = \frac{6380}{2\pi} = 1.015 \text{ kHz}$$

(c)   The period is $P = \dfrac{1}{f} = \dfrac{1}{1015} = 984.8 \ \mu s$

(d)   The phase angle is $-44°$.

(e)   The angular velocity is 6380 rad/s.

(f)   The graph is shown in Fig. 21-7.

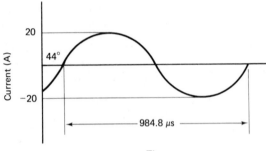

FIGURE 21-7 Example 21-8.

EXAMPLE 21-9   Find the voltage at $t = 200 \ \mu s$ for the following waveform:

$$v = 163 \sin (4000t)$$

SOLUTION

$$v = 163 \sin (4000 \times 200 \times 10^{-6})$$
$$= 163 \sin 0.8 \text{ rad}$$
$$= 116.9 \text{ V}$$

Note that the angle (0.8) is in radians.

EXAMPLE 21-10   Find the voltage at $t = 20 \ \mu s$ for the following waveform:

$$v = 1.312 \sin (6280t + 26°)$$

SOLUTION   The main difficulty here is that $6280t$ is in radians and 26° in degrees. However, we can convert the 26° to rad:

$$26° = 0.4538 \text{ rad}$$

(The way I did it was to find the sin of 26°, then change the calculator to the radian mode and find the $\sin^{-1}$ of this value). We can now complete the problem:

$$v = 1.312 \sin [(6280 \times 20 \times 10^{-6}) + 0.4538]$$
$$= 1.312 \sin 0.5794$$
$$= 0.7183$$
$$= 718.3 \text{ mV}$$

EXAMPLE 21-11 A certain 400-Hz current is 15 mA at a point 20 μs after $t = 0$. What is the maximum amplitude of this sinusoidal waveform?

SOLUTION

$$i = I \sin (2\pi ft)$$
$$0.015 = I \sin (2 \times \pi \times 400 \times 20 \times 10^{-6})$$
$$I = \frac{0.015}{\sin 0.016\pi}$$
$$= 298.5 \text{ mA}$$

Harmonics

A harmonic is a multiple of a given frequency. Thus, the second harmonic of 100 Hz is 200 Hz, the third harmonic 300 Hz, and the fifteenth harmonic 1500 Hz. The first harmonic is considered the fundamental frequency. All repetitive waveforms, regardless of shape, can be thought of as the sum of an infinite number of harmonics. For example, a square wave can be constructed from odd-numbered harmonics added in proper proportions (Fig. 21-8). The equation for this waveform is

$$v = V_m(\sin \omega t + \tfrac{1}{3} \sin 3\omega t + \tfrac{1}{5} \sin 5\omega t + \tfrac{1}{7} \sin 7\omega t + \ldots)$$

The fundamental is the term sin ωt. Each harmonic is indicated by the number preceding the ω. Thus, the seventh harmonic of the fundamental is sin 7 ωt. This seems reasonable since its angular velocity, 7ω, is seven times faster than that of the fundamental, ω. Note that the fractions in the equation indicate that the maximum amplitude of the third harmonic is $\tfrac{1}{3}$ that of the fundamental; the amplitude of the fifth harmonic is $\tfrac{1}{5}$ that of the fundamental; the amplitude of the nth harmonic is $1/n$ that of the fundamental, n being any odd integer.

Thus, the square wave is the sum of an infinite number of odd harmonics, each of proper amplitude. Other nonsinusoidal, repetitive waveforms are merely the sum of harmonics with the proper amplitude, frequency, and phase angle.

EXAMPLE 21-12 What are the second, fifth, tenth, and twentieth harmonics of the following frequencies?

(a) 500 Hz   (b) 2 kHz   (c) 4.3 MHz

SOLUTION

(a) 1 kHz, 2.5 kHz, 5 kHz, 10 kHz
(b) 4 kHz, 10 kHz, 20 kHz, 40 kHz
(c) 8.6 MHz, 21.5 MHz, 43 MHz, 86 MHz

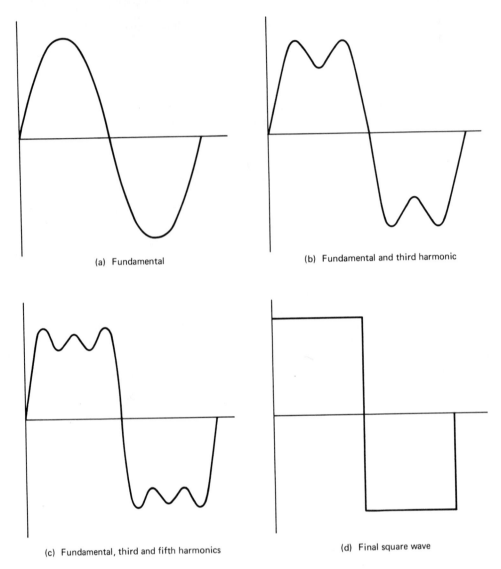

(a) Fundamental

(b) Fundamental and third harmonic

(c) Fundamental, third and fifth harmonics

(d) Final square wave

FIGURE 21-8 The square wave.

Wavelength

All radio waves travel through air or space at about the speed of light, 300 Mm/s. If we could take a picture of a 100-MHz FM radio wave traveling from the station to my home, we would find that a 100-MHz signal measures 3 m from the peak amplitude of one wave to the identical place on the succeeding wave. This distance is called the waveform's wavelength and may be expressed

$$v = f\lambda$$

where v is the velocity of the waveform (300 Mm/s for radio waves in space), f is the frequency, and λ (Greek lowercase lambda) is the wavelength.

EXAMPLE 21-13 What is the wavelength of a radio station operating on a frequency of 106 MHz?

SOLUTION

$$v = f\lambda$$
$$\lambda = \frac{v}{f} = \frac{300 \times 10^6}{106 \times 10^6}$$
$$= 2.83 \text{ m}$$

EXAMPLE 21-14 The velocity of a sound wave through dry air at 20°C is 344 m/s. What is the wavelength of a 2.5-kHz tone as it travels through air?

SOLUTION

$$\lambda = \frac{v}{f} = \frac{344}{2500} = 137.6 \text{ mm}$$

PROBLEMS

21-1. A wire loop is rotated in a magnetic field. If the maximum voltage is 60 V, compute the voltage at the following angle through the cycle: **(a)** 36° **(b)** 84° **(c)** 0.7 rad **(d)** 132°

21-2. A wire loop is rotated in a magnetic field producing a maximum of 435 V. Compute the voltage at the following angles through the cycle: **(a)** 1.5° **(b)** 44° **(c)** 6 rad **(d)** 210°

21-3. At a point 26° through its cycle a generator produces a voltage of 44 V. What is its maximum voltage?

21-4. At a point 133° through its cycle, a generator produces a voltage of −23 V. What is its maximum voltage?

21-5. If a generator produces 10 V at an angle of 16° through its cycle, what will it produce at 50°? at 110°?

21-6. At a point 44° through its cycle a generator produces 88 V. What voltage is it producing at 3°? at 175°?

21-7. Compute the frequency of waveforms if they complete the following number of cycles in the time indicated:
   (a)   27 cycles, 3 ms
   (b)   483 cycles, 5 μs
   (c)   5 Mc, 32.61 s

21-8. Repeat Prob. 21-7 for the following values:
   (a)   5 cycles, 3.16 s
   (b)   3 kc, 3.86 ms
   (c)   53 Gc, 49.31 s

21-9. How many cycles will a 4.31-kHz signal traverse in 50 ms? in 40 μs? in 1 h?

21-10. How many cycles will a 40-GHz signal traverse in 3 μs? in 40 ms? in 1 min?

21-11. Compute the period of the waveforms in Prob. 21-7.

21-12. Compute the period of the waveforms in Prob. 21-8.

21-13. Determine the **(a)** maximum amplitude, **(b)** frequency, **(c)** period, **(d)** phase angle, **(e)** angular velocity, and **(f)** graph for the equation $v = 40 \sin (377t + 64°)$.

21-14. Repeat Prob. 21-13 for $i = 60 \sin (5760t + 70°)$.

21-15. Repeat Prob. 21-13 for $v = 623 \sin (8623t - 60°)$.

21-16. Repeat Prob. 21-13 for $i = 0.023 \sin (7521t - 33°)$.

21-17. Find the voltage at the following times for the equation $v = 53 \sin (431t)$: **(a)** 200 μs **(b)** 20 ms **(c)** 4 s

21-18. Find the current for the equation $i = 0.031 \sin (87,640t)$ at the following times: **(a)** 1 μs **(b)** 5.63 μs **(c)** 7.31 ms

21-19. Find the current at the following times for the equation $i = 0.305 \sin (731t - 26°)$: **(a)** 4.6 μs **(b)** 82.1 ms **(c)** 463 ms

21-20. Find the voltage at the following times for the equation $v = 865 \sin (3463t + 70°)$: **(a)** 0 μs **(b)** 10 ms **(c)** 5.31 μs

21-21. What are the third, seventh, and twenty-sixth harmonics of the following fundamental waveforms? **(a)** 3 Hz **(b)** 46 kHz **(c)** 3.196 MHz

21-22. What are the third, sixth, and thirty-third harmonics of the following fundamental waveforms? **(a)** 4.31 Hz **(b)** 167 Hz **(c)** 3.159 GHz

21-23. What is the wavelength of a radio station operating at 590 kHz? at 146 MHz?

21-24. What is the wavelength of a radio station operating at 146 kHz? 3.61 GHz?

21-25. Sound travels through Mortimer Peck's wooden head at a rate of 4110 m/s. What is the wavelength of Mrs. Peck's shrill 3-kHz voice as it travels through Mortimer's head?

21-26. The velocity of sound through the Seine River at 15°C is 1437 m/s. What is the wavelength of a 400-Hz tone as it travels through this medium?

21-3 EXPRESSION OF AC VALUES

When we speak to one another about a waveform we can specify its frequency. But how can we specify the amplitude of a waveform if it is constantly changing? We shall see that there are four characteristics of amplitude that we can identify: peak amplitude, peak-to-peak amplitude, average amplitude, and rms amplitude.

Peak Measurements

There are two descriptions of amplitude involving the word "peak." The first, the peak amplitude, is the amplitude from the 0-V line to the maximum positive peak of the waveform. Thus, for Fig. 21-9(a) it would be 100 V.

The second, called peak to peak (p-p), is the amplitude from the signal's maximum negative excursion to its maximum positive excursion, in this case 200 V. Thus, the peak amplitude of the waveform in Fig. 21-9(a) is 100 V; the peak-to-peak amplitude, 200 V. In a similar manner, the peak amplitude of the waveform in Fig. 21-9(b) is 5 V; its p-p amplitude, 10 V.

Average Amplitude

The average amplitude of a waveform tells us something about its long-term characteristics. To find the average of the waveform in Fig. 21-10(a):

1. Find the total area underneath the waveform for one period.
2. Divide this by the period.

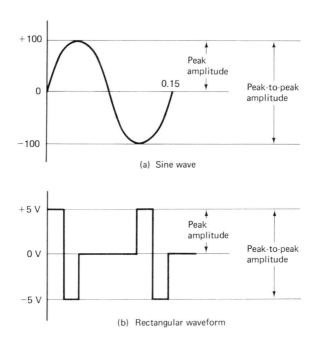

(a) Sine wave

(b) Rectangular waveform

FIGURE 21-9 Peak amplitudes.

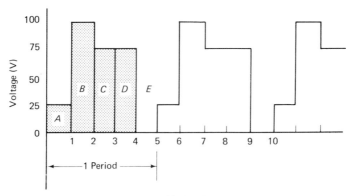

(a) Waveform

Portion	Height	Width	Area
A	25	1	25
B	100	1	100
C	75	1	75
D	75	1	75
E	0	1	0
Total			275

(b) Finding the area

FIGURE 21-10 Average voltage.

Thus, we have merely to find the area indicated by the shaded portion and divide this by the period. We can total the area as shown in Fig. 21-10(b). Since the width of the waveform is 5 (one complete period), we can find the average by dividing the total area by the width:

$$V_{av} = \frac{area}{width} = \frac{275}{5} = 55 \text{ V}$$

The average voltage is 55 V.

EXAMPLE 21-15 Find the average current of the waveform shown in Fig. 21-11.

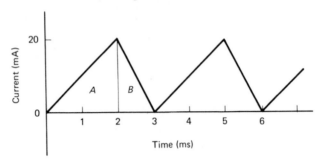

FIGURE 21-11 Example 21-15, average current.

SOLUTION The area of A is

$$A_A = \frac{1}{2}WH = \frac{1}{2} \times 2 \times 20 = 20$$

The area of B is

$$A_B = \frac{1}{2}WH = \frac{1}{2} \times 1 \times 20 = 10$$

The area is

$$A_T = A_A + A_B = 20 + 10 = 30$$

The average is the area divided by the width:

$$I_{av} = \frac{area}{W} = \frac{30}{3} = 10$$

Thus, the average amplitude is 10 mA.

The method for finding the average of a sinusoidal waveform is identical to that of the rectangular and the triangular waveforms. Note, however:

1. Just what is the average? Is it the algebraic average over the complete waveform (zero) or the average of the absolute value of the waveform? For the purpose of this discussion, we shall assume the latter definition.
2. Finding the area underneath a sine wave is no easy trick.

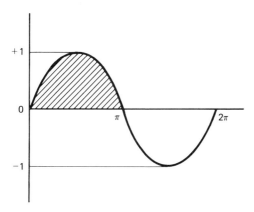

FIGURE 21-12 Sine wave.

It turns out that the area of the shaded part of Fig. 21-12 is, by calculus, equal to 2.000. Thus, the average is

$$V_{av} = \frac{\text{area}}{\text{period}} = \frac{2}{\pi} = 0.6366$$

Thus, we can state that, for sinusoidal waveforms only, the absolute average amplitude is equal to the peak amplitude multiplied by 0.6366:

$$V_{av} = 0.6366 \times V_{pk}$$

EXAMPLE 21-16 Find the absolute average for the following sinusoidal waveforms:

    (a) A peak current of 40 A
    (b) A p-p voltage of 60 V
    (c) The waveform $V = 160 \sin 377t$

SOLUTION

    (a) $I_{av} = 0.6366 I_{pk} = 0.6366 \times 40 = 25.46$ A

    (b) $V_{av} = 0.6366 V_{pk} = 0.6366 \times \dfrac{60}{2} = 19.10$ V

    (c) The peak voltage is 160.

$$V_{av} = 0.6366 \; V_{pk} = 0.6366 \times 160 = 101.9 \text{ V}$$

Root Mean Square

The root-mean-square (rms), or effective, value of an ac voltage is that value which would cause the same heating effect as an identical dc voltage. Thus, 10-V ac rms across a 10-kΩ resistor will produce the same heat as 10 V dc. This is, by far, the most common measurement of an ac waveform, for it relates ac to dc.

Root mean square means, literally, the root of the mean of the square. It can be found mathematically by:

1. Squaring the waveform.
2. Finding the mean (average) of this squared waveform.
3. Taking the square root of the result.

EXAMPLE 21-17 Find the average and rms of the waveform shown in Fig. 21-13(a).

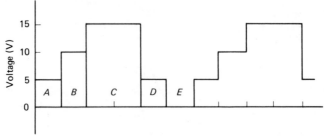

(a) Original waveform

Portion	Area ($l \times h$)
A	5
B	10
C	30
D	5
E	0
Total	50

(b) Area of original waveform

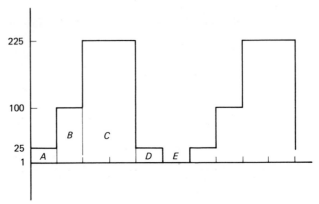

(c) Squared waveform

Portion	Area ($l \times h$)
A	25
B	100
C	450
D	25
E	0
Total	600

(d) Area of squared waveform

FIGURE 21-13 Example 21-17.

SOLUTION We must first find the area of the original [Fig. 21-13(b)]. Compute the average amplitude:

$$V_{av} = \frac{\text{area}}{\text{width}} = \frac{50}{6} = 8.333 \text{ V}$$

The rms is the root of the mean of the square. We must first square the waveform [Fig. 21-13(c)]. Next we find the area under this squared waveform [Fig. 21-13(d)]. We can now compute the mean (average) of the square:

$$V^2_{rms} = \frac{\text{area}}{\text{width}} = \frac{600}{6} = 100$$

We can now find the root of this mean:

$$V_{rms} = \sqrt{V^2_{rms}} = \sqrt{100} = 10 \text{ V}$$

Thus, the average is 8.333 V, but the rms is 10 V. If this waveform were applied to a resistor, it would produce the same heating effect as a 10-V dc source.

EXAMPLE 21-18 Compute the absolute average and rms of the current waveform of Fig. 21-14(a).

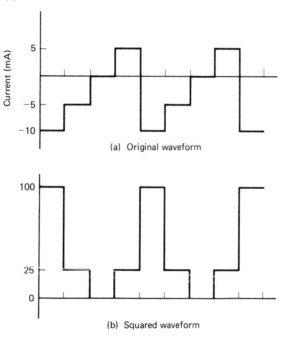

(a) Original waveform

(b) Squared waveform

FIGURE 21-14 Example 21-18.

SOLUTION When finding the absolute average, all areas are considered positive.

$$I_{av} = \frac{(10 \times 1) + (5 \times 1) + (0 \times 1) + (5 \times 1)}{4}$$

$$= 5 \text{ mA}$$

The squared waveform is shown in Fig. 21-14(b). Find the mean of the square:

$$I^2_{rms} = \frac{(100 \times 1) + (25 \times 1) + (0 \times 1) + (25 \times 1)}{4}$$

$$= 37.50$$

The root of the mean of the square is

$$I_{rms} = \sqrt{I^2_{rms}} = \sqrt{37.50} = 6.124 \text{ mA}$$

The absolute average current is 5 mA, and the rms current is 6.124 mA.

The rms of a sinusoidal waveform is found in the same manner [Fig. 21-15(a)]. First, the waveform is squared [Fig. 21-15(b)]. The shaded area turns out to be $\pi/2$. Thus,

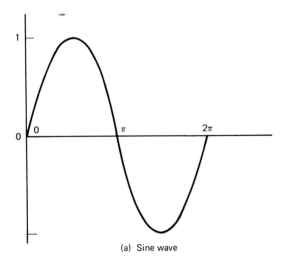

(a) Sine wave

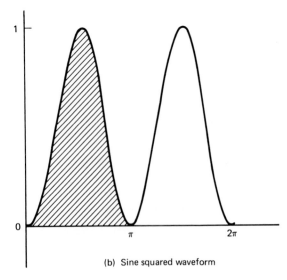

(b) Sine squared waveform

FIGURE 21-15 RMS of sine wave.

$$H_{av} = \frac{\pi/2}{\pi} = 0.5000$$
$$V_{rms} = \sqrt{0.5000} = 0.7071$$

Thus, the rms of a sinusoidal waveform is 0.7071 times its peak value.

EXAMPLE 21-19 Compute the rms value for the following sinusoidal waveforms:

    (a) A p-p voltage of 28.3 V
    (b) A peak current of 500 mA
    (c) $v = 160 \sin 377t$

SOLUTION In each case, rms is 0.7071 times peak.

    (a) $V_{rms} = 0.7071 V_{pk} = 0.7071 \times \dfrac{28.3}{2} = 10.01$ V

    (b) $I_{rms} = 0.7071 I_{pk} = 0.7071 \times 500 = 353.6$ mA
    (c) $V_{rms} = 0.7071 V_{pk} = 0.7071 \times 160 = 113.1$ V

PROBLEMS

21-27. Find V_{av} for Fig. 21-16.
21-28. Find I_{av} for Fig. 21-17.
21-29. Find the absolute average and the algebraic average for the waveform of Fig. 21-18.

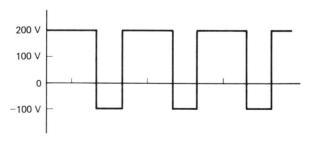

FIGURE 21-16 Problem 21-27.

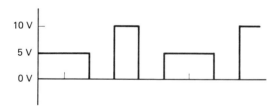

FIGURE 21-17 Problem 21-28.

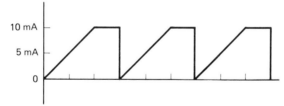

FIGURE 21-18 Problem 21-29.

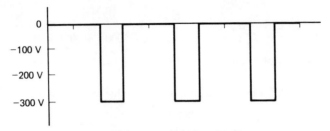

FIGURE 21-19 Problem 21-30.

21-30. Find V_{av} for Fig. 21-19.
21-31. Find V_{rms} for Fig. 21-16.
21-32. Find V_{rms} for Fig. 21-18.
21-33. Find V_{rms} for Fig. 21-19.
21-34. A certain rectangular waveform goes to +6 V for 10 ms, then remains at zero for 990 ms, then repeats the cycle. What is its rms value?
21-35. A certain rectangular current waveform goes to 1 A for 5.3 μs, then goes to 20 mA for 150 ms, then repeats the cycle. What is its rms value?
21-36. What are the absolute average and rms values of the following sinusoidal waveforms?
  (a) A p-p value of 20 A
  (b) A peak value of 465 V
  (c) $V = 33 \sin 377t$
21-37. What are the absolute average and rms values of the following sinusoidal waveforms?
  (a) A p-p value of 235 kV
  (b) A peak value of 27 μA
  (c) $i = (3.69 \times 10^{-3}) \sin (4780t + 26°)$

22

PHASORS

We have several times mentioned vectors and compared them to the spokes on a bicycle wheel. A vector is any quantity that has both magnitude and direction. However, the science of vectors has been borrowed by us in electronics to represent the difference in phase angle between electrical quantities. In this chapter we add the tool of phasors to our mathematical toolbox, enabling us to more deeply understand ac electronics.

22-1 *PHASORS AND SINE WAVES*

In Chapter 21 we noted that a sine wave can be represented as a rotating vector. Figure 22-1 illustrates such a case. The waveforms for two equations have been plotted:

$$H_A = V \sin \theta$$
$$H_B = V \sin (\theta - 45°)$$

On the figure, the large circle shows that H_A and H_B depend upon the length of vector V and the angle of rotation. H_A and H_B represent the vertical distance from the baseline (X axis) to the waveform and V represents the maximum amplitude attained. H_A and H_B depend upon the angle of rotation. The large circle details these values when the vector has rotated $45 + 23$ or $68°$ for H_A and $23°$ for H_B.

Next, examine carefully the small vectors, shown above the sine wave. Each of these represent the position of rotation of A and B vectors. Initially, the A vector is at $0°$ and the B vector is at $-45°$. Thus, H_A is 0 and H_B is $V \sin (-45°)$ or some negative value. Next, when the vectors have rotated $45°$, vector B is at $0°$ and its vertical value, H_B, is 0. Vector A, however, is at $+45°$ and its vertical value, H_A, is $V \sin (45°)$, or some positive value. At the next vector shown, vector A has rotated to $90°$ and B to $45°$. The vertical value of A, H_A, is now its maximum, $V \sin 90°$ or V. H_B is at $V \sin (45°)$, or 0.7071 V.

The intent of this drawing is to illustrate that sine waves can be thought of as rotating vectors. However, note that A always leads B by $45°$, along the entire waveform. Since vectors A and B represent phase differences, we refer to them as phasors in electronics.

Phasors can also be used to represent the sum and difference of sine waves (Fig. 22-2). In this figure, waveforms A and B have been added, obtaining another sine

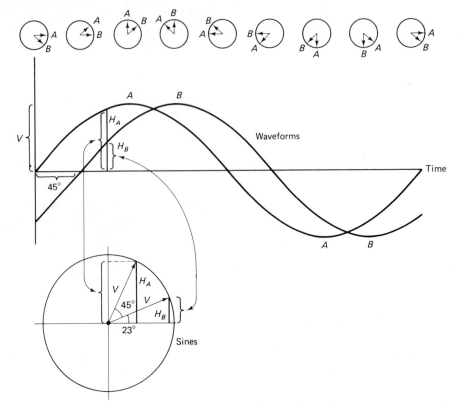

FIGURE 22-1 Instantaneous phase and amplitude.

wave, $A + B$. However, this one can also be represented as a rotating phasor as shown.

RMS Values

Since the phase difference is constant along the entire waveform and the phasor lengths are constant along the entire waveforms, this system can be used to represent rms values of voltage, current, and ac resistance (called impedance). However, to use them effectively, we must first study how phasors are expressed and manipulated by addition, subtraction, multiplication, division, and exponentiation. The rest of this chapter is devoted to this effort.

22-2 THE COMPLEX FIELD

Since phasors are borrowed by electronics from our mathematical friends, it seems only reasonable that we study why vectors are needed. It seems that every time mathematicians run into a brick wall, they invent a new system of mathematics to allow them to surmount the obstacle. Initially, human beings were interested only in counting things: dinosaur teeth, saber-toothed tigers, and caves, for example. How-

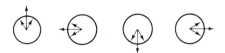

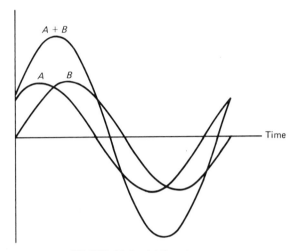

FIGURE 22-2 Adding sine waves.

ever, when Sir Arnold of the Lair wished to present his friend, Rock Granite, with one-half of his day's catch, he was forced to split a mammoth down the middle, and invented fractions to explain the incident to his wife. Next, he found that he had to express how many tyrannosaurs Rock owed him, so he invented negative numbers to solve this problem. His final dilemma occurred when he was amusing himself by finding square roots one day:

$$\sqrt{4} = \pm 2$$
$$\sqrt{3} = \pm 1.732$$
$$\sqrt{2} = \pm 1.414$$
$$\sqrt{1} = \pm 1$$
$$\sqrt{0} = 0$$
$$\sqrt{-1} = TILT$$

Being a very imaginative mathematician, Sir Arnold decided to define this number $\sqrt{-1}$ as i, meaning imaginary. This would allow him to find the square root of any negative number:

$$\sqrt{-100} = \sqrt{-1}\,\sqrt{100} = i10$$
$$\sqrt{-25} = \sqrt{-1}\,\sqrt{25} = i5$$
$$\sqrt{-3} = \sqrt{-1}\,\sqrt{3} = i1.732$$

This is the system that we borrowed from our mathematical friends. However, we already defined i as current, so we had to pick a different symbol. This, of course, put our imagination to the test, but, being the inventive individuals we are in electronics, we decided upon j. Thus,

$$\sqrt{-1} = j$$
$$\sqrt{-3} = j1.732$$
$$\sqrt{-25} = j5$$
$$\sqrt{-100} = j10$$

This j (also called a j factor) has some very interesting properties. Note what happens when we successively raise the power of j:

$$j = \sqrt{-1} = j$$
$$j^2 = \sqrt{-1}\,\sqrt{-1} = -1$$
$$j^3 = j \cdot j^2 = -j$$
$$j^4 = j^2 \cdot j^2 = (-1)(-1) = 1$$
$$j^5 = j \cdot j^4 = j$$

Note that we never have to express j to a higher power than 1, a trait we will find very useful in reducing algebraic expressions.

We now have two number systems: real (not having j factors) and imaginary (j factors). These can be placed upon an X-Y grid called a complex plane (Fig. 22-3). In this plane, real numbers are plotted on the horizontal axis and imaginary numbers on the vertical axis. The number 5 would be found by counting five spaces to the right of the origin. The number -3 would be found by counting 3 units to the left of the origin. The number $j6$, representing $6\sqrt{-1}$, would be found by counting up from the origin 6 units.

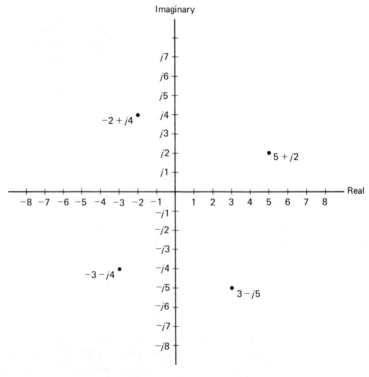

FIGURE 22-3 The complex plane.

A complex number is one that has both a real component and an imaginery component. The following are examples:

$$2 - j6 \qquad -5 + j3$$
$$7 + j13 \qquad -1 - j6$$

Furthermore, complex numbers can be plotted on the complex plane of Fig. 22-3. The number $5 + j2$ represents a point 5 units to the right and 2 units above the origin. The number $-2 + j4$ represents a number 2 units to the left and 4 units above the origin. The number $-3 - j4$ represents a point 3 units to the left and 4 units below the origin. The number $3 - j5$ represents a point 3 units to the right and 5 units below the origin.

EXAMPLE 22-1 Express the following as complex numbers:

(a) A number 10 units to the right of and 4 units above the origin.
(b) A number 7 units to the left of and 3 units above the origin.
(c) A number 60 units to the right of and 23 units below the origin.

SOLUTION

(a) $10 + j4$
(b) $-7 + j3$
(c) $60 - j23$

Complex numbers are considered as vectors originating at the origin and proceeding to a point on the complex plane [Fig. 22-4(a)]. Note that each phasor has both magnitude (expressed by its coordinates) and direction. The method we have been using to express these phasors is called rectangular coordinates. The following are examples of phasors expressed in rectangular coordinates:

$$-6 + j6 \qquad 2 + j5 \qquad 5 + j2 \qquad 3 - j4 \qquad -2 - j5$$

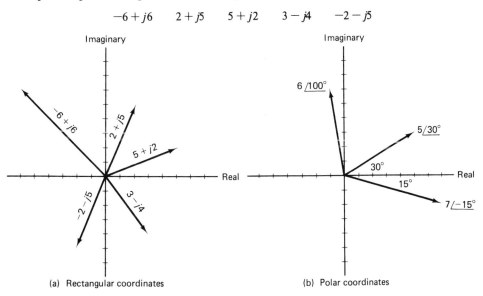

(a) Rectangular coordinates

(b) Polar coordinates

FIGURE 22-4 Phasors in the complex field.

Each has been plotted on Fig. 22-4(a).

We could, however, express phasors in another manner: their length and the angle they make with the +real axis [Fig. 22-4(b)]. This sytem is called polar coordinates. In this system, $5\underline{/30°}$ represents a phasor 5 units long at an angle of 30° above the +real axis (see the figure). The phasor $7\underline{/-15°}$ represents a phasor 7 units long at an angle of 15° below (clockwise to) the +real axis. We shall be manipulating phasors in both rectangular and polar coordinates.

EXAMPLE 22-2 Express the following phasors in polar coordinates:

(a) A phasor 32 units long and 28° above the +real axis.
(b) A phasor 77 units long and 82° below the +real axis.
(c) A phasor 1326 units long and 43.6° clockwise (cw) from the +real axis.
(d) A phasor 23,627 units long and 862° counterclockwise (ccw) from the +real axis.

SOLUTION

(a) $32\underline{/28°}$
(b) $77\underline{/-82°}$
(c) $1326\underline{/-43.6°}$
(d) $23,627\underline{/862°}$

PROBLEMS

22-1. Express the following points as complex numbers:
  (a) A point 10 units to the right of and 12 units above the origin.
  (b) A point 5 units to the left of and 13 units below the origin.
  (c) A point 53.1 units to the right of and 15.6 units below the origin.
22-2. Express the following points as complex numbers:
  (a) A point 32 units to the left of and 17 units above the origin.
  (b) A point 732 units to the right of and 569 units below the origin.
  (c) A point 33.69 units to the left of and 99.3 units below the origin.
22-3. Express the following phasors as polar coordinates:
  (a) A phasor 16 units long and 26° above the +real axis.
  (b) A phasor 23.1 units long and 65° below the +real axis.
  (c) A phasor 44.9 units long and 196° cw from the +real axis.
  (d) A phasor 839 units long and 342° ccw from the +real axis.
22-4. Express the following phasors as polar coordinates:
  (a) A phasor 62 units long and 47° above the +real axis.
  (b) A phasor 438 units long and 2.6° below the +real axis.
  (c) A phasor 3,625,721 units long and 43.9° cw to the +real axis.
  (d) A phasor 43,700 units long and 73.6° ccw to the +real axis.

22-3 RECTANGULAR COORDINATES

As noted in the preceding paragraph, rectangular coordinates can be used to specify any phasor in the form of $A + jB$, where A represents the distance from the origin along the horizontal axis and B represents the distance from the origin along the vertical axis.

Adding Phasors

Phasors can be added both graphically and numerically. To add two phasors graphically, merely form a parallelogram from the given phasors.

EXAMPLE 22-3 Graphically find the sum (also called the resultant) of phasors AB and AC in Fig. 22-5(a).

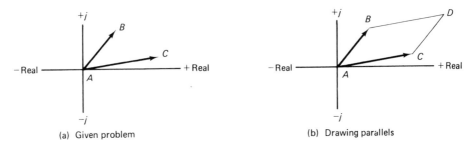

(a) Given problem   (b) Drawing parallels

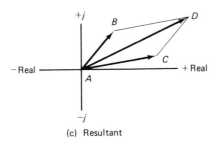

(c) Resultant

FIGURE 22-5 Example 22-3, phasor addition.

SOLUTION First, a line is drawn parallel to AC, passing through B [Fig. 22-5(b)]. A line is then drawn parallel to AB through C. The head of the resultant phasor is located at the intersection of these lines [Fig. 22-5(c)]. Thus, in phasor notation,

$$\overrightarrow{AB} + \overrightarrow{AC} = \overrightarrow{AD}$$

The symbol "$\longrightarrow$" identifies that line segment as a phasor.

There is a second way of graphically specifying phasor addition. Rather than connect the tails of the phasors together, they can be shown in a head-to-tail configuration [Fig. 22-6(a)]. The addition is actually the same as in the previous paragraph if you observe that BC is parallel to AD and AB parallel to DC [Fig. 22-6(b)]. The resultant is $\overrightarrow{AC}$.

Although graphical addition is very easy to perform, mathematical addition is even easier. Addition of phasors expressed in rectangular coordinates requires that the real numbers be summed and the j factors summed.

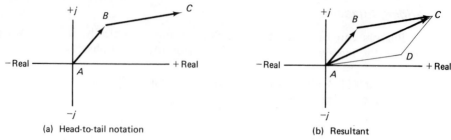

(a) Head-to-tail notation　　　　　　　(b) Resultant

FIGURE 22-6　Head to tail phasor addition.

EXAMPLE 22-4　Add the following phasors:

(a)　$3 + j4, 5 - j6$.
(b)　$-4 - j6, -10 - j1$.

SOLUTION

(a)　　　　　　　　　　$3 + j4$
　　　　　　　　　　　$5 - j6$
　　　　Resultant　$8 - j2$

Note that the sum of $j4$ and $-j6$ is $j2$. Graphically, this means that we traveled north 4 units on the j axis, then south 6 units, resting at $-j2$.

(b)　　　　　　　　　$-4 - j6$
　　　　　　　　　　$-10 - j1$
　　　　Resultant　$-14 - j7$

EXAMPLE 22-5　Add the following phasors:

$$23.1 + j6.7, 18.3 - j17.6, -31.7 + j93.2$$

SOLUTION　Sum the real components and the imaginary components.

　　　　　　　　　　　　　$23.1 + \ j6.7$
　　　　　　　　　　　　　$18.3 - j17.6$
　　　　　　　　　　　$-31.7 + j93.2$
　　　Sum:　　　　　　　$9.7 + j82.3$

Subtracting Phasors

Graphically, subtracting phasor M from phasor N can be analyzed as follows:

$$P = M - N$$

Thus,

$$M = P + N$$

Consequently, we can deduce that M is a resultant, N is one of its component phasors, and we must find the second component phasor. In Fig. 22-7, we are given M and N and must find P. We can find it by completing the parallelogram as shown.

To subtract one phasor from the other mathematically, merely subtract the real parts and the imaginary parts.

EXAMPLE 22-6　Subtract $12 - j6$ from $-6 + j20$.

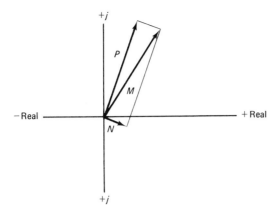

FIGURE 22-7   Subtracting phasors.

SOLUTION

$$\begin{array}{r} 12 - j6 \\ -(-6 + j20) \\ \hline 18 - j26 \end{array}$$

Subtracting -6 from 12 yields 18 (remember, this is algebraic subtraction). Subtracting $j20$ from $-j6$ yields $-j26$. Note that subtracting j factors is the same as subtracting real numbers, except for the inclusion of j in the result.

EXAMPLE 22-7   Subtract $-6 - j22$ from $-26 + j6$.

SOLUTION

$$\begin{array}{r} -26 + j6 \\ -(-6 - j22) \\ \hline -20 + j28 \end{array}$$

Multiplying Phasors

Although we could plunge right into the mathematics of multiplying phasors, let us first examine the process itself and the effect it has upon a phasor. To start,

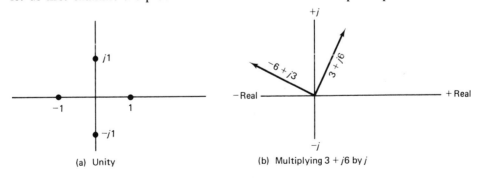

(a) Unity

(b) Multiplying 3 + j6 by j

FIGURE 22-8   Multiplying by j.

note that multiplying by j results in rotation of a phasor by 90°. Consider the number 1 (Fig. 22-8). Let us successively multiply by j:

$$1 = 1$$
$$1 \times j = j1$$
$$j1 \times j = j^2 1 = (-1)(1) = -1$$
$$-1 \times j = -j1$$
$$-j1 \times j = -j^2 1 = -(-1)(1) = 1$$

Thus, each time we have multiplied by j we have rotated the phasor 90°. We shall make use of this property extensively in reduction of phasor expressions.

Next, let us multiply the phasor $3 + j6$ by j:

$$j(3 + j6) = j3 + j^2 6 = (-1)(6) + j3 = -6 + j3$$

It can be shown by geometry that the phasors $3 + j6$ and $-6 + j3$ are also 90° apart. Thus, we can summarize:

The effect of multiplying any phasor by j is to rotate the phasor ccw by 90°.

Let us now examine the effect of multiplying a phasor by a scalar. A scalar is a number that has magnitude but no direction. It contrasts with a vector that has both magnitude and direction.

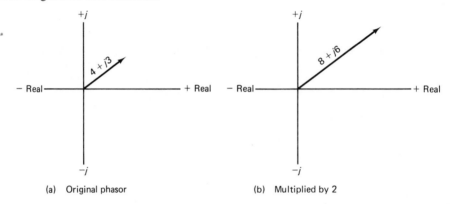

(a) Original phasor   (b) Multiplied by 2

FIGURE 22-9 Multiplying a phasor by a scalar.

Consider the phasor shown in Fig. 22-9(a). To multiply this phasor by the scalar 2, merely multiply both the real part by 2 and the imaginary part by 2. Thus,

$$(4 + j3) \times 2 = 8 + j6$$

Note that the effect, as shown in Fig. 22-9(b), is merely to lengthen the phasor. Next, let us combine the unity j factor with a scalar and note the effect. Multiplication is performed as if the j were an ordinary algebraic literal.

$$(3 + j4) \times (j2) = j6 + j^2 8 = -8 + j6$$

It turns out that this rotates the phasor by 90° and extends its length by a factor of 2 (Fig. 22-10).

We can now summarize the effect of scalar and j-factor multiplication:

1. Multiplying a phasor by a unity j factor rotates the phasor 90°.
2. Multiplying a phasor by a scalar changes the length of the phasor.
3. Multiplying a phasor by a nonunity j factor rotates the phasor by 90° and changes its length.

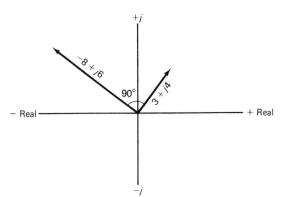

FIGURE 22-10 Multiplying a phasor by j2.

We have just one more case to examine, that of multiplying a phasor by a phasor. The feat is accomplished by treating each phasor as an ordinary binomial and performing the multiplication.

EXAMPLE 22-8 Multiply the phasors $3 - j1$ and $2 + j3$.

SOLUTION

$$
\begin{array}{r}
3 - j1 \\
2 + j3 \\
\hline
j9 - j^23 \\
6 - j2 \\
\hline
6 + j7 - j^23
\end{array}
$$

but $-j^23 = 3$.

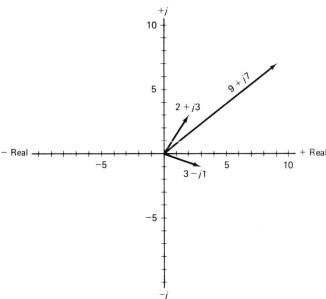

FIGURE 22-11 Example 22-7, multiplying phasors.

It follows, therefore, that

$$6 + j7 + 3 = 9 + j7$$

The result is graphed in Fig. 22-11. Note that changes in both magnitude and direction are reflected in the result.

EXAMPLE 22-9   Multiply $-3 + j6$ and $-7 - j8$.

SOLUTION

$$
\begin{array}{r}
-3 + j6 \\
-7 - j8 \\
\hline
j24 - j^2 48 \\
21 - j42 \\
\hline
21 - j18 - j^2 48 = 48 + 21 - j18 = 69 - j18
\end{array}
$$

Dividing Phasors

Division of phasors expressed in rectangular coordinates can be done by multiplying both numerator and denominator by a function of the denominator called a conjugate. A conjugate of a phasor is the original phasor with its j-factor sign reversed. Table 22-1 lists some examples. Now let us use this method in a few examples.

TABLE 22-1
Conjugates.

Phasor	Conjugate
$-3 - j6$	$-3 + j6$
$-4 + j12$	$-4 - j12$
$-6 - j23$	$-6 + j23$

EXAMPLE 22-10   Divide $6 - j4$ by $2 + j6$.

SOLUTION   The problem in fraction form is

$$\frac{6 - j4}{2 + j6}$$

Next, multiply both numerator and denominator by the conjugate of the denominator:

$$\frac{6 - j4}{2 + j6} \cdot \frac{2 - j6}{2 - j6} = \frac{12 - j36 - j8 + j^2 24}{4 - j12 + j12 - j^2 36} = \frac{-12 - j44}{40}$$

Note that multiplying the denominator by its conjugate results in the j factor dropping out. This is similar to the theorem in algebraic factoring:

$$(a + b)(a - b) = a^2 - b^2$$

Next, we must leave the answer in the form of a real number and an imaginary number:

$$\frac{-12 - j44}{40} = -\frac{12}{40} - j\frac{44}{40} = -0.3 - j1.1$$

This, then, is the final answer.

EXAMPLE 22-11   Divide $12.6 - j22.9$ by $-24.9 - j12.9$.

SOLUTION

Original problem:

$$\frac{12.6 - j22.9}{-24.9 - j12.9}$$

Multiplying by conjugates:

$$\frac{12.6 - j22.9}{-24.9 - j12.9} \cdot \frac{-24.9 + j12.9}{-24.9 + j12.9} = \frac{-313.74 + j162.54 + j570.21 - j^2295.41}{620.01 + 166.41}$$

$$= \frac{-18.33 + j732.75}{786.42}$$

$$= -0.02331 + j0.9318$$

Note that there is no need to multiply real by j factor in the denominator, for it will just drop out.

We have now discussed addition, subtraction, multiplication, and division of phasors using rectangular coordinates. Next, we shall do the same thing using polar notation.

PROBLEMS

Add the following phasors:

22-5. $4 + j3, -2 + j6$
22-6. $-5 + j2, -3 - j2$
22-7. $10 + j2, -5 + j3$
22-8. $15 + j16, -17 + j12$
22-9. $13.2 - j16.7, 5.2 - j12.8$
22-10. $-15.6 + j26.3, 77.2 - j16.2$
22-11. $0.031 - j0.076, -0.925 + j0.672$
22-12. $-13.6 + j17.3, -14.8 - j19.6$
22-13. $4 + j9, 5 - j6, -14 + j67$
22-14. $33 + j62, -46 - j7, 42 + j25$
22-15. $15 + j16, -12 + j13, 14 - j17$
22-16. $17.3 - j16.9, 4.3 + j12.6, -26.3 - j73.1$

Subtract the second phasor from the first:

22-17. $2 + j3, 3 + j6$
22-18. $4 - j6, 2 - j3$
22-19. $-18 + j13, -12 - j16$
22-20. $56 - j27, -126 + j48$
22-21. $88 - j29, -126 + j48$
22-22. $-77 + j92, -73 - j83$
22-23. $17.31 - j16.25, 42.67 - j18.72$
22-24. $-136.2 + j49.61, -83.77 + j78.96$
22-25. $-0.0357 + j0.0873, -0.0776 - j0.2314$
22-26. $12,000 - j16,723; -49,712 - j88,763$

Multiply the following phasors:

22-27. $4 + j6$, $2 + j3$
22-28. $3 + j7$, $-4 + j2$
22-29. $12 + j7$, $3 + j8$
22-30. $14 + j12$, $-12 + j10$
22-31. $-16 + j17$, $-22 - j18$
22-32. $-21 - j16$, $-12 - j15$
22-33. $14.3 - j16.2$, $-8.6 + j5.3$
22-34. $-56.21 - j25.63$, $12.15 - j25.63$
22-35. $-55.55 + j42.17$, $-31.62 - j73.13$
22-36. $123.6 + j93.7$, $-412.6 - j815.3$

Divide the first phasor by the second phasor:

22-37. $16 + j0$, $4 + j0$
22-38. $48 + j0$, $-3 + j0$
22-39. $7 + j3$, $2 - j4$
22-40. $-5 + j6$, $-3 - j4$
22-41. $15 - j5$, $-2 + j6$
22-42. $-43 - j60$, $5 - j4$
22-43. $17.3 - j16.2$, $-12.3 + j8.2$
22-44. $43.15 - j18.31$, $-12.63 - j17.31$

22-4 POLAR COORDINATES

As was stated earlier, we can also express a phasor by stating its length and its angle referred to the horizontal. In Fig. 22-12 the phasor $3 + j4$ can also be expressed as $5\underline{/53.13°}$. That is, it has a length of 5 units and its angle with reference to the horizontal is 53.13°. This is a very useful phasor form. However, we must be able to relate it to the rectangular form.

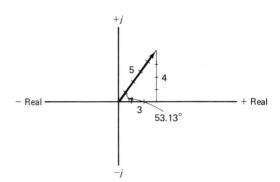

FIGURE 22-12 Polar coordinates.

Polar-to-Rectangular Conversion

Many calculators can convert directly from polar to rectangular coordinates. In addition, polar to rectangular conversion is easily accomplished by observing from Fig. 22-13 that:

1. Real $= V \cos \alpha$
2. Imaginary $= V \sin \alpha$

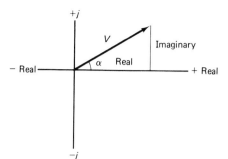

FIGURE 22-13 Polar-rectangular relationships.

EXAMPLE 22-12 Convert the following phasors to rectangular coordinates:

(a) $120\underline{/30°}$

(b) $60\underline{/212°}$

SOLUTION

(a) Real $= V \cos \alpha$
$= 120 \cos 30°$
$= 120 \times 0.8660$
$= 103.9$

Imaginary $= V \sin \alpha$
$= 120 \sin 30°$
$= 120 \times 0.5000$
$= 60.00$

Therefore, $120\underline{/30°} = 103.9 + j60.00$.

(b) Real $= V \cos \alpha$
$= 60 \cos 212°$
$= -50.88$

Imaginary $= V \sin \alpha$
$= 60 \sin 212°$
$= -31.80$

Therefore, $60\underline{/212°} = -50.88 - j31.80$.

Rectangular-to-Polar Conversion

Many calculators allow direct rectangular to polar conversion. The process can be described mathematically by the following formulas:

$$\alpha = \tan^{-1}\left(\frac{\text{imaginary}}{\text{real}}\right)$$

$$V = \frac{\text{imaginary}}{\sin \alpha}$$

This procedure can easily be done on any calculator with trigonometric functions.

EXAMPLE 22-13 Convert the following to polar coordinates:

(a) $3 - j4$

(b) $26 + j63$

SOLUTION

(a) To find the angle:

$$\alpha = \tan^{-1}\left(\frac{\text{imaginary}}{\text{real}}\right)$$

$$= \tan^{-1}\left(\frac{-4}{3}\right)$$

$$= \tan^{-1}(-1.333)$$

$$= -53.13°$$

To find the phasor length:

$$V = \frac{\text{imaginary}}{\sin \alpha}$$

$$= \frac{-4}{\sin(-53.13°)} = 5$$

Thus, $3 - j4 = 5.000 \underline{/-53.13°}$.

(b) $\alpha = \tan^{-1}\left(\frac{\text{imaginary}}{\text{real}}\right)$

$$= \tan^{-1}\left(\frac{63}{26}\right)$$

$$= 67.57°$$

$$V = \frac{\text{imaginary}}{\sin \alpha}$$

$$= \frac{63}{\sin(67.57°)}$$

$$= 68.15$$

Thus, $26 + j63 = 68.15 \underline{/67.57°}$.

Addition and Subtraction

Phasors in polar form can only be added or subtracted if their angles are the same, in which case the lengths are combined and the angle remains the same.

EXAMPLE 22-14 Perform the following phasor operations:

(a) $3\underline{/25°} + 6\underline{/25°}$
(b) $7\underline{/-35°} - 2\underline{/-35°}$

SOLUTION

(a) To add, merely add the phasor lengths, keeping the angles the same:

$$3\underline{/25°} + 6\underline{/25°} = (3 + 6)\underline{/25°} = 9\underline{/25°}$$

(b) To subtract, merely subtract the phasor lengths, keeping the angle the same:

$$7\underline{/-35°} - 2\underline{/-35°} = (7 - 2)\underline{/-35°} = 5\underline{/-35°}$$

If the angles are not identical, the phasors must be converted into rectangular form, and then added or subtracted.

EXAMPLE 22-15 Perform the following phasor operations:

(a) $4\underline{/26°} + 9\underline{/85°}$
(b) $6\underline{/-23°} - 7\underline{/64°}$

SOLUTION

(a) Since the angles are not identical, the phasors must be converted to rectangular form, then added:

$$
\begin{array}{rl}
4\underline{/26°} = & 3.595 + j1.753 \\
9\underline{/85°} = & 0.784 + j8.966 \\
\hline
\text{Sum:} & 4.379 + j10.719
\end{array}
$$

Converting back to polar: $11.58\underline{/67.78°}$. Thus, $4\underline{/26°} + 9\underline{/85°} = 11.58\underline{/67.78°}$.

(b) $\quad 6\underline{/-23°} = 5.523 - j2.344$
$\quad\quad\ \ 7\underline{/64°} = 3.069 + j6.292$

Subtracting: $2.454 - j8.636$
Converting back to polar: $8.978\underline{/-74.13°}$. Thus, $6\underline{/-23°} - 7\underline{/64°} = 8.978\underline{/-74.13°}$.

Multiplication

To multiply polar phasors, the lengths are multiplied and the angles added:

$$(A\underline{/\alpha_1})(B\underline{/\alpha_2}) = (A \cdot B)\underline{/\alpha_1 + \alpha_2}$$

Note that we are essentially using the same rules for the angles of these phasors as we would for exponents in algebra. That is, in algebra the exponents are added during multiplication. The reason for adding these angles is that they are, in fact, exponents. This can be shown by the following gymnastics:

The $\cos \alpha$ is represented by the series

$$\cos \alpha = 1 - \frac{\alpha^2}{2!} + \frac{\alpha^4}{4!} - \frac{\alpha^6}{6!} + \ldots$$

The $\sin \alpha$ is represented by the series

$$\sin \alpha = \alpha - \frac{\alpha^3}{3!} + \frac{\alpha^5}{5!} - \frac{\alpha^7}{7!} + \ldots$$

But $\epsilon^{j\alpha}$ is represented by the series

$$\epsilon^{j\alpha} = 1 + j\alpha - \frac{\alpha^2}{2!} - j\frac{\alpha^3}{3!} + \frac{\alpha^4}{4!} + j\frac{\alpha^5}{5!} - \frac{\alpha^6}{6!} - j\frac{\alpha^7}{7!} + \ldots$$

Therefore, $\epsilon^{j\alpha} = \cos \alpha + j \sin \alpha$. However, $V\underline{/\alpha} = V(\cos \alpha + j \sin \alpha)$. Therefore, $V\underline{/\alpha} = V\epsilon^{j\alpha}$, and α is, indeed, an exponent.

Now let us try this multiplication procedure on a few problems.

EXAMPLE 22-16 Multiply the following:

(a) $3\underline{/40°}$ by $26\underline{/-25°}$
(b) $3\underline{/-60°}$ by $5\underline{/-20°}$

SOLUTION

(a) $3\underline{/40°} \times 26\underline{/-25°} = (3 \times 26)\underline{/40° - 25°} = 78\underline{/15°}$

(b) $3\underline{/-60°} \times 5\underline{/-20°} = (3 \times 5)\underline{/-60° - 20°} = 15\underline{/-80°}$

Division

Division of polar phasors is accomplished by dividing the phasor lengths and subtracting the angles (as you would exponents).

EXAMPLE 22-17 Divide the following phasors:

(a) $26\underline{/25°}$ by $33\underline{/-75°}$

(b) $-33\underline{/-88°}$ by $27\underline{/45°}$

SOLUTION

(a) $\dfrac{26\underline{/25°}}{33\underline{/-75°}} = \dfrac{26}{33}\underline{/25° - (-75°)} = 0.788\underline{/100°}$

(b) $\dfrac{-33\underline{/-88°}}{27\underline{/45°}} = \dfrac{-33}{27}\underline{/(-88° - 45°)} = -1.222\underline{/-133°}$

Exponentiation

Polar phasors can be raised to any power by raising the phasor length to the desired power and multiplying the angle by the power. Note that, again, the angle is treated as an exponent.

$$(V\underline{/\alpha})^n = V^n\underline{/n\alpha}$$

EXAMPLE 22-18 Perform the following:

(a) $(12\underline{/25°})^3$

(b) $(6\underline{/-30°})^{2.6}$

SOLUTION Note that in each case the angle is multiplied by the exponent and the phasor length is raised to the exponent.

(a) $(12\underline{/25°})^3 = 12^3\underline{/25° \times 3} = 1728\underline{/75°}$

(b) $(6\underline{/-30°})^{2.6} = 6^{2.6}\underline{/-30° \times 2.6} = 105.5\underline{/-78.0°}$

PROBLEMS

Convert to rectangular coordinates:

22-45. (a) $30\underline{/45°}$ (b) $73\underline{/-26°}$

22-46. (a) $73\underline{/-14°}$ (b) $82\underline{/68°}$

22-47. (a) $123\underline{/160°}$ (b) $841\underline{/-215°}$

22-48. (a) $1231\underline{/47°}$ (b) $5.863\underline{/-17.3°}$

22-49. (a) $1769\underline{/-78°}$ (b) $88.31\underline{/41.65°}$

22-50. (a) $183.6\underline{/-143°}$ (b) $731\underline{/316.7°}$

22-51. (a) $46.83\underline{/98.1°}$ (b) $173.6\underline{/-173.6°}$

22-52. (a) $56.7\underline{/0°}$ (b) $35.6\underline{/-270°}$

Convert to polar coordinates:

22-53. (a) $3 + j4$ (b) $4 - j3$

22-54. (a) $-5 + j12$ (b) $12 - j5$

22-55. (a) $3 + j6$ (b) $-7 - j8$

22-56. (a) $-17.6 - j18.3$ (b) $-12.7 + j0.6$

22-57. (a) $831 + j1$ (b) $365 - j1873$

22-58. (a) $-55 - j156$ (b) $-837 - j3.2$

22-59. (a) $96 + j0$ (b) $0 - j65.2$

22-60. (a) $831.6 + j759.6$ (b) $-357.2 + j835.6$

Add or subtract as indicated. Leave the results in polar form:

22-61. $15\underline{/63°} + 27\underline{/63°}$

22-62. $48\underline{/-27°} + 75\underline{/-27°}$

22-63. $73.6\underline{/83.2°} + 13.3\underline{/83.2°}$

22-64. $735.3\underline{/47.6°} + 1572\underline{/47.6°}$

22-65. $3\underline{/27°} + 4\underline{/31°}$

22-66. $15\underline{/33} + 12\underline{/76°}$

22-67. $43\underline{/-7°} + 16\underline{/-20°}$

22-68. $73\underline{/42°} + 46\underline{/-77°}$

22-69. $12.16\underline{/13.2°} + 17.31\underline{/46.63°}$

22-70. $41.63\underline{/88.31°} + 73.16\underline{/47.63°}$

22-71. $136.2\underline{/-13.65°} + 98.6\underline{/33.46°}$

22-72. $256\underline{/183°} + 138\underline{/93.2°}$

22-73. $14\underline{/14°} - 5\underline{/14°}$

22-74. $112\underline{/75°} - 16\underline{/75°}$

22-75. $96\underline{/23°} - 48\underline{/23°}$

22-76. $33\underline{/26°} - 17\underline{/26°}$

22-77. $5\underline{/13°} - 6\underline{/27°}$

22-78. $14\underline{/63°} - 23\underline{/17°}$

22-79. $33\underline{/57°} - 28\underline{/-14°}$

22-80. $137\underline{/46°} - 73\underline{/-82°}$

22-81. $476.9\underline{/-73.6°} - 842.3\underline{/128.6°}$

22-82. $531.7\underline{/48.2°} - 292.3\underline{/66.7°}$

Multiply:

22-83. $3\underline{/45°}, 5\underline{/73°}$

22-84. $-6\underline{/3°}, 14\underline{/27°}$

22-85. $16\underline{/-12°}, 26\underline{/12°}$

22-86. $83.1\underline{/27.9°}, 77.3\underline{/-48.1°}$

22-87. $127.3\underline{/43.5°}, -173.6\underline{/26.89°}$

22-88. $66.31\underline{/23.68°}, 123\underline{/47°}$

22-89. $0.013\underline{/-43°}, 0.832\underline{/13.69°}$

22-90. $12.86\underline{/93°}, 14.6\underline{/375°}$

Divide the first phasor by the second phasor:

22-91. $4\underline{/17°}, 8\underline{/72°}$

22-92. $16\underline{/31°}, 12\underline{/33°}$

22-93. $17.3\underline{/-5°}, 18.6\underline{/-27°}$

22-94. $44.8\underline{/120°}, 86.2\underline{/-73°}$

22-95. $127.3\underline{/5°}, 3.7\underline{/-33.6°}$

22-96. $117.3\underline{/-46°}, 36.2\underline{/77.2°}$

22-97. $4375\underline{/23°}, 52.16\underline{/83.1°}$

22-98. $77.3\underline{/-76°}, 43.21\underline{/-7.6°}$

Exponentiate as indicated:

22-99. (a) $(3\underline{/4°})^2$ (b) $(17\underline{/2.6°})^3$

22-100. (a) $(14\underline{/23°})^3$ (b) $(33\underline{/63°})^2$

22-101. (a) $(36\underline{/27°})^4$ (b) $(13\underline{/-27°})^3$

22-102. (a) $(17.3\underline{/-14.3°})^3$ (b) $(0.013\underline{/-7.6°})^4$

22-103. (a) $(13.16\underline{/13.6°})^{1.2}$ (b) $(93.16\underline{/2.69°})^{4.6}$

23

LOGARITHMS

In 1614, John Napier, a distinguished Scottish mathematician, invented logarithms. Using this system, he was able to:

1. Reduce multiplication to an addition process
2. Reduce division to a subtraction process
3. Reduce exponentiation to a multiplication process

In electronics, many phenomena are best expressed in a logarithmic form. Audio intensity is measured in decibels, a logarithmic scale, because this best expresses the manner in which the ear responds to sound. In this chapter we examine logarithms: what they are, how they are manipulated, and where they are used.

23-1 SOME DEFINITIONS

A logarithm (or log) is an exponent—it is as simple as that. For example, we can express the number 100 as

$$10^2 = 100$$

Note that there are three parts to this statement: (a) the base, 10; (b) the exponent (or log), 2; and (c) the source, 100. In a similar manner, we can express any number as a base raised to an exponent.

EXAMPLE 23-1 Find the base, the log, and the source in each of the following:

(a) $3^{1.4650} = 5.000$
(b) $12^{2.9614} = 1570$
(c) $35^{0.3899} = 4.000$

SOLUTION

(a) Base = 3, log = 1.4650, source = 5.000
(b) Base = 12, log = 2.9614, source = 1570
(c) Base = 35, log = 0.3899, source = 4.000

The format we have been using for expressing this relationship is

$$(\text{base})^{\log} = \text{source} \qquad \textit{Example: } 10^3 = 1000$$

This is called the exponential form of expressing a log. However, a second form that is more frequently used is

$$\log_{\text{base}} \text{source} = \log \qquad \textit{Example:} \log_{10} 1000 = 3$$

read, "the log to the base 10 of 1000 is 3." What it is really asking is: "What exponent of 10 will result in a source of 1000?"

The following are some examples of both the exponential form and the logarithmic form:

Exponential	*Logarithmic*
$10^2 = 100$	$\log_{10} 100 = 2$
$10^3 = 1000$	$\log_{10} 1000 = 3$
$10^4 = 10,000$	$\log_{10} 10,000 = 4$
$10^{0.30103} = 2.000$	$\log_{10} 2 = 0.30103$
$10^{2.30103} = 200.0$	$\log_{10} 200 = 2.30103$
$10^{3.30103} = 2000$	$\log_{10} 2000 = 3.30103$

EXAMPLE 23-2 Express in logarithmic form:

(a) $10^{4.3} = 19953$ (b) $3^{6.9} = 1959.5$ (c) $8^{1.3} = 14.929$

SOLUTION

(a) $\log_{10} 19953 = 4.3$
(b) $\log_3 1959.5 = 6.9$
(c) $\log_8 14.929 = 1.3$

EXAMPLE 23-3 Express in exponential form:

(a) $\log_{10} 21 = 1.322$
(b) $\log_3 9 = 2.000$
(c) $\log_8 6.299 = 0.88504$

SOLUTION

(a) $10^{1.322} = 21$
(b) $3^{2.000} = 9$
(c) $8^{0.88504} = 6.299$

Calculating the Source

If we know the base and the exponent, we can calculate the source. If the base is 10, we can use the $\boxed{10^x}$ key of the calculator as follows:

1. Key in the exponent.
2. Depress the $\boxed{10^x}$ key.
3. The source will appear.

Many calculators have a $\boxed{y^x}$ key. This key has the advantage of being able to find the source from any base. However, the procedure depends upon whether your calculator has RPN notation or algebraic notation.

For RPN notation:

1. Key in the base.
2. Press ENTER.
3. Key in the exponent.
4. Press y^x.
5. The source will appear.

For algebraic notation:

1. Key in the base.
2. Press y^x.
3. Key in the exponent.
4. Press =.
5. The source will appear.

Try this procedure on the exponential forms above and those of Example 23-1.

Logarithms to Base 10

The above exponential forms illustrate that any number can be expressed as an exponent of 10. For example:

Exponential	Logarithmic
$10^0 = 1$	$\log_{10} 1 = 0$
$10^{0.30103} = 2$	$\log_{10} 2 = 0.30103$
$10^{0.47712} = 3$	$\log_{10} 3 = 0.47712$
$10^{0.60206} = 4$	$\log_{10} 4 = 0.60206$
$10^{0.69897} = 5$	$\log_{10} 5 = 0.69897$
$10^{0.77815} = 6$	$\log_{10} 6 = 0.77815$

These logs, which use a base of 10, are called common logs.

The only trick is finding the proper exponent (log). However, the calculator again rushes to our rescue, for we can find the log of any number to the base 10 by the following:

1. Key in the source.
2. Press LOG.
3. The log should appear.

Just for practice, find the log of the numbers 1 through 6, then use the exponential form to find the base. This will work for any log.

EXAMPLE 23-4 Find the logs of each of the following to base 10, then use these logs to recover the source:

(a) $\log_{10} 25 = ?$ (b) $\log_{10} 3.16 = ?$
(c) $\log_{10} 44 = ?$ (d) $\log_{10} 396 = ?$

SOLUTION

(a) Key in 25, then press LOG.
$\log_{10} 25 = 1.397940009$. Thus, the log of 25 to base 10 is 1.397940009. To recover the source, use this number as the exponent and find 10^x, giving 25.

The source can also be found by using 1.397940009 as the exponent and 10 as the base using the $\boxed{y^x}$ key.

(b) $\log_{10} 3.16 = 0.499687083$

$10^{0.499687083} = 3.16$

(c) $\log_{10} 44 = 1.643452676$

$10^{1.643452676} = 44$

(d) $\log_{10} 396 = 2.597695186$

$10^{2.597695186} = 396$

Logarithms to Base ϵ

We have discussed the number π. It is a number that occurs in nature and is always equal to 3.141592654 (approximately). There is another number that occurs frequently in nature and it is referred to by the Greek letter epsilon (ϵ), although we usually pronounce it "ee." This number has a value of 2.718281828459045 (approximately). Thus, it is slightly less than 3. This number is very frequently used as a base for logarithms, called natural logs. Whereas log refers to common logs (base 10), ln refers to natural logs (base ϵ). For example:

$$\epsilon^{0.69315} = 2 \qquad \ln 2 = 0.69315$$
$$\epsilon^{1.0986} = 3 \qquad \ln 3 = 1.0986$$
$$\epsilon^{1.3863} = 4 \qquad \ln 4 = 1.3863$$
$$\epsilon^{1.6094} = 5 \qquad \ln 5 = 1.6094$$

These logs can easily be found on the calculator using the following procedure:

1. Key in the source.
2. Press the $\boxed{LN}$ key.
3. The natural log will appear.

As in the case of the common logs, if the ln is known, the source can be found by using the $\boxed{\epsilon^x}$ key.

1. Enter the ln.
2. Press $\boxed{\epsilon^x}$.
3. The source will appear.

Try this on the examples above.

PROBLEMS

Find the base, the log, and the source in each of the following:

23-1. $5^6 = 15625$ **23-5.** $10^{0.5} = 3.1623$

23-2. $4^3 = 64$ **23-6.** $10^{0.8451} = 7$

23-3. $12^{2.3} = 303.47$ **23-7.** $46^{0.012} = 1.0470$

23-4. $7^{5.1} = 20417.4$ **23-8.** $2.7183^{3.4965} = 33.00$

Express the following in logarithmic form:

23-9. $10^{1.46} = 28.84$ **23-13.** $12^{1.82} = 92.07$

23-10. $10^{0.831} = 6.776$ **23-14.** $9^{0.99} = 8.804$

23-11. $4^5 = 1024$ **23-15.** $23^{0.63} = 7.209$

23-12. $2^{3.2} = 9.1896$ **23-16.** $7^{2.58} = 151.5$

Express in exponential form:

23-17. $\log_{10} 86 = 1.934$
23-18. $\log_8 77 = 2.089$
23-19. $\log_6 47 = 2.149$
23-20. $\log_5 82 = 2.738$

23-21. $\log_{15} 135 = 1.811$
23-22. $\log_{23} 77.88 = 1.389$
23-23. $\log_9 99 = 2.091$
23-24. $\log_{99} 8778 = 1.976$

Using a base of 10, calculate the source from the following exponents:

23-25. (a) 0.6 (b) 3.26 (c) 1.43
23-26. (a) 0.9 (b) 1.7 (c) 2.315
23-27. (a) -1.3 (b) 0.031 (c) 0
23-28. (a) -2.29 (b) 13 (c) 7.763

23-29. (a) 0.3 (b) 1.3 (c) 2.3
23-30. (a) 3.3 (b) 4.3 (c) 5.3
23-31. (a) 6.3 (b) 7.3 (c) 8.3
23-32. (a) -0.3 (b) -1.3 (c) -2.3

Calculate the following:

23-33. (a) 3^4 (b) 1.2^2 (c) 1.016^{16}
23-34. (a) 4^3 (b) $5^{1.2}$ (c) $6^{3.3}$
23-35. (a) $1.4^{1.6}$ (b) $3.4^{2.7}$ (c) $3.1^{0.15}$

23-36. (a) 5.9^{-2} (b) $3.15^{-0.16}$ (c) $4.3^{2.15}$
23-37. (a) $132^{0.69}$ (b) $13.2^{0.69}$ (c) $1.32^{0.69}$
23-38. (a) $132^{-0.69}$ (b) $13.2^{-0.69}$ (c) $1.32^{-0.69}$

Find the log to base 10. Also recover the source.

23-39. (a) 0.003 (b) 0.03 (c) 0.3
23-40. (a) 3 (b) 30 (c) 300
23-41. (a) 3000 (b) 30,000 (c) 300,000

23-42. (a) 2 (b) 4 (c) 8
23-43. (a) 3 (b) 6 (c) 12
23-44. (a) 13.6 (b) 136 (c) 1360

Find the natural log of each of the following. Also recover the source.

23-45. (a) 0.003 (b) 0.03 (c) 0.3
23-46. (a) 3 (b) 30 (c) 300
23-47. (a) 3000 (b) 30,000 (c) 300,000

23-48. (a) 2 (b) 4 (c) 8
23-49. (a) 3 (b) 6 (c) 12
23-50. (a) 13.6 (b) 136 (c) 1360

23-2 PROPERTIES OF LOGARITHMS

In this section we examine some of the properties of logs that make them useful in dealing with electronic quantities.

Log of Product

The log of a product is the sum of the individual logs:

$$\log XY = \log X + \log Y$$
$$\ln XY = \ln X + \ln Y$$

Let us consider the log (2×10). The log 2 is 0.30103 and the log 10 is 1.0. Thus, according to the foregoing rule, the log 20 is 1.30103, the sum of the individual logs. We can show this using the exponential form:

$$20 = 10 \times 2$$
$$10^{1.30103} = 10^1 \times 10^{0.30103}$$

Note that, according to the rules of exponents, we add the exponents when multiplying. But we already have shown that logs are exponents. Therefore, to multiply we add logs. Consider the following series of common logs:

$$2 = 10^0 \times 10^{0.30103} = 10^{0.30103}$$
$$20 = 10^1 \times 10^{0.30103} = 10^{1.30103}$$

$$200 = 10^2 \times 10^{0.30103} = 10^{2.30103}$$
$$2000 = 10^3 \times 10^{0.30103} = 10^{3.30103}$$

In log form:

$$\log 2 = 0.30103$$
$$\log 20 = 1.30103$$
$$\log 200 = 2.30103$$
$$\log 2000 = 3.30103$$

Examine this series closely. Note that the portion of the log to the right of the decimal point is always the same for the number 2, regardless of where the point is placed. This portion of the log is called the mantissa. The mantissa for any series of numerals will be identical, regardless of where the point is placed.

EXAMPLE 23-5   Find log 4.3, log 43, log 430, and log 4300 and compare the mantissas:

SOLUTION

$$\log 4.3 = 0.633468$$
$$\log 43 = 1.633468$$
$$\log 430 = 2.633468$$
$$\log 4300 = 3.633468$$

Note that the mantissas are identical. This mantissa principle occurs in common logs because every time the source is multiplied by 10, a whole number (1) must be added to the exponent. It does not apply to natural logs, because every time the source is multiplied by 10, 2.30259 (the ln 10) must be added to the exponent.

The portion of a common log to the left of the decimal point is called the characteristic, and determines where the decimal point occurs in the source.

The foregoing discussion illustrates the principle that multiplication becomes the addition of logs. This principle applies to all logs of identical bases.

EXAMPLE 23-6   Find the log 20 by the following methods:

(a) $\log 20 = \log 2 + \log 10$
(b) $\log 20 = \log 4 + \log 5$

SOLUTION

(a) $\log 20 = \log 2 + \log 10$
$1.30103 = 0.30103 + 1.00000$
(b) $\log 20 = \log 4 + \log 5$
$1.30103 = 0.60206 + 0.69897$

EXAMPLE 23-7   Find the ln 20 by the following methods:

(a) $\ln 20 = \ln 10 + \ln 2$
(b) $\ln 20 = \ln 4 + \ln 5$

SOLUTION

(a) $\ln 20 = \ln 10 + \ln 2$
$2.9957 = 2.3026 + 0.6931$

(b)   ln 20 = ln 4 + ln 5
         2.9957 = 1.3863 + 1.6094

Note that the ln of a product is the sum of the individual ln's.

The log of product law can be used very effective in solving logarithmic equations.

EXAMPLE 23-8   Solve for X:

$$\log X + 4 \log X = 12$$

SOLUTION

$$\log X + 4 \log X = 12$$
$$5 \log X = 12$$
$$\log X = \frac{12}{5} = 2.40$$
$$10^{2.40} = X$$
$$X = 251.19$$

EXAMPLE 23-9   Solve for P:

$$\log P + \log 36P = 2.61$$

SOLUTION

We can recognize that

$$\log 36P = \log 36 + \log P$$

Therefore,

$$\log P + \log 36P = 2.61$$
$$\log P + \log 36 + \log P = 2.61$$
$$2 \log P + 1.5563 = 2.61$$
$$2 \log P = 1.0537$$
$$\log P = 0.5268$$
$$10^{0.5268} = P$$
$$P = 3.3639$$

Log of Quotient

In the division process, logs obey the rule of exponents.

$$\log \left(\frac{X}{Y}\right) = \log X - \log Y$$

$$\ln \left(\frac{X}{Y}\right) = \ln X - \ln Y$$

EXAMPLE 23-10   Using the rule of logs for division, prove the following:

(a) $\log \left(\dfrac{100}{10}\right) = \log 100 - \log 10$

(b) $\log \left(\dfrac{365}{32.6}\right) = \log 365 - \log 32.6$

(c) $\ln \left(\dfrac{2753}{86}\right) = \ln 2753 - \ln 86$

SOLUTION

(a) $\log \dfrac{100}{10} = \log 100 - \log 10$

$\log 10 = \log 100 - \log 10$
$1.000 = 2.000 - 1.000$
$1.000 = 1.000$

(b) $\log \left(\dfrac{365}{32.6}\right) = \log 365 - \log 32.6$

$\log 11.1963 = \log 365 - \log 32.6$
$1.0491 = 2.5623 - 1.5132$
$1.0491 = 1.0491$

(c) $\ln \left(\dfrac{2753}{86}\right) = \ln 2753 - \ln 86$

$\ln 32.0116 = \ln 2753 - \ln 86$
$3.4661 = 7.9204 - 4.4543$
$3.4661 = 3.4661$

Note that in each case, the rule of logs for exponents applies.

The logarithmic rule for division is a very useful tool for solution of log equations.

EXAMPLE 23-11   Solve for Q:

$$\log \frac{Q}{10} = 1.5$$

SOLUTION

$$\log \frac{Q}{10} = 1.5$$
$$\log Q - \log 10 = 1.5$$
$$\log Q = 1.5 + 1 = 2.5$$
$$10^{2.5} = Q$$
$$Q = 316.23$$

EXAMPLE 23-12   Solve for R:

$$3 \log \frac{R+1}{6} = 4.31$$

SOLUTION

$$3 \log \frac{R+1}{6} = 4.31$$
$$\log \frac{R+1}{6} = \frac{4.31}{3} = 1.4367$$
$$\log (R+1) - \log 6 = 1.4367$$
$$\log (R+1) = 1.4367 + 0.7782$$
$$\log (R+1) = 2.2148$$
$$10^{2.2148} = R+1$$
$$R = 163.0$$

Log of Exponential

The law of exponents also applies to exponentiation.

$$\log X^N = N \log X$$

We can show this in the exponential form. Assume that $\log_B X^N = S$. Then

$$B^S = X^N$$

But X can be expressed as the base, B, raised to some power, L. Note that L is the log of X to base B:

$$B^S = (B^L)^N$$
$$= B^{LN}$$

For this to be true,

$$S = LN$$

must be true. Substitute $\log_B X^N$ for S and $\log_B X$ for L:

$$\log_B X^N = N \log_B X \qquad (22\text{-}1)$$

EXAMPLE 23-13   Use Eq. (22-1) to illustrate that

  (a) $\log 5^2 = 2 \log 5$
  (b) $\log 25^{1.31} = 1.31 \log 25$
  (c) $\ln 33.6^{2.69} = 2.69 \ln 33.6$

SOLUTION

  (a) $\log 5^2 = 2 \log 5$
       $\log 25 = 2 \log 5$
       $1.3979 = 2 \times 0.698970$
       $1.3979 = 1.3979$
  (b) $\log 25^{1.31} = 1.31 \log 25$
     $\log 68.8112 = 1.31 \log 25$
          $1.8313 = 1.31 \times 1.3979$
          $1.8313 = 1.8313$
  (c) $\ln 33.6^{2.69} = 2.69 \ln 33.6$
     $\ln 12760 = 2.69 \ln 33.6$
       $9.4541 = 2.69 \times 3.5145$
       $9.4541 = 9.4541$

Some seemingly impossible logarithmic equations are easily solved using these principles.

EXAMPLE 23-14   Solve for A:

$$2 \log A + \log A^3 = 4.713$$

SOLUTION

$$2 \log A + \log A^3 = 4.713$$
$$2 \log A + 3 \log A = 4.713$$

$$5 \log A = 4.713$$
$$\log A = 0.9426$$
$$10^{0.9426} = A$$
$$A = 8.7619$$

These principles also apply to roots. A root can be thought of as a fractional power, such as

$$\sqrt[2]{X} = X^{0.5}$$

EXAMPLE 23-15   Prove:

$$\log \sqrt{5} = 0.5 \log 5$$

SOLUTION

$$\log \sqrt{5} = \log 5^{0.5}$$

But we have just shown that

$$\log X^N = N \log X$$

Substitute:

$$\log 5^{0.5} = 0.5 \log 5$$
$$\log 2.2361 = 0.5 \log 5$$
$$0.3495 = 0.5 \times 0.6990$$
$$0.3495 = 0.3495$$

Changing Bases

What if a calculator only finds logs to base ϵ? Can logs to base 10 be found? Yes. The principle involved is

$$\log_A X = \log_A B \log_B X$$

We can show this as follows. Let

$$\log_A X = M \quad \text{and} \quad \log_B X = N$$

Then

$$A^M = X \quad \text{and} \quad B^N = X$$

Therefore,

$$A^M = B^N$$

But we can express B as A to some power, L. Note that this is the log of B to base A.

$$A^M = (A^L)^N$$
$$A^M = A^{LN}$$

For the equation to be equal, the exponents must be equal:

$$M = LN$$

But we originally defined

$$\log_A X = M$$
$$\log_B X = N$$

and

$$\log_A B = L \quad \text{because } A^L = B$$

Therefore,

$$\log_A X = \log_A B \log_B X$$

EXAMPLE 23-16 Assume that I can find logs only to the base 10. Find:

(a) $\log_8 35$
(b) $\log_\epsilon 263$

SOLUTION

(a) $\log_A X = \log_A B \log_B X$

$$\log_B X = \frac{\log_A X}{\log_A B}$$

$$\log_8 35 = \frac{\log_{10} 35}{\log_{10} 8} = \frac{1.5441}{0.9031}$$
$$= 1.7098$$

We can verify this:

$$8^{1.7098} = 35$$
$$35 = 35$$

(b) $\log_B X = \frac{\log_A X}{\log_A B}$

$$\log_\epsilon 263 = \frac{\log_{10} 263}{\log_{10} \epsilon}$$
$$= \frac{2.41996}{0.43429} = 5.5722$$

We can prove this by keying in 263 and pressing $\boxed{\text{LN}}$.

PROBLEMS

Prove the following:

23-51. $\log 30 = \log 3 + \log 10$ **23-53.** $\log 30 = \log 15 + \log 2$
23-52. $\log 30 = \log 5 + \log 6$

Knowing only the logs of 2, 3, 4, and 5, find the following logs:

23-54. $\log 8$ **23-59.** $\log 2.5$
23-55. $\log 15$ **23-60.** $\log 1.6667$
23-56. $\log 30$ **23-61.** $\log 1.3333$
23-57. $\log 20$ **23-62.** $\log 1.5$
23-58. $\log 24$

Solve the following equations:

23-63. $2 \log B + 3 \log B = 7$ **23-64.** $4 \log C - \log C = 3.15$

23-65. $2 \log M + \log 2M = 2.6$
23-66. $4 \log R - \log 5R = 3.69$

23-67. $6 \log R + 3 \log 2R = 1.31$
23-68. $3.1 \log 2T + 6.3 \log 3T = 11.62$

Prove the following:

23-69. $\log 30 = \log 15 + \log 2$
23-70. $\log 100 = \log 500 - \log 5$

23-71. $\log 64 = \log 32 + \log 2$
23-72. $\log 45 = \log 225 - \log 5$

Solve the following equations:

23-73. $\log \dfrac{X}{6} = 2.6$

23-74. $\log \dfrac{3X}{2} = 1.95$

23-75. $\log R + \log \dfrac{R}{2} = 1.36$

23-76. $5 \log \dfrac{T}{3} + 6 \log \dfrac{T}{2} = 3.66$

23-77. $\log \dfrac{4}{A} = 3.2$

23-78. $\log \dfrac{5}{A-6} = 2.1$

Use the law of logarithms concerning exponentials to prove that:

23-79. $\log 4^3 = 3 \log 4$
23-80. $\log 7^2 = 2 \log 7$
23-81. $\log 35^{1.2} = 1.2 \log 35$

23-82. $\log 15^{2.9} = 2.9 \log 15$
23-83. $\log 4.31^{5.2} = 5.2 \log 4.31$
23-84. $\log 17.36^{0.83} = 0.83 \log 17.36$

Find the following logs:

23-85. $\log_8 31 = ?$
23-86. $\log_7 44 = ?$
23-87. $\log_6 23 = ?$
23-88. $\log_{12} 15 = ?$

23-89. $\log_{13} 17 = ?$
23-90. $\log_{25} 459 = ?$
23-91. $\log_{20} 430 = ?$
23-92. $\log_{75} 1463 = ?$

23-3   THE DECIBEL

The decibel is a measure of comparing one signal to another. It has three definitions:

$$dB = 20 \log \left(\frac{V}{V_{\text{ref}}} \right) \qquad dB = 20 \log \left(\frac{I}{I_{\text{ref}}} \right) \qquad dB = 10 \log \left(\frac{P}{P_{\text{ref}}} \right)$$

It can be used to state the gain of an amplifier, for example. If we provide an input of 50 μW and the amplifier boosts this to 8 W, we have a gain in decibels of

$$dB = 10 \log \left(\frac{P}{P_{\text{ref}}} \right)$$
$$= 10 \log \left(\frac{8}{50 \times 10^{-6}} \right)$$
$$= 52.04 \text{ dB}$$

Note that we used the power form of the equation, for these were what were given. Note further that we used the input as our reference. We can also obtain losses from a circuit.

EXAMPLE 23-17   A certain circuit receives a 3.1-V signal and reduces this signal to 40 mV. What is its loss in decibels?

SOLUTION   We shall use the voltage form with the input being our reference:

$$dB = 20 \log \left(\frac{V}{V_{ref}} \right)$$

$$= 20 \log \left(\frac{0.04}{3.1} \right)$$

$$= -37.79$$

The minus sign indicates that signal V is less than signal V_{ref}.

The decibel is useful to us in acoustics because the ear hears in a logarithmic ratio. That is, an increase from 1 mW to 10 mW sounds to us the same as an increase from 10 mW to 100 mW. The decibel takes this into account.

EXAMPLE 23-18   Compute the increase in intensity of each of the following in decibels:

(a)   1 μW to 10 μW
(b)   10 μW to 100 μW
(c)   100 μW to 1 mW
(d)   1 kW to 10 kW

SOLUTION

(a) $dB = 10 \log \left(\frac{P}{P_{ref}} \right)$

$= 10 \log \left(\frac{10}{1} \right)$

$= 10$ dB

(b) $dB = 10 \log \left(\frac{P}{P_{ref}} \right)$

$= 10 \log \left(\frac{100}{10} \right)$

$= 10$ dB

(c) $dB = 10 \log \left(\frac{P}{P_{ref}} \right)$

$= 10 \log \left(\frac{1000}{100} \right)$

$= 10$ dB

(d) $dB = 10 \log \left(\frac{P}{P_{ref}} \right)$

$= 10 \log \left(\frac{10,000}{1000} \right)$

$= 10$ dB

Note that each computes to the same degree of increase in intensity.

PROBLEMS

Assuming the first value is the input and the second the output, compute the gain for each of the following in decibels.

23-93. 4 mV, 3 V	23-98. 4.65 A, 2.13 mA
23-94. 6 mW, 20 W	23-99. 1.31 μA, 4 A
23-95. 20 μV, 2 mV	23-100. 23.1 MV, 3 μV
23-96. 20 μW, 200 W	23-101. 1.31 mW, 3.46 W
23-97. 1.31 mV, 2.61 V	23-102. 439 mW, 3 W

24

INDUCTANCE IN SERIES *RL* CIRCUITS

In this chapter we study the mathematics of ac circuits having resistance and inductance. Because there is a phasor relationship between resistance and inductance when subjected to sine waves, these circuits deserve special attention.

24-1 INDUCTORS IN SERIES AND IN PARALLEL

We can define inductance as the property of a device that obeys the equation

$$v = L \frac{di}{dt} \tag{24-1}$$

where v is the instantaneous voltage, L the inductance measured in henries, and di/dt the rate of change of current per unit time. The most common inductor is a coil of wire. Applying Eq. (24-1) to the coil, the equation states that the amount of voltage across the coil is determined by the rate of change of current within the coil.

EXAMPLE 24-1 A coil has current increasing at a rate of 4 A every 3 sec. If the coil has a 5-H inductance, what is the voltage across the coil?

SOLUTION

$$v = L \frac{di}{dt}$$
$$= 5 \times \frac{4}{3} = 6.667 \text{ V}$$

Thus, the coil has a constant 6.667 V across it.

EXAMPLE 24-2 A voltmeter connected across a coil reads 23.2 V. If current is increasing at a rate of 4 mA every 100 μs, what is the inductance of the coil?

SOLUTION

$$v = L \frac{di}{dt}$$
$$23.2 = L \times \frac{0.004}{100 \times 10^{-6}}$$
$$L = 580.0 \text{ mH}$$

399

Inductors placed in series are additive. Note that this is similar to resistors connected in series.

$$L_t = L_1 + L_2 + L_3 + \ldots$$

EXAMPLE 24-3 Inductors of 3 H, 250 mH, and 7.2 mH are connected in series (Fig. 24-1). What is the total inductance of the circuit?

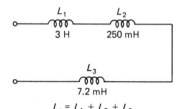

FIGURE 24-1 Example 24-3, inductors in series.

SOLUTION

$$
\begin{aligned}
L_t &= L_1 + L_2 + L_3 \\
&= 3 + 0.250 + 0.0072 \\
&= 3.2572 \text{ H}
\end{aligned}
$$

EXAMPLE 24-4 Compute the total inductance of the following inductors when connected in series: 1.5 H, 2.5 H, 780 mH, 430 mH, and 65 mH.

SOLUTION

$$
\begin{aligned}
L_t &= L_1 + L_2 + L_3 + L_4 + L_5 \\
&= 1.5 + 2.5 + 0.78 + 0.43 + 0.065 \\
&= 5.275 \text{ H}
\end{aligned}
$$

Inductors that are connected in parallel follow an inverse law similar to that of resistors connected in parallel.

$$\frac{1}{L_t} = \frac{1}{L_1} + \frac{1}{L_2} + \frac{1}{L_3} + \ldots$$

EXAMPLE 24-5 Compute total inductance of the following inductors connected in parallel: 3 H, 4 H, and 5 H (Fig. 24-2).

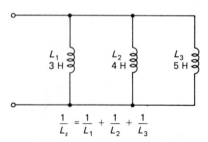

FIGURE 24-2 Example 24-5, inductors connected in parallel.

SOLUTION

$$\frac{1}{L_t} = \frac{1}{L_1} + \frac{1}{L_2} + \frac{1}{L_3}$$
$$= \frac{1}{3} + \frac{1}{4} + \frac{1}{5}$$
$$= 0.3333 + 0.250 + 0.200$$
$$= 0.7833$$
$$L_t = \frac{1}{0.7833} = 1.2766 \text{ H}$$

Note that the sum is less than any of the individual inductors.

EXAMPLE 24-6 Edgar Lectric's oscillator contains 760-μH and 700-μH coils connected in parallel. What is the total inductance of E. Lectric's circuit?

SOLUTION

$$\frac{1}{L_t} = \frac{1}{L_1} + \frac{1}{L_2}$$
$$= \frac{1}{760} + \frac{1}{700}$$
$$L_t = 364.4 \ \mu\text{H}$$

PROBLEMS

24-1. A coil has current steadily increasing at a rate of 50 mA every 100 ms. What is the voltage across the coil if the coil has an inductance of   **(a)** 10 μH,   **(b)** 100 μH,   **(c)** 1 mH,   **(d)** 10 mH,   **(e)** 100 mH,   **(f)** 1 H?

24-2. Compute the voltage across a 10-H coil if the current increases at a rate of   **(a)** 1 μA/s,   **(b)** 1 mA/s,   **(c)** 1 A/s.

24-3. A voltmeter connected across a 2-H coil reads 3.5 V. How fast is current increasing every second?

24-4. A voltmeter connected across a coil reads 20 mV. What is the inductance of the coil if current is increasing 15 mA every 2.5 ms?

Compute total inductance of the following inductors connected in series.

24-5. 2.5 H, 4.3 H

24-6. 3.6 mH, 12.7 mH

24-7. 77 mH, 120 mH, 460 mH

24-8. 4.3 μH, 9.6 μH, 43 μH

24-9. 7.3 H, 269 mH, 430 mH, 33 μH

24-10. 50 H, 740 mH, 320 μH, 40 μH

Compute total inductance of the following inductors connected in parallel:

24-11. 2.5 H, 4.3 H

24-12. 3.6 mH, 12.7 mH

24-13. 77 mH, 120 mH, 460 mH

24-14. 4.3 μH, 9.6 μH, 43 μH

24-15. 7.3 H, 269 mH, 430 mH, 33 μH

24-16. 50 H, 740 mH, 320 μH, 40 μH

24-17. What inductance must be placed in parallel with a 43-mH inductor to make 25 mH total?

24-18. What inductance must be placed in parallel with a 250-μH inductor to make 100 μH?

24-2 MUTUAL INDUCTANCE

Mutual inductance occurs any time the magnetic field of one coil links the magnetic field of another. It obeys the equation

$$L_t = L_1 + L_2 \pm 2M$$

where L_t is the total inductance, L_1 and L_2 the inductance of the individual inductors, and M the mutual inductance. All inductances are measured in henries (H). The sign of the $2M$ term is positive if the flux linkage opposes the flow of current, and is negative if the flux linkage aids current flow.

> **EXAMPLE 24-7** Two inductors, 350 mH and 600 mH, are connected in such a manner that there is 30 mH of mutual inductance opposing current flow. What is the total inductance?
>
> **SOLUTION**
>
> $$\begin{aligned} L_t &= L_1 + L_2 \pm 2M \\ &= 350 + 600 + 2(30) \\ &= 1.010 \text{ H} \end{aligned}$$

This mutual inductance can be computed according to the equation

$$M = k\sqrt{L_1 L_2}$$

where M is the mutual inductance, L_1 and L_2 are inductances of the individual inductors, and k is the coefficient of coupling, indicating how closely the coils are coupled. The constant k varies from 0 to 1. The closer the coils are linked, the higher the value of k. If 100% of the flux lines of one coil intersect the second coil, k has a value of 1.0; if 5% of the flux lines of one coil intersect the second, k is 0.05.

> **EXAMPLE 24-8** Inductors of 260 μH and 400 μH are connected in series such that 20% of the flux of one coil links the other in a direction that aids current flow. Compute the total inductance.
>
> **SOLUTION** We must first calculate the mutual inductance:
>
> $$\begin{aligned} M &= k\sqrt{L_1 L_2} \\ &= 0.2\sqrt{260 \times 10^{-6} \times 400 \times 10^{-6}} \\ &= 64.50 \ \mu\text{H} \end{aligned}$$
>
> Since the current flow is aided, the mutual inductance must be subtracted:
>
> $$\begin{aligned} L_t &= L_1 + L_2 - 2M \\ &= 260 + 400 - 2(64.5) \\ &= 531.0 \ \mu\text{H} \end{aligned}$$

EXAMPLE 24-9 Two inductors, 20 mH and 30 mH, are connected in series. What is the coefficient of coupling if total inductance is 55 mH?

SOLUTION We must first find the mutual inductance. Since total inductance is more than the sum of the two individual inductances, the sign of the $2M$ term must be positive.

$$L_t = L_1 + L_2 + 2M$$
$$55 = 30 + 20 + 2M$$
$$M = 2.5 \text{ mH}$$

We can now find the coefficient of coupling.

$$M = k\sqrt{L_1 L_2}$$
$$2.5 \times 10^{-3} = k\sqrt{20 \times 10^{-3} \times 30 \times 10^{-3}}$$
$$k = 0.1021$$

PROBLEMS

24-19. Two inductors, 40 mH and 60 mH, are connected in series such that there is 10 mH of mutual inductance aiding the current flow. Compute the total inductance.

24-20. Repeat Prob. 24-19 for a mutual inductance of 30 mH.

24-21. Two inductors, 4 H and 780 mH, are connected in series such that there is 300 mH of mutual inductance opposing the current flow. Compute the total inductance.

24-22. Repeat Prob. 24-21 for a mutual inductance of 100 μH.

24-23. Hank Link connected inductors of 20 mH and 40 mH in series such that k was 0.30 and opposed the current flow. What was the total inductance of Hank's circuit?

24-24. Repeat for a k of 0.70.

24-25. Series inductors of 750 mH and 920 mH are 15% flux-linked such that current is aided. What is the total inductance?

24-26. Series inductors of 420 μH and 670 μH are 3% flux-linked such that current is opposed. What is the total inductance?

24-27. Series inductors of 78 mH and 47 mH have a total inductance of 150 mH. What is the coefficient of coupling?

24-28. Repeat Prob. 24-27 if the total inductance is 100 mH.

24-3 TRANSFORMERS

A transformer (Fig. 24-3) is formed from two coils sharing the same core, usually air, powdered iron, or laminated iron plates. The input side of the transformer is

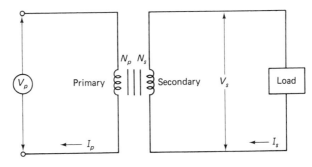

FIGURE 24-3 Transformer.

called its primary winding and the output side is called its secondary winding. The primary and secondary voltages and currents are related according to the equations

$$\frac{V_P}{V_s} = \frac{N_P}{N_s}$$ (24-2)

$$\frac{I_P}{I_s} = \frac{N_s}{N_P}$$ (24-3)

where V_P and V_s are primary and secondary voltages, I_P and I_s are primary and secondary currents, and N_P and N_s are the number of turns of the primary and secondary windings.

EXAMPLE 24-10 A transformer has 200 turns of primary winding and 400 turns of secondary windings. If its load draws 50 mA and has a voltage of 30 V across it, what is the primary voltage and current?

SOLUTION

$$\frac{V_P}{V_s} = \frac{N_P}{N_s}$$
$$\frac{V_P}{30} = \frac{200}{400}$$
$$V_P = 15 \text{ V}$$
$$\frac{I_P}{I_s} = \frac{N_s}{N_P}$$
$$\frac{I_P}{50} = \frac{400}{200}$$
$$I_P = 100 \text{ mA}$$

The impedance (ac resistance) looking into the transformer is related to the impedance of the load. We can develop this relationship by multiplying Eqs. (24-2) and (24-3):

$$\frac{V_P}{V_s}\left(\frac{I_s}{I_P}\right) = \frac{N_P}{N_s}\left(\frac{N_P}{N_s}\right)$$
$$\frac{V_P I_s}{V_s I_P} = \frac{N_P^2}{N_s^2}$$
$$\frac{V_P/I_P}{V_s/I_s} = \frac{N_P^2}{N_s^2}$$
$$\frac{Z_P}{Z_s} = \frac{N_P^2}{N_s^2}$$

where Z_P is the impedance looking into the primary winding and Z_s the impedance of the load, both measured in ohms.

EXAMPLE 24-11 A transformer has 400 primary turns and 1000 secondary turns. If 20 V at 40 mA is applied to the primary, what is the secondary current and voltage? What is the impedance of the load?

SOLUTION

$$\frac{V_P}{V_s} = \frac{N_P}{N_s}$$

$$\frac{20}{V_s} = \frac{400}{1000}$$

$$V_s = 50 \text{ V}$$

$$\frac{I_P}{I_s} = \frac{N_s}{N_P}$$

$$\frac{40}{I_s} = \frac{1000}{400}$$

$$I_s = 16 \text{ mA}$$

The impedance of the load is

$$Z_s = \frac{V_s}{I_s} = \frac{50}{16} = 3.125 \text{ k}\Omega$$

We could also have used the impedance equation:

$$\frac{Z_P}{Z_s} = \frac{N_P^2}{N_s^2} \qquad \frac{V_P/I_P}{Z_s} = \frac{N_P^2}{N_s^2}$$

$$\frac{20/40}{Z_s} = \frac{400^2}{1000^2}$$

$$Z_s = 3.125 \text{ k}\Omega$$

EXAMPLE 24-12 What turns ratio (N_P/N_s) must be selected so that an 8-Ω speaker load will look like 150 Ω of primary impedance?

SOLUTION

$$\frac{Z_P}{Z_s} = \frac{N_P^2}{N_s^2} = \left(\frac{N_P}{N_s}\right)^2$$

$$\frac{N_P}{N_s} = \sqrt{\frac{Z_P}{Z_s}} = \sqrt{\frac{150}{8}} = 4.330$$

Thus, the primary winding should have 4.330 times as many turns as the secondary winding.

PROBLEMS

Compute the missing quantities for these transformers:

	N_P	N_s	I_P	I_s	V_P	V_s
24-29.	1000	2000	20 mA		5 V	
24-30.	3000	100	10 mA		500 V	
24-31.	50	300	7 mA			75 V
24-32.	6000	100		3 mA	50 V	
24-33.	700			50 mA	25 V	150 V
24-34.		300	450 mA	30 mA	10 V	

24-35. A transformer has a primary winding of 250 turns. How many turns does its secondary winding have if its primary impedance is 4 kΩ and its secondary impedance is 16 Ω?

24-36. The transformer of Prob. 24-29 drives a 400-Ω load. What is its primary impedance?

24-37. The transformer of Prob. 24-30 has a primary input impedance of 3 kΩ. What is the impedance of its load?

24-38. What turns ratio is required so that the primary impedance is 5 kΩ and the load impedance is 500 Ω?

24-4 *RL* DISCHARGE CURVE

The inductor tends to prevent any change in current flow. Consider what happens when we close switch S_1 in Fig. 24-4. Before the switch is closed, the current through L_1 is zero. Thus, when the switch closes, the inductor, by our definition, tries to prevent the current from increasing. It does this by generating a counter EMF in opposition to V_s. Therefore, at a time immediately after the switch is closed, V_L is 10 V and I is 0 mA, as shown in Fig. 24-5. With 0 mA flowing through R_1, the voltage across the resistor is zero volts.

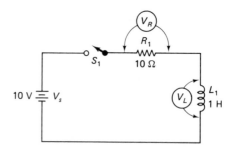

FIGURE 24-4 DC in an RL circuit.

However, as time progresses, the current does increase, resulting in a voltage drop V_R across the resistor and a drop of V_L across the inductor. Note that the sum of V_L and V_R must at all times be 10 V, the supply voltage. Finally, when sufficient time has progressed, the inductor will have a steady 1 A through it, generating a static magnetic field, resulting in no counter EMF, a drop of zero volts across L_1, and a drop of 10 V across R_1.

This curve is known as a time constant curve and V_L can be computed as

$$V_L = V_s \epsilon^{-Rt/L} \tag{24-4}$$

where V_L is the instantaneous voltage across the inductor, V_s the power supply voltage, ϵ the value 2.71828, R the resistance of R_1 in ohms, L the inductance of L_1 in henries, and t time in seconds. Using this equation, we can calculate the voltage at any time t.

EXAMPLE 24-13 Compute the voltage across L_1 and R_1 250 ms after closing switch S_1 in Fig. 24-4.

SOLUTION

$$V_L = V_s \epsilon^{-Rt/L}$$
$$= 10\epsilon^{(-10\times0.25)/1}$$
$$= 820.8 \text{ mV}$$
$$V_R = V_s - V_L = 10 - 0.8208 = 9.179 \text{ V}$$

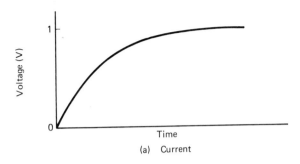

(a) Current

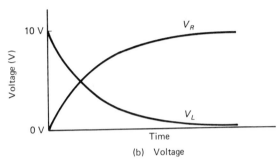

(b) Voltage

FIGURE 24-5 R/L time constants.

Equation (24-4) gives us the voltage across the inductor. We can derive an equation for the voltage across the resistor by subtracting it from the supply voltage:

$$V_R = V_s - V_L$$
$$= V_s - V_s\epsilon^{-Rt/L}$$
$$= V_s(1 - \epsilon^{-Rt/L})$$

We can also derive the resistor (and thus the inductor) current:

$$V_R = V_s(1 - \epsilon^{-Rt/L})$$
$$\frac{V_R}{R} = \frac{V_s}{R}(1 - \epsilon^{-Rt/L})$$
$$i = I_m(1 - \epsilon^{-Rt/L})$$

where i is instantaneous inductor or resistor current and I_m is the maximum possible current existing in the circuit. This occurs when the inductor offers no resistance to current flow and the current is dependent only upon the resistance. Note that, according to the equation, this will occur when time approaches infinity.

EXAMPLE 24-14 An inductor of 400 mH is in series with a 15-V supply and a 1-kΩ resistor. Compute the voltages across the inductor and resistor and the current at a time 237 μs after closing the circuit.

SOLUTION

$$V_L = V_s \epsilon^{-Rt/L}$$
$$= 15\epsilon^{-(1000\times237\times10^{-6}/0.4)}$$
$$= 8.294 \text{ V}$$
$$V_R = V_s - V_L$$
$$= 15 - 8.294$$
$$= 6.706 \text{ V}$$
$$i = I_m(I - \epsilon^{-Rt/L})$$
$$= \frac{V_s}{R}(1 - \epsilon^{-Rt/L})$$
$$= \frac{15}{1}(1 - \epsilon^{-(1000\times237\times10^{-6}/0.4)})$$
$$= 6.706 \text{ mA}$$

Note that, since we knew the resistance (1 kΩ) and the voltage across the resistance (6.706 V), we could have computed i by

$$i = \frac{v}{R} = \frac{6.706}{1} = 6.706 \text{ mA}$$

PROBLEMS

In the following problems, an inductor is connected in series with a resistor and a power supply. Compute V_L, V_R, and I at the time indicated after closing the switch.

	R	L	t	V_s
24-39.	27 Ω	4.5 H	200 ms	30 V
24-40.	100 Ω	3 H	20 ms	5 V
24-41.	1 kΩ	1 H	10 μs	100 V
24-42.	1 kΩ	1 H	100 μs	100 V
24-43.	1 kΩ	1 H	1 ms	100 V
24-44.	1 kΩ	1 H	1 s	100 V

24-5 INDUCTIVE REACTANCE

Inductive reactance is the opposition that an inductor offers to sinusoidal waveforms. It is defined according to the equation

$$X_L = 2\pi fL$$

where X_L is inductive reactance measured in ohms, π is the constant 3.14159265, f is frequency measured in hertz, and L is inductance in henries.

EXAMPLE 24-15 Compute inductive reactance for a 200-μH coil at:

(a) 1 Hz
(b) 1 kHz
(c) 1 MHz
(d) 1 GHz

SOLUTION

(a) $X_L = 2\,\pi fL = 2 \times \pi \times 1 \times 200 \times 10^{-6} = 1.257 \text{ m}\Omega$
(b) $X_L = 2\,\pi fL = 2 \times \pi \times 1000 \times 200 \times 10^{-6} = 1.257\ \Omega$
(c) $X_L = 2\,\pi fL = 2 \times \pi \times 10^6 \times 200 \times 10^{-6} = 1.257 \text{ k}\Omega$
(d) $X_L = 2\,\pi fL = 2 \times \pi \times 10^9 \times 200 \times 10^{-6} = 1.257 \text{ M}\Omega$

Note that inductive reactance increases as frequency increases.

Reactances in Series

We can derive an equation for inductive reactances in series by multiplying the equation for series inductors by $2\,\pi f$:

$$L_t = L_1 + L_2 + L_3 + \dots$$
$$2\,\pi fL_t = 2\,\pi fL_1 + 2\,\pi fL_2 + 2\,\pi fL_3 + \dots$$
$$X_{Lt} = X_{L1} + X_{L2} + X_{L3} + \dots$$

EXAMPLE 24-16 Inductive reactances of 40 kΩ and 7 kΩ are in series. What is the total reactance?

SOLUTION

$$X_{Lt} = X_{L1} + X_{L2} = 40 + 7 = 47 \text{ k}\Omega$$

EXAMPLE 24-17 Inductors having reactances of 2 kΩ and 3 kΩ are in series with 2-mH and 5.6-mH inductors at a frequency of 65 kHz. What is the total reactance?

SOLUTION We must first compute the reactances of the 2-mH and 5.6-mH coils:

$$X_{L1} = 2\,\pi fL = 2 \times \pi \times 65 \times 10^3 \times 2 \times 10^{-3} = 816.8\ \Omega$$
$$X_{L2} = 2\,\pi fL = 2 \times \pi \times 65 \times 10^3 \times 5.6 \times 10^{-3} = 2.287 \text{ k}\Omega$$

We can now compute total reactance.

$$X_{Lt} = X_{L1} + X_{L2} + X_{L3} + X_{L4}$$
$$= 816.8 + 2287.1 + 2000 + 3000$$
$$= 8.1039 \text{ k}\Omega$$

Since, in an inductor, voltage leads current by 90°, inductive reactance is always considered at an angle of $/90°$. Thus, the complete solution to Example 24-16 is $47/90°$ kΩ and the complete solution to Example 24-17 is $8.1039/90°$ kΩ. These phasor values should be used to express the voltages across reactances and the currents through these elements.

EXAMPLE 24-18 What is the current through a 2.5-mH coil when the voltage across it is 20 V at 5 kHz (Fig. 24-6)?

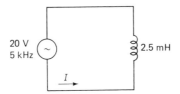

FIGURE 24-6 Example 24-18.

SOLUTION

$$X_L = 2 \pi fL = 2 \times \pi \times 5 \times 10^3 \times 2.5 \times 10^{-3} = 78.54 \; \Omega$$
$$= 78.54 \underline{/90°} \; \Omega$$

The angle was assigned because this was inductive reactance. We shall assume an angle of 0° for the supply voltage:

$$I = \frac{V}{X_L} = \frac{20 \underline{/0°}}{78.54 \underline{/90°}} = 254.6 \underline{/-90°} \; \text{mA}$$

Note that voltage leads current by 90°.

EXAMPLE 24-19 Compute the current through and voltage across the 2.9-mH coil of Fig. 24-7.

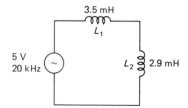

FIGURE 24-7 Example 24-19.

SOLUTION We can first find total reactance:

$$X_{Lt} = 2 \pi f(L_1 + L_2)$$
$$= 2 \times \pi \times 20 \times 10^3 \times (3.5 + 2.9) \times 10^{-3}$$
$$= 804.2 \underline{/90°} \; \Omega$$

The current can now be found:

$$I = \frac{V}{X} = \frac{5 \underline{/0°}}{804.2 \underline{/90°}} = 6.217 \underline{/-90°} \; \text{mA}$$

Find the reactance of L_2:

$$X_{L2} = 2 \pi fL = 2 \times \pi \times 20 \times 10^3 \times 2.9 \times 10^{-3}$$
$$= 364.42 \underline{/90°}$$

The voltage across L_2 can now be found:

$$V = IX_L = 6.217 \underline{/-90°} \times 0.36442 \underline{/90°}$$
$$= 2.266 \underline{/0°} \; \text{V}$$

Reactances in Parallel

We can develop an equation for reactances in parallel by dividing the equation for parallel inductors by $2 \pi f$:

$$\frac{1}{L_t} = \frac{1}{L_1} + \frac{1}{L_2} + \frac{1}{L_3} + \dots$$

$$\frac{1}{2 \pi fL_t} = \frac{1}{2 \pi fL_1} + \frac{1}{2 \pi fL_2} + \frac{1}{2 \pi fL_3} + \dots$$

$$\frac{1}{X_{Lt}} = \frac{1}{X_{L1}} + \frac{1}{X_{L2}} + \frac{1}{X_{L3}} + \dots$$

EXAMPLE 24-20   Inductive reactances of 3 kΩ, 5.6 kΩ, and 3.9 kΩ are connected in parallel. What is the total reactance?

SOLUTION   Since all phasors in the problem are at 90°, we shall suffix the angle after computation.

$$\frac{1}{X_t} = \frac{1}{X_1} + \frac{1}{X_2} + \frac{1}{X_3}$$
$$= \frac{1}{3} + \frac{1}{5.6} + \frac{1}{3.9}$$
$$X_t = 1.3015 \underline{/90°} \text{ k}\Omega$$

EXAMPLE 24-21   Three inductors are connected in parallel: a 3-mH coil, a 10-mH coil, and a coil having a reactance of 35.6 Ω. Compute the total reactance at 3.6 kHz.

SOLUTION   We must first find the reactance of the individual coils:

$$X_{L1} = 2\,\pi f L_1 = 2 \times \pi \times 3600 \times 0.003 = 67.858 \underline{/90°} \ \Omega$$
$$X_{L2} = 2\,\pi f L = 2 \times \pi \times 3600 \times 0.01 = 226.19 \underline{/90°} \ \Omega$$

We can now compute total reactance. Since all the phasors are at 90°, we shall affix the angles after computation:

$$\frac{1}{X_{Lt}} = \frac{1}{X_{L1}} + \frac{1}{X_{L2}} + \frac{1}{X_{L3}}$$
$$= \frac{1}{67.858} + \frac{1}{226.19} + \frac{1}{35.6}$$
$$X_{Lt} = 21.165 \underline{/90°} \ \Omega$$

These parallel circuits obey all the laws of parallel circuits we learned for dc circuits, provided that we drag the phasor along in our calculations. Thus, we can apply Ohm's law to any circuit.

EXAMPLE 24-22   Compute I_1, I_2, and I_t of Fig. 24-8 if V_s is 26 V.

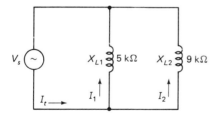

FIGURE 24-8   Examples 24-22 and 24-23.

SOLUTION   We can compute currents through the individual branches using Ohm's great and powerful law:

$$I_1 = \frac{V_s}{X_{L1}} = \frac{26 \underline{/0°}}{5 \underline{/90°}} = 5.200 \underline{/-90°} \text{ mA}$$
$$I_2 = \frac{V_s}{X_{L2}} = \frac{26 \underline{/0°}}{9 \underline{/90°}} = 2.889 \underline{/-90°} \text{ mA}$$

$$I_t = I_1 + I_2$$
$$= 5.200 \underline{/-90°} + 2.889 \underline{/-90°}$$
$$= 8.0889 \underline{/-90°} \text{ mA}$$

Note that, since no angle was given for V_s, we assumed it to be 0°.

EXAMPLE 24-23 If, in Fig. 24-8, I_1 is $20 \underline{/30°}$ mA, compute I_t.

SOLUTION Knowing I_1 and X_{L1}, we can compute voltage V_s:

$$V_s = I_1 X_{L1} = (20 \underline{/30°})(5 \underline{/90°})$$
$$= 100 \underline{/120°} \text{ V}$$

We can now find I_2:

$$I_2 = \frac{V_s}{X_{L2}} = \frac{100 \underline{/120°}}{9 \underline{/90°}} = 11.11 \underline{/30°} \text{ mA}$$

By adding I_1 and I_2, we can compute I_t:

$$I_t = I_1 + I_2 = 20 \underline{/30°} + 11.11 \underline{/30°}$$
$$= 31.11 \underline{/30°} \text{ mA}$$

Note that V at 120° still leads I at 30° by 90°.

PROBLEMS

24-45. Compute inductive reactance for a 3.5-H coil at (a) 1 Hz, (b) 1 kHz, (c) 1 MHz, (d) 1 GHz.

24-46. Compute inductive reactance for a 35-μH coil at (a) 3.9 Hz, (b) 3.9 kHz, (c) 3.9 MHz, (d) 3.9 GHz.

24-47. At what frequency will a 4.7-mH coil have a reactance of 256 Ω?

24-48. What is the inductance of a coil having a reactance of 5 kΩ at 29 kHz?

24-49. Compute the total inductive reactance of a series circuit composed of three inductive reactances: 3 kΩ, 5 kΩ, and 7 kΩ.

24-50. Compute the total reactance of a series circuit composed of the following inductive reactances: 40 kΩ, 35 kΩ, 70 kΩ, 3.5 kΩ, 7.7 kΩ.

24-51. Inductive reactances of 3.6 kΩ and 5.5 kΩ are in series with a 35-mH and a 72-mH inductor at 7.5 kHz. What is total reactance?

24-52. What inductance must be placed in series with a 4.5-kΩ inductive reactance to result in a total of 5.8 kΩ at 4.7 kHz?

24-53. What is the current through a 6-mH coil with 25 V, 10 kHz across it?

24-54. What is the current through a 4.7-H coil with 3 V, 500 Hz across it?

24-55. What is the reactance of a coil having 27 V across it and 30 mA through it?

24-56. What is the voltage across a 27-mH coil having a current 3.5 mA at 800 Hz?

24-57. Compute I of Fig. 24-9 if L_1 is 35 mH, L_2 is 50 mH, and V_s is 30 V at a frequency of 200 Hz.

24-58. Compute I of Fig. 24-9 if L_1 is 250 μH, L_2 is 730 μH, and V_s is 3.3 V at a frequency of 300 kHz.

24-59. Compute V_1 of Fig. 24-9 if I is $30 \underline{/20°}$ mA, L_1 is 12 H, L_2 is 5 H, and the frequency is 400 Hz.

24-60. Compute V_2 of Fig. 24-9 if V_1 is $4.53 \underline{/26°}$ V, L_1 is 880 mH, L_2 is 1.30 H, and the frequency is 750 Hz.

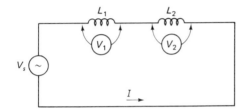

FIGURE 24-9 Problems 24-57 through 24-60.

24-61. Inductive reactances of 640 Ω and 430 Ω are connected in parallel. Compute the total inductive reactance.

24-62. Inductive reactances of 3.5 kΩ, 4.3 kΩ, 373 Ω, and 876 Ω are connected in parallel. Compute total reactance.

24-63. Three inductors are connected in parallel: a 40-mH coil, a 60-mH coil, and a coil having a reactance of 560 Ω. Compute the total reactance at 2 kHz.

24-64. Four inductors are connected in parallel: a 700-mH coil, a 3.6-H coil, a coil having 3460-Ω reactance, and a coil having 5361-Ω reactance. Compute the total inductive reactance at 500 Hz.

24-65. In Fig. 24-10, $V_s = 20$ V at 5 kHz, $L_1 = 3.5$ mH, and $L_2 = 5.5$ mH. Compute I_1, I_2, and I_t.

24-66. In Fig. 24-10, $V_s = 5$ V at 30 kHz, $L_1 = 8.6$ mH, and $L_2 = 1.3$ mH. Compute I_1, I_2, and I_t.

24-67. In Fig. 24-10, compute I_2 and V_s if I_1 is $3\underline{/27°}$ mA, L_1 is 45 mH, L_2 is 55 mH, and the frequency is 5 kHz.

24-68. In Fig. 24-10, compute L_1 if I_2 is $3.65\underline{/-22°}$ mA, I_1 is $5.73\underline{/-22°}$ mA, L_2 is 66 mH, and the frequency is 4.3 kHz.

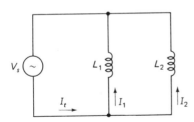

FIGURE 24-10 Problems 24-65 through 24-68.

24-6 SERIES RL CIRCUITS

In this section we consider circuits that have both inductance and resistance in series. In such a circuit, inductive reactance (X_L) and resistance (R) are phasor-added to obtain impedance (Z). Figure 24-11 illustrates this relationship. Inductive reactance is always considered as having a phase angle of 90°, resistance always has a phase angle of 0°, and the impedance result will have a phase angle somewhere between 0° and 90°.

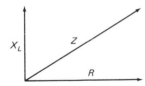

FIGURE 24-11 AC phasor relationship.

EXAMPLE 24-24 A resistor of 6.8 kΩ is in series with a 20-mH inductor. What is the total circuit impedance at 42 kHz?

SOLUTION We must first compute the reactance:

$$X_L = 2\pi fL = 2 \times \pi \times 42 \times 10^3 \times 0.02$$
$$= 5278\ \Omega$$

Note that, since this is inductive reactance, it has a phase angle of 90°. In j notation, it is 5278 units in the $+j$ direction. We now have only to phasor add to obtain total impedance:

$$Z = R + jX_L = 6800 + j5278$$
$$= 8.608 \underline{/37.82°}\ k\Omega$$

Figure 24-12 illustrates the result.

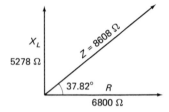

FIGURE 24-12 Example 24-24.

EXAMPLE 24-25 Repeat Example 24-24 for a frequency of 4.8 kHz.

SOLUTION

$$X_L = 2\pi fL = 2 \times \pi \times 4800 \times 0.02 = 603.2\ \Omega$$
$$Z = R + jX_L = 6800 + j603 = 6.827 \underline{/5.069°}\ k\Omega$$

EXAMPLE 24-26 Resistors of 4.7 kΩ and 7.5 kΩ are in series with an inductive reactance of 5 kΩ and a 200-mH coil at 3 kHz. What is the total circuit impedance?

SOLUTION

$$Z = R_1 + R_2 + jX_{L1} + jX_{L2}$$
$$= 4700 + 7500 + j5000 + j(2 \times \pi \times 3000 \times 0.2)$$
$$= 12{,}200 + j5000 + j3770$$
$$= 12{,}200 + j8770$$
$$= 15.03 \underline{/35.71°}\ k\Omega$$

Theoretically, a coil offers no resistance to direct current, according to the equation $X_L = 2\pi fL$. However, because a coil is made up of wire and because that wire has resistance, the coil does not have a perfect 90° phase angle.

EXAMPLE 24-27 A 2.5-H coil has a dc resistance of 326 Ω. What is its impedance at 300 Hz?

SOLUTION

$$X_L = 2\,\pi f L = 2 \times \pi \times 300 \times 2.5 = 4712\ \Omega$$
$$X = R + jX_L = 326 + j4712$$
$$= 4724\underline{/86.04^\circ}\ k\Omega$$

Note that it is not perfect, for it has a phase angle of less than 90°.

PROBLEMS

24-69. A resistor of 10 kΩ is in series with a 40-mH inductor at 50 kHz. What is the circuit's impedance?

24-70. A resistor of 8.6 kΩ is in series with a 200-μH inductor at 2 MHz. What is the circuit's impedance?

24-71. Resistors of 4 MΩ and 3 MΩ are in series with a 200-mH inductor at 860 kHz. What is the total impedance of the circuit?

24-72. Resistors of 36 kΩ, 12 kΩ, and 15 kΩ are in series with an inductive reactance of 49 kΩ and an inductor of 2 H at a frequency of 10 kHz. What is the circuit's impedance?

24-73. Resistors of 4.9 kΩ, 13 kΩ, and 7.5 kΩ are in series with inductive reactances of 1.2 kΩ and 5.8 kΩ and inductors of 450 mH and 720 mH at a frequency of 1 kHz. What is the total impedance of the circuit?

24-74. Resistors of 3.8 kΩ and 1.2 kΩ are in series with inductive reactances of 7.2 kΩ and 1.6 kΩ and inductors of 200 mH and 330 mH at a frequency of 20 kHz. Compute total impedance.

24-75. A 220-μH inductor has a dc resistance of 25 Ω. What is its impedance at 790 kHz?

24-76. A 5-mH inductor has a dc resistance of 14 Ω. What is its impedance at 1 MHz?

24-77. A 100-mH inductor has a dc resistance of 473 Ω. What is its impedance at 2.5 MHz?

24-78. A 68-μH inductor has a dc resistance of 216 mΩ. What is its impedance at 10 kHz?

24-7 OHM'S LAW IN SERIES REACTIVE CIRCUITS

As long as all the calculations are done in phasor algebra, all the techniques used in dc analysis are applicable to ac analysis.

EXAMPLE 24-28 A 260-mH coil is in series with a 5.3-kΩ resistor at 2 kHz across a 20-V supply. Compute total current and the voltage drops across the inductor and the resistor [Fig. 24-13(a)].

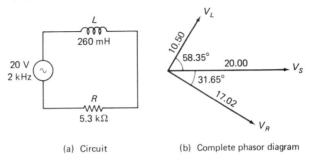

(a) Circuit   (b) Complete phasor diagram

FIGURE 24-13 Example 28-28.

SOLUTION

$$X_L = 2\,\pi fL = 2 \times \pi \times 2000 \times 0.26 = 3.267 \text{ k}\Omega$$
$$Z = R + jX_L = 5.3 + j3.267 = 6.226\underline{/31.65°} \text{ k}\Omega$$

The current can now be computed:

$$I = \frac{V}{Z} = \frac{20\underline{/0°}}{6.226\underline{/31.65°}} = 3.212\underline{/-31.65°} \text{ mA}$$

Knowing the current, we can now calculate the voltage drops.

$$V_L = IX_L = (3.212\underline{/-31.65°})(3.267\underline{/90°})$$
$$= 10.50\underline{/58.35°}$$
$$V_R = IR = (3.212\underline{/-31.65°})(5.3\underline{/0°})$$
$$= 17.02\underline{/-31.65°} \text{ V}$$

Note that the supply voltage is the phasor sum of the voltage across the resistor and the voltage across the inductor.

$$V_s = V_R + V_L$$
$$= 17.02\underline{/-31.65°} + 10.50\underline{/58.35°}$$
$$= 14.49 - j8.934 + 5.508 + j8.934$$
$$= 20.00 + j0$$
$$= 20\underline{/0°} \text{ V}$$

The complete phasor diagram is shown in Fig. 24-13(b).

EXAMPLE 24-29   A 36-mH inductor is in series with a 4.7-kΩ resistor and a 25-kHz voltage source. If the voltage across the inductor is $28\underline{/0°}$ V, what is the voltage across the resistor?

SOLUTION   The reactance of the inductor is

$$X_L = 2\,\pi fL = 2 \times \pi \times 25{,}000 \times 0.036 = 5.655 \text{ k}\Omega$$

The current through the inductor is

$$I = \frac{V_L}{X_L} = \frac{28\underline{/0°}}{5.655\underline{/90°} \text{ k}\Omega} = 4.951\underline{/-90°} \text{ mA}$$

We can now compute the voltage drop across the resistor:

$$V_R = IR = (4.951\underline{/-90°})(4.7\underline{/0°})$$
$$= 23.27\underline{/-90°} \text{ V}$$

PROBLEMS

24-79.   An 8-V 3-kHz supply is in series with a 9.8-kΩ resistance and an 880-mH inductor. Compute the total current and the voltage drops across each of the elements. Draw phasor diagrams for the result.

24-80.   Repeat Prob. 24-79 for 300 Hz.

24-81.   Repeat Prob. 24-79 for 30 Hz.

24-82.   Repeat Prob. 24-79 for 30 kHz.

24-83. A 300-Hz voltage supply is in series with a 26-mH coil and an 80.0-Ω resistor. If the resistor has 3 V across it, what is the supply voltage?

24-84. A 4.5-MHz voltage source is in series with a 20-μH coil and a 3.9-kΩ resistor. If the coil drops 0.63 V, what is the voltage across the resistor?

24-85. A 3-kHz supply delivers 20/0° V to a series *RL* circuit. What are the values of resistance and inductance if the current is 5/−30° mA?

24-86. A 6-kHz supply delivers 30/0° V to a series *RL* circuit. If the voltage across the inductor is 25/17° V, what is the voltage across the resistor?

25

CAPACITANCE IN SERIES *RC* CIRCUITS

The capacitor and the inductor complement one another, for the voltage leads the current in an inductive circuit, whereas it lags the current in a capacitive circuit. This chapter, therefore, complements Chapter 24. Here, we first examine capacitors when they are connected in series and parallel. We then investigate how capacitors and resistors can be used to form timing circuits. Finally, we examine the capacitor in the sinusoidal environment.

25-1 CAPACITORS IN SERIES AND IN PARALLEL

A capacitor is formed from two conductors separated by an insulator. The capacitor is capable of storing an electrical charge in its dielectric (the insulator), enabling this energy to be recovered later. It can best be defined as:

$$Q = CV$$

where Q is the charge stored in the dielectric in coulombs, C is the value of the capacitor in farads, and V is the voltage across the capacitor. Let us now assume that the charge across the capacitor is increasing at a rate of Q coulombs per second. We should also expect the voltage to increase by a rate of V volts per second. We can express this as

$$\frac{dQ}{dt} = C \frac{dV}{dt}$$

This equation states that the increase of coulombs per unit time is equal to the capacitance multiplied by the increase of voltage per unit time. However, we recognize that dQ/dt is defined as current. Therefore, the fundamental definition of capacitance is

$$i = C \frac{dV}{dt}$$

where i is the instantaneous current, C the capacitance in farads, and dV/dt the rate of change in voltage per unit time.

EXAMPLE 25-1 What is the current through a 100-μF capacitor if the voltage is steadily increasing at a rate of 5 V every 30 ms?

SOLUTION

$$i = C\frac{dV}{dt} = 100 \times 10^{-6} \times \frac{5}{0.030} = 16.67 \text{ mA}$$

Note that the term dV/dt represents the increase (or decrease) of voltage per unit time. In this example

$$\frac{dV}{dt} = \frac{5}{0.030} = 166.7 \text{ V/s}$$

The voltage is increasing at a rate of 166.7 V/s.

Capacitors in Parallel

Capacitors connected in parallel are additive:

$$C_t = C_1 + C_2 + C_3 + \ldots$$

This seems reasonable for, by connecting them in parallel, we are increasing plate area, and the larger the plate area, the larger the capacitance.

EXAMPLE 25-2 Capacitors of 3 μF, 5 μF, and 100 μF are connected in parallel. What is the total capacitance?

SOLUTION

$$\begin{aligned} C_t &= C_1 + C_2 + C_3 \\ &= 3 + 5 + 100 \\ &= 108 \ \mu\text{F} \end{aligned}$$

Note that we did all our work in microfarads. Figure 25-1 illustrates the circuit.

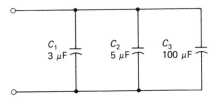

C_1 3 μF C_2 5 μF C_3 100 μF

FIGURE 25-1 Example 25-2, capacitors in parallel.

EXAMPLE 25-3 Capacitors of 1.5 μF, 0.015 μF, 0.47 μF, and 2000 pF are connected in parallel. What is the total capacitance?

SOLUTION We shall use farads as our units:

$$\begin{aligned} C_t &= C_1 + C_2 + C_3 + C_4 \\ &= 1.5 \times 10^{-6} + 0.015 \times 10^{-6} + 0.47 \times 10^{-6} + 2000 \times 10^{-12} \\ &= 1.987 \times 10^{-6} \\ &= 1.987 \ \mu\text{F} \end{aligned}$$

Capacitors in Series

Capacitors connected in series obey an inverse equation:

$$\frac{1}{C_t} = \frac{1}{C_1} + \frac{1}{C_2} + \frac{1}{C_3} + \ldots$$

Note that capacitors and resistors are opposites as far as these equations are concerned. Capacitors in series obey an inverse law, whereas resistors in series are additive. Capacitors in parallel are additive, whereas resistors in series obey the inverse law.

EXAMPLE 25-4 Capacitors of 10 μF, 1.5 μF, and 4.7 μF are connected in parallel (Fig. 25-2). What is the total capacitance?

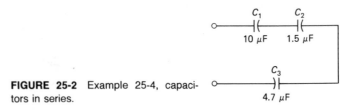

FIGURE 25-2 Example 25-4, capacitors in series.

SOLUTION

$$\frac{1}{C_t} = \frac{1}{C_1} + \frac{1}{C_2} + \frac{1}{C_3}$$
$$= \frac{1}{10} + \frac{1}{1.5} + \frac{1}{4.7}$$
$$C_t = 1.021 \ \mu F$$

EXAMPLE 25-5 Capacitors of 0.47 μF, 0.068 μF, and 2000 pF are connected in parallel. What is the total capacitance?

SOLUTION

$$\frac{1}{C_t} = \frac{1}{C_1} + \frac{1}{C_2} + \frac{1}{C_3}$$
$$= \frac{1}{0.47} + \frac{1}{0.068} + \frac{1}{0.002}$$
$$C_t = 1935 \ pF$$

PROBLEMS

Compute total capacitance of the following parallel-connected capacitors.

25-1. 10 μF, 15 μF, 50 μF

25-2. 100 μF, 200 μF, 800 μF

25-3. 0.47 μF, 1 μF, 0.1 μF

25-4. 0.047 μF, 0.001 μF, 5000 pF, 800 pF

25-5. 1.5 μF, 0.47 μF, 0.68 μF, 15,000 pF

25-6. 0.08 μF, 0.15 μF, 5000 pF, 1800 pF

Compute total capacitance of the following series-connected capacitors:

25-7. 10 μF, 15 μF, 50 μF

25-8. 100 μF, 200 μF, 800 μF

25-9. 0.47 μF, 1 μF, 0.1 μF

25-10. 0.047 μF, 0.001 μF, 5000 pF, 800 pF

25-11. 1.5 μF, 0.47 μF, 0.68 μF, 15,000 pF

25-12. 0.08 μF, 0.15 μF, 5000 pF, 1800 pF

One of the most commonly used circuits for timing is a capacitor charging through a resistor [Fig. 25-3(a)]. Assume that the capacitor is fully discharged. Next, close

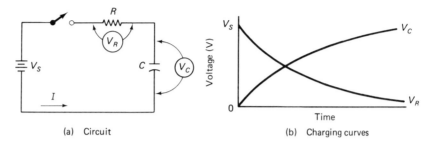

(a) Circuit   (b) Charging curves

FIGURE 25-3 RC timing circuit.

the switch. The voltages will follow the curves shown in Fig. 25-3(b). These curves have the following equations:

$$V_C = V_s(1 - \epsilon^{-t/RC}) \tag{25-1}$$
$$V_R = V_s\epsilon^{-t/RC} \tag{25-2}$$
$$I = I_M\epsilon^{-t/RC} \tag{25-3}$$

where V_C is the voltage across the capacitor in volts, V_s the supply voltage in volts, ϵ the constant 2.718281828, t the time measured in seconds, R the resistance in ohms, C the capacitance in farads, V_R the voltage across the resistor in volts, I the current at time t in amperes, and I_M the maximum current. This maximum current occurs at time 0 (when the switch is closed) and can be calculated as V_s/R.

These equations have several interesting characteristics:

1. The sum of V_C and V_R is the supply voltage.

$$\begin{aligned}V_C + V_R &= V_s(1 - \epsilon^{-t/RC}) + V_s\epsilon^{-t/RC} \\ &= V_s - V_s\epsilon^{-t/RC} + V_s\epsilon^{-t/RC} \\ &= V_s\end{aligned}$$

Therefore, we can find V_R and subtract this value from V_s to find V_C.
2. V_C is 0 and V_R is V_s at time 0.

$$\begin{aligned}V_R &= V_s\epsilon^{-0/RC} = V_s\epsilon^0 = V_s \\ V_C &= V_s - V_R = V_s - V_s = 0 \text{ V}\end{aligned}$$

3. V_C is V_s volts and V_R is 0 V when t is infinite.

$$\begin{aligned}V_R &= V_s\epsilon^{-\text{large}/RC} = V_s\epsilon^{-\text{large}} = V_s(0) = 0 \\ V_C &= V_s - V_R = V_s - 0 = V_s\end{aligned}$$

These facts are also apparent from Fig. 25-3(b).

EXAMPLE 25-6 If R is 2 MΩ, C is 0.47 μF, and V_s is 20 V, compute the voltages across the resistor and capacitor and the current at a time:

(a) 50 μs after closing the switch

(b) 50 ms after closing the switch

(c) 50 s after closing the switch

SOLUTION

(a) $V_R = V_s \epsilon^{-t/RC}$

$\qquad = 20 \times \epsilon^{-(50\times10^{-6})/(2\times10^6\times0.47\times10^{-6})}$

$\qquad = 19.99893620 \text{ V}$

$V_C = V_s - V_R$

$\qquad = 20 - 19.99893620$

$\qquad = 1.0638 \text{ mV}$

$I = I_m \epsilon^{-t/RC} = \dfrac{V_S}{R} \epsilon^{-t/RC}$

$\qquad = \dfrac{20}{2 \times 10^6} \epsilon^{-(50\times10^{-6})/(2\times10^6\times0.47\times10^{-6})}$

$\qquad = 9.999468100 \ \mu A$

(b) This time, to make things easier, let us calculate the term $-t/RC$.

$$-\frac{t}{RC} = \frac{-50 \times 10^{-3}}{2 \times 10^6 \times 0.47 \times 10^{-6}} = -53.19 \times 10^{-3}$$

$\qquad V_R = V_s \epsilon^{-t/RC}$

$\qquad\qquad = 20 \times \epsilon^{-53.19\times10^{-3}}$

$\qquad\qquad = 18.96 \text{ V}$

$\qquad V_C = V_s - V_R$

$\qquad\qquad = 20 - 18.96$

$\qquad\qquad = 1.04 \text{ V}$

$$I = \frac{V_s}{R} \epsilon^{-t/RC}$$

$\qquad\qquad = \dfrac{20}{2 \times 10^6} \epsilon^{-53.19\times10^{-3}}$

$\qquad\qquad = 9.482 \ \mu A$

(c) $-\dfrac{t}{RC} = \dfrac{-50}{2 \times 10^6 \times 0.47 \times 10^{-6}} = -53.1915$

$\qquad V_R = V_s \epsilon^{-t/RC}$

$\qquad\qquad = 20\epsilon^{-53.1915}$

$\qquad\qquad = 158.6 \times 10^{-24} \text{ V}$

$\qquad V_C = V_s - V_R$

$\qquad\qquad = 20 - (158.6 \times 10^{-24})$

$\qquad\qquad = 20.00 \text{ V}$

$$I = \frac{V_s}{R} \epsilon^{-t/RC}$$

$\qquad\qquad = \dfrac{20}{2 \times 10^6} \epsilon^{-53.1915}$

$\qquad\qquad = 79.29 \times 10^{-30} \text{ A}$

Thus, the voltage across the resistor is essentially 0, the current is 0, and the capacitor has been charged to 20 V.

EXAMPLE 25-7 A resistor must be placed in series with a 0.015-μF capacitor so that the voltage across the capacitor will be 1 V 30 ms after closing the switch. Find the value of the resistor if the supply voltage is 5 V.

SOLUTION

$$V_C = V_s(1 - \epsilon^{-t/RC})$$
$$1 = 5(1 - \epsilon^{-(30 \times 10^{-3})/(R \times 0.015 \times 10^{-6})})$$
$$\frac{1}{5} = 1 - \epsilon^{-(30 \times 10^{-3})/(R \times 0.015 \times 10^{-6})}$$
$$0.8 = \epsilon^{-(30 \times 10^{-3})/(R \times 0.015 \times 10^{-6})}$$

We now shall take the $\log_\epsilon$ of both sides. The $\log_\epsilon$ of the right side is the exponent itself.

$$-0.223144 = \frac{-30 \times 10^{-3}}{R \times 0.015 \times 10^{-6}}$$
$$R = 8.963 \text{ M}\Omega$$

EXAMPLE 25-8 A 2000-pF capacitor is in series with a 47-kΩ resistor. How long will it take to charge the capacitor to 10 V after applying 25 V to the series circuit?

SOLUTION

$$V_C = V_s(1 - \epsilon^{-t/RC})$$
$$10 = 25(1 - \epsilon^{-t/(47 \times 10^3 \times 2000 \times 10^{-12})})$$
$$0.6 = \epsilon^{-t/(47 \times 10^3 \times 2000 \times 10^{-12})}$$

Take the $\log_\epsilon$ of both sides:

$$-0.51083 = \frac{-t}{47 \times 10^3 \times 2000 \times 10^{-12}}$$
$$t = 48.02 \text{ } \mu\text{s}$$

Discharge

Once a capacitor has been charged, it may be discharged by the circuit shown in Fig. 25-4(a). Note that the voltage across the capacitor and resistor are equal after closing the switch. The equations for voltage and current are

$$V = V_s\epsilon^{-t/RC}$$
$$I = \frac{V_s}{R}\epsilon^{-t/RC}$$

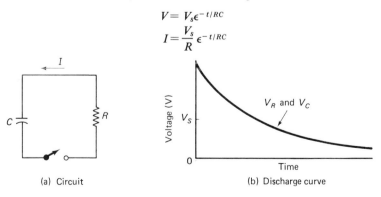

(a) Circuit   (b) Discharge curve

FIGURE 25-4 RC discharge.

EXAMPLE 25-9 A 1-μF capacitor has been charged to 25 V. If R is 390 kΩ, what will the voltage and current of the capacitor be 200 ms after closing the switch [Fig. 25-4(a)]?

SOLUTION

$$V = V_s \epsilon^{-t/RC}$$
$$= 25 \times \epsilon^{-0.2/(390 \times 10^3 \times 1 \times 10^{-6})}$$
$$= 14.97 \text{ V}$$

Since the instantaneous voltage is 14.97 V, the instantaneous current is this value divided by the resistance:

$$I = \frac{V}{R} = \frac{14.97}{390,000} = 38.38 \ \mu\text{A}$$

PROBLEMS

25-13. A discharged 100-μF capacitor is in series with a 25-kΩ resistor. Compute the resistor voltage, capacitor voltage, and current **(a)** 1 s, **(b)** 10 s, and **(c)** 100 s after applying 30 V to the circuit.

25-14. A discharged 0.02-μF capacitor is in series with a 470-kΩ resistor. Compute the resistor voltage, capacitor voltage, and current **(a)** 1 ms, **(b)** 10 ms, and **(c)** 100 ms after applying 100 V to the circuit.

25-15. Repeat Prob. 25-14 for a 0.05-μF capacitor.

25-16. Repeat Prob. 25-13 for a 47-μF capacitor.

25-17. What value of resistor must be placed in series with a 0.83-μF capacitor to cause it to charge to 20 V from a 30-V supply in 60 μs?

25-18. Repeat Prob. 25-17 for 300 ms.

25-19. What is the time required to charge a 0.1-μF capacitor through 27 kΩ to 30 V from a 60-V supply?

25-20. Repeat Prob. 25-19 for a 160-V supply.

25-21. A 2.7-μF capacitor has been charged to 35 V. What will its current be 400 ms after placing a 120-kΩ resistor in parallel with it?

25-22. Select a resistor to discharge a 100-μF capacitor from 28 V to 20 V in 100 ms.

25-3 CAPACITIVE REACTANCES IN SERIES AND IN PARALLEL

Capacitive reactance can be defined as the opposition a capacitor offers to a sinusoidal waveform. It is expressed by the equation

$$X_C = \frac{1}{2\pi fC}$$

where X_C is capacitive reactance in ohms, f is frequency in hertz, and C is capacitance in farads. Capacitive reactance has a phase angle of $-90°$.

EXAMPLE 25-10 Compute the capacitive reactance of a 0.47-μF capacitor at:

(1) 100 Hz (b) 100 kHz (c) 100 MHz

SOLUTION

(a) $X_C = \dfrac{1}{2\pi fC} = \dfrac{1}{2 \times \pi \times 100 \times 0.47 \times 10^{-6}}$

$= 3.386\underline{/-90°}\ k\Omega$

The angle was assigned as $-90°$ because this is capacitive.

(b) $X_C = \dfrac{1}{2\pi fC} = \dfrac{1}{2 \times \pi \times 100 \times 10^3 \times 0.47 \times 10^{-6}}$

$= 3.386\underline{/-90°}\ \Omega$

(c) $X_C = \dfrac{1}{2\pi fC} = \dfrac{1}{2 \times \pi \times 100 \times 10^6 \times 0.47 \times 10^{-6}}$

$= 3.386\underline{/-90°}\ m\Omega$

Note that capacitive reactance goes down as frequency goes up.

We can apply Ohm's law principles to capacitive reactance.

EXAMPLE 25-11 A 0.002-μF capacitor is across a 20-V 400-Hz source. What is the current?

SOLUTION We must first compute the reactance:

$$X_C = \frac{1}{2\pi fC}$$

$$= \frac{1}{2 \times \pi \times 400 \times 0.002 \times 10^{-6}}$$

$$= 198.9\underline{/-90°}\ k\Omega$$

We can now calculate current. We shall assume that the power supply has a phase angle of $0°$.

$$I = \frac{V}{X} = \frac{20\underline{/0°}}{198.9 \times 10^3\ \underline{/-90°}}$$

$$= 100.5\underline{/90°}\ \mu A$$

EXAMPLE 25-12 A 0.047-μF capacitor has $2\underline{/30°}$ mA of 1 kHz current flowing through it. What is the voltage across the capacitor?

SOLUTION

$$X_C = \frac{1}{2\pi fC} = \frac{1}{2 \times \pi \times 1000 \times 0.047 \times 10^{-6}}$$

$$= 3.386\underline{/-90°}\ k\Omega$$

$$V = IX_C$$

$$= (2 \times 10^{-3}\underline{/30°})(3386\underline{/-90°})$$

$$= 6.773\underline{/-60°}\ V$$

Capacitive Reactances in Series

Capacitive reactances in series are additive:

$$X_{Ct} = X_{C1} + X_{C2} + X_{C3} + \ldots$$

EXAMPLE 25-13 Capacitive reactances of 1 kΩ, 400 Ω, and 5 kΩ are in series. What is the total reactance?

SOLUTION

$$\begin{aligned} X_{Ct} &= X_{C1} + X_{C2} + X_{C3} \\ &= 1000 + 400 + 5000 \\ &= 6400\underline{/-90°} \ \Omega \end{aligned}$$

EXAMPLE 25-14 Capacitors of 2.4 μF, 1 μF, and 0.47 μF are in series with a 1-kΩ capacitive reactance at 3 kHz. What is the total reactance of the circuit?

SOLUTION We can find the reactance of the capacitors in two different ways:

1. Find the individual reactances.
2. Find the total capacitance, then the total reactance.

We shall use the latter method.

$$\frac{1}{C_t} = \frac{1}{C_1} + \frac{1}{C_2} + \frac{1}{C_3}$$
$$\frac{1}{C_t} = \frac{1}{2.4} + \frac{1}{1} + \frac{1}{0.47}$$
$$C_t = 0.2821 \ \mu F$$
$$X_C = \frac{1}{2 \pi f C} = \frac{1}{2 \times \pi \times 3000 \times 0.2821 \times 10^{-6}}$$
$$= 188.0 \ \Omega$$

We can now find the total reactance:

$$\begin{aligned} X_{Ct} &= X_{C1} + X_{C2} \\ &= 188.0 + 1000 \\ &= 1188\underline{/-90°} \ \Omega \end{aligned}$$

Capacitive Reactances in Parallel

Capacitive reactances in parallel obey the inverse rule:

$$\frac{1}{X_{Ct}} = \frac{1}{X_{C1}} + \frac{1}{X_{C2}} + \frac{1}{X_{C3}} + \ldots$$

EXAMPLE 25-15 Capacitive reactances of 1 kΩ and 400 Ω are in parallel with a 0.05-μF capacitor at 300 Hz. What is the total reactance?

SOLUTION

$$X_{C3} = \frac{1}{2 \pi f C} = \frac{1}{2 \times \pi \times 300 \times 0.05 \times 10^{-6}}$$
$$= 10.61\underline{/-90°} \ k\Omega$$

$$\frac{1}{X_{Ct}} = \frac{1}{X_{C1}} + \frac{1}{X_{C2}} + \frac{1}{X_{C3}}$$

$$= \frac{1}{1000} + \frac{1}{400} = \frac{1}{10,610}$$

$$X_{Ct} = 278.22 \underline{/-90°} \ \Omega$$

PROBLEMS

Compute the capacitive reactance:

25-23. (a) 3.5 μF, 100 Hz (b) 3.5 μF, 100 kHz

25-24. (a) 450 pF, 1 kHz (b) 450 pF, 1 MHz

25-25. (a) 0.05 μF, 1 Hz (b) 0.05 μF, 2Hz

25-26. (a) 0.05 μF, 3 Hz (b) 0.05 μF, 4Hz

25-27. A 4.7-μF capacitor is across a 60-Hz 120-V source. Compute the current through the capacitor.

25-28. Repeat Prob. 25-27 for 600 Hz.

25-29. A 50-pF capacitor is across a 1-V 1-MHz source. Compute the capacitor current.

25-30. Repeat Prob. 25-29 for 10 MHz.

25-31. A 2000-pF capacitor has 1 μA of 200 kHz current through it. Compute the capacitor voltage.

25-32. A 0.02-μF capacitor has 1.6 mA of 20 kHz current through it. Compute its voltage.

25-33. Capacitive reactances of 1 kΩ, 10 kΩ, and 5.5 kΩ are in series. Compute total reactance.

25-34. Capacitive reactances of 2 kΩ, 5 kΩ, and 500 Ω are in series. Compute total reactance.

25-35. Capacitive reactances of 750 Ω and 1.2 kΩ are in series with a 0.75-μF capacitor at 2.5 kHz. Compute the total reactance.

25-36. Capacitive reactances of 200 Ω, 500 Ω, and 800 Ω are in parallel. Compute the total reactance.

25-37. Capacitive reactances of 2 kΩ and 750 Ω are in parallel with a 0.1-μF capacitor at 100 Hz. Compute the total reactance.

25-38. Capacitive reactances of 100 Ω and 780 Ω are in parallel with a 3.5-μF capacitor and a 6.9-μF capacitor at 250 Hz. Compute the total reactance.

25-4 *SERIES* RC *CIRCUITS*

In our study of inductance, we learned that impedance is the phasor sum of reactance and resistance. This also applies to capacive reactance.

EXAMPLE 25-16 A 0.1-μF capacitor is in series with a 500-Ω resistor at 8 kHz. What is the total impedance?

SOLUTION The capacitive reactance is

$$X_C = \frac{1}{2\pi fC}$$

$$= \frac{1}{2 \times \pi \times 8000 \times 0.1 \times 10^{-6}}$$

$$= 198.9 \underline{/-90°} \ \Omega$$

We can now find the impedance.

$$Z_t = R - jX_C$$
$$= 500 - j198.9$$
$$= 538.1\underline{/-21.70°} \ \Omega$$

Note that since capacitive reactance has a phase angle of $-90°$, we designated it $-j$ in rectangular coordinates. The phasor diagram is shown in Fig. 25-5.

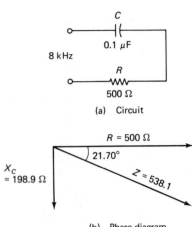

(a) Circuit

$$R = 500 \ \Omega$$
$$21.70°$$
$$X_C = 198.9 \ \Omega$$
$$Z = 538.1$$

FIGURE 25-5 Example 25-16, series RC impedance.

(b) Phase diagram

EXAMPLE 25-17 Capacitors of 0.05 μF and 0.1 μF are in series with resistors of 1400 Ω and 2100 Ω. Compute impedance for a frequency of 900 Hz.

SOLUTION

$$X_{C1} = \frac{1}{2\pi fC_1} = \frac{1}{2 \times \pi \times 900 \times 0.05 \times 10^{-6}}$$
$$= 3.537\underline{/-90°} \ k\Omega$$
$$X_{C2} = \frac{1}{2\pi fC_2} = \frac{1}{2 \times \pi \times 900 \times 0.1 \times 10^{-6}}$$
$$= 1.768\underline{/-90°} \ k\Omega$$

We can now compute the impedance.

$$Z = R_1 + R_2 - jX_{C1} - jX_{C2}$$
$$= 1400 + 2100 - j3537 - j1768$$
$$= 6.356\underline{/-56.59°} \ k\Omega$$

PROBLEMS

Compute the impedance of series circuits with the following elements:

25-39. $C = 0.7$ μF, $R = 900$ Ω, $f = 7500$ Hz
25-40. $C = 0.31$ μF, $R = 270$ Ω, $f = 10$ kHz
25-41. $C = 3500$ pF, $R = 1$ kΩ, $f = 80$ kHz
25-42. $C = 33$ pF, $R = 100$ kΩ, $f = 40$ kHz
25-43. $C_1 = 27$ μF, $C_2 = 43$ μF, $R_1 = 120$ Ω, $R_2 = 80$ Ω, $f = 10$ Hz
25-44. $C_1 = 0.05$ μF, $C_2 = 0.15$ μF, $R_1 = 8$ kΩ, $R_2 = 15$ kΩ, $R_3 = 7.5$ kΩ, $f = 120$ Hz

By applying the Ohm's law principles we learned for dc problems, we can solve ac problems. We have only to observe that all ac quantities are phasors.

EXAMPLE 25-18 A 510-Ω resistance is in series with a 0.08-μF capacitor and a 10-kHz 5-V supply. Compute the voltages across the capacitor and resistor [Fig. 25-6(a)].

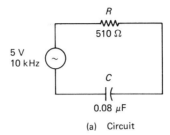

(a) Circuit

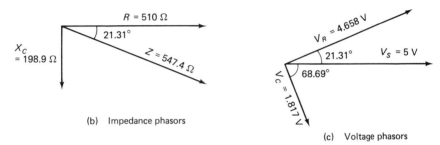

(b) Impedance phasors

(c) Voltage phasors

FIGURE 25-6 Example 25-18, series RC circuit.

SOLUTION We must first compute the reactance:

$$X_C = \frac{1}{2\pi f C}$$
$$= \frac{1}{2 \times \pi \times 10^4 \times 0.08 \times 10^{-6}}$$
$$= 198.9 \underline{/-90°}\ \Omega$$

We can now find the impedance:

$$Z = R - jX_C$$
$$= 510 - j198.9$$
$$= 547.4 \underline{/-21.31°}\ \Omega$$

The impedance phasor diagram is shown in Fig. 25-6(b). The total current can now be found:

$$I = \frac{V_s}{Z}$$
$$= \frac{5 \underline{/0°}}{547.4 \underline{/-21.31°}}$$
$$= 9.134 \underline{/21.31°}\ \text{mA}$$

We can now find the voltage drops.

$$V_R = IR$$
$$= (9.134\,\underline{/21.31°})(0.510\,\underline{/0°})$$
$$= 4.658\,\underline{/21.31°}\ \text{V}$$
$$V_C = IX_C$$
$$= (9.134\,\underline{/21.31°})(0.1989\,\underline{/-90°})$$
$$= 1.817\,\underline{/-68.69°}\ \text{V}$$

The voltage phasor diagram is shown in Fig. 25-6(c).

EXAMPLE 25-19 A series circuit consists of a 400-Hz supply, a 0.24-μF capacitor, a 0.67-μF capacitor, and a 1.8-kΩ resistor. If the current is $3\,\underline{/30°}$ mA, what is the supply voltage?

SOLUTION

$$X_{C1} = \frac{1}{2\,\pi fC}$$
$$= \frac{1}{2 \times \pi \times 400 \times 0.24 \times 10^{-6}}$$
$$= 1.658\,\underline{/-90°}\ \text{k}\Omega$$
$$X_{C2} = \frac{1}{2\,\pi fC}$$
$$= \frac{1}{2 \times \pi \times 400 \times 0.67 \times 10^{-6}}$$
$$= 593.9\,\underline{/-90°}\ \text{k}\Omega$$
$$Z = R - jX_{C1} - jX_{C2}$$
$$= 1800 - j1658 - j593.9$$
$$= 2.883\,\underline{/-51.36°}\ \text{k}\Omega$$

We can now find the total voltage.

$$V = IZ$$
$$(3\,\underline{/30°})(2.883\,\underline{/-51.36°})$$
$$= 8.648\,\underline{/-21.36°}\ \text{V}$$

PROBLEMS

25-45. An 860-Ω resistor is in series with a 0.15-μF capacitor and a 10-V 1200-Hz supply. Compute the voltage across the capacitor and resistor.

25-46. A 370-Ω resistor is in series with a 0.015-μF capacitor and an 18-kHz 3-V supply. Compute the voltage across the resistor and the capacitor.

25-47. A resistance of 4 kΩ is in series with an 800-pF capacitor at 25 kHz. If the current is $30\,\underline{/40°}$ μA, compute the supply voltage.

25-48. A 3.9-kΩ resistor, a 5.1-kΩ resistor, a 0.01-μF capacitor, and a 0.02-μF capacitor are in series. If the voltage across the 0.01-μF capacitor is $3\,\underline{/10°}$ V at 1000 Hz, compute the supply voltage.

26

RLC CIRCUITS

The *RLC* circuit has resistance, inductance, and capacitance. In this chapter we first examine series *RLC* circuits, then parallel *RLC* circuits, and finally, circuits having both series and parallel elements. We then examine a special type of *RLC* circuit, the resonant circuit.

26-1 SERIES CIRCUITS

Those circuits having inductance, capacitance, and resistance in series have impedance phasors of between $-90°$ and $+90°$. Note that inductive reactance cancels some of the capacitive reactance in the following example.

EXAMPLE 26-1 An inductive reactance of 2 kΩ is in series with a capacitive reactance of 1500 Ω and a resistance of 500 Ω. Compute the total impedance (Fig. 26-1).

SOLUTION We have only to phasor-add the quantities, bearing in mind that inductive reactance has a phase angle of $+90°$ *(+j)*, resistance has a phase angle of $0°$ (real number), and capacitive reactance has a phase angle of $-90°$ *(−j)*.

$$Z = R + jX_L - jX_C$$
$$= 500 + j2000 - j1500$$
$$= 500 + j500$$
$$= 707.1 \underline{/45°} \ \Omega$$

EXAMPLE 26-2 A 0.35-H inductor, a 0.05-μF capacitor, and a 1.2-kΩ resistor are in series. What is the impedance at 1 kHz?

SOLUTION

We must first compute the reactances:

$$X_L = 2 \pi fL$$
$$= 2 \times \pi \times 1000 \times 0.35$$
$$= 2.199 \ k\Omega$$
$$X_C = \frac{1}{2 \pi fC}$$
$$= \frac{1}{2 \times \pi \times 1000 \times 0.05 \times 10^{-6}}$$
$$= 3.183 \ k\Omega$$

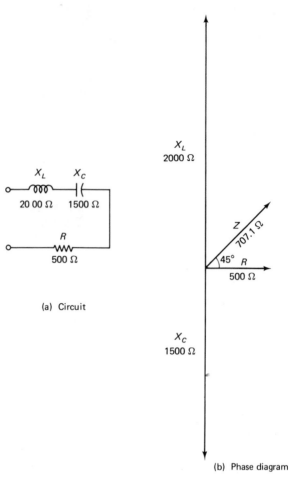

FIGURE 26-1 Example 26-1, series RLC circuit.

(a) Circuit

(b) Phase diagram

We can now find the impedance.

$$Z = R + jX_L - jX_C$$
$$= 1200 + j2199 - j3183$$
$$= 1200 - j984.0$$
$$= 1.552\underline{/-39.35°} \text{ k}\Omega$$

By applying the principles of Ohm's law, we can compute currents, voltages, and resistances. However, all quantities are phasors.

EXAMPLE 26-3 A series circuit consists of a 0.1-μF capacitor, a 0.5-H inductor, a 1.8-kΩ resistor, and an 800-Hz 12-V supply. Compute the voltage across each of the elements [Fig. 26-2(a)].

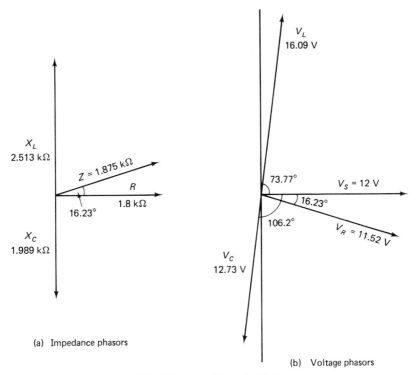

(a) Impedance phasors

(b) Voltage phasors

FIGURE 26-2 Example 26-3.

SOLUTION Computing the reactances:

$$X_C = \frac{1}{2\pi fC}$$

$$= \frac{1}{2 \times \pi \times 800 \times 0.1 \times 10^{-6}}$$

$$= 1.989 \text{ k}\Omega$$

$$X_L = 2\pi fL$$

$$= 2 \times \pi \times 800 \times 0.5$$

$$= 2.513 \text{ k}\Omega$$

We can now compute the impedance.

$$Z = R + jX_L - jX_C$$

$$= 1800 + j2513 - j1989$$

$$= 1800 + j523.8$$

$$= 1875\underline{/16.23°} \ \Omega$$

Find the total current:

$$I = \frac{V}{Z} = \frac{12\underline{/0°}}{1875\underline{/16.23°}} = 6.401\underline{/-16.23°} \text{ mA}$$

Compute the voltage drops across each of the elements:

$$V_R = IR$$
$$= (6.401\underline{/-16.23°})(1.8\underline{/0°})$$
$$= 11.52\underline{/-16.23°} \text{ V}$$
$$V_C = IX_C$$
$$= (6.401\underline{/-16.23°})(1.989\underline{/-90°})$$
$$= 12.73\underline{/-106.2°} \text{ V}$$
$$V_L = IX_L$$
$$= (6.401\underline{/-16.23°})(2.513\underline{/90°})$$
$$= 16.09\underline{/73.77°} \text{ V}$$

Note that V_L exceeds V_s. The phasor diagram is shown in Fig. 26-2(b).

PROBLEMS

26-1. An inductive reactance of 12 kΩ is in series with a capacitive reactance of 5 kΩ and a resistance of 7 kΩ. Compute the impedance.

26-2. An inductive reactance of 3.2 MΩ, a capacitive reactance of 4.6 MΩ, and a resistance of 5 MΩ are in series. Compute the impedance.

26-3. If $X_L = 260$ Ω, $X_C = 440$ Ω, and $R = 300$ Ω, compute the impedance for this series combination.

26-4. If $X_L = 4.2$ kΩ, $X_C = 260$ Ω, and $R = 4300$ Ω, compute the impedance for this series combination.

26-5. A 350-mH coil, a 0.35-μF capacitor, and a 500-Ω resistor are in series at 1 kHz. Compute the impedance.

26-6. Repeat Prob. 26-5 for $C = 20$ μF, $L = 30$ mH, $R = 300$ Ω, and $f = 410$ Hz.

26-7. Compute the voltage drops across each of the elements if 20 V at 350 Hz is applied to a series circuit consisting of a 5-μF capacitor, a 50-mH coil, and a 110-Ω resistor.

26-8. Compute the voltage drops across the following series elements connected across a 30-V 10-kHz supply: 0.05-μF capacitor, 7.5-mH coil, and 750-Ω resistor.

26-9. A series circuit consists of a 0.005-μF capacitor, a 50-mH coil, and a 2.7-kΩ resistor. If the capacitor has a voltage of $1.5\underline{/30°}$ V at 6 kHz across it, compute the supply voltage.

26-10. A series circuit consists of a 0.001-μF capacitor, a 370-μH inductor, and a 2.4-kΩ resistor across a 100-kHz supply. If the current is $8.5\underline{/-45°}$ μA, compute the supply voltage.

26-2 PARALLEL CIRCUITS

The approach to solving parallel ac *RLC* circuits is the same as that of dc circuits. The only difference is that all quantities are phasors.

EXAMPLE 26-4 A 3-kΩ inductive reactance, a 5-kΩ capacitive reactance, and a 5-kΩ resistor are in parallel. Compute the impedance (Fig. 26-3).

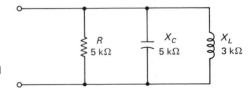

FIGURE 26-3 Example 26-4, parallel RLC circuit.

SOLUTION

$$\frac{1}{Z} = \frac{1}{R} + \frac{1}{-jX_C} + \frac{1}{jX_L} \qquad (26\text{-}1)$$

$$= \frac{1}{5\underline{/0^\circ}} + \frac{1}{5\underline{/-90^\circ}} + \frac{1}{3\underline{/90^\circ}}$$

$$= 0.2\underline{/0^\circ} + 0.2\underline{/90^\circ} + 0.3333\underline{/-90^\circ}$$

$$= 0.2 + j0.2 - j0.3333$$

$$= 0.2 - j0.1333$$

$$= 0.2404\underline{/-33.69^\circ}$$

$$Z = \frac{1}{0.2404\underline{/-33.69^\circ}} = 4.160\underline{/33.69^\circ} \text{ k}\Omega$$

Example 26-4, although correct, was a little awkward because of all the inverting that had to be done. We can simplify this procedure by defining the terms shown in Table 26-1. We now can develop a new inverse equation.

TABLE 26-1   Definitions of inverse AC quantities.

Fraction	Symbol	Name	Unit of Measure
$\frac{1}{Z}$	Y	Admittance	Siemens (S)
$\frac{1}{R}$	G	Conductance	Siemens (S)
$\frac{1}{X}$	B	Susceptance	Siemens (S)

Starting with Eq. (26-1), we can move the j's to the numerators by multiplying both numerator and denominator by j:

$$\frac{1}{Z} = \frac{1}{R} + \frac{1}{-jX_C} + \frac{1}{jX_L}$$

$$= \frac{1}{R} + \frac{(j)}{(j)} \cdot \frac{1}{(-j)X_C} + \frac{j}{j} \cdot \frac{1}{jX_L}$$

$$= \frac{1}{R} + j\frac{1}{X_C} - j\frac{1}{X_L}$$

Note that $j^2 = -1$. Substitute the inverse symbols:

$$Y = G - jB_L + jB_C$$

where B_L is inductive susceptance and B_C is capacitive susceptance.

EXAMPLE 26-5   A 5-kΩ inductive reactance, a 2-kΩ capacitive reactance, and a 4-kΩ resistance are in parallel. Compute the impedance.

SOLUTION

$$Y = G - jB_L + jB_C$$

$$= \frac{1}{4} - j\left(\frac{1}{5}\right) + j\left(\frac{1}{2}\right)$$

$$= 0.25 - j0.2 + j0.5$$
$$= 0.25 + j0.3$$
$$= 0.3905 \underline{/50.19°}$$
$$Z = \frac{1}{Y} = \frac{1}{0.3905 \underline{/50.19°}} = 2.561 \underline{/-50.19°} \text{ k}\Omega$$

These parallel techniques also apply to problems involving current and voltage.

EXAMPLE 26-6 A parallel circuit has a 0.01-μF capacitor, a 70-mH coil, and a 10-kΩ resistor across a 2.5-V 15-kHz supply. Compute the total current and branch currents.

SOLUTION We must first compute the reactances:

$$X_C = \frac{1}{2 \pi fC} = \frac{1}{2 \times \pi \times 15,000 \times 0.01 \times 10^{-6}}$$
$$= 1.061 \text{ k}\Omega$$
$$X_L = 2 \pi fL = 2 \times \pi \times 15,000 \times 0.07 = 6.597 \text{ k}\Omega$$

We can now compute the admittance:

$$Y = G - jB_L + jB_C$$
$$= \frac{1}{10} - j\frac{1}{6.597} + j\frac{1}{1.061}$$
$$= 0.100 + j0.7909$$
$$= 0.7972 \underline{/82.79°}$$
$$Z = \frac{1}{Y} = \frac{1}{0.7972 \underline{/82.79°}} = 1.254 \underline{/-82.79°} \text{ k}\Omega$$

We can now find the current:

$$I = \frac{V}{Z} = \frac{2.5 \underline{/0°}}{1.254 \underline{/-82.79°}} = 1.993 \underline{/82.79°} \text{ mA}$$

We can also find the branch currents:

$$I_L = \frac{V}{X_L} = \frac{2.5 \underline{/0°}}{6.597 \underline{/90°}} = 378.9 \underline{/-90°} \text{ }\mu\text{A}$$
$$I_C = \frac{V}{X_C} = \frac{2.5 \underline{/0°}}{1.061 \underline{/-90°}} = 2.356 \underline{/90°} \text{ mA}$$
$$I_R = \frac{V}{R} = \frac{2.5 \underline{/0°}}{10 \underline{/0°}} = 250.0 \underline{/0°} \text{ }\mu\text{A}$$

EXAMPLE 26-7 Compute the branch currents of Fig. 26-4.

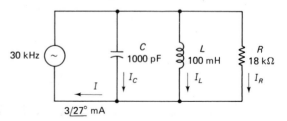

FIGURE 26-4 Example 26-7.

SOLUTION

$$X_L = 2\pi fL = 2 \times \pi \times 30 \times 10^3 \times 0.1 = 18.85\,\underline{/90°}\ \text{k}\Omega$$

$$X_C = \frac{1}{2\pi fC} = \frac{1}{2 \times \pi \times 30 \times 10^3 \times 1000 \times 10^{-12}} = 5.305\,\underline{/-90°}\ \text{k}\Omega$$

Compute the impedance:

$$\begin{aligned}
Y &= G - jB_L + jB_C \\
&= \frac{1}{18} - j\frac{1}{18.85} + \frac{1}{5.305} \\
&= 0.0556 + j0.1354 = 0.1464\,\underline{/67.70°} \\
Z &= \frac{1}{Y} = 6.831\,\underline{/-67.70°}\ \text{k}\Omega
\end{aligned}$$

The supply voltage is

$$\begin{aligned}
V_s &= IZ = (3.00\,\underline{/27°})(6.831\,\underline{/-67.70°}) \\
&= 20.49\,\underline{/-40.70°}\ \text{V}
\end{aligned}$$

We can now compute the branch currents.

$$I_C = \frac{V_s}{X_C} = \frac{20.49\,\underline{/-40.70°}}{5.305\,\underline{/-90°}} = 3.863\,\underline{/49.30°}\ \text{mA}$$

$$I_L = \frac{V_s}{X_L} = \frac{20.49\,\underline{/-40.70°}}{18.85\,\underline{/90°}} = 1.087\,\underline{/-130.7°}\ \text{mA}$$

$$I_R = \frac{V_s}{R} = \frac{20.49\,\underline{/-40.70°}}{18\,\underline{/0°}} = 1.138\,\underline{/-40.70°}\ \text{mA}$$

PROBLEMS

Compute the impedance of the following elements in parallel:

26-11. $X_L = 5$ kΩ, $X_C = 10$ kΩ, $R = 7$ kΩ

26-12. $X_L = 40$ kΩ, $X_C = 29$ kΩ, $R = 30$ kΩ

26-13. $C = 0.05$ μF, $L = 100$ mH, $R = 7$ kΩ, $f = 20$ kHz

26-14. $C = 3$ μF, $L = 270$ mH, $R = 500$ Ω, $f = 400$ Hz

Compute the branch currents of the following parallel-connected circuits:

26-15. $X_L = 7.8$ kΩ, $X_C = 12$ kΩ, $R = 6.8$ kΩ, $V_s = 30$ V

26-16. $X_L = 15$ kΩ, $X_C = 25$ kΩ, $R = 45$ kΩ, $V_s = 10$ V

26-17. $C = 0.02$ μF, $L = 60$ mH, $R = 4.7$ kΩ, $V_s = 5$ V, $f = 15$ kHz

26-18. $C = 20$ μF, $L = 10$ H, $R = 30$ Ω, $V_s = 20$ V, $f = 60$ Hz

26-19. A circuit similar to Fig. 26-4 has $C = 0.02$ μF, $L = 750$ mH, $R = 1$ kΩ, and $f = 1$ kHz. If the total current, I_t, is $47\,\underline{/0°}$ mA, compute the branch currents.

26-20. A circuit similar to Fig. 26-4 has $C = 0.08$ μF, $L = 500$ mH, $R = 5$ kΩ, and $f = 800$ Hz. If the current through the capacitor is $5\,\underline{/0°}$ mA, compute the total current.

26-3 SERIES–PARALLEL CIRCUITS

The analysis of series–parallel circuits uses a combination of series analysis techniques and parallel analysis techniques. The following example demonstrates this.

EXAMPLE 26-8 Compute the supply current of Fig. 26-5.

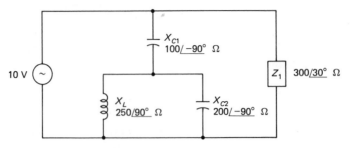

FIGURE 26-5 Example 26-8.

SOLUTION Note that Z_1 is a complex impedance in parallel with the rest of the network. We shall find the total impedance and then the total current.

$$X_{EQ1} = X_L \| X_{C2}$$
$$= \frac{(250 \underline{/90°})(200 \underline{/-90°})}{250 \underline{/90°} + 200 \underline{/-90°}}$$
$$= 1000 \underline{/-90°}$$

But this is in series with X_{C1}:

$$X_{EQ2} = X_{EQ1} + X_{C1}$$
$$= 1000 \underline{/-90°} + 100 \underline{/-90°}$$
$$= 1100 \underline{/-90°}$$

We can now parallel Z_1:

$$Z_t = X_{EQ2} \| Z_1$$
$$= \frac{(1100 \underline{/-90°})(300 \underline{/30°})}{1100 \underline{/-90°} + 300 \underline{/30°}}$$
$$= \frac{330,000 \underline{/-60°}}{-j1100 + 259.8 + j150}$$
$$= \frac{330,000 \underline{/-60°}}{259.8 - j950}$$
$$= \frac{330,000 \underline{/-60°}}{984.9 \underline{/-74.70°}}$$
$$= 335.1 \underline{/14.70°}$$

We can now compute the current:

$$I = \frac{V}{Z_t} = \frac{10 \underline{/0°}}{335.1 \underline{/14.70°}} = 29.85 \underline{/-14.70°} \text{ mA}$$

All the dc network theorems also apply to ac as long as phasors are used. The following example demonstrates the use of Thévenin's theorem as applied to ac.

EXAMPLE 26-9 Compute the resistor current of Fig. 26-6.

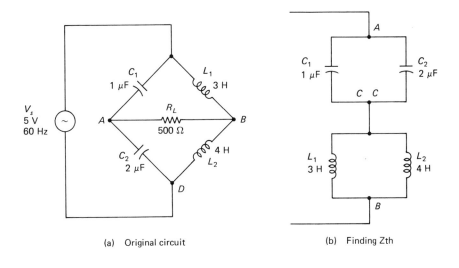

(a) Original circuit

(b) Finding Zth

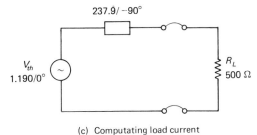

(c) Computating load current

FIGURE 26-6 Example 26-9.

SOLUTION We shall use Thévenin's Theorem. Discarding the resistor, the voltage at A can be computed:

$$X_{C1} = \frac{1}{2\pi fC} = \frac{1}{2 \times \pi \times 60 \times 10^{-6}} = 2.653\,\underline{/-90°}\text{ k}\Omega$$

$$X_{C2} = \frac{1}{2\pi fC} = \frac{1}{2 \times \pi \times 60 \times 2 \times 10^{-6}} = 1.326\,\underline{/-90°}\text{ k}\Omega$$

The left branch impedance is

$$Z_{LH} = X_{C1} + X_{C2} = 3.979\,\underline{/-90°}\text{ k}\Omega$$

The left branch current is

$$I = \frac{V}{Z_{LH}} = \frac{5\,\underline{/0°}}{3.979\,\underline{/-90°}} = 1.257\,\underline{/90°}\text{ mA}$$

The voltage drop across C_2 is, therefore,

$$V_A = IX_C = (1.257\,\underline{/90°})(1.326\,\underline{/-90°})$$
$$= 1.667\,\underline{/0°}\text{ V}$$

Note that this is reasonable, considering the capacitor values. Compute the voltage to B:

$$X_{L1} = 2 \pi fL = 2 \times \pi \times 60 \times 3 = 1.131\underline{/90°} \text{ k}\Omega$$
$$X_{L2} = 2 \pi fL = 2 \times \pi \times 60 \times 4 = 1.508\underline{/90°} \text{ k}\Omega$$

The voltage at V is

$$V_B = \frac{X_{L2}}{L_{L1} + X_{L2}} \times V_s$$
$$= \frac{1.508\underline{/90°}}{1.131\underline{/90°} + 1.508\underline{/90°}} \times 5\underline{/0°}$$
$$= 2.857\underline{/0°} \text{ V}$$

The thevenized voltage is the difference between A and B:

$$V_{th} = V_B - V_A$$
$$= 2.857\underline{/0°} - 1.667\underline{/0°}$$
$$= 1.190\underline{/0°}$$

We must now find Z_{th}. We can do this by recognizing that the circuit becomes as shown in Fig. 26-6(b).

$$C_t = C_1 + C_2 = 1 + 2 = 3 \ \mu\text{F}$$
$$L_t = L_1 \| L_2 = \frac{3 \times 4}{3 + 4} = 1.714 \text{ H}$$
$$X_{Ct} = \frac{1}{2 \pi fC_t} = \frac{1}{2 \times \pi \times 60 \times 3 \times 10^{-6}}$$
$$= 884.2\underline{/-90°}$$
$$X_{Lt} = 2 \pi fL = 2 \times \pi \times 60 \times 1.714$$
$$= 646.27\underline{/90°} \ \Omega$$
$$Z_{th} = X_{Ct} + X_{Lt}$$
$$= 884.2\underline{/-90°} + 646.27\underline{/90°}$$
$$= 237.9\underline{/-90°} \ \Omega$$

We can now replace the load and figure its current [Fig. 26-6(c)].

$$I = \frac{V_{th}}{Z_{th} + R_L} = \frac{1.190\underline{/0°}}{500 - j237.9} = 2.150\underline{/25.45°} \text{ mA}$$

$$V_L = IR_L = (2.150\underline{/25.45°})(0.500\underline{/0°})$$
$$= 1.075\underline{/25.45°} \text{ V}$$

PROBLEMS

26-21. In a circuit similar to Fig. 26-5, $X_{C1} = 2000 \ \Omega$, $X_{C2} = 5000 \ \Omega$, $X_L = 1000 \ \Omega$, and $Z_1 = 4000 - j2000$, and $V_s = 26$ V. Compute the total current.

26-22. Compute the supply current of Fig. 26-7.

26-23. If $f = 600$ Hz, compute the current through the 400-Ω resistor of Fig. 26-7.

26-24. Compute the current through R_2 of Fig. 26-8.

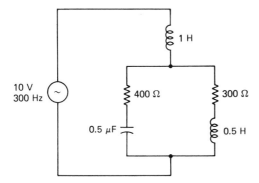

FIGURE 26-7 Problem 26-22.

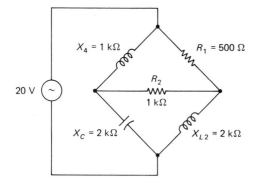

FIGURE 26-8 Problem 26-24.

26-4 SERIES RESONANCE

Series resonance can be defined as a condition of a series RLC circuit at which the circuit acts as a pure resistance. Thus, it is a point at which X_L and X_C cancel each other, leaving only the resistance. We can show this mathematically as

$$Z = R + jX_L - jX_C$$

Therefore, for $Z = R$,

$$X_L = X_C$$

$$2\pi fL = \frac{1}{2\pi fC}$$

Solve for f:

$$f = \frac{1}{2\pi\sqrt{LC}}$$

This equation will tell us at what frequency X_L and X_C are equal.

EXAMPLE 26-10 A circuit has a 0.01-μF capacitor, a 50-mH indicator, and a 300-Ω resistor in series at the resonant frequency. If 5 V at the resonant frequency is placed across the circuit, compute the current and the voltage drops across the inductor, capacitor, and resistor.

SOLUTION The resonant frequency is found by

$$f = \frac{1}{2\pi\sqrt{LC}} = \frac{1}{2 \times \pi \times \sqrt{0.05 \times 0.01 \times 10^{-6}}}$$
$$= 7.118 \text{ kHz}$$

We can now find the reactance of the capacitor and inductor:

$$X_L = 2\pi fL = 2 \times \pi \times 7118 \times 0.05$$
$$= 2.236 \text{ k}\Omega$$

$$X_C = \frac{1}{2\pi fC} = \frac{1}{2 \times \pi \times 7118 \times 0.01 \times 10^{-6}}$$
$$= 2.236 \text{ k}\Omega$$

Note that the impedance becomes

$$Z = R + jX_L - jX_C$$
$$= 300 + j2236 - j2236$$
$$= 300 + j0$$

Therefore, it is purely resistive. We can now find the current:

$$I = \frac{V}{Z} = \frac{5\underline{/0°}}{300\underline{/0°}} = 16.67\underline{/0°} \text{ mA}$$

The various voltage drops are

$$V_R = IR = (16.67\underline{/0°})(0.300\underline{/0°})$$
$$= 5.000\underline{/0°} \text{ V}$$

Note that the voltage across the resistance is equal to the applied voltage.

$$V_L = IX_L$$
$$= (16.67\underline{/0°})(2.236\underline{/90°})$$
$$= 37.27\underline{/90°} \text{ V}$$
$$V_C = IX_C$$
$$= (16.67\underline{/0°})(2.236\underline{/-90°})$$
$$= 37.27\underline{/-90°} \text{ V}$$

Note that the voltages across the capacitor and inductor are much higher than the supply voltage. However, the algebraic sum is

$$V_s = V_R + V_L + V_C$$
$$= 5.000 + j37.27 - j37.27$$
$$= 5.000\underline{/0°} \text{ V}$$

The *Q* of a Circuit

If we were to vary the frequency of the power supply across a series resonant circuit and measure the resultant current, we would have the resonance curve shown in Fig. 26-9. The *Q* of a circuit is a measure of how wide or how narrow the curve is and can be defined as

$$Q = \frac{f_r}{f_{BW}} \quad \text{or} \quad Q = \frac{X_L}{R}$$

where Q is the quality factor (figure of merit), f_r the resonant frequency measured in Hz, and f_{BW} the width of the band of frequencies having amplitudes of greater than $0.7071 I_{max}$.

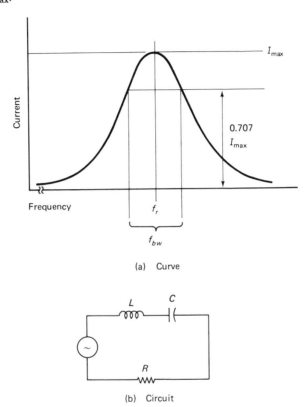

(a) Curve

(b) Circuit

FIGURE 26-9 Resonance curve.

EXAMPLE 26-11 What is the resonant frequency and bandwidth of a series circuit consisting of a 10-Ω resistor, a 1000-pF capacitor, and a 180-μH inductor?

SOLUTION The resonant frequency is

$$f_r = \frac{1}{2\pi\sqrt{LC}} = \frac{1}{2 \times \pi \times \sqrt{180 \times 10^{-6} \times 1000 \times 10^{-12}}}$$
$$= 375.1 \text{ kHz}$$

The Q can be determined by

$$Q = \frac{X_L}{R} = \frac{2\pi fL}{R}$$
$$= \frac{2 \times \pi \times 375.1 \times 10^3 \times 180 \times 10^{-6}}{10}$$
$$= 42.43$$

Now we can find the bandwidth:

$$Q=\frac{f_r}{f_{BW}} \qquad f_{BW}=\frac{f_r}{Q}$$

$$f_{BW}=\frac{375.1 \times 10^3}{42.43}$$

$$=8.842 \text{ kHz}$$

EXAMPLE 26-12 A capacitor must be selected to resonate with a 20-mH coil at a frequency of 84 kHz. What size must the capacitor be? If the coil has a 200-Ω resistance, what is the Q of the circuit?

$$f=\frac{1}{2\pi\sqrt{LC}}$$

$$C=\frac{1}{4\pi^2 f^2 L}$$

$$=\frac{1}{4\times\pi^2\times(84{,}000^2)\times 0.02}$$

$$=179.5 \times 10^{-12}$$

$$=179.5 \text{ pF}$$

The Q is:

$$Q=\frac{X}{R}=\frac{2\pi fL}{R}=\frac{2\times\pi\times 84{,}000\times 0.02}{200}$$

$$=52.78$$

PROBLEMS

Compute the resonant frequency:

26-25. 0.05-μF capacitor, 250-mH inductor
26-26. 1-μF capacitor, 8-H inductor
26-27. 2500-pF capacitor, 250-μH inductor
26-28. 0.015-μF capacitor, 20-mH inductor

Compute V_R, V_L, V_C, and Q of the following series resonant circuits:

26-29. $C=0.05\ \mu$F, $L=45$ mH, $R=100\ \Omega$, $V_s=14$ V
26-30. $C=180$ pF, $L=200\ \mu$H, $R=50\ \Omega$, $V_s=2.5$ V
26-31. Select a capacitor to resonate with a 200-mH coil at 2 kHz. If the circuit has a resistance of 100 Ω, what is the Q of the circuit? What is its bandwidth?
26-32. Repeat Prob. 26-31 for a frequency of 4 kHz.

26-5 PARALLEL RESONANCE

Parallel resonance can be defined in much the same manner as series resonance. The equation for resonance of the parallel circuit of Fig. 26-10 is

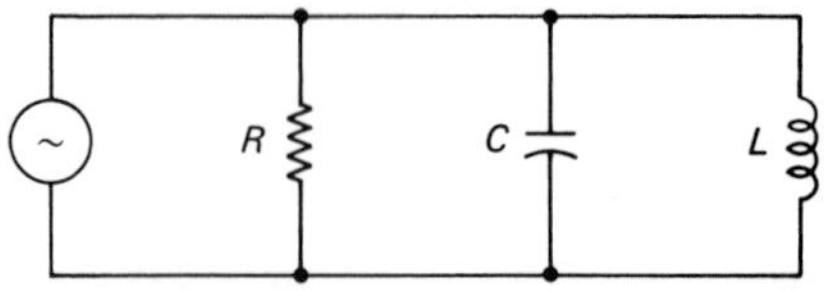

FIGURE 26-10 Parallel resonant circuit.

$$f_r = \frac{1}{2\pi\sqrt{LC}}$$

where f_r is the resonant frequency. However, in the parallel resonant circuit, the line current is minimum at resonance. The mathematical treatment of the circuit is as in any parallel circuit.

EXAMPLE 26-13 A 4.3-μF capacitor, a 50-mH inductor, and a 10-kΩ resistor are connected in parallel. What is the resultant resonant frequency (Fig. 26-10)?

SOLUTION

$$f = \frac{1}{2\pi\sqrt{LC}} = \frac{1}{2\times\pi\times\sqrt{0.05\times4.3\times10^{-6}}}$$
$$= 343.2 \text{ Hz}$$

This equation applies to circuits such as those of Fig. 26-10 or to those having equal resistances in the capacitance and inductive branch. For those having unequal resistances in the branches, such as in Fig. 26-11, the definition of resonance diverges to the following three definitions:

1. That frequency at which the circuit acts as a pure resistance. This is called the antiresonant frequency. This definition obeys the equation

$$f_{ar} = \frac{1}{2\pi}\sqrt{\frac{1}{LC} - \frac{R^2}{L^2}}$$

2. That frequency at which X_L and X_C are equal. This definition obeys the equation

$$f_r = \frac{1}{2\pi\sqrt{LC}}$$

3. That frequency at which the line current is minimum. This definition is of no practical importance.

EXAMPLE 26-14 In a circuit similar to Fig. 26-11, $L = 3$ mH, $R = 50$ Ω, and $C = 0.05$ μF. Compute the resonant and antiresonant frequencies.

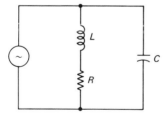

FIGURE 26-11

SOLUTION

$$f_r = \frac{1}{2\pi\sqrt{LC}}$$
$$= \frac{1}{2\times\pi\times\sqrt{0.003\times0.05\times10^{-6}}}$$
$$= 12.99 \text{ kHz}$$

$$f_{ar} = \frac{1}{2\pi} \sqrt{\frac{1}{LC} - \frac{R^2}{L^2}}$$

$$= \frac{1}{2\pi} \sqrt{\frac{1}{0.003 \times 0.05 \times 10^{-6}} - \frac{50^2}{0.003^2}}$$

$$= 12.72 \text{ kHz}$$

The Q of a Parallel Circuit

The Q of a parallel circuit can be defined as

$$Q = \frac{R_p}{X} = \frac{f_r}{f_{BW}}$$

where R_p is a resistor that is assumed to be in parallel with the capacitor and inductor as in Fig. 26-10.

> **EXAMPLE 26-15** In a circuit similar to Fig. 26-10, $R_p = 10$ kΩ, $L = 5$ mH, and $C = 2500$ pF. Determine the resonant frequency, bandwidth, and Q.

SOLUTION

$$f_r = \frac{1}{2\pi\sqrt{LC}} = \frac{1}{2 \times \pi \times \sqrt{0.005 \times 2500 \times 10^{-12}}}$$

$$= 45.02 \text{ kHz}$$

$$Q = \frac{R}{2\pi f L} = \frac{10,000}{2 \times \pi \times 45,020 \times 0.005}$$

$$= 7.071$$

$$f_{BW} = \frac{f_r}{Q} = \frac{45.02}{7.071} = 6.366 \text{ kHz}$$

PROBLEMS

Determine the resonant frequency for the following circuits, similar to Fig. 26-10.

26-33. $L = 20$ mH, $C = 126$ μF, $R = 5$ kΩ
26-34. $L = 460$ mH, $C = 0.078$ μF, $R = 10$ kΩ
26-35. $L = 10$ H, $C = 100$ μF, $R = 300$ Ω
26-36. $L = 10$ μH, $C = 30$ pF, $R = 47$ kΩ

Determine the resonant and antiresonant frequencies for the following circuits, similar to Fig. 26-11.

26-37. $L = 5$ mH, $C = 0.15$ μF, $R = 75$ Ω
26-38. $L = 50$ mH, $C = 0.05$ μF, $R = 91$ Ω
26-39. $L = 500$ mH, $C = 0.86$ μF, $R = 56$ Ω
26-40. $L = 380$ μH, $C = 0.031$ μF, $R = 250$ Ω

Determine the resonant frequency, Q, and bandwidth for the following parallel circuits similar to Fig. 26-10.

26-41. $L = 30$ mH, $C = 13$ μF, $R = 5$ kΩ
26-42. $L = 400$ mH, $C = 0.015$ μF, $R = 10$ kΩ

PART III

ACTIVE-DEVICE MATHEMATICS

In this part of the book we examine mathematics needed to understand the computer. We look at the number system of the computer, then the peculiar algebra that is used to build such computer systems.

27

COMPUTER ARITHMETIC

As each day passes, computers become a greater part of our lives. This continual advancement requires that we have a knowledge of both the number systems used in such systems and the boolean algebra that forms the basis for the system hardware. We investigate these two subjects in this chapter.

27-1 THE DECIMAL NUMBER SYSTEM

Before examining other number systems in detail, the reader should observe several important characteristics of the decimal system. We have become so accustomed to its use that its orderliness can easily be overlooked. The decimal number system contains 10 unique symbols: 0, 1, 2, 3, 4, 5, 6, 7, 8, and 9. To many readers this is not an earth-shattering revelation. Note that, although we call it a decimal (tens) system, it does not have a symbol for 10, but expresses it and any number above 10 as a combination of these symbols. Because it has 10 symbols, its radix (or base) is said to be 10.

Even though the decimal system has only 10 symbols, any number of any magnitude can be expressed by using our system of positional weighting. Consider the number 3472. It can be broken down as follows:

$$3472 = 3000 \quad + 400 \quad + 70 \quad + 2$$
$$= 3 \times 10^3 + 4 \times 10^2 + 7 \times 10^1 + 2 \times 10^0$$

Note that, because the 3 is positioned four places to the left of the decimal point, it has a much greater weight than the 2. If the 3 were next to the point, it would only be worth 3 instead of 3000. Thus, the position of the digit with reference to the decimal point determines its weight. This can be illustrated further by rearranging the digits of the previous example; the value of the number changes.

$$4237 = 4000 \quad + 200 \quad + 30 \quad + 7$$
$$= 4 \times 10^3 + 2 \times 10^2 + 3 \times 10^1 + 7 \times 10^0$$

The principle of positional weighting can be extended to any number system. Any number can be represented by the equation

$$Y = d_n \times r^n + d_{n-1} \times r^{n-1} + \ldots + d_1 \times r^1 + d_0 \times r^0 \tag{27-1}$$

where Y is the value of the entire number, d_n the value of the nth digit from the point, and r the radix, or base. This equation has already been applied to the base 10 number system. Next, we apply it to the binary (base 2) system.

27-2 THE BINARY NUMBER SYSTEM

Table 27-1 illustrates the binary number system. This system of numbering is called a binary system, because each number position can take on only two values: zero and one. These positions are called "bits," a contraction of the words "binary digits." Look closely at this binary counting sequence. It is very similar to the decimal system of counting. Starting at zero, the next number is 1. The right-hand bit has now assumed the highest value within the system, 1. The next number in the sequence will require that a 1 be added to the column two-bit and that the column one-bit be set to zero. This binary sequence continues until the decimal number fifteen (1111) is reached, the highest number that can be represented by four bits.

TABLE 27-1 Binary numbers.

Binary				Decimal
8	4	2	1	
0	0	0	0	0
0	0	0	1	1
0	0	1	0	2
0	0	1	1	3
0	1	0	0	4
0	1	0	1	5
0	1	1	0	6
0	1	1	1	7
1	0	0	0	8
1	0	0	1	9
1	0	1	0	10
1	0	1	1	11
1	1	0	0	12
1	1	0	1	13
1	1	1	0	14
1	1	1	1	15

Binary numbers are used extensively throughout all digital systems, because of the very nature of electronics. A 1 can be represented by a saturated transistor, a light turned on, a relay energized, or a magnet magnetized in a particular direction. A zero can be represented as a cutoff transistor, a light turned off, a deenergized relay, or the magnet magnetized in the opposite direction. In each of these, there are only two values that the device can assume. For this reason, we study the binary system of numbers in detail.

Converting Binary to Decimal

Like the decimal system, the binary system is positionally weighted. Each position represents a particular n of 2^n. Table 27-2 illustrates both binary and decimal positional weighting. The base is indicated by the subscripts. Because of this property of positional weighting, the procedure for converting a binary number to decimal is very similar to that of breaking a decimal number into its weighted values, as discussed in Section 27-1. It merely requires substitution of numbers into Eq. (27-1). The d's will all be 0's or 1's, the r's will all be 2 (the radix), and the n's various powers of 2, depending upon the position of the digit with reference to the binary point. (In binary it is referred to as a binary point, not a decimal point.)

TABLE 27-2 Binary and decimal positional weighting.

Binary	Decimal
$1_2 = 1 \times 2^0 = 1_{10}$	$1_{10} = 1 \times 10^0 = 1_{10}$
$10_2 = 1 \times 2^1 = 2_{10}$	$10_{10} = 1 \times 10^1 = 10_{10}$
$100_2 = 1 \times 2^2 = 4_{10}$	$100_{10} = 1 \times 10^2 = 100_{10}$
$1000_2 = 1 \times 2^3 = 8_{10}$	$1000_{10} = 1 \times 10^3 = 1000_{10}$
$10,000_2 = 1 \times 2^4 = 16_{10}$	$10,000_{10} = 1 \times 10^4 = 10,000_{10}$

EXAMPLE 27-1 Convert 10111_2 to decimal.

SOLUTION

$$Y = d_4 \times r^4 + d_3 \times r^3 + d_2 \times r^2 + d_1 \times r^1 + d_0 \times r^0$$
$$= 1 \times 2^4 + 0 \times 2^3 + 1 \times 2^2 + 1 \times 2^1 + 1 \times 2^0$$
$$= 16 + 0 + 4 + 2 + 1$$
$$10111_2 = 23_{10}$$

EXAMPLE 27-2 Convert 1011101001_2 to decimal.

SOLUTION

$$Y = d_9 \times r^9 + d_8 \times r^8 + d_7 \times r^7 + d_6 \times r^6$$
$$+ d_5 \times r^5 + d_4 \times r^4 + d_3 \times r^3 + d_2 \times r^2$$
$$+ d_1 \times r^1 + d_0 \times r^0$$
$$= 1 \times 2^9 + 0 \times 2^8 + 1 \times 2^7 + 1 \times 2^6 + 1 \times 2^5 + 0 \times 2^4$$
$$+ 1 \times 2^3 + 0 \times 2^2 + 0 \times 2^1 + 1 \times 2^0$$
$$= 512 + 0 + 128 + 64 + 32 + 0 + 8 + 0 + 0 + 1$$
$$1011101001_2 = 745_{10}$$

This method will always provide the correct decimal representation of a binary number. There is a second method, called the dibble-dobble method, that will also provide the solution. To use this method, start with the left-hand bit. Multiply this value by 2 and add the next bit to the right. Multiply this value by 2 and add the next bit to the right. Stop when the binary point is reached.

EXAMPLE 27-3 Convert 110111_2 to decimal.

SOLUTION

Copy down the left bit:		1
Multiply by 2, add next bit	$(2 \times 1) + 1 =$	3
Multiply by 2, add next bit	$(2 \times 3) + 0 =$	6
Multiply by 2, add next bit	$(2 \times 6) + 1 =$	13
Multiply by 2, add next bit	$(2 \times 13) + 1 =$	27
Multiply by 2, add next bit	$(2 \times 27) + 1 =$	55

Therefore, $110111_2 = 55_{10}$.

The two methods described above provide the means for converting binary numbers to decimal numbers. However, in digital circuits, it is also necessary to convert from decimal to binary.

Converting Decimal to Binary

One method of converting decimal to binary is the opposite of the dibble-dobble method. Using this method, the number is successively divided by 2 and its remainders recorded. The final binary result is obtained by assembling all the remainders, with the last remainder being the most significant bit (MSB).

EXAMPLE 27-4 Convert 43_{10} to binary.

SOLUTION

Successive division	Remainders
2)43	
2)21	1
2)10	1
2) 5	0
2) 2	1
2) 1	0
2) 0	1

Read the remainders from the bottom to the top:

$$43_{10} = 101011_2$$

EXAMPLE 27-5 Convert 200_{10} to binary.

SOLUTION

Successive division	Remainders
2)200	
2)100	0
2) 50	0
2) 25	0
2) 12	1
2) 6	0
2) 3	0
2) 1	1
2) 0	1

Reading the remainders up, the result is

$$200_{10} = 11001000_2$$

PROBLEMS

Convert from binary to decimal:

27-1. **(a)** 1011 **(b)** 0111 **(c)** 1001

27-2. **(a)** 1110 **(b)** 1111 **(c)** 1101

27-3. **(a)** 11,011 **(b)** 10,100 **(c)** 10,111

27-4. **(a)** 10,001 **(b)** 10,110 **(c)** 11,101

27-5. **(a)** 110,011,101 **(b)** 101,110,001

27-6. **(a)** 100,001,101 **(b)** 100,111,110

Convert from decimal to binary:

27-7. **(a)** 21 **(b)** 31 **(c)** 26

27-8. **(a)** 18 **(b)** 22 **(c)** 25

27-9. **(a)** 86 **(b)** 72 **(c)** 92

27-10. **(a)** 88 **(b)** 95 **(c)** 49

27-3 THE OCTAL NUMBER SYSTEM

Binary numbers are long—not opinion, fact. To represent the number 4096_{10}, 12 bits must be used. These numbers are fine for stupid machines, but for human beings they are simply too bulky.

Consider a binary register three bits long. The largest number that can be expressed by these three bits is 7, and there are eight unique numbers. Table 27-3 compares the binary, octal, and decimal number systems. Note that there are no eights or nines in the octal system, but that it is merely a representation of the decimal value of each three-bit group of binary bits.

TABLE 27-3 Binary, octal, and decimal numbers.

Binary	Octal	Decimal
0	0	0
1	1	1
10	2	2
11	3	3
100	4	4
101	5	5
110	6	6
111	7	7
1 000	10	8
1 001	11	9
1 010	12	10
1 011	13	11
1 100	14	12
1 101	15	13
1 110	16	14
1 111	17	15
10 000	20	16

Binary–Octal Conversions

Conversion from binary to octal proceeds from the foregoing definitions. Simply divide the number into groups of three bits each, starting at the binary point. Then express each group by its decimal (or octal) equivalent. Note that the highest digit in the octal system is a 7. (Remember that the highest digit in the decimal system is a nine?)

EXAMPLE 27-6   Convert 11111011110101_2 to octal.

SOLUTION

Divide into groups of three	11,111,011,110,101
Express each group in decimal	3 7 3 6 5

Therefore, $11111011110101_2 = 37{,}365_8$.

EXAMPLE 27-7   Convert 1011110100011000111_2 to octal.

SOLUTION

Divide into groups	1,011,110,100,011,000,111
Express groups in decimal	1 3 6 4 3 0 7

Therefore, $1011110100011000111_2 = 1{,}364{,}307_8$.

Conversion from octal to binary is the reverse of this process. Express each octal number in its appropriate binary notation.

EXAMPLE 27-8   Convert 3674_8 to binary.

SOLUTION

Copy the octal number	3 6 7 4
Convert each to binary	011 110 111 100

Therefore, $3674_8 = 11{,}110{,}111{,}100_2$.

Converting Octal to Decimal

Since the octal number system is an extension of the binary system, these numbers must frequently be converted to decimal. The octal system is a positional number system, as are the binary and decimal systems, obeying Eq. (27-1) with a radix of eight. As in the case of the binary system, this equation forms the basis for converting from octal to decimal.

EXAMPLE 27-9   Convert 357_8 to decimal.

SOLUTION

$$357 = d_2 \times r^2 + d_1 \times r^1 + d_0 \times r^0$$
$$= 3 \times 8^2 + 5 \times 8^1 + 7 \times 8^0$$
$$= 192 + 40 + 7$$
$$= 239$$

Therefore, $357_8 = 239_{10}$. Note that all the numbers on the right of the equation are decimal.

EXAMPLE 27-10   Convert 6421_8 to decimal.

SOLUTION

$$6421_8 = 6 \times 8^3 + 4 \times 8^2 + 2 \times 8^1 + 1 \times 8^0$$
$$= 3072 + 256 + 16 + 1$$
$$= 3345_{10}$$

Converting Decimal to Octal

Decimal numbers can be converted to octal by successive division. The method requires that the decimal number be successively divided by 8 and the remainders collected, from the last to the first. Note the similarity between this and the decimal to binary method discussed in Section 27-2.

EXAMPLE 27-11   Convert 1359_{10} to octal.

SOLUTION

Successive Division	Remainders
8)1359	
8) 169	7
8) 21	1
8) 2	5
8) 0	2

Reading the remainders from the bottom to the top, the result is 2517_8. Therefore, $1359_{10} = 2517_8$.

EXAMPLE 27-12   Convert 7777_{10} to octal.

SOLUTION

Successive Division	Remainders
8)7777	
8) 972	1
8) 121	4
8) 15	1
8) 1	7
8) 0	1

Therefore, $7777_{10} = 17,141_8$.

PROBLEMS

Convert from binary to octal:

27-11. (a) 101,101,111 (b) 111,011,010

27-12. (a) 111,110,101 (b) 011,100,001

27-13. 111,101,110,111,111,000

27-14. 011,101,011,010,111,101

Convert from octal to binary:

27-15. (a) 765 (b) 542 (c) 676

27-16. (a) 1356 (b) 7215 (c) 7642

Convert from octal to decimal:

27-17. (a) 346 (b) 762 (c) 431

27-18. (a) 7625 (b) 1356 (c) 7265

27-19. (a) 1234 (b) 4362 (c) 1726

27-20. (a) 77,321 (b) 14,365 (c) 132,711

Convert from decimal to octal:

27-21. (a) 963 (b) 210 (c) 469

27-22. (a) 186 (b) 343 (c) 592

27-23. (a) 1,865 (b) 12,347 (c) 27,226

27-24. (a) 77,169 (b) 123,642 (c) 962,431

Like the octal system, the hexadecimal system was borne out of a need to express binary numbers concisely. However, computer number (word) lengths come in eight bits, 16 bits, 32 bits, and so on. Thus, grouping in four-bit groups instead of three-bit groups was a reasonable step. If these four-bit groups are used, 16 unique numbers can be defined: zero through 15_{10}. However, unlike the octal system, there are no symbols within our Arabic number system that represent 10_{10} through 15_{10} without using more than one symbol. So, drawing on American genius, individuality, and bravery, the symbols A, B, C, D, E, and F were selected. Table 27-4 compares the binary, hexadecimal, octal, and decimal symbols.

TABLE 27-4   Computer number systems.

Binary	Hexadecimal	Octal	Decimal
0000	0	0	0
0001	1	1	1
0010	2	2	2
0011	3	3	3
0100	4	4	4
0101	5	5	5
0110	6	6	6
0111	7	7	7
1000	8	10	8
1001	9	11	9
1010	A	12	10
1011	B	13	11
1100	C	14	12
1101	D	15	13
1110	E	16	14
1111	F	17	15
1 0000	10	20	16

Converting Hexadecimal to Decimal

The hexadecimal number system is, like its neighbors, octal, decimal, and binary, a positionally weighted system obeying Eq. (27-1), where the radix is 16. To convert from hexadecimal (usually called hex for short) to decimal, merely substitute the digits into the equation.

EXAMPLE 27-13   Convert $2C9_{16}$ to decimal.

SOLUTION

$$2C9_{16} = d_2 \times r^2 + d_1 \times r^1 + d_0 \times r^0$$
$$= 2 \times 16^2 + 12 \times 16^1 + 9 \times 16^0$$
$$= 512 + 192 + 9$$
$$= 713_{10}$$

Note that a C represents a decimal 12 as indicated in Table 27-4.

EXAMPLE 27-14   Convert EB4A$_{16}$ to decimal.

SOLUTION

$$EB4A_{16} = d_3 \times r^3 + d_2 \times r^2 + d_1 \times r^1 + d^0 \times r^0$$
$$= 14 \times 16^3 + 11 \times 16^2 + 4 \times 16^1 + 10 \times 16^0$$
$$= 57,344 + 2816 + 64 + 10$$
$$= 60,234_{10}$$

Converting Decimal to Hexadecimal

Any decimal number can be converted to hex by successively dividing by 16. The remainders can then be converted to hex and read up from the bottom to obtain the hexadecimal results.

EXAMPLE 27-15   Convert 423_{10} to hexadecimal.

SOLUTION

Successive Division	Remainders	Hex Notation
16)423		
16) 26	7	7
16) 1	10	A
0	1	1

Reading the remainders up, the result is $1A7_{16}$.

EXAMPLE 27-16   Convert 72905_{10} to hexadecimal.

SOLUTION

Successive Division	Remainders	Hex Notation
16)72905		
16) 4556	9	9
16) 284	12	C
16) 17	12	C
16) 1	1	1
0	1	1

Reading the remainders up, the result is $11,CC9_{16}$.

PROBLEMS

Convert the following hexadecimal numbers to decimal:

27-25. (a) 1B6   (b) 4C7   (c) 123

27-26. (a) BB3   (b) 86B   (c) 14A

27-27. (a) 99B   (b) 8CA   (c) CA7

27-28. (a) 1C1B   (b) 8ABC   (c) F035

Convert the following decimal numbers to hexadecimal:

27-29. (a) 131   (b) 323   (c) 462

27-30. (a) 398   (b) 401   (c) 275

27-31. (a) 1369   (b) 1053   (c) 1762

27-32. (a) 13,642   (b) 12,941   (c) 98,364

27-5 INTRODUCTION TO BOOLEAN ALGEBRA

Boolean algebra is a system of mathematical logic. It differs from both ordinary algebra and the binary number system. As an illustration, in boolean, $1 + 1 = 1$; in ordinary algebra and in binary arithmetic the result is 10. Thus, although there are similarities, boolean algebra is a unique system.

There are two constants within the boolean system: 0 and 1. Every number is either a 0 or a 1. There are no negative or fractional numbers. Thus:

$$\text{If } X = 1, \quad \text{then} \quad X \neq 0 \tag{27-2}$$
$$\text{If } X = 0, \quad \text{then} \quad X \neq 1 \tag{27-3}$$

27-6 THE AND OPERATOR

The AND function is defined in boolean algebra by use of the dot ($\cdot$), or no operator symbol at all. Thus, it is similar to multiplication in ordinary algebra. For example: $A \cdot B = C$ means that if A is true AND B is true, then C will be true. Under any other condition, C will be false. There are four different combinations of A and B that must be considered, as shown in Table 27-5. Note that only when both A and B are 1 (or true) will the output, C, be 1 (or true).

TABLE 27-5 AND functions.

$A \cdot B = C$
$0 \cdot 0 = 0$
$0 \cdot 1 = 0$
$1 \cdot 0 = 0$
$1 \cdot 1 = 1$

AND Gates

The AND function is also used in electronics. Figure 27-1 shows the symbol for an AND gate and its associated truth table. This particular gate has two inputs and one output. Since many logic gates that are on the market today define a 0-V level as false and a plus 5-V level as true, we shall make this assumption in this example. Thus, this gate will provide +5 V on its output, C, if and only if both A and B inputs are at +5 V. Under any other set of input conditions, the output will be 0-V.

AND gates may have any number of inputs. Figure 27-2 shows a four-input AND gate and its associated truth table. Note that the output will be a 1 only if all four inputs have 1's on them.

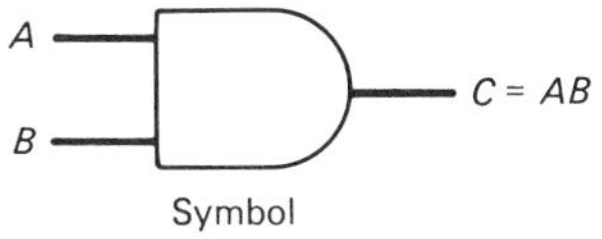

```
              0 1 2 3
A ___         A  0 0 1 1
      )----  C = AB    B  0 1 0 1
B ___                  C  0 0 0 1
    Symbol
                       Truth table
```

FIGURE 27-1 The AND gate.

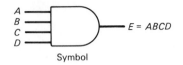

Symbol

	0 1 2 3	4 5 6 7	8 9 10 11	12 13 14 15
A	0 0 0 0	0 0 0 0	1 1 1 1	1 1 1 1
B	0 0 0 0	1 1 1 1	0 0 0 0	1 1 1 1
C	0 0 1 1	0 0 1 1	0 0 1 1	0 0 1 1
D	0 1 0 1	0 1 0 1	0 1 0 1	0 1 0 1
E	0 0 0 0	0 0 0 0	0 0 0 0	0 0 0 1

FIGURE 27-2   The four input AND gate.   Truth table

AND Laws[1]

From the foregoing discussion, there are three boolean algebraic laws that shall be examined closely:

$$A \cdot 1 = A \qquad (27\text{-}4)$$
$$A \cdot 0 = 0 \qquad (27\text{-}5)$$
$$A \cdot A = A \qquad (27\text{-}6)$$

All three of these can be verified by remembering what the AND symbol means. Consider Eq. (27-4), and apply it to a two-input AND gate as shown in Fig. 27-3. If A equals 0 and the other input is 1, the output is 0 [Fig. 27-3(a)]. If A equals 1 and the other input is 1, the output is 1 [Fig. 27-3(b)]. Thus, the output is always equal to the A input. Note that we could just as well replace the gate with a length of wire from the A input to the output.

(a)   (b)

FIGURE 27-3   Verifying $A \cdot 1 = A$.

Next, consider Eq. (27-5) and apply it to a two-input gate, as shown in Fig. 27-4. Note that no matter what value A takes on, 1 or 0, the output will always be a zero.

Consider Eq. (27-6) and its realization in Fig. 27-5. Note that the output always takes on the value of A. These illustrations demonstrate the truth of the three laws.

FIGURE 27-4   Verifying $A \cdot 0 = 0$.

[1]   Although not mathematically precise, all axioms, postulates, lemmas, and theorems are called laws in this chapter, for simplicity and ease of discussion.

FIGURE 27-5   Verifying A·A = A.

27-7   THE OR OPERATOR

The OR operator is indicated by using a plus sign. Thus, $A + B = C$ means that if A is true OR B is true, C will be true. Under any other set of conditions (there is only one other set of conditions—when both A and B are false), C will be false. There are four different combinations of A and B that must be examined in the light of the OR function, as shown in Table 27-6.

TABLE 27-6   The OR function.

$A + B = C$
$0 + 0 = 0$
$0 + 1 = 1$
$1 + 0 = 1$
$1 + 1 = 1$

OR Gates

The symbol for the OR gate as used in electronics in shown in Fig. 27-6, together with its associated truth table. The OR gate may have any number of inputs. A four-input gate is shown in Fig. 27-7, together with its truth table. Note that any input being a 1 causes the output to be a 1.

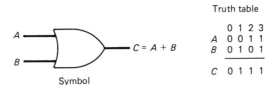

FIGURE 27-6   The OR gate.

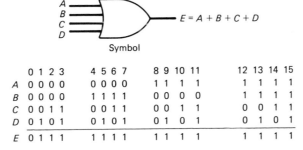

FIGURE 27-7   The four input OR gate.

OR Laws

As with the AND gate, there are several OR laws that become apparent by studying the OR gate.

$$A + 1 = 1 \qquad (27\text{-}7)$$
$$A + 0 = A \qquad (27\text{-}8)$$
$$A + A = A \qquad (27\text{-}9)$$

The two possible cases of Eq. (27-7) are shown in Fig. 27-8. Regardless of what value A takes on, the output is always 1.

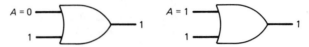

FIGURE 27-8   Verifying A ± 1 = 1.

Figure 27-9 illustrates Eq. (27-8). When A is set to 0, the output is 0. When A is set to 1, the output is 1. Therefore, the output assumes the value of A.

FIGURE 27-9   Verifying A + 0 = A.

Figure 27-10 illustrates Eq. (27-9). With A set to 0, the output is 0. With A set to 1, the output is 1. Thus, the output always equals A.

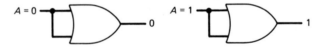

FIGURE 27-10   Verifying A + A = A.

27-8   THE NOT OPERATOR

The symbol for the NOT operator is the vinculum, also called an overscore or bar. It represents an inversion of logic levels.

Inverters

The logic diagrams for the NOT operator are called inverters, and are shown with the truth table in Fig. 27-11. The circle actually represents the inversion and

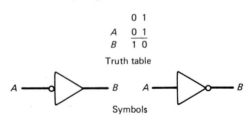

FIGURE 27-11   The inverter.

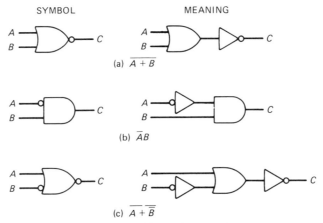

SYMBOL   MEANING

(a) $\overline{A + B}$

(b) $\overline{A}B$

(c) $\overline{A + \overline{B}}$

FIGURE 27-12 The circle as an inverter.

the triangle an amplifier. When a single circuit is used for inversion alone, the triangle is included with the symbol. When the circuit is used for both gating and inversion, the circle is shown in series with a gate input or output lead. Some examples of this use of the circle are shown in Fig. 27-12, together with their meanings. In each case the circle represents an inversion.

NOT Laws

There are several laws of boolean algebra that become apparent when examining the inverter.

$$\overline{0} = 1 \tag{27-10}$$
$$\overline{1} = 0 \tag{27-11}$$
$$\text{If } A = 0, \quad \text{then} \quad \overline{A} = 1 \tag{27-12}$$
$$\text{If } A = 1, \quad \text{then} \quad \overline{A} = 0 \tag{27-13}$$

Equations (27-10) through (27-13) can be verified by examining Fig. 27-13. There are only two possibilities for input A on the inverter: either the input is a 0 or it is a 1. If the input is 0, the output, $\overline{A}$, must be 1. If the input is 1 the output, $\overline{A}$, must be 0.

A fourth law that emerges from the definition of an inverter is

$$\overline{\overline{A}} = A \tag{27-14}$$

This law can be verified by examining Fig. 27-14. Note that when a 0 appears on the input to the first inverter, a 1 appears at the $\overline{A}$ and a 0 at the output of the second inverter. Similarly, when a 1 appears at A, a 0 appears at $\overline{A}$ and a 1 at the

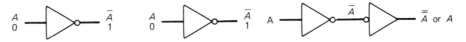

FIGURE 27-13 Verifying NOT relationships.

FIGURE 27-14 Verifying $\overline{\overline{A}} = A$.

output of the second inverter. Therefore, through the method of perfect induction, we can reason that A inverted twice, $\bar{\bar{A}}$, is identical to A.

27-9 LAWS OF BOOLEAN ALGEBRA

Boolean algebra is a system of mathematics. As with any such system, there are fundamental laws that are used to build a workable, cohesive framework upon which are placed the theorems proceeding from these laws. Many of these laws have been already discussed in earlier sections of this chapter. They will be repeated here, together with others that will provide all the tools necessary for manipulating boolean expressions.

Laws of Complementation

The term "complement" simply means to invert; to change 1's to 0's and 0's to 1's. The five laws of complementation were discussed in Section 27-8, where inverters were introduced. They are as follows:

Law 1	$\bar{0} = 1$
Law 2	$\bar{1} = 0$
Law 3	If $A = 0$, then $\bar{A} = 1$
Law 4	If $A = 1$, then $\bar{A} = 0$
Law 5	$\bar{\bar{A}} = A$

AND Laws

The four AND laws are:

Law 6	$A \cdot 0 = 0$
Law 7	$A \cdot 1 = A$
Law 8	$A \cdot A = A$
Law 9	$A \cdot \bar{A} = 0$

Laws 6 through 8 were discussed in Section 27-6 in the discussion of AND gates. Figure 27-15 illustrates Law 9. If A were 0, then $\bar{A}$ would be 1. The AND gate would then have a 0 on one of its inputs and a 1 on the other input, causing a 0 to appear at the output. If A were a 1, $\bar{A}$ would be a 0 and the AND would still output a 0. Thus, regardless of the value of A, $A \cdot \bar{A} = 0$.

OR Laws

The four OR laws are as follows:

Law 10	$A + 0 = A$
Law 11	$A + 1 = 1$
Law 12	$A + A = A$
Law 13	$A + \bar{A} = 1$

Laws 10 through 12 were discussed in Section 27-7 when OR gates were considered. Law 13 is illustrated in Fig. 27-16. If A were a 1, then $\bar{A}$ would be 0. Therefore, the expression of the OR gate would be $1 + 0 = 1$. If A were 0, then $\bar{A}$ would be a

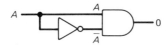

FIGURE 27-15 Verifying A · $\overline{A}$ = 0.

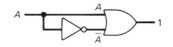

FIGURE 27-16 Verifying A + A = 1.

1. The OR gate function would receive $0 + 1$ and the output would be a 1. As can be seen, regardless of the value of A, $A + \overline{A} = 1$.

Commutative Laws

The commutative laws allow the change in position of an AND or OR variable:

Law 14 $A + B = B + A$
Law 15 $A \cdot B = B \cdot A$

Table 27-7 is a truth table illustrating Law 14. By perfect induction, every case of $A + B$ yields an identical result to $B + A$. Thus, this law has been verified.

TABLE 27-7   Verifying that $A + B = B + A$.

	0	1	2	3
A	0	0	1	1
B	0	1	0	1
$A + B$	0	1	1	1
$B + A$	0	1	1	1

Table 27-8 illustrates a truth table for A, B, $A \cdot B$, and $B \cdot A$. Note that, after all cases have been examined, $A \cdot B = B \cdot A$.

TABLE 27-8   Verifying that $A \cdot B = B \cdot A$.

	0	1	2	3
A	0	0	1	1
B	0	1	0	1
$A \cdot B$	0	0	0	1
$B \cdot A$	0	0	0	1

Associative Laws

The associative laws allow the grouping of variables. The two associative laws are:

Law 16 $A + (B + C) = (A + B) + C$
Law 17 $A \cdot (B \cdot C) = (A \cdot B) \cdot C$

Table 27-9 is a truth table for Law 16. ORing A and B yields the result shown by $(A + B)$. ORing B and C yields the result shown by $(B + C)$. The next step is to OR the row $(A + B)$ with row C, yielding the results shown by row $(A + B) + C$. Note that this result is in every case the same as ORing row A with row $(B + C)$. This verifies the law.

TABLE 27-9 Verifying that $A + (B + C) = (A + B) + C$.

	0	1	2	3	4	5	6	7
A	0	0	0	0	1	1	1	1
B	0	0	1	1	0	0	1	1
C	0	1	0	1	0	1	0	1
$(A + B)$	0	0	1	1	1	1	1	1
$(B + C)$	0	1	1	1	0	1	1	1
$(A + B) + C$	0	1	1	1	1	1	1	1
$A + (B + C)$	0	1	1	1	1	1	1	1

Table 27-10 is a truth table illustrating Law 17. Again, the results of $(A \cdot B) \cdot C$ are, in every case, identical with the results of $A \cdot (B \cdot C)$. This verifies the law.

TABLE 27-10 Verifying that $(A \cdot B) \cdot C = A \cdot (B \cdot C)$.

	0	1	2	3	4	5	6	7
A	0	0	0	0	1	1	1	1
B	0	0	1	1	0	0	1	1
C	0	1	0	1	0	1	0	1
$(A \cdot B)$	0	0	0	0	0	0	1	1
$(B \cdot C)$	0	0	0	1	0	0	0	1
$(A \cdot B) \cdot C$	0	0	0	0	0	0	0	1
$A \cdot (B \cdot C)$	0	0	0	0	0	0	0	1

Distributive Laws

The distributive laws allow the factoring or multiplying out of expressions. Three distributive laws will be considered:

Law 18	$A \cdot (B + C) = (A \cdot B) + (A \cdot C)$
Law 19	$A + (B \cdot C) = (A + B) \cdot (A + C)$
Law 20	$A + (A \cdot B) = A + B$

Table 27-11 is a perfect induction proof for Law 18, since, in every case, the left side of the identity is the same as the right side.

TABLE 27-11 Verifying that $A \cdot (B + C) = (A \cdot B) + (A \cdot C)$.

	0	1	2	3	4	5	6	7
A	0	0	0	0	1	1	1	1
B	0	0	1	1	0	0	1	1
C	0	1	0	1	0	1	0	1
$(B + C)$	0	1	1	1	0	1	1	1
$(A \cdot B)$	0	0	0	0	0	0	1	1
$(A \cdot C)$	0	0	0	0	0	1	0	1
$A \cdot (B + C)$	0	0	0	0	0	1	1	1
$(A \cdot B) + (A \cdot C)$	0	0	0	0	0	1	1	1

Law 19 differs from the algebra we struggled with in high school. This boolean law can be proven as follows:

$$A + BC = A + BC$$
$$= A \cdot 1 + BC \qquad \text{Law 7}$$
$$= A(1 + B) + BC \qquad \text{Laws 11 and 14}$$
$$= A + AB + BC \qquad \text{Law 18}$$
$$= A(1 + C) + AB + BC \qquad \text{Law 11}$$
$$= AA + AC + AB + BC \qquad \text{Laws 8 and 13}$$
$$= A(A + C) + BA + BC \qquad \text{Laws 18 and 15}$$
$$= A(A + C) + B(A + C) \qquad \text{Law 18}$$
$$= (A + C)A + (A + C)B \qquad \text{Law 15}$$
$$= (A + C)(A + B) \qquad \text{Law 18}$$
$$= (A + B)(A + C) \qquad \text{Law 15}$$

Law 20 is also departure from numerical algebra. It can be proven as follows:

$$A + \overline{A}B = A + \overline{A}B$$
$$= A \cdot 1 + \overline{A}B \qquad \text{Law 7}$$
$$= A(1 + B) + \overline{A}B \qquad \text{Law 11 and 14}$$
$$= A \cdot 1 + AB + \overline{A}B \qquad \text{Law 18}$$
$$= A + AB + \overline{A}B \qquad \text{Law 7}$$
$$= A + BA + B\overline{A} \qquad \text{Law 15}$$
$$= A + B(A + \overline{A}) \qquad \text{Law 18}$$
$$= A + B \cdot 1 \qquad \text{Law 13}$$
$$= A + B$$

De Morgan's Theorem

One of the most powerful identities used in boolean algebra is De Morgan's theorem. It provides two tools to the designer:

1. It allows removal of individual variables from under a NOT sign. For example, $\overline{A + BC}$ can be transformed into $\overline{A}(\overline{B} + \overline{C})$.
2. It allows transformation from a sum of products form to a product of sum form. For example, $A\overline{B}C + A\overline{B}\,\overline{C}$ can be transformed into $(\overline{A} + B + \overline{C})(\overline{A} + B + C)$.

The theorem can be expressed by the following identity:

Law 21 $$\overline{A} \cdot \overline{B} = \overline{A + B}$$

To verify the identity, refer to Table 27-12. Note that for each assigned value of A and B, the theorem is true. Therefore, the identity has been proven by perfect induction.

TABLE 27-12 Truth table, De Morgan's theorem.

	0	1	2	3
A	0	0	1	1
B	0	1	0	1
$\overline{A}$	1	1	0	0
$\overline{B}$	1	0	1	0
$\overline{A} \cdot \overline{B}$	1	0	0	0
$A + B$	0	1	1	1
$\overline{A + B}$	1	0	0	0

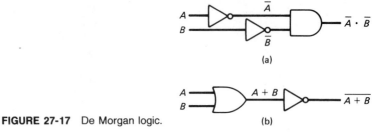

FIGURE 27-17 De Morgan logic. (b)

All of this is very interesting but, as the monkey said to the dog concerning the dog's tail: "What good is it?" Let us examine the hardware implementation for each side of De Morgan's theorem. Figure 27-17 compares the logic required for each. Observe that the basic logic function involved can be either an OR gate or an AND gate, depending upon what is available to the designer. Both forms execute the same function. This allows the designer to use two tools for implementing any function, rather than one, and he can convert it to whatever form is convenient.

Although the identity above represents De Morgan's theorem, the transformation is more easily performed by following these steps:

(a) Complement the entire function.
(b) Change all the ANDs to ORs and all the ORs to ANDs.
(c) Complement each of the individual variables.

This procedure is called demorganization.

EXAMPLE 27-17 Demorganize the function $\overline{A\overline{B} + C}$.

SOLUTION

Given:	$\overline{A\overline{B} + C}$
Complement function:	$A\overline{B} + C$
Change operators:	$(A + \overline{B})(C)$
Complement variables:	$(\overline{A} + B)(\overline{C})$

This transformation can be performed on any part of a function. However, it only applies to two levels of logic: a product of sums or a sum of products. For example, it could only be used to transform $A + B(C + D)$ if $C + D$ were considered one variable. Setting $C + D$ to R, the function becomes $A + BR$, a sum of the products $A \cdot 1$ and BR. To illustrate transformation of a multiple level function, consider the following example:

EXAMPLE 27-18 Reduce the expression $A + B(\overline{C + \overline{DE}})$.

SOLUTION First, demorganize $\overline{DE}$.

Complement:	DE
Change operators:	$D + E$
Complement variables:	$\overline{D} + \overline{E}$
Substitute back into expression:	$A + B(\overline{C + \overline{D} + \overline{E}})$
Next, demorganize $(\overline{C + \overline{D} + \overline{E}})$:	

Complement:	$(C + \overline{D} + \overline{E})$
Change operators:	$(C \cdot \overline{D} \cdot \overline{E})$
Complement variables:	$(\overline{C} \cdot D \cdot E)$
Substitute back into expression:	$A + B(\overline{C} \cdot D \cdot E)$
Final result:	$A + B\overline{C}DE$

From this example, it may seem that several demorganizations may be necessary. However, each should only involve two levels of logic.

PROBLEMS

Demorganize:

27-33. (a) $\overline{A + B}$ (b) $\overline{A} + \overline{B}$

27-34. (a) $\overline{AB} + C$ (b) $\overline{A}\overline{B} + C$

27-35. (a) $\overline{A + B + C}$ (b) $\overline{A} + \overline{B} + \overline{C}$

27-36. (a) $\overline{AB}\ \overline{CD}$ (b) $\overline{ABCD}$

Reduce by demorganizing:

27-37. (a) $\overline{AB + \overline{CDE}}$ (b) $\overline{\overline{A}\overline{B} + \overline{CDE}}$

27-38. (a) $\overline{ABC}\ \overline{DE}$ (b) $\overline{ABCD\overline{E}}$

27-39. (a) $A + \overline{BC} + \overline{DE}$ (b) $A + \overline{BC} + \overline{\overline{D}\ \overline{E}}$

27-40. (a) $(\overline{\overline{A} + \overline{B}})(\overline{C + D})\overline{E}$ (b) $A + \overline{\overline{B} + C + \overline{D} + \overline{\overline{E}}}$

27-10 *REDUCING BOOLEAN EXPRESSIONS*

Since every logic operator represents a corresponding element of hardware, the designer must reduce every boolean equation to as simple form as possible in order to reduce cost. This will require the use of the 21 laws discussed in Section 27-9. The techniques used for these reductions are similar to those used in ordinary algebra. The following procedure can be used as a general approach:

(a) Multiply all variables necessary to remove parentheses.

(b) Look for identical terms. Using Law 12, one of these can be dropped.

(c) Look for a variable and its negation in the same term. This term can be dropped. For example,

$$A\overline{A}C = 0 \cdot C \qquad \text{Law 9}$$
$$= 0 \qquad \text{Law 6}$$

(d) Look for pairs of terms that are identical except for one variable. If one variable is missing, the larger term can be dropped. For example,

$$ABCD + ABD = ABD(C + 1) \qquad \text{Law 18}$$
$$= ABD \cdot 1 \qquad \text{Law 11}$$
$$= ABD \qquad \text{Law 7}$$

If the one variable is present but negated in the second term, it can be reduced. For example,

$$ABCD + A\overline{B}CD = ACD(B + \overline{B}) \qquad \text{Law 18}$$
$$= ACD \cdot 1 \qquad \text{Law 13}$$
$$= ACD \qquad \text{Law 7}$$

The preceding procedure can be used for most cases. However, each expression must be examined for combinations that would permit reduction.

EXAMPLE 27-19 Reduce the expression $\overline{\overline{AB} + \overline{A}} + AB$.

SOLUTION

Demorganize $\overline{AB}$:	$\overline{\overline{A} + \overline{B} + \overline{A}} + AB$	Law 21
Reduce:	$\overline{\overline{A} + \overline{B}} + AB$	Law 12
Reduce:	$\overline{\overline{A} + \overline{B}} + A$	Law 20
Rearrange:	$A + \overline{A} + \overline{B}$	Law 14
Reduce:	$1 + \overline{B}$	Law 13
Reduce:	$\overline{1}$	Law 11
Convert:	0	Law 2

EXAMPLE 27-20 Reduce the expression $AB + \overline{AC} + A\overline{B}C(AB + C)$.

SOLUTION

Multiply:	$AB + \overline{AC} + AAB\overline{B}C + A\overline{B}CC$	Law 18
Reduce:	$AB + \overline{AC} + A\overline{B}C$	Laws 9, 6, 8
Demorganize $\overline{AC}$:	$AB + \overline{A} + \overline{C} + A\overline{B}C$	Law 21
Rearrange:	$AB + \overline{C} + \overline{A} + A\overline{B}C$	Law 14
Reduce:	$AB + \overline{C} + \overline{A} + \overline{B}C$	Law 20
Rearrange:	$\overline{A} + AB + \overline{C} + \overline{B}C$	Law 14
Reduce:	$\overline{A} + B + \overline{C} + \overline{B}$	Law 20
Reduce:	1	Laws 11, 13

PROBLEMS

Reduce:

27-41. $\overline{AB} + B$

27-42. $\overline{A}\overline{B} + B$

27-43. $AB\overline{B} + \overline{A}B$

27-44. $AB + A\overline{B}$

27-45. $ABC + A\overline{B}C$

27-46. $\overline{A}BC + \overline{A}\overline{B}C$

27-47. $A\overline{B}\overline{C} + \overline{A}\overline{B}C$

27-48. $ABB + AB + \overline{A}$

27-49. $A\overline{B}\overline{C} + AB + AC$

27-50. $ABC(AB + B)$

27-51. $A(A + B) + B(AB + B)$

27-52. $AA(ABC + AC) + AB(A + B)$

27-53. $AA(A\overline{A} + BC + AC)(B + C)$

27-54. $\overline{AB}(AB + BC\overline{C} + AC)(B + \overline{A})$

27-55. $\overline{ABC}(\overline{A + B + C})$

27-56. $\overline{AB}(AB + \overline{AC})$

27-57. $(\overline{ABC} + \overline{AB} + \overline{AC})(\overline{AB} + \overline{A})$

27-58. $(AB\overline{B} + A\overline{B} + \overline{AB}C)(\overline{AB} + A)$

APPENDIX I

TABLE OF SYMBOLS

I-1. Electrical Symbols

Quantity	Quantity Symbol	Unit	Unit Symbol
admittance	Y	siemen	S
capacitance	C	farad	F
charge	Q	coulomb	C
conductance	G	siemen	S
current	I	ampere	A
electric field strength	E	volt per meter	V/m
electric flux	ψ	coulomb	C
electric flux density	D	coulomb per square meter	C/m²
energy, work	W	joule	J
impedance	Z	ohm	Ω
inductance	L	henry	H
permittivity	ϵ	farad per meter	F/m
power, active	P	watt	W
power, apparent	S	voltampere	VA
power, reactive	Q	var	var
reactance	X	ohm	Ω
relative permittivity	ϵ_r	(numeric)	
resistance	R	ohm	Ω
resistivity	ρ	ohm-meter	$\Omega \cdot m$
susceptance	B	siemen	S
voltage	V, E	volt	V

I-2. Magnetic Symbols

Quantity	Quantity Symbol	Unit	Unit Symbol
magnetic field strength	H	ampere per meter	A/m
magnetic flux	ϕ	weber	Wb
magnetic flux density	B	tesla	T
magnetomotive force	F	ampere (amp turn)	
permeability	μ	henry per meter	H/m
permeance	P	weber per ampere	Wb/A
relative permeability	μ_r	(numeric)	
reluctance	R	ampere per weber	A/Wb

I-3. Other Symbols

Quantity	Quantity Symbol	Unit	Unit Symbol
angular velocity	ω	radian per second	rad/s
force	F	newton	N
frequency	f	hertz	Hz
length	l	meter	m
mass	m	kilogram	kg
temperature, absolute	T	kelvin	K
temperature, customary	T	degree Celsius	°C
time	t	second	s

APPENDIX II

GREEK ALPHABET

Name	English Equivalent	Lowercase Capital	Designates
alpha	a	α A	angles, coefficients, attenuation constant
beta	b	β B	angles, coefficients, phase constant, current gain
gamma	g	γ Γ	angles, conductivity, specific gravity complex propagation constant
delta	d	δ Δ	increment or decrement, angle, density increment or decrement
epsilon	e	ϵ E	2.718281828, permittivity, dielectric constant
zeta	z, dz	ζ Z	
eta	ē, H	η H	efficiency, hysteresis
theta	th	θ Θ	angles, phase
iota	i	ι I	
kappa	k	κ K	susceptibility, coupling coefficient
lambda	l	λ Λ	wavelength, attenuation constant
mu	m	μ M	permeability, amplification factor, prefix micro
nu	n	ν N	
omicron	o	o O	
pi	p	π Π	3.14159265 product
rho	r	ρ P	resistivity, coordinates
sigma	s	σ Σ	deviation, leakage coefficient summation
tau	t	τ T	time constant, transmission factor
upsilon	u	υ Y	
phi	ph	ϕ Φ	angles, magnetic flux

chi	ch	χ X	
psi	ps	ψ Ψ	dielectric flux, phase difference
omega	ō	ω Ω	angular velocity resistance

APPENDIX III

INTERNATIONAL SYSTEM OF UNITS (SI) CONVERSION FACTORS

Symbol	Given:	Multiply by:	To Obtain:	Symbol
		Length		
in.	inches	25.4*	millimeters	mm
ft	feet	30.48*	centimeters	cm
yd	yards	0.9144*	meters	m
mi	miles (statute)	1.609	kilometers	km
nmi	miles (nautical)	1.852*	kilometers	km
μ	micron	1.0*	micrometers	m
Å	angstrom	0.1*	nanometers	nm
		Area		
cmil	circular mils	0.0005067	square millimeters	mm²
in.²	square inches	6.452	square centimeters	cm²
ft²	square feet	0.09290	square meters	m²
yd²	square yards	0.8361	square meters	m²
mi²	square miles (statute)	2.590	square kilometers	km²
		Volume		
fl oz	fluid ounces (U.S.)	29.57	cubic centimeters, milliliters	cm³ or ml
gal	gallons (U.S.)	3.785	liters	⁄
gal	gallons (Canada)	4.546	liters	⁄
in.³	cubic inches	16.39	cubic centimeters	cm³
ft³	cubic feet	0.02832	cubic meters	m³
yd³	cubic yards	0.7646	cubic meters	m³
		Speed		
ft/min	feet per minute	5.080*	millimeters per second	mm/s
mi/h	miles per hour	0.4470	meters per sec	m/s
km/h	kilometers per hour	0.2778	meters per sec	m/s
kn	knots	0.5144	meters per sec	m/s
		Mass		
oz	ounces (avdp)	28.35	grams	g
lb	pounds (avdp)	0.4536	kilograms	kg
ton	short tons (2000 lb)	0.9072	metric tons (1000 kg)	t
		Density		
lb/ft³	pounds per cubic foot	16.02	kilograms per cubic meter	kg/m³

Force

ozf	ounces-force	0.2780	newtons	N
lbf	pounds-force	4.448	newtons	N
kgf	kilograms-force	9.807	newtons	N
dyn	dynes	10^{-5}*	newtons	N

Work, Energy, Power

ft-lbf	foot pounds-force	1.356	joules	J
cal	calorie	4.184*	joules	J
Btu	British thermal units	1055	joules	J
hp	horsepower	746*	watts	W
ft-lbf/s	foot pounds-force per second	1.356	watts	W
Btu/h	British thermal units per hour	0.2931	watts	W

Light

| fc | footcandles | 10.76 | lux | lx |
| fL | footlamberts | 3.426 | candelas per square meter | cd/m² |

Temperature

| °F | degrees Fahrenheit | $(°F - 32) \cdot \dfrac{5}{9}$ | degrees Celsius | °C |
| °C | degrees Celsius | $°C \cdot \dfrac{5}{9} + 32$ | degrees Fahrenheit | °F |

Notes:

1. * means exact value.
2. The values on the left may be obtained by dividing the quantities on the right by the conversion factor.
3. To form fractional conversion factors, put the conversion factor with the quantity on the right. Examples:

$$\frac{1 \text{ in.}}{25.4 \text{ mm}} \qquad \frac{1 \text{ gal (U.S.)}}{3.785 \; \ell}$$

ANSWERS TO ODD-NUMBERED PROBLEMS

CHAPTER 2

2-1. $26 \cdot V \qquad 26 \times V \qquad 26\,V$
2-3. $100 \cdot C \qquad 100 \times C \qquad 100\,C$
2-5. $f \cdot L \qquad L \cdot f \qquad f \times L \qquad L \times f \qquad fL \qquad Lf$
2-7. $L \cdot C \qquad C \cdot L \qquad L \times C \qquad C \times L \qquad LC \qquad CL$

2-9.	64	**2-33.**	16	**2-57.**	30
2-11.	216	**2-35.**	88	**2-59.**	403.3
2-13.	243	**2-37.**	10	**2-61.**	1,063,142
2-15.	46,656	**2-39.**	87	**2-63.**	260.51
2-17.	10	**2-41.**	275	**2-65.**	37
2-19.	20	**2-43.**	58,880	**2-67.**	2.465
2-21.	17	**2-45.**	27	**2-69.**	28.847
2-23.	6	**2-47.**	99.71	**2-71.**	1
2-25.	63	**2-49.**	864.7	**2-73.**	2.60
2-27.	16	**2-51.**	22.53	**2-75.**	3516.59
2-29.	3072	**2-53.**	20,160	**2-77.**	53.976
2-31.	12,348	**2-55.**	107,833	**2-79.**	15

2-81.	21
2-83.	36
2-85.	11
2-87.	2
2-89.	7
2-91.	1264
2-93.	187
2-95.	522
2-97.	8
2-99.	16
2-101.	315,792

CHAPTER 3

3-1. (a) 8 (b) 4 (c) -5 (d) -123
3-3. (a) -6.23 (b) -2.44 (c) -14.38
  (d) 16.01
3-5. -74
3-7. 5.64
3-9. (a) -3 (b) -10 (c) -88 (d) 56
3-11. (a) 10.99 (b) -6.60 (c) 4.86
  (d) -0.79
3-13. $6R$
3-15. $-2.1QP$
3-17. $2R + 3T$
3-19. $-1.7LM - 3.9L^2M$
3-21. $15Z + 11T$
3-23. $4.5Z_1Z_2 - 2.1Z_1Z_3 - 11.2Z_2Z_3$
3-25. $10A - B + C + 4AB + CD$

3-27. $13L_1^4L_2^2L_3 + 6L_1^2L_2^2 - 4L_1^2L_2^4$
  $+ 7L_1L_2L_3^5$
3-29. $8I_1^2R_1 + 8I_2^2R_2 - 12I_1^2R_2$
3-31. $4f^2L^2 + 7f^2L + 4f^2L^2$
3-33. $4X^2$
3-35. $4A^2B$
3-37. $26A^2Z - 75AZ$
3-39. $-8.216P^2Q^2RS^2 - 7.313P^2Q^2RS$
3-41. $-61VI + 36I^2R$
3-43. $35I^2X_c + 19I^2Z - 7I^2R + 6I^2X_L$
3-45. $5M^5 - 7M^4N^2 + 5M^3N^2$
  $+ 2M^2N^3 - 9N^5$
3-47. $3.153f_1^3 - 1.315f_1^2f_0 + 4.928f_1f_0^2 - f_0^3$
3-49. 12
3-51. $5AB - 2C$

3-53. $5VI + P + 31Q$

3-55. 15

3-57. 2.50

3-59. $2X_L^2 - 7X_C^2 + Z^2$

3-61. $10L_1 - 2L_2 - 6M_1$

3-63. $-W - X + Y + Z$

3-65. $14L_1 - 15L_2 - 7L_3$

3-67. $6R_1^2 + 3R_2^2$

3-69. $9C_1 + 8C_2 - 15C_3$

CHAPTER 4

4-1. (a) 432   (b) -315 (c) -945 (d) 2336

4-3. (a) 1188   (b) -196 (c) 323   (d) -3240

4-5. (a) 4.654   (b) -46.07 (c) -406.6 (d) -9.322

4-7. -630

4-9. 47,424

4-11. 286.90

4-13. (a) 3^8 (b) 4^{12} (c) 5^{13} (d) 7^{14}

4-15. (a) 4^2 (b) 7^1 (c) 9^{-22} (d) 12^{-5}

4-17. (a) A^5 (b) X_1^{10} (c) P^{11} (d) V^{11}

4-19. (a) L_1^2 (b) R_1^{-3} (c) M_3^{-11}

4-21. (a) $56A^3B^3$ (b) $8L_1^2L_2^8$ (c) $765N_p^6N_s^3$

4-23. (a) $31.49a^4bC$ (b) $1.094L_1^{-1}L_2^3L_3$ (c) $68.11Q_1^{-2}Q_2^2$

4-25. (a) $21R_1^5R_2 + 14R_1^2R_2R_3^3$

(b) $17P_1^2P_2^2 + 17P_1^2P_3^3$

(c) $105A^2B^2R + 105A^2R^3$

4-27. (a) $2.1AWX^4Y^3 - 6.3AX^4Y^2Z$

(b) $1.350L_1^4L_2^2 - 18.00L_1^2L_2^4$

4-29. (a) $AR + AS + BR + BS$

(b) $A^2R^2 + A^2B^2 - B^2R^2 - B^4$

(c) $12A^6 - 21A^3B + 24A^3B^2 - 42B^3$

4-31. (a) $35.77Q^3P^2 - 53.29BQP^3 - 12.74BQ^2P^3 + 18.98B^2P^4$

(b) $8.61X_L^3X_C - 31.00X_LX_C^2 - 2.96X_L^3X_C + 10.66X_L^2X_C^2$

(c) $7.409V_1^3 + 6.453V_1^2V_2 - 20.77V_1V_2^2 - 18.09V_2^3$

4-33. $A^2 + 2AB + B^2 + AC + BC$

4-35. $R^3 - R^2S + 2RS - 2S^2 - RT + ST$

4-37. $12X_L^3 - 10X_L^2X_C + 20X_LX_C^2 - 6X_C^3$

4-39. $0.319I_LI_R^2 - 1.4674I_L^2I_C + 14.855I_LI_C^2 - 3.16I_CI_R^2 - 3.16I_C^3$

4-41. $X_1^3 + X_1^2X_2 - X_1X_2^2 - X_2^3$

4-43. $21R^2 + 21R^2T + 58R + 109RT + 42RT^2 + 36T^2 + 60T + 21$

4-45. $3V_1^2 - 27V_1V_2 - 14V_1V_3 + 91V_2V_3 - 42V_2^2 - 49V_3^2$

4.47. $3C_T^2C_LC_R - 21C_MC_T + 3C_L^4C_T - C_TC_L^2C_R + C_LC_M - C_L^5 + C_TC_LC_R^2 - 7C_MC_R + C_L^4C_R$

4-49. $97.94W^3 + 235.565W^2Y + 94.695WYZ + 43.86WY^2 + 19.26Y^2Z + 24.5348W^2Z + 54.0209WYZ + 23.7219YZ^2$

4-51. $12R^2 - 36R^2S - 8RS + 6RS^2 + 12S^3$

4-53. $4X_1 - 4X_2 - 2X_1^2 + 2X_2^2 + 6XX_1 - 6XX_2$

4-55. $-5Y - WX + 6WZ + 4W^2X - 4WY$

4-57. (a) $\dfrac{1}{4}$ (b) $\dfrac{1}{3}$ (c) $\dfrac{1}{2}$

4-59 (a) $\dfrac{12}{19}$ (b) $\dfrac{1}{30}$ (c) $\dfrac{1}{56}$

4-61. (a) $\dfrac{7}{16}$ (b) $\dfrac{2}{7}$ (c) $\dfrac{4}{49}$

4-63. (a) $\dfrac{48}{119}$ (b) $\dfrac{128}{207}$ (c) $\dfrac{49}{289}$

4-65. (a) $\dfrac{2268}{4477}$ (b) $\dfrac{3072}{3417}$ (c) $\dfrac{10,285}{51,233}$

4-67. (a) $\dfrac{21}{16}$ (b) $\dfrac{7}{6}$ (c) $\dfrac{12}{7}$

4-69. (a) $\dfrac{7}{15}$ (b) $\dfrac{9}{40}$ (c) $\dfrac{11}{4}$

4-71. (a) $\dfrac{17}{13}$ (b) $\dfrac{299}{306}$ (c) $\dfrac{6}{5}$

4-73. (a) $\dfrac{5}{6}$ (b) $\dfrac{5}{6}$ (c) $\dfrac{9}{10}$

4-75. (a) $\dfrac{23}{30}$ (b) $\dfrac{38}{33}$ (c) $\dfrac{139}{176}$

4-77. (a) $\dfrac{113}{100}$ (b) $\dfrac{367}{432}$ (c) $\dfrac{41}{36}$

4-79. (a) $\dfrac{1}{16}$ (b) $\dfrac{1}{12}$ (c) $\dfrac{3}{8}$

4-81. (a) $\dfrac{17}{45}$ (b) $\dfrac{5}{6}$ (c) $\dfrac{3}{4}$

4-83. (a) $6\dfrac{5}{8}$ (b) $17\dfrac{3}{7}$

4-85. (a) $4\dfrac{13}{24}$ (b) $3\dfrac{41}{60}$

4-87. (a) $10\dfrac{25}{32}$ (b) $10\dfrac{26}{27}$

4-89. (a) $2\dfrac{7}{24}$ (b) $1\dfrac{103}{285}$

4-91. (a) 0.6190 (b) -1.1905 (c) 1.6447

4-93. (a) 0.4500 (b) -2.6667 (c) -3.3333

4-95. (a) 3^3 (b) 4^6 (c) 5 (d) 0

4-97. (a) 4^6 (b) 7^{13} (c) 9^{-1} (d) 12^{-19}

4-99. (a) A (b) X_1^6 (c) P^3 (d) V^3

4-101. (a) L_1^{-8} (b) R_1^{17} (c) M_3^3 (d) Q_1

4-103. (a) $3AB^2$ (b) $6A^3$

4-105. (a) $1.4769RS^2TV^{-1}$
  (b) $2.084M^2P^{-2}Q^3$

4-107. (a) $3^1 + 3^2 = 12$ (b) $2^4 + 2^3 - 2 = 22$

4-109. (a) $3X + 4$ (b) $Y^2 - 3Y + 1$

4-111. $1.5XY + 2Y^2 + X^{-1}Y^3 - 0.5X^{-2}Y^4$

4-113. $1.732RS^{-1} + 3.4319 - 1.732R^{-1}S$

4-115. $X^2 - X + 3$ Rem: 0

4-117. $A^3B + 2A^2B^2 + 2AB^3$ Rem: 0

4-119. $3X^2 + 2XY - 2Y^2$ Rem: 0

CHAPTER 5

5-1. $V = 1$

5-3. $X = 3.667$

5-5. $S = 2.250$

5-7. $A = 0.1003$

5-9. $I = 0.9571$

5-11. $V = 18$

5-13. $P = 20.33$

5-15. $R = -14.75$

5-17. $R_1 = 46$

5-19. $G = -4.687$

5-21. $R = -13.80$

5-23. $I = -18$

5-25. $X_L = 5.105$

5-27. $L = 0.4800$

5-29. $R = 3.668$

5-31. $P = -6.400$

5-33. $I = -1.962$

5-35. $R = 1.578$

5-37. $P = 0.07397W$

5-39. $X = 70$

5-41. $C = \$0.25$
  $2C = \$0.50$

5-43. $4001\,\mu\text{A}$

5-45. $\dfrac{1}{C_T} = \dfrac{1}{C_1} + \dfrac{1}{C_2} + \dfrac{1}{C_3} + \ldots$

5-47. $I = \dfrac{V}{R}$

5-49. $C_m = \dfrac{X^2}{2V}$

5-51. $h_{21} = -\dfrac{1 + h_{fe}}{h_{fe}}$

5-53. $E_a = \dfrac{E_d\,Ll}{2AD}$

5-55. $n = \dfrac{f_1 - f_1m - f_2}{f_2}$

5-57. $S = \dfrac{Y - b}{X}$

5-59. $T^2 = \dfrac{K - G\omega}{G\omega^3}$

5-61. $L_k = \dfrac{L_1(1 - m^2)}{m}$

5-63. $a = \dfrac{A_i}{A_i + 1}$

5-65. $Y_l = \dfrac{A_i h_{22}}{h_{21} - A_i}$

5-67. $r_b = \dfrac{a r_l - A_v r_e}{A_v (1 - a)}$

5-69. $L_s = \dfrac{L}{1 - \omega^2 L C_d}$

5-71. $r_m = \dfrac{(r_c + r_e + r_o)(r_g + r_b + r_e) - r_e^2}{r_g + r_b}$

5-73. $A_1 = \dfrac{0.049\,V}{a}\left(\dfrac{1}{T_M} - \dfrac{1}{T_E}\right) - A_2 - A_3$

5-75. Increase by 20%.

5-77. Twice the original rate.

5-79. $L \propto 1.563 L_0$

5-81. Decrease by 27%.

5-83. Quadrupled.

5-85. $K = 3$

5-87. 207 peanuts

5-89. $A_2 = 12.5$ in.^{2}

5-91. $P_2 = 133.8$

CHAPTER 6

6-1. 1.235 kΩ, 1.365 kΩ, 1.30 kΩ, 1% rel.

6-3. 712.5 Ω, 787.5 Ω, 750 Ω

6-5. 1.76 kΩ, 2.64 kΩ, 2.2 kΩ

6-7. 58.9 kΩ, 65.1 kΩ, 62 kΩ

6-9. 3.135 Ω, 3.465 Ω, 3.3 Ω

6-11. 0.312 Ω, 0.468 Ω, 0.39 Ω

6-13. Brown, orange, brown, gold

6-15. Yellow, orange, orange, gold, red

6-17. Orange, blue, gold, gold

6-19. Green, blue, gold, gold, yellow

6-21. (a) 3.169×10^1 31.69×10^0
(b) 2.1673×10^4 21.673×10^3

6-23. (a) 1.390×10^{-2} 13.90×10^{-3}
(b) -1.364×10^{-5} -13.64×10^{-6}

6-25. (a) -7.65317×10^5 -765.317×10^3
(b) 8.472×10^6 8.472×10^6

6-27. (a) 5.367×10^7 53.67×10^6
(b) 3.4672×10^4 34.672×10^3

6-29. (a) 37,600   (b) 734.1

6-31. (a) $-7,316,000$ (b) 38,700

6-33. (a) -79.6 (b) 0.0876

6-35. (a) 7.705×10^6 (b) 73.4777×10^6
(c) 69.30×10^{-3} (d) 166.3×10^3

6-37. (a) 72.329×10^3 (b) 625.7×10^3
(c) 7.76384×10^0 (d) 128.6×10^6

6-39. (a) 7.7737×10^{-3} (b) 639.8×10^0
(c) -10.487×10^{-6} (d) 554.5×10^{-3}

6-41. (a) 7.298×10^{-9} (b) 59.008×10^{-9}
(c) -3.0988×10^0 (d) 170.84×10^6

6-43. (a) 4.306×10^{33} (b) 25.93×10^{12}
(c) 3.544×10^3 (d) 895.1×10^{-21}

6-45. (a) 4   (b) 3   (c) 5

6-47. (a) 4   (b) 3   (c) 6

6-49. (a) 3   (b) 3   (c) 4

6-51. (a) 1.25, 1.35 (b) 1.295, 1.305
  (c) 1.2995, 1.3005
6-53. (a) 438.075, 438.085
  (b) 438.0825, 48.0835
  (c) 438.08355, 48.08365
6-55. (a) 251.25×10^4, 251.35×10^4
  (b) 6.4345×10^{-5}, 6.4355×10^{-5}
  (c) 5.7625×10^2, 5.7635×10^2
6-57. (a) 3.2 (b) 4.8 (c) 7.2
6-59. (a) 3.15 (b) 0.743 (c) 7.77
6-61. (a) -1.26 (b) 9.92 (c) 0.812
6-63. (a) ± 68.17 abs, $\pm 8\%$ rel.
  (b) ± 0.3145 abs., $\pm 5\%$ rel.
6-65. (a) ± 0.27 abs., $\pm 8.23\%$ rel.
  (b) ± 0.003 abs., $\pm 1.038\%$ rel.
6-67. (a) ± 4 abs., $\pm 103.3\%$ rel.
  (b) ± 0.1548 abs., $\pm 4\%$ rel.
6-69. $1.000000003 \times 10^{10}$ adding left to right.
  $1.000000000 \times 10^{10}$ adding right to left.
  Left to right is more accurate.
6-71. $5.000000002 \times 10^{-1}$
  $5.000000001 \times 10^{-1}$ (more accurate)
6-73. -0.31 ± 0.02 ($\pm 6.45\%$)
6-75. -0.011 ± 0.0143 ($\pm 130\%$)
6-77. 0.086 ± 0.3917 ($\pm 455\%$)
6-79. 13.86, 0.9, 6.494%
6-81. 0.01486, 0.00014, 0.9421%
6-83. 9.53, 0.2502, 2.625%
6-85. 8.990, 0.02229, 0.2480%
6-87. 67.92, 0.08000, 0.1178%
6-89. -938, 542.2, 57.81%
6-91. 208152, 8421, 4.046%
6-93. 1.685, 0.01121, 0.6652%
6-95. 87.42, 5.245, 6.000%
6-97. 10211, 612.6, 6.000%
6-99. 0.03317, 0.0002595, 0.7825%
6-101. 189.33, 2.007, 1.060%
6-103. (a) 136.59 MΩ (b) 17.57 kV
6-105. (a) 8.29 mA (b) 93.7 μA
6-107. (a) 22.9 GHz (b) 71pF
6-109. (a) 4,960,000 Ω (b) 5420 Ω
6-111. (a) 0.000,000,08135 s
  (b) 0.000,000,000,105 F
6-113. 273.375 ft²/rod²
6-115. 14.000×10^6 grain/ton
6-117. 28.16×10^3 BN/mi
6-119. 30,000 ms
6-121. 3.5×10^{12} nA
6-123. Equations (c), (d), (f), and (g) are invalid.

CHAPTER 7

7-1. 130 mΩ

7-3. 7.333 Ω

7-5. 17.66 A

7-7. 876.1 mA

7-9. 355.3 kV

7-11. 514.7 V

7-13. 57.90 kΩ

7-15. 982.9 kΩ

7-17. 1.103 mA

7-19. 12.62 μA

7-21. 2.547 kV

7-23. 1.008 kV

7-25. 5 μA

7-27. 3.333 mΩ

7-29. 1.186 mA

7-31. 58.50 mA

7-33. (a) 427.4 mA, 273.8 Ω
   (b) 512.8 mA, 228.2 Ω
   (c) 854.7 mA, 136.9 Ω
   (d) 4.274 A, 27.38 Ω

7-35. 80 mW, 312.5 Ω

7-37. $I = 31.62$ mA $V = 31.62$ V

7-39. $P = 150$ μW
   $V = 150$ mV

7-41. $I_5 = 100$ mA
   $I_{10} = 50$ mA
   $I_{15} = 33.33$ mA

7-43. $R = 114.7$ Ω, $P = 75.42$ W

7-45. $I = 9.154$ mA, $P = 336.9$ mW

7-47. $I = 538.8$ A, $R = 382.3$ μΩ

7-49. $V = 130.13$ kV, $P = 118.4$ kW

7-51. $V = 903.6$ mV, $R = 229.3$ mΩ

7-53. $V = 80.00$ V, $I = 9.313$ mA

7-55. $3.00

7-57. $63.07

7-59. $R_T = 1.1608$ kΩ

7-61. $R_T = 5.064$ kΩ

7-63. $R_T = 8.489$ MΩ

7-65. $R_T = 49.30$ kΩ

7-67. $V_1 = 2.360$ V, $V_2 = 8.656$ V,
   $V_3 = 4.984$ V

7-69. $V_1 = 10.38$ V, $V_2 = 6.210$ V,
   $V_3 = 25.41$ V

7-71. $V_1 = 21.14$ V, $V_2 = 1.664$ V,
   $V_3 = 196.0$ mV

7-73. $V_5 = 4.800$ V, $V_{7.5} = 7.200$ V

7-75. $P_1 = 31.28$ mW, $P_2 = 114.7$ mW,
   $P_3 = 66.07$ mW, $P_T = 212.1$ mW

7-77. $P_1 = 2.747$ mW, $P_2 = 1.643$ mW,
   $P_3 = 6.723$ mW, $P_T = 11.11$ mW

7-79. $I = 755.2$ μA

7-81. $I = -1.055$ mA

7-83. $I = 383.7$ mA

7-85. $I = -244.9$ μA, $V_{R1} = -16.16$ V,
   $V_{R2} = -21.31$ V,
   $V_{R3} = -10.53$ V, $P = 11.76$ mW

7-87. $I = -727.3$ μA, $V_{R1} = -10.84$ V,
   $V_{R2} = -2.684$ V,
   $V_{R3} = -989.2$ mV, $P = 10.55$ mW

7-89. $R_1 = 312.5$ Ω, $P_1 = 80.00$ mW
   $R_2 = 187.5$ Ω, $P_2 = 48.00$ mW
   $R_3 = 375.0$ Ω, $P_3 = 96.00$ mW
   $R_4 = 375.0$ Ω, $P_4 = 96.00$ mW

7-91. $R_1 = 1.350$ kΩ, $P_1 = 540.0$ mW
   $R_2 = 1.600$ kΩ, $P_2 = 640.0$ mW
   $R_3 = 300.0$ Ω, $P_3 = 120.0$ mW
   $R_4 = 500.0$ Ω, $P_4 = 200.0$ mW

7-93. $R_1 = 50.00$ Ω, $P_1 = 500.0$ mW
   $R_2 = 50.00$ Ω, $P_2 = 500.0$ mW
   $R_3 = 50.00$ Ω, $P_3 = 500.0$ mW
   $R_4 = 100.0$ Ω, $P_4 = 1.000$ W

CHAPTER 8

8-1. (a) $9A^2B^2$ (b) $16C^2D^4$

8-3. (a) $27A^3C^3$ (b) $16A^4B^6$

8-5. (a) $25R^4S^6$ (b) $-343R_4^3R_2^6$

8-7. (a) $308.2x^8y^{12}$ (b) $-5606W^{20}P^{10}$

8-9. (a) $-6.800V_1^7V_2^{14}$
   (b) $-2.23731X_1^{18}X_2^{27}$

8-11. (a) $\pm2AB$ (b) $\pm7C^2D$

8-13. (a) $\pm4C_1^5C_2^2$ (b) $\pm6D_1^4D_2^3D_3^2$

8-15. (a) $\pm1.732AB^2$ (b) $\pm2.646AB^2CD^3$

8-17. (a) $2 \cdot 2 \cdot 5 \cdot 7$ (b) $2 \cdot 2 \cdot 3 \cdot 5$
   (c) $2 \cdot 2 \cdot 2 \cdot 2 \cdot 3 \cdot 5$

8-19. (a) $2 \cdot 2 \cdot 2 \cdot 3 \cdot 11$
   (b) $2 \cdot 2 \cdot 2 \cdot 3 \cdot 3 \cdot 13$
   (c) $2 \cdot 2 \cdot 3 \cdot 17$

8-21. (a) $2 \cdot 2 \cdot 2 \cdot 2 \cdot 2 \cdot 2 \cdot 3 \cdot 19$
   (b) $3 \cdot 3 \cdot 3 \cdot 3 \cdot 17$
   (c) $2 \cdot 2 \cdot 3 \cdot 3 \cdot 11 \cdot 13$

8-23. (a) $3 \cdot 5 \cdot 5 \cdot 11 \cdot 17 \cdot 17$

(b) $2 \cdot 2 \cdot 2 \cdot 2 \cdot 3 \cdot 11 \cdot 11 \cdot 19$

(c) $3 \cdot 3 \cdot 5 \cdot 11 \cdot 17 \cdot 23$

8-25. $2(A + B)$

8-27. $4(R_1 + 2R_2)$

8-29. $5S_1(5 + 33S_2)$

8-31. $7X^2 Y^2(2X^3 Y^2 + 1)$

8-33. $9P_L(P_L + 2)$

8-35. $3V_1(3 + V_2)$

8-37. $6R_1R_2(3R_1 + 2R_2)$

8-39. $3(A + B + C)$

8-41. $2V_1V_2(8V_2 + 9V_1 + 6)$

8-43. $4E_2^2(12E_1^2E_2V_1 + 5E_2V_1^3V_2^2 + 10V_2^2)$

8-45. $B^2(20A^2BC + 4ABC^2D + 6A^3E + C^2)$

8-47. $(A + 5B)^2$

8-49. $(5B + 2C)^2$

8-51. $(5E_1 + 3E_2)^2$

8-53. $(4A_1 + 3A_2)^2$

8-55. $(5AB^2 + 3CD)^2$

8-57. $(3R_1R_2 + 4R_3)^2$

8-59. $A(X + Y)^2$

8-61. $2XY(9X + 6Y + Y^2)$

8-63. $7MN(2M + N)^2$

8-65. $2AB(9A^2B^3 + 5C^2D)^2$

8-67. $(P - 5Q)^2$

8-69. $(5Z_1 - 2Z_2)^2$

8-71. $4X^2(X - 1)^2$

8-73. $4L_1^2L_3^2(4L_1L_2^2 - 3L_3)^2$

8-75. $(B + N)(B - N)$

8-77. $(2L_1 + 3L_2)(2L_1 - 3L_2)$

8-79. $(E_1 + E_2)(E_1 - E_2)$

8-81. $(10Z_1Z + 9Z_3^2)(10Z_1Z - 9Z_3^2)$

8-83. $(6P_1P_2^5 + 11P^{13})(6P_1P_2^5 - 11P^{13})$

8-85. $5X_1^2(8X_1 + 5X_2)(8X_1 - 5X_2)$

8-87. $7L_1^2L_2L_3^6(7L_1L_2^2 + 8L_3)(7L_1L_2^2 - 8L_3)$

8-89. $(a + b + c + d)(a + b - c - d)$

8-91. $\left(\dfrac{1}{a + b} + \dfrac{1}{c + d}\right)\left(\dfrac{1}{a + b} - \dfrac{1}{c + d}\right)$

8-93. $(3V_1 + V_2)(2V_1 + V_2)$

8-95. $(4I + J)(2I + J)$

8-97. $(12P + Q)(2P + Q)$

8-99. $(10Q + R)(3Q + R)$

8-101. $(12V_s + V_p)(3V_s + V_p)$

8-103. $(5D_1 + D_2)(12D_1 + D_2)$

8-105. $(3X - 1)(2X - 1)$

8-107. $(4L_1 - L_2)(3L_1 - L_2)$

8-109. $(18P - Q)(P - Q)$

8-111. $(10Y_1 - Y_2)(3Y_1 - Y_2)$

8-113. $(10M_A - M_B)(5M_A - M_B)$

8-115. $(3W - X)(2W + X)$

8-117. $(10R_1 - R_2)(R_1 + R_2)$

8-119. $(12L_1 - L_2)(2L_1 + L_2)$

8-121. $(6L_1 - L_2)(5L_1 + L_2)$

8-123. $(4Z_1 + Z_2)(Z_1 - Z_2)$

8-125. $(5Q + P)(4Q - P)$

8-127. $(8MN - P)(2MN + P)$

8-129. $(12C_1^2C_2^3 + C_3)(2C_1^2C_2^3 - C_3)$

8-131. $4R_1R_2^2(4R_1 + R_2)(2R_1 - R_2)$

8-133. $(4N + 3P)(7N + 2P)$

8-135. $(9V_1 + 5V_2)(6V_1 + 5V_2)$

8-137. $(5W + 6N)(3W + 5N)$

8-139. $(11V_1 + 12V_2)(V_1 + V_2)$

8-141. $(3Z + 5R)(2Z - 5R)$

8-143. $(6L - 7M)(2L + 3M)$

8-145. $(9T_1 - 2T_2)(7T_1 + 5T_2)$

8-147. $(13L - 4M)(L + 4M)$

8-149. $(9A_3 - 7A_4)(3A_3 - 5A_4)$

8-151. $2B^2C(B + C)$

8-153. $4V_1^2V_2(2V_1^4 + V_2V_3^3)$

8-155. $7AB(2A + B)^2$

8-157. $3R_1^2R_2R_3(R_1^2 + R_2)^2$

8-159. $(6A_1A_2^2 - 5A_3)^2$

8-161. $M(M - N)(M + N)$

8-163. $F(G_1^2 - 2F)(G_1^2 + 2F)$

8-165. $4B(2B + C)(3B + C)$

8-167. $2R_1R_2(3R_1 + R_2)(2R_1 + R_2)$

8-169. $3MT^2(M^2T^2 - U)(4M^2T^2 - 3U)$

8-171. $2Z_1(3Z_1 - Z_2)(2Z_1 - 3Z_2)$

8-173. $9R_1^2R_L(R_1 - R_L)(6R_1 + 5R_L)$

8-175. $21E_1E_2(6E_1 + 7E_2)(3E_1 - 4E_2)$

8-177. $(L_S - L_P)(L_S + L_T)(L_S - L_T)$

8-179. $(A^2 + AB + B^2)(A + B)$

8-181. $4A^2C(AC^2 + 2B)(AC^2 - 2B)$

8-183. $2V_1(V_1 + V_2)(V_1 - V_2)^2$

CHAPTER 9

9-1. $\dfrac{A}{2B^2}$

9-3. $\dfrac{16N_1N_2^2}{13N_3}$

9-5. $\dfrac{3R_1(R_1 + R_2)}{R_2}$

9-7. $\dfrac{2Q(2P + 3Q)}{P(3P - 2Q)}$

9-9. $\dfrac{4I_1^3(I_1 - I_2)}{3I_2^2(3I_1 + I_2)}$

9-11. $\dfrac{A - B}{A + B}$

9-13. $\dfrac{X - Y}{X + Y}$

9-15. $\dfrac{3}{2R_2^2}$

9-17. $\dfrac{12}{AB}$

9-19. $\dfrac{R_1^3 R_2}{R_3^2}$

9-21. $\dfrac{3A + 3B}{2A - 2B}$

9-23. $\dfrac{V_0^2 - V_S^2}{6V_0^2}$

9-25. $\dfrac{12I_1^2 + I_1 I_2 - 6I_2^2}{2I_1^2 + 2I_1 I_2}$

9-27. $\dfrac{2L_1^2 + 2L_1 M}{L_1^2 - M^2}$

9-29. $\dfrac{9.89R_1^2 + 40.49R_1 R_2 + 37.82R_2^2}{R_1^2 + R_1 R_2}$

9-31. $\dfrac{6A^2 + 11AB + 3B^2 + 2AC + 3BC}{AB + BC}$

9-33. $\dfrac{7}{2AB}$

9-35. $\dfrac{9L_2^2}{4L_1 L_3}$

9-37. $\dfrac{V_1 - V_3}{3}$

9-39. $\dfrac{12L}{2L + 3}$

9-41. $\dfrac{R - 3}{3}$

9-43. $\dfrac{9(R + 1)}{16}$

9-45. $\dfrac{(Q - 5)(Q + 1)}{Q(Q - 1)^3}$

9-47. $\dfrac{6}{7}$

9-49. 1

9-51. $\dfrac{3A}{14}$

9-53. $\dfrac{3T}{5LM}$

9-55. $\dfrac{P}{3}$

9-57. $\dfrac{3}{(L + M)(L - M)}$

9-59. $\dfrac{1}{M_1 - M_2}$

9-61. (a) 6   (b) 24   (c) 12

9-63. (a) 56   (b) 24   (c) 8

9-65. (a) 24   (b) 252   (c) 660

9-67. (a) 420   (b) 660

9-69. (a) 58,080   (b) 15,120

9-71. (a) X^3 (b) XY^2

9-73. $X^2(X + Y)(X - Y)$

9-75. WAN^3

9-77. $L(L + M)^3$

9-79. $\dfrac{5R_1}{3R_2}$

9-81. $\dfrac{2V^2}{3R}$

9-83. $\dfrac{33M + 24N}{44N}$

9-85. $\dfrac{11R_1 R_2}{72R_3}$

9-87. $\dfrac{-6ab + 5ac + bc}{abc}$

9-89. $\dfrac{-7L_1 + 6L_2 - 3}{L_1 L_2}$

9-91. $\dfrac{-7V_1^2 R_1 R_2 + 6V_1^2 R_1 R_3 + 3V_1^2 R_2 R_3}{R_1 R_2 R_3}$

9-93. $\dfrac{27R_1 + R_2}{30}$

9-95. $\dfrac{-8P_N^2 + 14P_N P_R - 7P_R^2}{6P_N P_R}$

9-97. $\dfrac{P_L}{R_1 - 3}$

9-99. $\dfrac{V_1(V + 6)}{(V_1 + 3)(V_1 - 3)}$

9-101. $\dfrac{-16}{I_1(I_1 + 2)(I_1 - 2)}$

9-103. $\dfrac{3M}{M - 3}$

9-105. $\dfrac{13}{3(T - 4)}$

9-107. $\dfrac{4A + 3B}{5AB}$

9-109. $\dfrac{1}{LM}$

9-111. $R + 1$

9-113. $\dfrac{C^2 + C + 1}{1 - C - C^2}$

9-115. $\dfrac{B + 3}{(B - 2)^2}$

9-117. $3V = 4\pi r^3$

9-119. $3V = \pi r^2 h$

9-121. $48\text{EIN} = \text{DPB}(3L^2 - 4B^2) - 3M_1 L^2$

9-123. $X = 1.400$

9-125. $L = -3.727$

9-127. $M = 7.154$

9-129. $k = -38$

9-131. $R_T = -0.4578$

9-133. $C = 1.333F$

9-135. $C_T = \dfrac{R_1 R_2 R_T (C_1 - C_2)}{R_1 + R_2}$ $R_1 = -\dfrac{C_T R_2}{C_T - C_1 R_2 R_T + C_2 R_2 R_T}$

9-137. $C_1 = \dfrac{-(C_2 C_3 + C_2 C_3^2)}{C_3 - C_3^2 - C_2 C_3}$

CHAPTER 10

	I_1	I_2	I_3	I_T
10-1.	1.000 mA	240 μA	470.6 μA	1.711 mA
10-3.	4.800 A	2.400 A	3.158 A	10.36 A
10-5.	2.400 μA	5.106 μA	16.00 μA	23.51 μA

10-7. 763.9 mA, 572.9 mA, 1.337 A

10-9. 6.000 kΩ

10-11. 5.546 kΩ

10-13. 984.4 Ω

10-15. 158.7 Ω

10-17. 188.8 Ω

10-19. 818.2 Ω

10-21. 14.85 Ω

10-23. 1.112 kΩ

10-25. 2.7 kΩ

10-27. 410 Ω

10-29. $I_1 = 1.200$ mA, $I_2 = 1.000$ mA, $I_3 = 800.0$ μA
$P_1 = 14.40$ mW, $P_2 = 12.00$ mW, $P_3 = 9.600$ mW
$I_t = 3.000$ mA, $P_t = 36.00$ mW

10-31. $R_3 = 2.000$ kΩ, $I_1 = 5.000$ mA, $I_2 = 3.333$ mA
$P_1 = 50.00$ mW, $P_2 = 33.33$ mW, $P_3 = 50.00$ mW
$I_t = 13.33$ mA, $P_t = 133.3$ mW

10-33. $R_3 = 450.0$ Ω, $I_1 = 7.500$ mA, $I_2 = 15.00$ mA
$I_3 = 33.33$ mA, $P_1 = 112.5$ mW, $P_2 = 225.0$ mW
$I_t = 55.83$ mA, $P_t = 837.5$ mW

10-35. $R_2 = 5.000$ kΩ, $R_3 = 2.012$ kΩ, $I_1 = 6.061$ mA
$I_3 = 9.939$ mA, $P_1 = 121.2$ mW, $P_2 = 80.00$ mW
$P_3 = 198.8$ mW, $P_t = 400$ mW

10-37. $R_1 = 15.00$ kΩ, $R_3 = 7.500$ kΩ, $I_2 = 1.500$ mA
$P_1 = 15.00$ mW, $P_2 = 22.50$ mW, $P_3 = 30.00$ mW
$I_t = 4.500$ mA, $P_t = 67.50$ mW

10-39. $R_t = 6.545$ Ω, $I_1 = 166.7$ mA, $I_2 = 333.3$ mA
$I_3 = 416.7$ mA, $I_t = 916.7$ mA

10-41. $R_x = 12.00$ kΩ

10-43. $R_x = 12.00$ kΩ

10-45. $R_x = 120.0$ kΩ

10-47. $R_x = 1.200$ kΩ

10-49. $D = 30{,}490$ ft or 5.775 mi

10-51. $D = 30{,}800$ ft or 5.833 mi

10-53. 22,810 ft or 4.320 mi

10-55. 45.53 mi

10-57. 37.33 mi

10-59. $R_1 = 1.333$ kΩ, $P_1 = 1.200$ W
$R_2 = 500$ Ω, $P_2 = 200$ mW
$R_3 = 2.000$ kΩ, $P_3 = 450$ mW
$R_4 = 2.000$ kΩ, $P_4 = 200$ mW

10-61. $R_1 = 393.9$ Ω, $P_1 = 10.73$ W
$R_2 = 115.4$ Ω, $P_2 = 1.950$ W
$R_3 = 300.0$ Ω, $P_3 = 3.000$ W
$R_4 = 200.0$ Ω, $P_4 = 500.0$ mW

10-63. $R_1 = 44.94$ kΩ, $P_1 = 142.4$ mW
$R_2 = 254.2$ kΩ, $P_2 = 354.0$ mW
$R_3 = 36.36$ kΩ, $P_3 = 44.00$ mW
$R_4 = 80.00$ kΩ, $P_4 = 80.00$ mW

10-65. $R_1 = 6.667$ kΩ, $P_1 = 60.00$ mW
$R_2 = 10.34$ kΩ, $P_2 = 87.00$ mW
$R_3 = 10.00$ kΩ, $P_3 = 10.00$ mW

10-67. $R_1 = 90.91$ Ω, $P_1 = 275$ mW
$R_2 = 400.0$ Ω, $P_2 = 1.000$ W
$R_3 = 166.7$ Ω, $P_3 = 150.0$ mW
$R_4 = 222.2$ Ω, $P_4 = 450.0$ mW

CHAPTER 11

11-1. (a) 23.85 V (b) 22.61 V (c) 14.86 V
11-3. 171.4 A, 2.571 V, $P_C = 440.8$ W, $P_B = 587.8$ W
11-5. 4.332 A
11-7. 6.720 A, 40.32 V, 271.0 W
11-9. 1.340 A, 13.40 V, $P_{BA} = 359.0$ mW, $P_{BB} = 448.7$ mW, $P_L = 17.95$ W
11-11. 774.2 mA, 5.806 V
11-13. (a) 62,500 cmils (b) 1 Mcmil (c) 155,000 cmils (d) 1.048 Mcmils
11-15. 2.624 Ω
11-17. 19.65 mΩ per wire, 39.31 Ω total
11-19. AWG 14
11-21. 2.195 A, 101.0 V
11-23. $51.33 \dfrac{\Omega\text{-cmil}}{\text{ft}}$
11-25. 1.095 Ω
11-27. 27.71 Ω
11-29. 1.529 Ω

CHAPTER 12

12-1. Graph
12-3. Graph
12-5. Graph
12-7. Graph

		Independent	Dependent	Functional
12-9.	(a)	B	A	$A = f(B)$
	(b)	R_1, R_2, R_3	R_T	$R_T = f(R_1, R_2, R_3)$
	(c)	r	A	$A = f(r)$
	(d)	f_0, Q	B	$B = f(f_0, Q)$
	(e)	C_1, C_2, C_3	C_T	$C_T = f(C_1, C_2, C_3)$

12-11. Graphs
12-13. Graphs
12-15. (a) $X = -3$ (b) $X = -2$ (c) $X = 3$
(d) $R = -2$ (e) $R = -18$ (f) $R = -4$

12-17. (a) $X = \pm 2$ (b) $X = -4, -3$ (c) $X = 2, 4$

	Y Intercept	Slope	Equation
12-19. A	-3	-1.5	$2Y = -3X - 6$
B	5.5	0.5	$2Y = X + 11$
C	21	-3	$Y = -3X + 21$

12-21. Graph, $Y = 0.7910X + 21.13$

CHAPTER 13

13-1. $4A, 1A$

13-3. $1.765, -0.8235$

13-5. \$0.25, \$0.05

13-7. $X = 1, Y = 1$

13-9. $A = 2, B = 5$

13-11. $X = 6, Z = -3$

13-13. $M = 14.6, N = -4.40$

13-15. $W = 20.00, X = -10.00$

13-17. $M = 4.106, N = 2.617$

13-19. $M_1 = 1.143, M_2 = 25.24$

13-21. $X = 4.308, Y = 0.4615$

13-23. $Y_1 = 1.507, Y_2 = 3.603$

13-25. $M = 2.137, N = 0.3642$

13-27. $W = 2, X = 3, Y = -1$

13-29. $C_1 = 3, C_2 = 4, C_3 = -1$

13-31. $P = 4, Q = -4, R = 0$

13-33. $X = \dfrac{CE - BF}{AE - BD}$

$Y = \dfrac{AF - CD}{AE - BD}$

CHAPTER 14

14-1. 7

14-3. 14

14-5. 165.80

14-7. 0

14-9. 104.152

14-11. 217.45×10^6

14-13. $T = 3.750, U = 5.250$

14-15. $X = 0.5432, Y = -0.5895$

14-17. $X_1 = -3.357, X_2 = -11.087$

14-19. $R_1 = 1610.9, R_2 = 731.66$

14-21. $A = 2.000, B = -2.000,$
$C = 3.000$

14-23. $V_1 = 1.215, V_2 = 6.425, V_3 = 3.190$

14-25. $X_{L1} = 26.00, X_{L2} = 44.20,$
$X_{L3} = 2.155$

14-27. $A = 4, B = -6, C = 0, D = 3$

14-29. $R = 1.182, S = 2.201,$
$T = 8.583, U = -2.409$

14-31. (a) 0 (b) 0 (c) 20 (d) 0

14-33. -36

14-35. (a) -168 (b) 80

14-37. (a) 44 (b) 375

14-39. $X = 2, Y = 3$

14-41. $L_1 = 2.500, L_2 = 1.500$

14-43. $I_1 = 3.000, I_2 = 6.000, I_3 = 1.000$

14-45. $R_1 = 150.0, R_2 = 270.0,$
$R_3 = 51.00$

14-47. $L_1 = 3.519, L_2 = 4.318, L_3 = 8.430$

14-49. $A = 1, B = 2, C = 3, D = 4$

14-51. $L_1 = 27.59, L_2 = -14.09,$
$L_3 = -1.632, L_4 = -6.618$

CHAPTER 15

15-1. $I_1 = 903.2$ mA
$I_2 = 645.2$ A

15-3. $I_1 = -875.0$ μA
$I_2 = 1.125$ mA

15-5. $I_1 = -500.0$ μA
$I_2 = 0.000$ A
$I_3 = 500.0$ μA

15-7. $I_1 = -10.00$ mA
$I_2 = 0.000$ mA
$I_3 = 10.00$ mA

15-9. $I_1 = -128.6$ mA
$I_2 = 242.9$ mA
$I_3 = -114.3$ mA

15-11. $V_1 = 7.059$ V

15-13. $V_1 = 19.51$ V

15-15. $V_s = 49.97$ V

15-17. $V_3 = 7.692$ V

15-19. $V_4 = 6.923$ V

15-21. $V_{th} = 10$ V, $R_{th} = 15$ kΩ

15-23. $V_1 = 19.51$ V

15-25. $V_{th} = 6$ V, $R_{th} = 1.5$ Ω

15-27. $I_N = 1.5$ mA, $R_N = 6.667$ kΩ   15-33. $I_L = 119.93$ mA
15-29. $I_N = 120$ A, $R_N = 0.1$ Ω   15-35. $R_T = 7.0909$ kΩ
15-31. $V_{L1} = V_{L2} = 99.006$ V   15-37. $R_T = 6.917$ kΩ

CHAPTER 16

16-1. (a) $R = 498.0$ kΩ in series
     (b) $R = 998.0$ kΩ in series
     (c) $R = 1.998$ MΩ in series
     (d) $R = 4.998$ MΩ in series

16-3. $R_1 = 500.0$ kΩ
     $R_2 = 300.0$ kΩ
     $R_3 = 100.0$ kΩ
     $R_4 = 50.00$ kΩ
     $R_5 = 30.00$ kΩ
     $R_6 = 10.00$ kΩ
     $R_7 = 8.00$ kΩ

16-5. $R_1 = 5.000$ MΩ
     $R_2 = 4.000$ MΩ
     $R_3 = 500.0$ kΩ
     $R_4 = 400.0$ kΩ
     $R_5 = 50.00$ kΩ
     $R_6 = 40.00$ kΩ
     $R_7 = 9.850$ kΩ

16-7. (a) $R_s = 15.00$ Ω
     (b) $R_s = 6.667$ Ω
     (c) $R_s = 606.1$ mΩ
     (d) $R_s = 120.2$ mΩ
     (e) $R_s = 60.06$ mΩ

16-9. $R_1 = 160.0$ Ω
     $R_2 = 20.00$ Ω
     $R_3 = 16.00$ Ω
     $R_4 = 2.000$ Ω
     $R_5 = 2.000$ Ω

16-11. $R_1 = 45.00$ Ω
     $R_2 = 15.00$ Ω
     $R_3 = 13.50$ Ω
     $R_4 = 1.350$ Ω
     $R_5 = 150.0$ mΩ

16-13. % Error $= 25\%$
16-15. % Error $= 10\%$
16-17. % Error $= 1.7\%$

CHAPTER 17

17-1. (a) X^6 (b) $3X^8$
17-3. (a) $30R^{M+N}$ (b) $15A^{P+2}$
17-5. $18L^{7N}M^{-4N+4}$
17-7. $-48Z_1^{5.5A-4.35}Z_2^{11.00A-0.7}$
17-9. (a) X^{-2} (b) $1.5\,A$
17-11. (a) $3R^{N-2}$ (b) $2A^{4N-3}$
17-13. $0.6667A^6B^{-3N-9}$
17-15. $5R_1^{-11.5}R_2^{-1.30}$
17-17. (a) $9A^2$ (b) $16A^{12}C^8$
17-19. (a) $27A^{-6}B^9$ (b) $9.766 \times 10^{-4}L_1^{10}L_2^{20}$
17-21. (a) $L^{3P+6}M^{3P-6}$
     (b) $R_1^{12R+4T}R_2^{8R-4T}$
17-23. $4^{A+B}R_1^{3A^2+4AB+B^2}R_2^{2A^2-AB-3B^2}$
17-25. $9A^{12}B^{-6}$
17-27. $3.375L_1^{-3N+18}L_2^{-30N-27}$
17-29. (a) 64 (b) 25 (c) 27
17-31. (a) 1.732 (b) 1.710 (c) 1.710
17-33. (a) 43.06 (b) 16.982 (c) 166.47
17-35. (a) 5.099 (b) 5.916 (c) 4.899
17-37. (a) 5.992 (b) 4.148 (c) 9.017
17-39. (a) $WN\sqrt{N}$ (b) $W^2P^2\sqrt{P}$
     (c) $Z_1^5Z_2^5\sqrt{Z_2}$

17-41. (a) $A^2B^3C^{-5}\sqrt{C}$
     (b) $R_1R_2^{-2}R_3^{-2}\sqrt[3]{R_2}$
     (c) $M_1M_2^5M_3^{-8}\sqrt[4]{M_2^2M_3}$
17-43. (a) $4\sqrt{2}$ (b) $\sqrt{74}$ (c) $2\sqrt{15}$
17-45. (a) $20\sqrt{3}$ (b) $3\sqrt[3]{15}$ (c) $4\sqrt[4]{3}$
17-47. (a) $A_1A_2\sqrt{10A_2}$ (b) $2Z_1Z_2^{-3}\sqrt{5}$
17-49. (a) $6AB^3$ (b) $5LM^{-2}\sqrt{2LM^{-1}}$
17-51. (a) $15L_1^8L_2^6\sqrt{L_2}$
     (b) $25R_1^8R_2^{-7}\sqrt{R_2-1}$
17-53. (a) $4Z_1^5Z_2^{-7}\sqrt[3]{12Z_2^{-1}}$
     (b) $2J_1^4J_2^{-7}J_3^8\sqrt[4]{45}$
17-55. (a) $10\sqrt{2}$ (b) $15\sqrt{3}$
17-57. (a) $-\sqrt{2}$ (b) $17\sqrt{3}$
17-59. (a) $5\sqrt{2}$ (b) $8\sqrt{3}$
17-61. (a) $13A\sqrt{3}$ (b) $28T_1T_2\sqrt{2}$
17-63. $8 - 2\sqrt{2}$
17-65. $14 + 40\sqrt{2}$
17-67. $2\sqrt{2}$
17-69. $22\sqrt{2}$
17-71. $4L_1L_2\sqrt{5}$
17-73. $-2 + 5\sqrt{2}$
17-75. (a) $24\sqrt{2}$ (b) $-600\sqrt{3}$

17-77. $-63 + 31\sqrt{2}$

17-79. $258 + 110\sqrt{3}$

17-81. -3090

17-83. (a) $\dfrac{\sqrt{2}}{2}$ (b) $\dfrac{\sqrt{5}}{15}$ (c) $\dfrac{\sqrt{7}}{119}$

17-85. (a) $\dfrac{1}{18}$ (b) $\dfrac{\sqrt{3}}{84}$ (c) $\dfrac{\sqrt{3}}{36}$

17-87. (a) $\dfrac{3 + 6\sqrt{5}}{-171}$ (b) $\dfrac{8 - 2\sqrt{3}}{52}$

17-89. (a) $\dfrac{69 - 24\sqrt{5}}{-171}$ (b) $\dfrac{90 + 68\sqrt{3}}{-111}$

17-91. (a) $\dfrac{42 + 21\sqrt{5} + 36\sqrt{3} + 18\sqrt{15}}{-59}$ (b) $\dfrac{56 + 49\sqrt{3} - 48\sqrt{5} - 42\sqrt{15}}{-131}$

CHAPTER 18

18-1. $X = -6.541, -0.4586$

18-3. $X = -52.63, -1.368$

18-5. $X = 0.1161, -0.08612$

18-7. $X = 2.000, 4.000$

18-9. $Y = 4.000, -2.000$

18-11. $N = 5.000, -5.000$

18-13. $R_2 = 8.000, -7.000$

18-15. $P = 1.500, -0.6667$

18-17. $B = -1.500, 0.5000$

18-19. $X = 7.000, 6.000$

18-21. $R_1 = -4.000, 1.000$

18-23. $A = -8.170, -0.1632$

18-25. $A = 0.3333 \pm j0.9428$

CHAPTER 19

19-1. (a) $6°40'$
(b) $19°29'23''$
(c) $-12°10'30''$

19-3. (a) $3°55'$
(b) $508°30'42''$
(c) $221°6'53''$

19-5. (a) $25.533°$
(b) $63.7333°$

19-7. $14.6075°$

19-9. $75.2506°$

19-11. $123.0036°$

19-13. (a) 0.4712 rad (b) 1.3439 rad
(c) -0.5880 rad

19-15. 1.11219 rad

19-17. 0.665033 rad

19-19. (a) 36.096 deg (b) -4.0107 deg
(c) 17.762 deg

19-21. (a) 50.42 deg (b) -41.83 deg
(c) 246.9 deg

19-23. (a) 3.989 grad (b) 51.44 grad
(c) 84.33 grad

19-25. (a) 37.56 grad (b) 3.820 grad
(c) -200.5 grad

19-27. $DE = 5$ m, $BD = 2$ m

19-29. $AB = 0.8889$ cm, $CD = 90$ cm

19-31. $\gamma = 30°$

19-33. $\alpha = 20°$

19-35. $AC = 20.00$ ft

19-37. 292.4 ft

19-39. $X = 908.4 \ \Omega$

CHAPTER 20

20-1. Adj. $= 323.3$ cm; hyp. $= 538.8$ cm

	Sin α	Cos α	Tan α
20-3.	0.6508	0.7592	0.8571
20-5.	0.4546	0.8906	0.5105
20-7.	0.9791	0.2018	4.8519
20-9.	0.6169	0.7871	0.7837

	Sin α	Cos α	Tan α
20-11.	0.9887	0.1501	6.5883
20-13.	1.737×10^{-9}	1.000	1.737×10^{-9}
20-15. (a)	0.5000	0.8660	0.5774
(b)	0.4586	0.8886	0.5161
(c)	0.6833	0.7302	0.9358
(d)	0.9995	0.0314	31.8205
(e)	0.7923	−0.6101	−1.2985
(f)	0.9278	−0.3730	−2.4876
(g)	0.7808	0.6247	1.2499
(h)	0.0210	0.9998	0.0210

20-17. $\beta = 80°$, $B = 2.116$, $C = 12.19$
20-19. $\beta = -17.84°$, $B = 118.7$, $C = 124.7$
20-21. $\alpha = 6°$, $A = 304.46$, $C = 306.14$
20-23. $\alpha = 87°$, $A = 0.4927$, $C = 8.211$
20-25. $\beta = 43°$, $A = 22.92$, $B = 24.57$
20-27. $\beta = 7°$, $A = 0.9408$, $B = 7.6625$
20-29. 314.3 ft
20-31. $\alpha = 45°$, $\beta = 45°$, $C = 42.43$
20-33. $\alpha = 64.10°$, $\beta = 25.90°$, $C = 5.403$
20-35. $\alpha = 35.52°$, $\beta = 54.48°$, $B = 558.9$
20-37. $\alpha = 81.63°$, $\beta = 8.374°$, $B = 1.019$
20-39. $\alpha = 9.339°$, $\beta = 80.66°$, $A = 8301$
20-41. $\alpha = 59.90°$, $\beta = 30.10°$, $A = 1.078$
20-43. Rise = 7 in., angle = 34.99°, length = 122.1 in.
20-45. $X = 299.6\ \Omega$, $\theta = 48.51°$

	Sin	Cos	Tan
20-47. (a)	0.5592	−0.8290	−0.6745
(b)	0.7547	−0.6561	−1.1504
(c)	−0.2250	−0.9744	0.2309
20-49. (a)	−0.5000	0.8660	−0.5774
(b)	0.3907	0.9205	0.4245
(c)	−0.8090	−0.5878	1.3764

20-51. −1 to +1
20-53. −∞ to +∞
20-55. Solution 1: $\beta = 61.25°$, $\gamma = 92.75°$, $C = 22.79$
Solution 2: $\beta = 118.75°$, $\gamma = 35.25°$, $C = 13.17$
20-57. $\alpha = 56.24°$, $\beta = 20.76°$, $A = 104.61$
20-59. $\alpha = 48.74°$, $\gamma = 54.26°$, $C = 3.974$
20-61. $\beta = 106.0°$, $A = 43.63$, $C = 52.55$
20-63. $\beta = 62°$, $A = 96.71$, $B = 90.31$
20-65. $C = 49.82$
20-67. $C = 825.9$
20-69. $B = 6.004$
20-71. $B = 14,717$

	Sec	Cosec	Cotan
20-73. **(a)**	1.113	2.281	2.050
(b)	−1.836	1.192	−0.6494
20-75. **(a)**	1.012	6.392	6.314
(b)	−1.970	−1.161	−0.5891

20-77. $\sin 4X = 2 \sin 2X \cos 2X$
$$= 2(2 \sin X \cos X)(\cos^2 X - \sin^2 X)$$
$$= 4 \sin X \cos^3 X - 4 \sin^3 X \cos X$$
$$= 4 \sin X \cos^3 X - 4 \sin X \cos X(1 - \cos^2 X)$$
$$= 4 \sin X \cos^3 X - 4 \sin X \cos X + 4 \sin X \cos^3 X$$
$$= 8 \sin X \cos^3 X - 4 \sin X \cos X$$
$$= 8 \cos^3 X \sin X - 4 \sin X \cos X$$

20-79. $[\cos(X + Y)][\cos(X - Y)]$
$$= (\cos X \cos Y - \sin X \sin Y)(\cos X \cos Y + \sin X \sin Y)$$
$$= \cos^2 X \cos^2 Y - \sin^2 X \sin^2 Y$$
$$= \cos^2 X(1 - \sin^2 Y) - \sin^2 Y(1 - \cos^2 X)$$
$$= \cos^2 X - \cos^2 X \sin^2 Y - \sin^2 Y + \sin^2 Y \cos^2 X$$
$$= \cos^2 X - \sin^2 Y$$

CHAPTER 21

21-1. **(a)** 35.27 V   **(b)** 59.67 V
   (c) 38.65 V   **(d)** 44.59 V
21-3. 100.37 V
21-5. **(a)** 27.79 V   **(b)** 34.09 V
21-7. **(a)** 9.000 kHz   **(b)** 96.60 MHz   **(c)** 153.3 kHz
21-9. **(a)** 215.5 c   **(b)** 0.1724 c   **(c)** 15.516 Mc
21-11. **(a)** 111.1 μs   **(b)** 10.35 ns   **(c)** 6.562 μs
21-13. **(a)** 40 V   **(b)** 60.00 Hz   **(c)** 16.67 ms
   (d) 64.00°   **(e)** 377 rad/s
21-15. **(a)** 623 V   **(b)** 1.372 kHz   **(c)** 728.7 μs
   (d) −60°   **(e)** 8623 rad/s
21-17. **(a)** 4.563 V   **(b)** 38.20 V   **(c)** 35.52 V
21-19. **(a)** −132.8 mA   **(b)** 32.22 mA   **(c)** −293.3 mA
21-21. **(a)** 9 Hz, 21 Hz, 78 Hz   **(b)** 138 kHz, 322 kHz, 1.196 MHz
   (c) 9.588 MHz, 22.372 MHz, 83.096 MHz
21-23. 508.5 m, 2.055 m
21-25. 1.370 m
21-27. 4.000 V
21-29. $V_{ABS} = 166.7$ V, $V_{ALG} = 100.0$ V
21-31. $V_{RMS} = 5.477$ V
21-33. $V_{RMS} = 173.2$ V
21-35. $I_{RMS} = 20.86$ mA
21-37. **(a)** $V_{AVG} = 74.80$ kV, $V_{RMS} = 83.085$ kV
   (b) $I_{AVG} = 17.19$ μA, $I_{RMS} = 19.09$ μA
   (c) $I_{AVG} = 2.349$ mA, $I_{RMS} = 2.609$ mA

CHAPTER 22

22-1. (a) $10 + j12$ (b) $-5 - j13$
(c) $53.1 - j15.6$

22-3. (a) $16\,\angle 26°$ (b) $23.1\,\angle{-65°}$
(c) $44.9\,\angle{-196°}$ (d) $839\,\angle 342°$

22-5. $2 + j9$

22-7. $5 + j5$

22-9. $18.4 - j29.5$

22-11. $-0.894 + j0.596$

22-13. $-5 + j70$

22-15. $17 + j12$

22-17. $-1 - j3$

22-19. $-6 + j29$

22-21. $214 - j77$

22-23. $-25.36 + j2.47$

22-25. $0.0419 + j0.3187$

22-27. $-10 + j24$

22-29. $-20 + j117$

22-31. $658 - j86$

22-33. $-37.12 + j215.11$

22-35. $4840.4 + j2729.0$

22-37. $4 + j0$

22-39. $0.1000 + j1.700$

22-41. $-1.500 - j2.000$

22-43. $-1.582 + j0.2627$

22-45. (a) $21.21 + j21.21$
(b) $65.61 - j32.00$

22-47. (a) $-115.6 + j42.07$
(b) $-688.9 + j482.4$

22-49. (a) $367.8 - j1730$
(b) $65.99 + j58.69$

22-51. (a) $-6.598 + j46.36$
(b) $-172.5 - j19.35$

22-53. (a) $5\,\angle 53.13°$ (b) $5\,\angle{-36.87°}$

22-55. (a) $6.708\,\angle 63.43°$ (b) $10.63\,\angle{-131.2°}$

22-57. (a) $831.0\,\angle 0.06895°$
(b) $1908\,\angle{-78.97°}$

22-59. (a) $96\,\angle 0°$ (b) $65.2\,\angle{-90°}$

22-61. $42\,\angle 63°$

22-63. $86.9\,\angle 83.2°$

22-65. $6.996\,\angle 29.29°$

22-67. $58.70\,\angle{-10.52°}$

22-69. $28.26\,\angle 32.92°$

22-71. $215.8\,\angle 5.912°$

22-73. $9\,\angle 14°$

22-75. $48\,\angle 23°$

22-77. $1.668\,\angle{-106.5°}$

22-79. $35.66\,\angle 104.9°$

22-81. $1296\,\angle{-59.39°}$

22-83. $15\,\angle 118°$

22-85. $416\,\angle 0°$

22-87. $-22{,}099\,\angle 70.39°$

22-89. $0.010816\,\angle{-29.31°}$

22-91. $0.5000\,\angle{-55.00°}$

22-93. $0.9301\,\angle 22.00°$

22-95. $34.41\,\angle 38.60°$

22-97. $83.88\,\angle{-60.10°}$

22-99. (a) $9\,\angle 8°$ (b) $4913\,\angle 7.80°$

22-101. (a) $1.680 \times 10^6\,\angle 108.0°$
(b) $2197\,\angle{-81.00°}$

22-103. (a) $22.03\,\angle 16.32°$
(b) $1.144 \times 10^9\,\angle 12.37°$

CHAPTER 23

	Base	Log	Source
23-1.	5	6	15,625
23-3.	12	2.3	303.47
23-5.	10	0.5	3.1623
23-7.	46	0.012	1.0470

23-9. $\log_{10} 28.84 = 1.46$

23-11. $\log_4 1024 = 5$

23-13. $\log_{12} 92.07 = 1.82$

23-15. $\log_{23} 7.209 = 0.63$

23-17. $10^{1.934} = 86$

23-19. $6^{2.149} = 47$

23-21. $15^{1.811} = 135$

23-23. $9^{2.091} = 99$

23-25. (a) 3.981 (b) 1819.7 (c) 26.92

23-27. (a) 0.05012 (b) 1.074 (c) 1.000

23-29. (a) 1.995 (b) 19.95 (c) 199.5

23-31. (a) 1.995×10^6 (b) 19.95×10^6
(c) 199.5×10^6

23-33. (a) 81 (b) 1.440 (c) 1.289

23-35. (a) 1.713 (b) 27.23 (c) 1.185

23-37. (a) 29.05 (b) 5.932 (c) 1.211

23-39. (a) -2.523 (b) -1.523 (c) 0.5229

23-41. (a) 3.4771 (b) 4.4771 (c) 5.4771

23-43. (a) 0.4771 (b) 0.7782 (c) 1.0792

23-45. (a) -5.8091 (b) -3.5066
(c) -1.2040

23-47. (a) 8.0064 (b) 10.309 (c) 12.61
23-49. (a) 1.0986 (b) 1.7918 (c) 2.4849
23-51. $1.4771 = 0.4771 + 1$
23-53. $1.4771 = 1.1761 + 0.3010$
23-55. $\log 15 = \log 3 + \log 5 = 1.1761$
23-57. $\log 20 = \log 4 + \log 5 = 1.3010$
23-59. $\log 2.5 = \log 5 - \log 2 = 0.3979$
23-61. $\log 1.3333 = \log 4 - \log 3$
$= 0.1249$
23-63. $B = 25.12$
23-65. $M = 5.8388$
23-67. $R = 1.1097$
23-69. $1.4771 = 1.1761 + 0.3010$
23-71. $1.8062 = 1.5051 + 0.3010$
23-73. $X = 2389$

23-75. $R = 6.7689$
23-77. $A = 0.002524$
23-79. $1.8062 = 1.8062$
23-81. $1.8529 = 1.8529$
23-83. $3.2993 = 3.2993$
23-85. 1.6514
23-87. 1.7500
23-89. 1.1046
23-91. 2.0241
23-93. 57.50 dB
23-95. 40.00 dB
23-97. 65.99 dB
23-99. 129.7 dB
23-101. 34.22 dB

CHAPTER 24

24-1. (a) 5 μV (b) 50 μV (c) 500 μV
(d) 5 mV (e) 50 mV (f) 500 mV
24-3. 1.75 V/s
24-5. 6.8 H
24-7. 657 mH
24-9. 7.999 H
24-11. 1.581 H
24-13. 42.564 mH
24-15. 32.99 μH
24-17. 59.72 μH
24-19. 80 mH
24-21. 5.38 H
24-23. 76.97 mH
24-25. 1.4208 H
24-27. 0.2064
24-29. 10 mA, 10 V
24-31. 1.1667 mA, 12.5 V
24-33. 4200 turns, 300.0 mA
24-35. 15.81 turns
24-37. 3.333 Ω

	V_L	V_R	I
24-39.	9.036 V	20.96 V	776.5 mA
24-41.	99.00 V	995.0 mV	995.0 μA
24-43.	36.79 V	63.21 V	63.21 mA

24-45. (a) 21.99 Ω (b) 21.99 kΩ
(c) 21.99 MΩ (d) 21.99 GΩ
24-47. 8.669 kHz

24-49. 15 kΩ
24-51. 14.14 kΩ
24-53. 66.31 mA
24-55. 900.0 Ω
24-57. 280.9 mA
24-59. $904.8\underline{/110.0°}$ V
24-61. $257.2\underline{/90°}$ Ω
24-63. $196.0\underline{/90°}$ Ω
24-65. $I_1 = 181.9\underline{/-90°}$ mA
$I_2 = 115.7\underline{/-90°}$ mA
$I_t = 297.6\underline{/-90°}$ mA
24-67. $V_s = 4.241\underline{/117.0°}$ V
$I_2 = 2.455\underline{/27.00°}$ mA
24-69. $Z = 16.06\underline{/51.49°}$ kΩ
24-71. $Z = 7.083\underline{/8.776°}$ MΩ
24-73. $Z = 29.17\underline{/29.47°}$ kΩ
24-75. $Z = 1.092\underline{/88.69°}$ kΩ
24-77. $Z = 1.571\underline{/89.98°}$ MΩ
24-79. $I = 415.2\underline{/-59.43°}$ μA
$V_R = 4.069\underline{/-59.43°}$ V
$V_L = 6.888\underline{/30.57°}$ V
24-81. $I = 816.2\underline{/-0.9697°}$ μA
$V_R = 7.999\underline{/0.9697°}$ V
$V_L = 135.4\underline{/89.03°}$ mV
24-83. $3.518\underline{/31.49°}$ V
24-85. $R = 3.464$ kΩ
$L = 106.1$ mH

CHAPTER 25

25-1. 75 μF
25-3. 1.57 μF
25-5. 2.665 μF

25-7. 5.357 μF
25-9. 0.07618 μF
25-11. 0.01410 μF

25-13. (a) $V_R = 20.11$ V
$V_c = 9.890$ V
$I = 804.4$ μA
(b) $V_R = 549.5$ mV
$V_c = 29.45$ V
$I = 21.98$ μA
(c) $V_R = 127.5 \times 10^{-18}$ V
$V_c = 30.00$ V
$I = 5.098 \times 10^{-21}$ A

25-15. (a) $V_R = 95.83$ V
$V_c = 4.166$ V
$I = 203.9$ μA
(b) $V_R = 65.34$ V
$V_c = 34.66$ V
$I = 139.0$ μA
(c) $V_R = 1.419$ V
$V_c = 98.58$ V
$I = 3.019$ μA

25-17. $R = 65.80$ Ω
25-19. 1.871 ms
25-21. $I = 84.86$ μA
25-23. (a) 454.7 Ω (b) 454.7 mΩ
25-25. (a) 3.183 MΩ (b) 1.592 MΩ
25-27. $212.6 \underline{/90.00°}$ mA
25-29. $314.2 \underline{/90.00°}$ μA
25-31. $397.9 \underline{/-90.00°}$ μA
25-33. 16.5 kΩ
25-35. 2.035 kΩ
25-37. 527.38 Ω
25-39. $900.5 \underline{/-1.929°}$ Ω
25-41. $1.150 \underline{/-29.61°}$ kΩ
25-43. $980.2 \underline{/-78.23°}$ Ω
25-45. $V_R = 6.972 \underline{/45.79°}$ V
$V_c = 7.168 \underline{/-44.21°}$ V
25-47. $V_s = 267.2 \underline{/-23.31°}$ mV

CHAPTER 26

26-1. $9.899 \underline{/45.00°}$ kΩ
26-3. $349.9 \underline{/-30.96°}$ Ω
26-5. $1.815 \underline{/74.01°}$ kΩ
26-7. $V_R = 19.71 \underline{/-9.805°}$ V
$V_L = 19.70 \underline{/80.20°}$ V
$V_c = 16.29 \underline{/-99.80°}$ V
26-9. $V_s = 1.232 \underline{/68.29°}$
26-11. $5.735 \underline{/34.99°}$ kΩ
26-13. $161.2 \underline{/-88.68}$ Ω
26-15. $I_R = 4.412 \underline{/0.00°}$ mA
$I_L = 3.846 \underline{/-90.00°}$ mA
$I_c = 2.500 \underline{/90.00°}$ mA
26-17. $I_R = 1.064 \underline{/0.00°}$ mA
$I_L = 884.2 \underline{/-90.00°}$ μA
$I_c = 9.425 \underline{/90.00°}$ mA
26-19. $I_R = 46.82 \underline{/9.496°}$ mA
$I_L = 9.936 \underline{/-85.05°}$ mA
$I_c = 5.884 \underline{/94.95°}$ mA
26-21. $I_L = 37.63 \underline{/82.06°}$ mA

26-23. $I_{400} = 3.719 \underline{/-59.99°}$ mA
26-25. 1.424 kHz
26-27. 201.3 kHz
26-29. $V_R = 14.00 \underline{/0°}$ V
$V_L = 132.8 \underline{/90°}$ V
$V_c = 132.8 \underline{/-90°}$ V
$Q = 9.487$
26-31. $C = 0.03166$ μF
$Q = 25.13$
$f_{BW} = 79.58$ Hz
26-33. 100.3 Hz
26-35. 5.033 Hz
26-37. $f_r = 5.812$ kHz
$f_{ar} = 5.299$ kHz
26-39. $f_r = 242.7$ Hz
$f_{ar} = 242.1$ Hz
26-41. $f_r = 254.9$ Hz
$Q = 104.1$
$f_{BW} = 2.449$ Hz

CHAPTER 27

27-1. (a) 11 (b) 7 (c) 9
27-3. (a) 27 (b) 20 (c) 23
27-5. (a) 413 (b) 369
27-7. (a) 10,101 (b) 11,111 (c) 11,010
27-9. (a) 1,010,110 (b) 01,001,000
(c) 1,011,100

27-11. (a) 557 (b) 732
27-13. 756,770
27-15. (a) 111,110,101
(b) 101,100,010
(c) 110,111,110
27-17. (a) 230 (b) 498 (c) 281

27-19. (a) 668 (b) 2290 (c) 982
27-21. (a) 1703 (b) 322 (c) 725
27-23. (a) 3511 (b) 30,073 (c) 65,132
27-25. (a) 438 (b) 1223 (c) 291
27-27. (a) 2459 (b) 2250 (c) 3239
27-29. (a) 83 (b) 143 (c) 1CE
27-31. (a) 559 (b) 41D (c) 6E2
27-33. (a) $\overline{A}\,\overline{B}$ (b) $\overline{AB}$
27-35. (a) $\overline{A}\,\overline{B}\,\overline{C}$ (b) $\overline{ABC}$
27-37. (a) $\overline{A} + \overline{B} + \overline{C} + DE$
　　　(b) $\overline{A}\,\overline{B} + CD + \overline{E}$

27-39. (a) $A + BCDE$
　　　(b) $\overline{A}(B + C) + D + E$
27-41. 1
27-43. $\overline{A}B$
27-45. AC
27-47. $A\overline{B} + \overline{A}C$
27-49. A
27-51. $A + B$
27-53. AC
27-55. $\overline{A}\,\overline{B}\,\overline{C}$
27-57. $\overline{A} + \overline{B}$

INDEX

INDEX

Abscissa, 182
Acoustics, 398
Acute angle, 311
Addition, 7
Addition, errors in, 90–91
Addition, evaluating, 14
Addition, of negative numbers, 19–22
Addition, notation, 7
Addition of algebraic expressions, 24
Addition of decimal numbers, 81–82
Addition of fractions, 42–43
Admittance, 435
Alternating current, 283
Ammeter, 270–74
Ampacity, 177–78
Amplitude, 350, 358–61
AND operator, 457
Angle, 311–18, 333–41
Angle, phase, 351–52
Angular velocity, 352
Arabic symbols, 6
Arc cosine, 333
Arc sine, 333
Arc tangent, 333
Associative law, 463–64
Average amplitude, 358–61
Axes, 182, 188
Ayrton shunt, 271–75

Bandwidth, 443, 446
Base, 386, 388–89, 395–96
Battery, 169–74, 254
Binary number, 448
Binary number system, 449–52
Binomial, definition, 12
Binomials, adding, 25
Binomials, factoring, 120–27
Binomials, multiplication by, 36–37

Bit, 449
Boolean algebra, 457–68
Brackets, 16
Bridge, balanced, 160–63
Bridge, unbalanced, 247–49, 262

Calculator, 2
Calculator, angles, 318, 328
Calculator, EEX function, 83
Calculator, functions, 4
Calculator, log functions, 387–88, 396
Calculator, trig functions, 330, 334, 338–40, 380
Calculators, errors, 86–89
Capacitance, 418–20
Capacitor, 418–20
Circle, 300, 313, 316–17
Circular mil, 176–77
Coefficient, 12
Coefficients in addition, 24
Common factor, 120–22
Common log, 388
Commutative law, 463
Complementary angle, 312
Complementation, 462
Complex field, 368–72
Conductance, 435
Conductors, 174–81
Conjugate, 378–79
Constant, 11, 291–92
Conversion, coordinate, 380–82
Cosecant, 344–45
Cosine, 327–28, 336–41
Cosines, law of, 343–44
Cotangent, 344–45
Cramer's rule, 219–23
Current in a parallel circuit, 154–55
Current in a series circuit, 107

Decibel, 397–98
Decimal numbers, 81–83
Degrees, 313–16
DeMorgan's Theorem, 465–67
Denominator, 39
Denominator, lowest common, 144–46
Determinants, 217–36
Dibble-dobble, 450
Dielectric, 418
Difference, 11
Digits, significant, 84–85
Dimensional analysis, 98–99
Direct current, 1
Discharge, 406–8, 423–24
Distribution systems, 241–43
Distributive law, 464–65
Division, errors in, 93–94
Division, evaluating, 13
Division, notation, 7
Division, signs in, 51–52
Division of algebraic expressions, 53–55
Division of decimal numbers, 82–83
Division of fractions, 42

Ellipse, 300–301
Energy, 103–6
Engineering notation, 80
Equation, conditional, 57, 150–51
Equation, identity, 57
Equation, sine wave, 352–55
Equations, Cramer's rule, 219–23
Equations, forming, 65–67
Equations, fractional, 62–63, 149–52
Equations, graphing, 194–96, 198–201, 204–6, 299–302
Equations, higher order, 193–94
Equations, literal, 67–69, 152, 215–16
Equations, quadratic, 299–310
Equations, rules in, 58
Equations, simultaneous, 203–15, 219–35
Equations, solving, 59–64, 303–4
Equilateral triangle, 319
Error, absolute, 86
Error, relative, 86
Errors, meter loading, 278–81
Errors, propagation of, 90–94
Exponentiation, decimal numbers, 83
Exponentiation, evaluating, 13
Exponentiation, integer, 7
Exponents, 52–53, 284–89, 386
Exponents, grouping, 17
Expression, 12
Expressions, adding, 24–26

Expressions, evaluating, 13–15
Expressions, subtracting, 26–28

Factors, 116–35, 304–6
Factors, determinants, 229
Fixed notation, 79
Fractional exponents, 288–89
Fractions, addition of, 42–43, 144–46
Fractions, algebraic, 137–52
Fractions, arithmetic, 39–49
Fractions, complex, 147–49
Fractions, division of, 42, 141
Fractions, improper, 47–49
Fractions, multiplication of, 41–42, 140–41
Fractions, proper, 47–49
Fractions, subtraction of, 46, 145–46
Frequency, 350–51
Full-scale current, 265–66, 270–72
Functions, trigonometric, 325–28

Gate, 457, 459
Generator, 347–48
Grad, 317
Grand Coulee Dam, 347
Graphs, 182–201, 299–302
Graphs, phasor, 373
Graphs, plotting, 182
Grouping, signs of, 16–17, 28–30, 38, 51–52
Guage, wire, 175–76

Harmonic, 355
Heirarchy of operations, 8–9
Henry, 399
Hexadecimal number system, 455–56
Hyperbola, 301
Hypotenuse, 320

Identities, trigonometric, 345–46
Identity, 57
Imaginary number, 369
Imaginary roots, 308–10
Impedance, 335, 404
Inductance, 399
Inductance, mutual, 402–3
Inductive reactance, 408–12
Inductor, 399
Insertion loss, 280–81
Intercept, 198
Inverse ac quantities, 435–37
Inverse function, 333
Inverter, 460–61

Irrational number, 290
Isoceles triangle, 319

j-factor, 370
Kirchhoff's Law, 112, 237–39, 245

Lamp, 277
Law of cosines, 343–44
Law of sines, 342–43
LCD, 44
Literals, 7, 11, 68, 290–91
Loading, meter, 277–80
Logarithms, 386–95
Loop equation, 112–13, 237–43
Loop resistance, 161
Loss, insertion, 280–81
Lowest common denominator, 44, 144–46
Lowest common multiple, 142–43

Matrix, 217, 232
Memory, calculator, 4
Mesh, 237–43
Meter, 265–81
Meter loading, 277–80
Metric units, 95–96
Minor, 223–24
Minutes, 313–16
Monomials, division by, 54–55
Monomials, multiplication by, 35–36
Monomials, powers of, 116
Monomials, roots of, 117, 292–93
Most significant bit, 451
Multiple, lowest common, 142–43
Multiplication, errors in, 92–93
Multiplication, evaluating, 13
Multiplication, exponents, 33–34
Multiplication, notation, 7
Multiplication, signs in, 31–33
Multiplication of decimal numbers, 82
Multiplication of fractions, 41–42
Murray bridge method, 162
Mutual inductance, 402–3

Natural log, 389
Negative numbers, 19–24
Network theorems, 237–63
Node, 244–45
Norton's Theorem, 255–57
Notation, algebraic, 2
Notation, engineering, 80
Notation, fixed, 79

Notation, RPN, 2
Notation, scientific, 79
NOT operator, 460–61
Number line, 19–23
Number system, binary, 449–52
Number system, decimal, 448–49
Number system, hexadecimal, 455–56
Numerator, 39

Obtuse angle, 311
Octal number system, 452–53
Ohm, Georg Simon, 101
Ohmmeter, 275–76
Ohm's Law, 101
Ohms per volt, 266
Operations, heirarchy, 8–9
Ordinate, 182
OR operator, 459

Parabola, 301–2
Parallel batteries, 254
Parallel circuits, 154–60
Parentheses, 16
Peak-to-peak value, 358
Peak value, 358
Period, 351
Phase angle, 351–52, 367–68
Phasor, 367–84
Pivotal reduction, 232–36
Polar coordinates, 380
Polynomial, definition, 12
Polynomials, adding, 25
Polynomials, division of, 54–55
Polynomials, multiplication by, 35, 37
Positional weighting, 448
Power, 103–5, 110, 159
Powers, (*See* Exponents)
Primary winding, 404
Prime numbers, 119–20
Proportionality, 70–74
Pythagorean Theorem, 321, 334

Q, 442–46
Quadrants, 183
Quadratic formula, 306–10
Quotient, 11

Radian, 316, 352
Radicals, 289–98
Radix, 448
Rational number, 290

Ray, 312
Reactance, 335
Reactance, inductive, 408–12
Rectangular coordinates, 372–73
Reduction, fractional, 40–42, 137–39
Reduction, pivotal, 232–36
Reduction theorem, 230–31
Reference voltage, 166
Resistance, 335
Resistance, battery internal, 169–70
Resistance, wire, 177
Resistivity, 178–80
Resistor, color code, 76–78
Resistors, parallel, 156–58
Resistors, series, 107–11
Resistor string, 268, 272
Resonance, 441–46
Reverse Polish notation, 2
Right angle, 311
Right triangle, 319, 325–26
RMS, 361–65, 368
Root, 289, 308, 369
Root-mean-square, 361–65, 368
Rounding off, 85
RPN notation, 3

Scalene triangle, 319
Scientific notation, 79
Secant, 344–45
Secondary winding, 404
Seconds, 313–16
Sensitivity, meter, 266
Sexagesimal system, 313–16
Shunt, 271–75
Significant digits, 84–85
Signs, 16–17, 28–30, 31–33, 38
Signs in factoring, 132
Similar triangles, 319–20, 325–26
Sine, 327–28, 336–41
Sines, law of, 342–43
Sine wave, 347–57, 367–68
Square, difference of, 126–27
Square, perfect, 122–24
Square of differences, 124–26
Square wave, 355
Subscripts, 12–13
Subtraction, errors in, 91–92
Subtraction, evaluating, 14
Subtraction, notation, 7
Subtraction of algebraic expressions, 26–28
Subtraction of decimal numbers, 82

Subtraction of fractions, 46
Subtraction of negative numbers, 22–24
Sum, 11
Superposition theorem, 257–61
Superscript, 13
Supplementary angle, 312
Susceptance, 435
Switch, 271

Tangent, 327–28, 336–41
Temperature, effect upon wire, 180–81
Term, 12
Thévenin's Theorem, 251–53, 439–40
Time constant, 406–8, 421–24
Transformer, 403–5
Triangle, 311
Trigonometric identities, 345–46
Trigonometry, 325–46
Trinomial, definition, 12
Trinomials, adding, 25
Trinomials, factoring, 127–36

Unary, 8
Units, Conversion of, 96–98
Units of measure, 95–99
Universal shunt, 271–75

Variable, 11
Variable, dependent, 190–91, 299
Variable, independent, 190–91, 299
Varley bridge method, 162
Vector, 349, 367
Velocity, angular, 352
Vinculum, 16, 39
Voltage distribution, 241–43
Voltage divider, 114–15, 164–68
Voltage in a parallel circuit, 154–55
Voltage in a series circuit, 109–10
Voltmeter, 265–69
Volt-ohm-milliameter, 276
VOM, 276

Wavelength, 355–56
Wheatstone bridge, 160
Wire, 174–81
Wire table, 175
Wye-delta transformation, 261–63